A Century of Mathematics in America

Part III

A Century of Mathematics in America

Part III

Edited by Peter Duren
with the assistance of Richard A. Askey
Harold M. Edwards
Uta C. Merzbach

NON CIRCULATING

NATIONAL UNIVERSITY
LIBRARY
SAN DIEGO

American Mathematical Society • Providence, Rhode Island

Library of Congress Cataloging-in-Publication Data
(Revised for vol. 3)

A century of mathematics in America.
(History of mathematics; v. 2-)
1. Mathematics—United States—History—20th century. I. Duren, Peter L., 1935- .
II. Askey, Richard. III. Edwards, Harold M., 1936- . IV. Merzbach, Uta C., 1933- .
QA27.U5C46 1989 510′.973 88-22155
ISBN 0-8218-0124-4 (v. 1)
ISBN 0-8218-0130-9 (v. 2)
ISBN 0-8218-0136-8 (v. 3)

Copying and reprinting. Individual readers of this publication, and nonprofit libraries acting for them, are permitted to make fair use of the material, such as to copy an article for use in teaching or research. Permission is granted to quote brief passages from this publication in reviews, provided the customary acknowledgment of the source is given.

Republication, systematic copying, or multiple reproduction of any material in this publication (including abstracts) is permitted only under license from the American Mathematical Society. Requests for such permission should be addressed to the Manager of Editorial Services, American Mathematical Society, P.O. Box 6248, Providence, Rhode Island 02940-6248.

Copyright ©1989 by the American Mathematical Society. All rights reserved.
Printed in the United States of America.
The American Mathematical Society retains all rights
except those granted to the United States Government.

The paper used in this book is acid-free and falls within the guidelines
established to ensure permanence and durability. ♾

10 9 8 7 6 5 4 3 2 95 94 93 92 91

Contents

Surveys and Recollections

Preface

A Century of Mathematics in America was originally envisaged as a single-volume collection of newly written and reprinted historical articles issued to mark the Centennial of the American Mathematical Society. The mathematical community greeted the project with such enthusiasm, however, and with so many good ideas, that the collection quickly expanded to fill two and then three volumes. This is the third and final volume.

Part II featured historical articles on departments of mathematics at leading American universities. Part III now continues with histories or partial histories of Johns Hopkins, Clark, Columbia, MIT, Michigan, Texas, and the Institute for Advanced Study. As before, the selection of institutions was largely governed by the willingness of qualified people to undertake the task of producing the articles.

Other contributions include histories of American mathematical activity during the nineteenth century, some surveys of the work of individual mathematicians, further accounts of mathematicians' participation in the war effort, and a collection of articles on probability and statistics. We are greatly indebted to all of the writers for giving us these marvelous views of America's mathematical past.

As the project comes to a close, the editors feel moved to say that they have found it gratifying and highly educational. After reading so many interesting and informative articles, in which characters and themes continually reemerge in various guises, they have gained a more profound understanding of American mathematical history. Inevitably there are serious omissions in the collection, especially in the portrayal of individual mathematicians and their work. Nevertheless, the editors believe that the three-part collection is a valuable addition to the historical record.

To all who have given assistance, offered suggestions, or criticized articles in the making, the editors want to express their gratitude. In particular, they want to acknowledge the continued participation of Mary Lane and Donna Harmon in the editorial work, and the expert technical assistance of the AMS editorial staff.

Peter Duren
Richard Askey
Harold Edwards
Uta Merzbach

A Century of Mathematics in America

Part III

Karen Hunger Parshall majored in mathematics at the University of Virginia, then studied mathematics and history at the University of Chicago, where she received her Ph.D. in history in 1982 under Allen G. Debus and I. N. Herstein. She is now an Assistant Professor of Mathematics and History at the University of Virginia. Her research interests center on the history of nineteenth- and early twentieth-century mathematics and its institutions. David E. Rowe took his Ph.D. in mathematics from the University of Oklahoma in 1981, specializing in topology under Leonard Rubin. From 1983 to 1985, he studied in Göttingen as a Humboldt Foundation Fellow. He is now an Associate Professor of Mathematics at Pace University (Pleasantville, N.Y.) and is completing a dissertation in history entitled "Felix Klein and the Göttingen mathematical tradition" under Joseph Dauben at the Graduate Center, CUNY. The present article is a prelude to the authors' forthcoming book, The Emergence of an American Mathematical Research Community: J. J. Sylvester, Felix Klein, and E. H. Moore.

American Mathematics Comes of Age: 1875–1900

KAREN HUNGER PARSHALL* AND DAVID E. ROWE†

Within the history of science in general and the history of mathematics in particular, issues such as the beginning of American research mathematics and the subsequent founding of a mathematical community have been conspicuously ignored. In the last fifteen years, historians of American science have generated quite a number of new books on the subject of the development of science in America. With few exceptions, however, none of them

*Departments of Mathematics and History, University of Virginia, Charlottesville, VA 22903-3199. This research was partially supported by National Science Foundation Scholars Award #SES-8509795.

†Department of Mathematics, Pace University, Pleasantville, NY 10570. The present paper represents a merging of the talks that the two authors gave in the Special Session on the History of Mathematics at the annual meeting of the American Mathematical Society in Phoenix, Arizona in January 1989. We would like to thank Professors Peter Duren and Richard Askey for asking us to prepare this joint work.

Congress of Mathematicians, World's Columbian Exposition, 1893. Bottom row, left to right, James E. Oliver and William E. Story; second row, William B. Smith, Henry S. White, Felix Klein, Harry W. Tyler, and Thomas F. Holgate; third row, Arthur G. Webster, C. A. Waldo, E. Study, J. M. Van Vleck, H. T. Eddy, J. B. Shaw, James McMahon, and Professor of Mathematics at Hope College (John Kleinheksel); top row, E. M. Blake, H. G. Keppel, Frank Loud, Henry Taber, Oskar Bolza, E. H. Moore, and Heinrich Maschke.

deals with the period between 1875 and 1900, and none of them deals, to any extent, with mathematics.[1]

It is surely true that the years from roughly 1800 to 1875 witnessed a steady organization of the American scientific community and an increase in the overall level of scientific research being pursued by Americans. This certainly justifies a concentration on the first three quarters of the nineteenth century. Yet, it was during the last quarter of that century and through the first quarter of the twentieth century that the seeds of this earlier developmental period bore fruit. Furthermore, as John Servos has recently pointed out, mathematics, both as the handmaiden of the sciences and as an independent intellectual endeavor in its own right, was at the heart of advances, first in physics and later in chemistry and biology.[2] Since the developments of the other sciences hinged on the development of mathematics, it thus becomes crucial to the understanding of the entire history of American science to come to terms with the emergence of a mathematical research community in the United States.

To say that American mathematics came of age between 1875 and 1900 implies that it did not spring up *ex nihilo.* As an integral part of the curriculum at all levels, mathematics had come to America with the first educational institutions. Until the latter part of the nineteenth century, though, instruction had remained woefully elementary.[3] Prior to the 1820s, the curriculum of America's colleges had followed the eighteenth-century English model, concentrating primarily on Latin, Greek, philosophy, the rudiments of Newtonian mechanics, a little trigonometry and a bit of mathematics from Euclid, that is, arithmetic, elementary algebra, and some geometry. The War of 1812, however, symbolized a shift in focus from things English to things French, and Americans in higher education saw a country in which science and mathematics were highly respected and flourishing. As a result, many American colleges had established professorships in science by the 1820s to complement their extant mathematical chairs, and French texts in translation had

---

[1]See, for example, Stanley M. Guralnick, *Science and the Antebellum American College* (Philadelphia: American Philosophical Society, 1975); Sally Gregory Kohlstedt, *The Formation of the American Scientific Community: The Association for the Advancement of Science 1848–60* (Urbana: University of Illinois Press, 1976); Nathan Reingold, ed., *The Sciences in the American Context: New Perspectives* (Washington, D.C.: Smithsonian Institution Press, 1976); Charles E. Rosenberg, *No Other Gods: On Science and American Social Thought* (Baltimore: Johns Hopkins University Press, 1976); Daniel J. Kevles, *The Physicists: The History of a Scientific Community in Modern America* (New York: Alfred A. Knopf, 1978); John C. Greene, *American Science in the Age of Jefferson* (Ames: The Iowa State University Press, 1984); and Robert V. Bruce, *The Launching of Modern American Science, 1846–1876* (New York: Alfred A. Knopf, 1987).

[2]John Servos, "Mathematics and the Physical Sciences in America, 1880–1930," *Isis* 77 (1986): 611–629.

[3]On early nineteenth-century American mathematics education, see Florian Cajori, *The Teaching and History of Mathematics in the United States* (Washington, D.C.: Government Printing Office, 1890).

come to set new standards for scientific and mathematical learning.[4] Thus, in mathematics the stakes were raised, and calculus was introduced into a curriculum which became more and more science-oriented. By mid-century, in fact, the number of science professors amounted to almost half of many faculties, and a third of the courses which students took were scientific or mathematical in nature.

Of importance to the present discussion, however, is the fact that this rise in science teaching did not imply an increase in basic scientific research. Prior to 1875, although research was considered prestigious within the growing scientific community, there were no institutional mandates and few institutional facilities for research.[5] Furthermore, since there was little training in science beyond the undergraduate level, anyway, few people were able to reach the research level in their chosen discipline. Virtually only those who chose to study abroad, although there were notable exceptions to this, could get the extra training they needed to become productive researchers.[6] All of this began to change after 1875 with the founding of the Johns Hopkins University.

What made Johns Hopkins, as conceived and implemented by its first president, Daniel Coit Gilman, so different?[7] Unlike the presidents of long extant colleges and universities such as Harvard, Yale, and Princeton, Gilman labored under neither an unbending tradition nor a firmly entrenched philosophy of education. He realized that for his new university to survive and prosper, it had to offer something different within the context of American education. As a result of his observations abroad, Gilman recognized that the United States trailed far behind the European countries in offering advanced training in the theoretical as well as in the practical sciences. Thus, in contrast to the pre-1875 American college and university, Johns Hopkins stressed graduate education, but not at the expense of undergraduate studies, and it made research and publication institutionally sanctioned and supported activities. One of its goals was to make the United States competitive with Europe at the research level. In mathematics, it achieved this goal by appointing the then sixty-one-year-old British mathematician, James Joseph Sylvester.

---

[4] *Ibid.*, and Stanley M. Guralnick, "The American Scientist in Higher Education, 1820–1910," in Nathan Reingold, ed., *The Sciences in the American Context: New Perspectives*, pp. 99–141. The figures which follow come from Guralnick, op. cit., pp. 107–108.

[5] Rosenberg, p. 146.

[6] Among these exceptions were Benjamin Peirce and Josiah Willard Gibbs.

[7] On the history of the Johns Hopkins University, see John C. French, *A History of the University Founded by Johns Hopkins* (Baltimore: The Johns Hopkins University Press, 1946), Hugh Hawkins, *Pioneer: A History of The Johns Hopkins University 1874–1889* (Ithaca: Cornell University Press, 1960), and Francesco Cordasco, *Daniel Coit Gilman and the Protean Ph.D.: The Shaping of American Graduate Education* (Leiden: E. J. Brill, 1960).

Born into a Jewish family in London in 1814, Sylvester went to St. John's College, Cambridge in 1831.[8] There, in spite of a second place finish in 1837 on the prestigious Mathematical Tripos, his failure to submit to the Thirty-Nine Articles of the Church of England prevented him from taking his British degree. In 1841, he did earn his B.A. and M.A., but from Trinity College, Dublin. By 1846, he had met Arthur Cayley, while both were studying for the Bar, and had struck up a friendship and mathematical association which would end only with Cayley's death in 1895. Together, these two mathematicians launched the field of invariant theory, one of the most active research areas of nineteenth-century mathematics, and made far-reaching contributions to higher geometry, to the theory of matrices, and to combinatorics.[9] Once again, though, Sylvester's reputation and prodigious research proved inconsequential in the broader sphere. The longstanding Tests Act denied him, on religious grounds, the sort of prestigious university position he merited, and so, from 1855 to 1870, he was professor of mathematics at the Royal Military Academy at Woolwich. He finally left academe in 1870 in the wake of a pension dispute with the Academy and remained unemployed until 1876.

Knowing of this sad state of affairs, the Harvard mathematician Benjamin Peirce greeted the news of the new university to be founded in Baltimore as a potential godsend both for his British friend and for American mathematics. In the most eloquent of terms, he urged Gilman to choose Sylvester for the mathematics professorship and assured him of the wisdom of such a choice. On September 18, 1875, Peirce wrote:

> Hearing that you are in England, I take the liberty to write you concerning an appointment in your new university, which I think would be greatly for the benefit of our country and of American science if you could make it. It is that of one of the two greatest geometers of England, J. J. Sylvester. If you enquire about him, you will hear his genius universally recognized but his power of teaching will probably be said to be quite deficient. Now there is no man living who is more luminary in his language, to those who have the capacity to comprehend him than Sylvester, provided the

[8]There are many short, biographical sources on Sylvester. See, for example, H. F. Baker's notice in *The Collected Mathematical Papers of James Joseph Sylvester*, H. F. Baker, ed., 4 vols. (Cambridge: University Press, 1904–1912; reprint ed., New York: Chelsea Publishing Co., 1973), 4:xv–xxxvii (hereinafter cited as *Math. Papers J.J.S.*).

[9]On the mathematics of Cayley and Sylvester, see Tony Crilly, "The Rise of Cayley's Invariant Theory (1841–1862)," *Historia Mathematica* **13** (1986): 241–254; Tony Crilly, "The Decline of Cayley's Invariant Theory (1863–1895)," *Historia Mathematica* **15** (1988): 332–347; Karen Hunger Parshall, "America's First School of Mathematical Research: James Joseph Sylvester at The Johns Hopkins University 1876–1883," *Archive for History of Exact Sciences* **38** (1988): 153–196; and Karen Hunger Parshall, "Toward a History of Nineteenth-Century Invariant Theory," in David E. Rowe and John McCleary, eds., *The History of Modern Mathematics*, 2 vols. (Boston: Academic Press, 1989), 1: to appear.

> hearer is in a lucid interval. But as the barn yard fowl cannot understand the flight of the eagle, so it is the eaglet only who will be nourished by his instruction .... Among your pupils, sooner or later, there must be one, who has a genius for geometry. He will be Sylvester's special pupil—the one pupil who will derive from his master, knowledge and enthusiasm—and that one pupil will give more reputation to your institution than the ten thousand, who will complain of the obscurity of Sylvester, and for whom you will provide another class of teachers .... I hope that you will find it in your heart to do for Sylvester—what his own country has failed to do—place him where he belongs—and the time will come, when all the world will applaud the wisdom of your selection.[10]

Sylvester was indeed appointed and officially assumed his duties in the fall of 1876. He began by choosing the first class of graduate fellows in mathematics, a class of two: George Bruce Halsted, who would become a controversial professor at the University of Texas at Austin, and Thomas Craig, who would eventually succeed Sylvester at Johns Hopkins. Later that summer, Sylvester stole Hopkins' first associate for undergraduate teaching, William E. Story, from Harvard. By 1878, he had founded the *American Journal of Mathematics*; he had brought out its first number with the help of Story, whom he had enlisted as managing editor; he had published over twenty papers on invariant theory; and he had gathered around him almost a dozen graduate students and assistants. By 1881, he and his assembled students and associates, Craig, Story, Fabian Franklin, Christine Ladd Franklin, William P. Durfee, and Charles S. Peirce, among others, were realizing Gilman's goal.[11]

Doctoral Dissertations Written under Sylvester at Johns Hopkins

1. Thomas Craig, "The representation of one surface upon another, and some points in the theory of the curvature of surfaces," 1878.
2. George Bruce Halsted, "Basis for a dual logic," 1879.
3. Fabian Franklin, "Bipunctual coordinates," 1880.
4. Washington Irving Stringham, "Regular figures in n-dimensional space," 1880.

[10]Benjamin Peirce to Daniel C. Gilman, September 18, 1875, Daniel C. Gilman Papers, Ms. 1, Special Collections Division, Milton S. Eisenhower Library, The Johns Hopkins University (hereinafter cited as Gilman Papers). As quoted in Parshall, "America's First School of Mathematical Research," pp. 167–168. We thank the Johns Hopkins University for permission to quote from its archives.

[11]Parshall, "America's First School of Mathematical Research," pp. 165–172. Although women were not formally admitted to Johns Hopkins in the 1870s and 1880s, Christine Ladd (later Mrs. Franklin) asked for and got permission to attend Sylvester's lectures. Sylvester even persuaded Gilman to grant her a fellowship.

5. Oscar Howard Mitchell, "Some theorems in numbers," 1882.
6. William Pitt Durfee, "Symmetric functions," 1883.
7. George Stetson Ely, "Bernouilli's numbers," 1883.
8. Ellery William Davis, "Parametric representations of curves," 1884.

Sylvester had indeed founded a mathematical school engaged in, and even competing with one another in, producing and publishing results which were recognized as significant and original both in England and on the Continent. This school centered around what Sylvester called his "Mathematical Seminarium." With Sylvester as its director and the students as his assistants, the mathematical seminarium operated as a sort of laboratory for the production of new mathematics. Basing his lectures on whatever research problems engaged him at the moment, the director offered his assistants the opportunity to join with him in creating mathematics. By posing open questions and suggesting possible attacks on difficult points, he coaxed his students into proving new results. Sylvester captured well the cooperative spirit of this mathematical laboratory in his farewell address to the Johns Hopkins University on December 20, 1883. In his words:

> I have written a great deal, and almost every paper I have written in the course of the last seven years, has originated either in the work of the Lecture room, or in private communication with my own pupils; and there are few papers in which their names do not appear. Now I remember a considerable Memoir, which you may say I have the bad taste to entitle "A Constructive Theory of Partitions"—there is no fault to be found in that part of the title, but now comes the objectionable part,— "arranged in three Acts, an Interact and an Exodion".... That paper, extending over 85 pages of the *American Journal of Mathematics*, originated with one of my students .... Mr. Durfee, in response to a question I propounded to him, brought me an answer, in less than 24 hours, founded upon a principle, vast and fertile, due to a method discovered more than 30 years ago, but which remained sterile and abortive until the discovery of Durfee gave it vitality and energy. Except for that method and the improvement made by Durfee, this long paper in three acts, an interact and an exodium [sic] would never have been written.[12]

In this paper which appeared in 1883, the Sylvester school proved that eager American mathematical neophytes responded very favorably to Sylvester's

[12]Remarks of Professor Sylvester, at the Farewell Reception tendered to him by the Johns Hopkins University, December 20, 1883 reported by Arthur S. Hathaway, typescript, p. 20, Gilman Papers. The paper referred to is James Joseph Sylvester, "A Constructive Theory of Partitions, Arranged in Three Acts, an Interact, and an Exodion," *American Journal of Mathematics* **5** (1882): 251–330, or *Math. Papers J.J.S.*, 4: 1–83.

idiosyncratic teaching techniques. In fact, this joint effort presents what George Andrews has termed "monumental" contributions to combinatorics.[13]

As Sylvester's correspondence reveals, he had taken as his conscious goal the task of creating a successful, research-level school of mathematics in America. On May 12, 1881, he wrote to his friend Cayley in England: "I firmly believe that there is a better opportunity for creating a great mathematical school here than exists in England and the young men of the Country are fired with the love of science and seem to me to be especially gifted with a genius for Mathematics which has never before now had a chance of showing itself."[14] With their results, particularly of 1883, Sylvester and his school succeeded in putting America on the "mathematical map." Of Sylvester's students, Thomas Craig did important work on the theory of differential equations culminating in his book, *A Treatise on Linear Differential Equations*; George Bruce Halsted distinguished himself in non-Euclidean geometry as well as in the history of mathematics; Fabian Franklin published a new proof of Euler's pentagonal number theorem in addition to his work on invariant theory; W. Irving Stringham advanced the theory of elliptic and theta functions; William Durfee worked on the theory of symmetric functions and its connections with invariant theory; and George Ely and Oscar Mitchell excelled in number theory. These students put forth their ideas, not only in journals published in the United States, but also in such foreign journals as the *Comptes rendus* of the French Académie des Sciences and the *Journal für die reine und angewandte Mathematik* (*Crelle*). Furthermore, the *American Journal of Mathematics*, where they published most frequently, was widely subscribed to abroad. As Sylvester's correspondence shows, the mathematical results which he and his students published there served to awaken Europe to America's growing mathematical sophistication.[15]

With Sylvester's accomplishments clearly in evidence, the following question naturally arises: how had the direction of American mathematics changed by December 1883 when Sylvester left Johns Hopkins to assume the Savilian Chair of Mathematics at New College, Oxford? The establishment at Hopkins of a graduate school which engaged in properly graduate education, that is, in the training of future researchers, forced other institutions which saw more advanced education as part of their mission, to establish similar

[13]George P. Andrews, *The Theory of Partitions*, Encyclopedia of Mathematics and its Applications, vol. 2 (Reading: Addison Wesley Publishing Co., 1976), p. 14.

[14]James Joseph Sylvester to Arthur Cayley, May 12, 1881, Sylvester Papers, St. John's College, Cambridge, Box 11. We thank the Master and Fellows of St. John's for permission to quote from their archives.

[15]For excerpts from letters to Sylvester from Charles Hermite testifying to this growing esteem, see Parshall, "America's First School of Mathematical Research," pp. 189–190.

schools.[16] With the increase in the number of graduate schools, the level of mathematical research in the United States gradually rose, and American students no longer had to look abroad for training. But American mathematical output increased only gradually. By and large, Sylvester's students failed to transport their research ethic directly to other institutions of higher education around the country. Still, the guiding philosophy of the Johns Hopkins, with its emphasis on graduate training and research, was transferred to extant universities like Harvard, Princeton, and Yale and to newly forming ones such as Clark and Chicago.[17] By the 1890s, a dozen or more American universities could boast able research mathematicians, and by 1910, several of these schools had native-son professors who enjoyed, or would soon enjoy, sustained international reputations.

The ten years that immediately followed Sylvester's departure from Johns Hopkins, however, marked a brief interlude in the training of American mathematicians on American soil. During this period, many students opted to pursue their graduate education abroad. Indeed, they were largely compelled to do so, in view of the fact that the leading universities in the United States were still in a state of transition and not yet equipped to prepare a first generation of productive research mathematicians. The quality and number of American aspirants studying overseas from 1884 to 1894 clearly reflected the mathematical coming of age underway on this side of the Atlantic. As their mentors at this crucial stage of the maturation process, the young itinerants favored the mathematicians of Germany.

In 1904, Thomas Fiske, the founder of the American Mathematical Society, estimated that about twenty percent of the Society's membership had undertaken doctoral or post-doctoral studies in Germany.[18] Impressive as this figure may seem, among the elite mathematicians of the country the percentage of those who studied at one or more of the German universities was even higher than this. Some of these prominent Americans went to Berlin to hear the lectures of Weierstrass, Kronecker, and Fuchs. Others studied with Sophus Lie in Leipzig. Several were drawn to Hilbert in Göttingen, especially after 1900, the year in which he delivered his famous Paris lecture. During the critical period from 1880 to 1895, however, the most popular and influential teacher of American mathematicians was Felix Klein. Klein attracted a handful of prominent Americans while he was in Leipzig from 1880 to

[16]See, for example, Charles Eliot's remarks on behalf of Harvard University in *Johns Hopkins University Celebration of the Twenty-Fifth Anniversary of the Founding of the University and Inauguration of Ira Remsen, LL.D. As President of the University* (Baltimore: The Johns Hopkins Press, 1902).

[17]Laurence R. Veysey, *The Emergence of the American University* (Chicago: University of Chicago Press, 1965), pp. 95–96.

[18]Thomas S. Fiske, "Mathematical Progress in America," *Bulletin of the American Mathematical Society* **11** (1905): 238–246; in Peter Duren et. al., eds. *A Century of Mathematics in America, Part I* (Providence: American Mathematical Society, 1988), pp. 3–12, on p. 5.

1885, but at Göttingen in the following decade, they came to him in droves. No less than six of these former students went on to become presidents of the American Mathematical Society, and thirteen served as its vice president.[19]

STUDENTS OF KLEIN WHO SERVED AS PRESIDENT OF THE AMERICAN MATHEMATICAL SOCIETY

1. W. F. Osgood (1905–1906)
2. H. S. White (1907–1908
3. M. Bôcher (1909–1910)
4. H. B. Fine (1911–1912)
5. E. B. Van Vleck (1913–1914)
6. V. Snyder (1927–1928)

STUDENTS OF KLEIN WHO SERVED AS VICE PRESIDENT OF THE AMERICAN MATHEMATICAL SOCIETY

1. H. B. Fine (1892–1893)
2. H. S. White (1901)
3. M. Bôcher (1902)
4. W. F. Osgood (1903)
5. A. Ziwet (1903)
6. O. Bolza (1904)
7. I. Stringham (1906)
8. H. Maschke (1907)
9. E. B. Van Vleck (1909)
10. M. W. Haskell (1913)
11. V. Snyder (1916)
12. F. N. Cole (1921)
13. H. W. Tyler (1923)

Klein first began to take a serious interest in American mathematics late in 1883 when he was offered Sylvester's chair at Johns Hopkins. In retrospect, it seems likely that Klein would have actually made the move to Baltimore if President Gilman had extended a sufficiently attractive offer.[20] When negotiations failed, however, Gilman lost the chance to secure Klein as Sylvester's successor, and Hopkins soon fell from its preeminent position among American universities in the field of mathematics. In 1889, it also lost the services of the highly respected William Story, a Leipzig Ph.D. and one of the first American mathematicians to study abroad. Story went to newly founded Clark University , whose original faculty included Henry Seeley White, Oskar Bolza, and Henry Taber.[21] Clark not only had more depth than Hopkins, its mathematics faculty was, for a brief time, the strongest in the country.

[19]Here the phrase "Klein's students" means those who studied with him at Leipzig and Göttingen whether or not they wrote their doctoral dissertation under him. For example, Osgood took his Ph.D. under Max Noether at Erlangen, and Cole returned to Harvard for his doctorate. A complete list of the American Mathematical Society presidents and vice presidents up to 1938 can be found in Raymond Clare Archibald, ed., *A Semicentennial History of the American Mathematical Society, 1888–1938*, 2 vols. (New York: American Mathematical Society, 1938) 1: 106–107.

[20]See the documents in Klein Nachlass XXII L:7, "Berufung nach Baltimore," Niedersächsische Staats- und Universitätsbibliothek, Göttingen (hereinafter abbreviated NSUB); and Constance Reid, "The Road Not Taken," *Mathematical Intelligencer* **1** (1978): 21–23.

[21]Roger Cooke and V. Frederick Rickey discuss the Clark University mathematics department in detail in "W. E. Story of Hopkins and Clark," in this volume, pp. 29–76.

Another school that rose to prominence in mathematics during the 1890s was Harvard University. Felix Klein's profound influence on Harvard mathematics can be traced back to his student Frank Nelson Cole, who came to Leipzig on a Parker Fellowship in 1883. When Cole returned to Harvard to complete his degree he took the "new math," that is, group theory and Riemann surfaces, with him. As one of his auditors, William Fogg Osgood, later recalled:

> [Cole] had just returned from Germany and was aglow with the enthusiasm which Felix Klein inspired in his students. Cole was not the first to give formal lectures at Harvard on the theory of functions of a complex variable, Professor James Mills Peirce having lectured on this subject in the seventies. That presentation was, however, solely from the Cauchy standpoint, being founded on the treatise of Briot and Bouquet *Fonctions Elliptiques.* Cole brought home with him the geometric treatment which Klein had given in his noted Leipsic [sic] lectures of the winter of 1881–1882. Cole also gave a course in Modern Higher Algebra, with its applications to geometry. The enthusiasm which he felt for his subject was contagious. Interesting as were the other courses I have mentioned, they stood as the Old over against the New and of the latter Cole was the apostle. The students felt that he had seen a great light. Nearly all the members of the Department attended his lectures. It was the beginning of a new era in graduate education at Harvard, and mathematics has been taught here in that spirit ever since.[22]

About this latter point, Osgood was certainly in a position to know, for as Garrett Birkhoff once remarked, Osgood's "... course on functions of a complex variable remained the key course for Harvard graduate students until World War II."[23] Clearly struck by Cole's lectures, Osgood decided to journey to Göttingen and seek out for himself the "great light" that Cole had seen. One year later, his future Harvard colleague, Maxime Bôcher followed his lead.

The German mathematical experience left lasting impressions on both of these young men. Osgood, who took his inspiration from Klein's approach to function theory, also infused his work with a precision reminiscent of Weierstrass' school. This combination yielded impressive results in 1900 when

[22]William F. Osgood on Cole in Thomas S. Fiske, "Frank Nelson Cole," *Bulletin of the American Mathematical Society* **33** (1927): 773–777 on pp. 773–774.

[23]Garrett Birkhoff, "Some Leaders in American Mathematics: 1891–1941," in Dalton Tarwater, ed., *The Bicentennial Tribute to American Mathematics, 1776–1976* (n.p.: Mathematical Association of America, 1977), pp. 25–78 on p. 34. Birkhoff's emphasis.

Osgood published the first truly rigorous proof of the Riemann mapping theorem. By 1907, he had written over a dozen research papers, a lengthy survey article on function theory for Klein's *Encyklopädie der mathematischen Wissenschaften*, and his own *Lehrbuch der Funktionentheorie*, a work which eventually went through five editions. Like his friend Osgood, Maxime Bôcher also distinguished himself as a mathematician. Bôcher earned his Göttingen Ph.D. in 1891 with a prize-winning dissertation in which he developed certain ideas on Lamé functions presented by Klein during the course of his lectures on the subject. Returning to the United States and a teaching position at Harvard, Bôcher expanded his thesis into the classic volume *Ueber die Reihenentwicklungen der Potentialtheorie* in 1894.[24]

As for Cole, he took his Ph.D. at Harvard in 1886 and stayed on as a lecturer before leaving for an instructorship at the University of Michigan in 1888. While at Michigan, he introduced George A. Miller to what had become his own primary field of expertise, the theory of finite groups.[25] Miller went on to study with the two leading group theorists of the era, Sophus Lie in Leipzig and Camille Jordan in Paris, and during a long career at the University of Illinois, he published over 400 papers on the theory of finite groups.[26] His teacher, Cole, finally left Michigan in 1895 for the professorship at Columbia he would hold until his death in 1926. For twenty-five of his thirty years at Columbia, Cole also served faithfully as the secretary of the American Mathematical Society.

As evidenced by the succession of mathematicians, Cole, Osgood, Bôcher, Harvard clearly functioned as an important focal point for Klein's influence on American mathematics. Another such focus was Princeton. During the summer of 1884, the young Henry Burchard Fine made his way to Leipzig, where Klein had just completed the first half of a two-semester course on elliptic and hyperelliptic functions. In spite of the fact that Fine had missed the first part of the course, Klein advised him to attend the second half even if he could not follow it completely.[27] This was just contrary to the advice he normally gave his new American students. Given their usually woeful state of readiness for advanced mathematics, Klein tended to urge them to start at the beginning and to build from there. But Klein also had a very keen eye for talent, and he sensed in Fine a student equal to the work. Fine did enroll in the course and, restudying his notes after a few weeks had passed,

[24]On Osgood, see Archibald, *Semicentennial of the AMS*, **1**: 153–158; on Bôcher, see William F. Osgood, "The Life and Services of Maxime Bôcher," *Bulletin of the American Mathematical Society* **25** (1919): 337–350.

[25]Archibald, *Semicentennial of the AMS*, **1**: 100–103.

[26]See George A. Miller, *The Collected Works of George Abram Miller*, 3 vols. (Urbana, Ill.: University of Illinois Press, 1935, 1938, 1946).

[27]On Fine, see Oswald Veblen, "Henry Burchard Fine—In Memoriam," *Bulletin of the American Mathematical Society* **35** (1929): 726–730; and Archibald, *Semicentennial of the AMS*, **1**: 167–170.

found that the entire subject was perfectly clear. Fine's beautifully written lecture notes, housed today in the Princeton Archives, testify to the acuteness of Klein's sixth sense. Furthermore, by the end of the 1885–1886 academic year, Fine had completed his doctoral thesis under Klein's supervision on a topic suggested by Eduard Study. This was the first of nine dissertations written by American students under Klein's direction.[28]

American Doctoral Dissertations Written under Klein

1. H. B. Fine, "On the singularities of curves of double curvature," Leipzig, 1886.
2. M. W. Haskell, "Ueber die zu der Kurve $\mu^3\nu + \nu^3\lambda + \lambda^3\mu = 0$ im projektiven Sinne gehörende mehrfache Ueberdeckung der Ebene," Göttingen, 1890.
3. M. Bôcher, "Ueber die Reihenentwicklungen der Potentialtheorie," Göttingen, 1891.
4. H. S. White, "Abelsche Integrale auf singularitätenfreien einfach überdeckten, vollständigen Schnittkurven eines beliebig ausgedehnten Raumes," Göttingen, 1891.
5. H. D. Thompson, "Hyperelliptische Schnittsysteme und Zusammenordnung der algebraischen und transzendenten Thetacharakteristiken," Göttingen, 1892.
6. E. B. van Vleck, "Zur Kettenbruchentwicklung Lamścher und ähnlicher Integrale," Göttingen, 1893.
7. F. S. Woods, "Ueber Pseudominimalflächen," Göttingen, 1895.
8. V. Snyder, "Ueber die linearen Komplexe der Lieschen Kugelgeometrie," Göttingen, 1895.
9. M. F. Winston, "Ueber den Hermiteschen Fall der Laméschen Differentialgleichung," Göttingen, 1897.

It was Fine's prompting that brought another Princeton graduate, Henry Dallas Thompson, to Felix Klein in Göttingen. Thompson spent six semesters there and finished with a dissertation dealing with a topic in hyperelliptic functions. He joined the Princeton faculty in 1888 and taught there for over thirty years. Both Thompson and Fine were present in 1896 when Klein was awarded an honorary doctorate at the Princeton sesquicentennial celebration. On this occasion, their former mentor also delivered a series of four lectures on the mathematical analysis of a spinning top.[29]

For many years, Fine guided science at Princeton from his position as Dean of the Science Faculty. Like Cole, he was not a top-flight research

[28] A (nearly) complete list of Klein's Ph.D. students can be found in Felix Klein, *Gesammelte Mathematische Abhandlungen*, 3 vols. (Berlin: Springer-Verlag, 1923), 1: pp. 11–13 (hereinafter cited as *Klein G.M.A.*).

[29] Felix Klein, "The Mathematical Theory of the Top," *Klein G.M.A.*, 2: 618–654.

mathematician, but he nevertheless played an important role in directing this country's mathematical development. During the academic year 1911–1912 he served as president of the American Mathematican Society, and his popular textbooks on algebra and the calculus were considered unsurpassed for their clarity of exposition. Fine made his most lasting accomplishments in his role as an administrator, however. Not only was he an excellent fund-raiser, but he also succeeded in attracting figures like Luther P. Eisenhart, Oswald Veblen, Gilbert A. Bliss, George D. Birkhoff, and Joseph H. M. Wedderburn to Princeton. Largely as a result of his appointments, Princeton became, after 1900, one of the three leading centers for mathematics in the United States, alongside Chicago and Harvard.[30]

These elite institutions did not provide the only sources of mathematical talent in turn-of-the-century America, though. At Wesleyan College, the astronomer and later American Mathematical Society vice president, John Monroe Van Vleck sent three of his undergraduates on to Göttingen: Henry Seeley White, Frederick Shenstone Woods, and his own son, Edward Burr Van Vleck.[31] Each of these students wrote a doctoral dissertation under Felix Klein before returning to teach mathematics in the United States. White, who had originally gone to Leipzig to study under Lie and Study, left there for Göttingen after one semester. On earning his degree, he took a position first at Clark, next at Northwestern, and finally at Vassar. Van Vleck brought his Göttingen degree back to the University of Wisconsin in 1893. Moving on to Wesleyan from 1895 to 1906, he returned to Wisconsin in 1906 and remained there for essentially the rest of his career. He succeeded White, Bôcher, and Fine as president of the American Mathematical Society in 1913. Woods came home to play an important role in upgrading mathematics instruction at his own institution, the Massachusetts Institute of Technology, as well as other technical schools through the widely adopted Woods and Bailey calculus text. His MIT colleague, Harry W. Tyler, who served as an American Mathematical Society vice president, also studied with Klein from 1887–1888 before going on to take his doctorate under Paul Gordan at Erlangen.[32]

Klein's first prominent American student, however, was Washington Irving Stringham, who came to Leipzig in 1880 immediately after taking his doctorate under Sylvester at Hopkins. He arrived at a most opportune time,

[30]In recognition of Fine's many contributions to his alma mater, Fine Hall, the present-day home of Princeton mathematics, was named in his honor. The building at Princeton presently called Fine Hall, however, was constructed many years after Fine's death.

[31]Robert A. Rosenbaum, "There were Giants in those Days: Van Vleck and his Boys," *Wesleyan University Alumnus* (Nov. 1956): 2–3.

[32]On White and Van Vleck, see Archibald, *Semicentennial of the AMS*, **1**: 158–161, 170–173; G. D. Birkhoff, "Edward Burr Van Vleck in Memoriam," *Bulletin of the American Mathematical Society* **50** (1944): 37–41. On Woods, Dirk Struik wrote an unpublished memoir that can be found in the archives of Wesleyan University in Middletown, Connecticut.

as Klein was just beginning a two-semester course on "Funktionentheorie in geometrischer Behandlungsweise." Klein's lectures from the second semester of this course formed the basis for his famous booklet *Ueber Riemann's Theorie der algebraischen Funktionen und ihrer Integrale.* Stringham left Leipzig in 1882 to accept a position at the University of California where he remained for the rest of his career. In 1890, he was joined at Berkeley by another Klein pupil, Mellon Woodman Haskell, who spent more time in Göttingen than any of Klein's other American students. On returning to the United States, Haskell prepared an English translation of Klein's "Erlangen Program," which was published in the second volume of the newly founded *Bulletin of the New York Mathematical Society.* One of Klein's last American students was Virgil Snyder, who went on to become a leading figure in algebraic geometry at Cornell University. Snyder was a fixture at Cornell, where he taught for more than forty years producing thirty-nine doctoral students along the way.[33]

Obviously, Klein could not have enjoyed such striking success with these Americans had he not possessed certain extraordinary qualities as a teacher. Among these were an unusual breadth of knowledge and a quick eye for fertile new ideas, characteristics that made him an unusually effective *Doktorvater.* One need only consider the diverse themes chosen by his students for their doctoral theses, many of which were undertaken as an elaboration of ideas presented by Klein in his lectures. During the late 1880s and early 1890s, Klein focused both on mathematical physics and on a geometric approach to elliptic, hyperelliptic, and Abelian integrals and functions. Since his lectures were highly informal compared to those of most German mathematics professors, the assistants charged with the task of writing them up for circulation in the *Lesezimmer* ended up burning a lot of midnight oil. For Klein's three-semester course on Abelian functions, several Americans lost sleep in a collaborative effort to produce the *Ausarbeitung.*[34]

As Fritz König has pointed out, Klein preferred to illustrate the key motivating principles of a given theory by choosing representative examples rather than by developing a comprehensive presentation of the theory itself. Furthermore, Klein peppered his lectures with numerous references to great nineteenth-century figures whose work was otherwise difficult or impossible for students to understand. He often colored his remarks on Cayley, Lie, Riemann, Plücker, Clebsch, Kronecker, Weierstrass, and others with personal assessments of the individuals and their work. Since such pronouncements were rarely heard in conventional mathematics lectures, those with a thirst for

[33]On Snyder, see Archibald, *Semicentennial of the AMS*, **1**: 218–223.

[34]Six students attended the first semester of this course, and five of them, Haskell, Osgood, Thompson, Tyler, and White, were Americans.

a broad, semi-historical approach to mathematical ideas knew where to go. For similar reasons, Klein's seminars also drew respectably sized audiences.[35]

Beginning with the winter semester of 1893–1894, several women also came to Göttingen to study with Felix Klein. The first were Mary "May" Winston, a student from the University of Chicago, and an Englishwoman named Grace Chisholm. Both went on to complete dissertations under Klein, effectively opening the door for foreign women to attend the Prussian universities. (Ironically enough, German women had to wait another fifteen years for this privilege.) Grace Chisholm, who later married the English analyst W. H. Young, was said to have been Klein's favorite student. The letters she wrote home during this time vividly conveyed the excitement she felt as one of the first women to attend classes at a German university and under a great German professor. Consider, for example, her account of the first day of classes:

> Klein had his first lecture on the hypergeometric functions.... Miss Winston and I made for the Sanctum and found Klein there working till lecture time. Klein, instead of beginning with his usual "Gentlemen!" began "Listeners!" ["Meine Zuhörer"] with a quaint smile; he forgot once or twice and dropped into "Gentlemen!" again, but afterwards he corrected himself with another smile. He has the frankest, pleasantest smile and his whole face lights up with it. He spoke very slowly and distinctly and used the blackboard very judiciously. Mr. Woods said he never heard anyone lecture so well and neither have I. I found my notes afterwards perfectly clear though queerly spelt; but I understood as well as at an English lecture.[36]

The following semester, Chisholm described her lecture before Klein's seminar, a daunting experience for any aspiring doctoral student:

> The lecture came off yesterday, and if it is a success to interest one twelfth of one's audience I may be said to have achieved one. As to the other eleven I do not know what they thought about it, but May Winston says they were all quite wide awake, which is something that cannot be said for all the preceding lectures .... It took a little over an hour to deliver and there were a good many interruptions, which is always a good sign. Once... Professor Klein asked for an explanation of certain facts, a thing he is very fond of doing. I had been more frightened than anything of his questions,

[35]Fritz König's remarks "zum didaktischen Vorgehen Kleins" are in Felix Klein, *Funktionentheorie in geometrischer Behandlungsweise*, Teubner-Archiv zur Mathematik, vol. 7 (Leipzig: B. G. Teubner, 1987), pp. 255–256.

[36]As quoted by Ivor Grattan-Guinness in "A Mathematical Union: William Henry and Grace Chisholm Young," *Annals of Science* **29** (1972): 105–183 on p. 123.

> it is so difficult to think on an occasion like that, and although the same thing happens to nearly every one I always think it looks foolish not to be able to answer. The Gods willed on this occasion that my brain should work, and I gave the explanation to my own astonishment, and I fancy, to his too.[37]

May Winston also lectured in Klein's seminar on two separate occasions. In 1894, she spoke on "Die Kugelfunktion als spezielle Fälle der hypergeometrischen Funktion," and the following semester she lectured in a seminar on the foundations of real analysis. Most of the Americans who studied in Göttingen made at least one presentation in Klein's seminar, which was clearly one of the focal points of his teaching activity. Unlike Sylvester's highly improvised laboratory for concocting new ideas, however, Klein preferred a tightly structured setting for exploring a wide variety of mathematical subjects, many of which were far removed from his own research interests. This proved a useful vehicle for introducing students to the vast body of literature that poured from journals like Klein's own *Mathematische Annalen*.[38]

Yet, despite Klein's unprecedented influence on American mathematics, none of his American students developed into a close mathematical disciple by carrying on the distinctively Kleinian geometric approach to function theory and other branches of mathematics. For example, none compares in this regard with his German students, Robert Fricke, Walter von Dyck, Ferdinand Lindemann, or even Arnold Sommerfeld. While Klein's *Gedankenwelt* undeniably inspired nearly all of the dissertations written by his American students, its impact on these young mathematicians proved short-lived. Even where its influence was most striking, as in the cases of Bôcher, White, Snyder, and Haskell, the Americans wandered from their mentor's path after returning to the United States. Relative to his transatlantic students, Klein's influence simply lay more in his ability to train them as research mathematicians than in the specific ideas they researched under his supervision. Indeed, to many Americans, Klein served as an emissary for and a symbol of the rich expanse of mathematical culture, something they very much wanted to transplant to their own country.

Among Klein's many outstanding German students, two actually played a decisive role both in this transplantation and in the emergence of American mathematics. Oskar Bolza and Heinrich Maschke, both former Gymnasium teachers, came to the United States because neither had reasonable prospects for breaking into the German system of higher education. Bolza, whose

---

[37] *Ibid.*, pp. 123–124.

[38] After a student gave a lecture before the seminar, he or she entered a synopsis of the presentation in a protocol book which Klein kept for each of his seminars over a period of more than forty years. Today, these protocol books may be found in the so-called "Giftschrank" (the "poison cabinet") in the library of the Mathematics Institute in Göttingen.

principal research interests lay in function theory and in the calculus of variations, had a substantial background in physics as well, having studied with Kirchhoff and Helmholtz in Berlin. There, he also came under the influence of Weierstrass, but eventually took his degree in Göttingen under Klein in 1886 with a dissertation on the reduction of hyperelliptic to elliptic integrals. His friend, Maschke, known primarily as a geometer, was actually a very versatile mathematician conversant with practically all major fields of research. Studying first with Koenigsberger in Heidelberg, Maschke spent three years in Berlin before taking his Göttingen doctorate in 1880.[39]

During the academic year 1886–1887, the two friends studied together privately with Felix Klein, who met with them weekly in his home. According to Bolza, "Maschke, ... whose gifts were more in line with Klein's approach, won great and lasting rewards from this year with Klein."[40] As for Bolza himself, the experience proved a near catastrophe. From his point of view, "... Klein's brilliant genius, supported by a wonderful capability for geometric visualization that enabled him to divine the results and his sovereign command of almost every area of mathematics, which provided him with the richest abundance of techniques for handling any task" clashed with his own "... purely analytic gifts, deficient of all fantasy and lying in an entirely different direction."[41] The result was a nearly total breakdown in his confidence. Ironically enough, Klein had gone through just this same sort of crisis three years earlier when he found himself stranded in the wake of Poincaré's genius.[42]

Neither Bolza nor Maschke relished the idea of spending his life teaching mathematics in the secondary schools, but it was Bolza who took the first leap. In so doing, he had Klein's support and the encouragement of his American students, Cole and Haskell. Thus, in April of 1888, Bolza arrived in Baltimore with nothing more than a letter of introduction from Felix Klein to Simon Newcomb. Unlike Sylvester and Cayley, Newcomb was often rather pessimistic about the future of mathematics in the United States. As he wrote Klein, "I never advise a foreign scientific investigator to come to this country, but always tell him that the difficulties in the way of immediate success are the same that a foreigner would encounter in any other country."[43] He went on to say that there was little opportunity to teach higher mathematics: "We have indeed several hundred so-called colleges; but I doubt that ... one half of the professors of mathematics in them could tell what a determinant is. All they want in their professors is an elementary knowledge of the branches

[39]See Bolza's autobiography, *Aus meinem Leben* (München: Verlag Ernst Reinhardt, 1936); on Maschke, see O. Bolza, "Heinrich Maschke: His Life and Work," *Bulletin of the American Mathematical Society* **15** (1908): 85–95.

[40]Bolza, *Aus meinem Leben*, p. 18.

[41]*Ibid.*

[42]See Jeremy Gray, *Linear Differential Equations and Group Theory from Riemann to Poincaré* (Boston: Birkhäuser, 1986), pp. 273–309.

[43]Simon Newcomb to Felix Klein, April 23, 1888, Klein Nachlass XI, NSUB.

they teach and the practical ability to manage a class of boys, among whom many will be unruly."[44] Thus, Bolza counted himself lucky when Newcomb supported his appointment as "Reader in Mathematics" at Johns Hopkins. In January 1889, he taught a one-semester course there on substitution theory, relying on notes from one of Klein's lectures on the subject. He followed his short stint at Hopkins with a three-year associateship at Clark University, which was about to open for instruction in the fall of 1889.

Unfortunately, the situation at Clark rapidly deteriorated during the three years Bolza spent there, a circumstance he attributed primarily to politics rather than to financial difficulties. In a letter to Klein, he described how President G. Stanley Hall had embittered the faculty with his "endless lies."[45] Yet, in all fairness, Hall was in an impossible situation. At this time, presidents at most other leading universities held nearly complete control over the procurement and disbursement of their institution's funds. In his role as benefactor, however, Jonas Clark made sure that he had Hall's hands tied relative to finances. Unfortunately, Mr. Clark apparently thought that running a university was little different from running a business firm. In the end, his frugal business sense more than Hall's incompetence, caused the university's undoing. In January 1892, all but two of the school's faculty members signed a document in which they collectively tendered their resignation. Although this was eventually withdrawn, discontent continued to rule the campus. Seizing this opportunity, William Rainey Harper, president of the newly founded University of Chicago, raided the Clark campus and offered its faculty the chance to abandon their sinking ship for his new luxury liner backed by Rockefeller money. Not surprisingly, his pitch worked, and he eventually walked away with most of Clark's outstanding scholars, including the physicist A. A. Michelson, the anthropologist Franz Boas, and the mathematician Oskar Bolza.[46]

Like Hall at Clark, Harper was also interested in hiring prominent German scholars whenever he could. Shortly before the job of putting together a faculty at Chicago began, Heinrich Maschke had finally followed his friend, Bolza, to the United States and had taken a job as an electrician for the Weston Electrical Company in Newark. Prior to Maschke's departure, Klein had predicted that, like Odysseus, after many wanderings he would end up in Ithaca (i.e., at Cornell).[47] As it turned out, Maschke did even better. Bolza managed to negotiate a position for them both at Chicago. Thus, when the University of Chicago opened its doors in the fall 1892, two of its three mathematicians were students of Felix Klein.

[44] *Ibid.*

[45] Oskar Bolza to Felix Klein, May 15, 1892, Klein Nachlass VIII, NSUB.

[46] The relationship between Hall and Clark is chronicled in Orwin Rush, ed., *Letters of G. Stanley Hall to Jonas Gilman Clark* (Worcester, Mass.: Clark University Library, 1948).

[47] Heinrich Maschke to Felix Klein, July 5, 1892, Klein Nachlass X, NSUB.

Even with the founding of universities like Clark and Chicago, American mathematicians continued to study in Göttingen. Although they went in ever decreasing numbers through the 1920s, Americans such as Earle R. Hedrick, Max Mason, Charles Noble, and William D. Cairns went to Göttingen to study not under Klein but under the then reigning star, David Hilbert. Hilbert's arrival there in 1895 allowed Klein a free hand to pursue the various organizational and administrative projects he had long had in view. In fact, during his visit to Chicago in 1893, Klein already sensed that American mathematics was about to enter a new era. In the closing remarks of his Evanston Colloquium lectures he suggested that it was time for him to relinquish his role as the premier teacher of American mathematicians:

> ... I do not regard it as at all desirable that all students should confine their mathematical studies to my courses or even to Göttingen. On the contrary, it seems to me far preferable that the majority of the students attach themselves to other mathematicians for certain special lines of work. My lectures may then serve to form the wider background on which these special lectures are projected. It is in this way, I believe, that my lectures will prove of the greatest benefit.[48]

Even as Klein spoke, Eliakim Hastings Moore and his colleagues at the newly founded University of Chicago stood ready to assume the responsibility of educating American mathematicians.

E. H. Moore was born in 1862 in Marietta, Ohio, a small town on the Ohio-West Virginia border.[49] At the age of seventeen, he had and took the opportunity to go to Yale where he fell under the influence of the mathematician-astronomer, Hubert Anson Newton. In 1883, the year Sylvester left Hopkins, Moore received the A.B. degree as valedictorian of his class and earned his Ph.D. in mathematics two years later under Newton for a thesis on the algebra of n-dimensional geometry.[50] Realizing that his student had advanced as far as an American education at the time allowed, Newton encouraged Moore to continue his studies in Germany.

[48]Felix Klein, *The Evanston Colloquium: Lectures on Mathematics* (New York: Macmillan, 1894), p. 98.

[49]On the details of E. H. Moore's life, see Gilbert A. Bliss, "Eliakim Hastings Moore," *Bulletin of the American Mathematical Society*, 2d. ser., **39** (1933): 831–838. On E. H. Moore at the University of Chicago, see Karen Hunger Parshall, "Eliakim Hastings Moore and the Founding of a Mathematical Community in America, 1892–1902," *Annals of Science* **41** (1984): 313–333; reprinted in Peter Duren et al., eds., *A Century of Mathematics in America, Part II* (Providence: American Mathematical Society, 1988), pp. 155–175.

[50]E. H. Moore, "Extensions of Certain Theorems of Clifford and Cayley in the Geometry of n Dimensions," *Transactions of the Connecticut Academy of Arts and Sciences* 7 (1885): 9–26.

Traveling to Göttingen in the summer of 1885, Moore spent one semester there studying German and mathematics before moving on to Berlin for the winter of 1886. In Berlin, he fell under the influence not only of Karl Weierstrass but also of Leopold Kronecker before returning to the United States to begin his career at the end of the summer. After serving first as a high school instructor and then as a tutor at his alma mater, Moore held his first permanent university job at Northwestern in 1888. Soon thereafter, though, he had a much more attractive option to consider.

Harper, the president-elect of the University of Chicago, approached Moore with an offer of a full professorship and the acting headship of the Department of Mathematics at his new university.[51] After relatively painless negotiations, Moore accepted the position and made the short move from Evanston to Hyde Park. As with the choice of Sylvester at Hopkins, the selection of Moore at Chicago benefited the university as well as American mathematics. During his forty years on the faculty there, Moore not only succeeded in building a first rate department but also proved instrumental in organizing a self-sustaining American mathematical community.

When the University of Chicago opened in the fall of 1892, E. H. Moore and his two colleagues, Oskar Bolza and Heinrich Maschke, began their instruction of mathematics at both the graduate and undergraduate levels. A priori, it was not at all clear that these three men would be able to work together as a like-minded, mathematical team. Reflecting back on the situation many years later, Bolza explained that Moore ". . . was almost five years younger that I, even more than eight years younger than Maschke and was at that time little known. In addition to that, Maschke and I were foreigners who for many years had been close friends and who had lived in the absolute freedom of the German university. All of these were factors which could have risked the inner peace of the department."[52] Could have, but did not risk that all-important inner peace, for by all accounts, these three mathematicians complemented one another perfectly as teachers and as scholars.[53] In fact, the first evidence of their ability to work together successfully came very early on in their association and centered on the World's Columbian Exposition.

Held in Chicago in 1893 to commemorate the four-hundredth anniversity of the discovery of America, the Columbian Exposition involved, in addition to the displays, amusements, and cultural activities associated with a world's

[51]Richard J. Storr has chronicled the founding of the University of Chicago in *A History of the University of Chicago: Harper's University The Beginnings* (Chicago: University of Chicago Press, 1966).

[52]Bolza, *Aus Meinem Leben*, p. 26.

[53]See, for example, Bliss' remarks in "Eliakim Hastings Moore," p. 833.

fair, a series of congresses which reflected the then current intellectual endeavors of the world.[54] Relative to mathematics, Moore organized a committee consisting of Bolza, Maschke, Henry Seeley White, and himself which extended invitations on behalf of the Congress to mathematicians from the United States and Europe. The venture proved quite successful, attracting forty-five mathematicians from Austria, Germany, Italy, and nineteen states of the Union, as well as contributed papers from mathematical representatives of France, Russia, and Switzerland. Furthermore, Felix Klein, who had longed to lecture in the United States ever since the negotiations over Sylvester's chair at Hopkins failed, readily accepted the invitation to participate in the Congress as the keynote speaker. Considering the fact that three of the members of the organizing committee, Bolza, Maschke, and White, had studied under Klein in Germany, this was an obvious choice. Yet, it also underscored the enormous debt that American mathematics owed to Germany. The American participation at all levels of the Congress proved, however, that mathematics in this country was beginning to stand on its own two feet.

After the formal close of the Congress, Moore and his Chicago colleagues took further advantage of Klein's presence in the United States by attending the Evanston Colloquium lectures. Hosted by Henry Seeley White, by then at Northwestern, Klein gave a two-week-long series of special lectures to roughly two dozen auditors before returning to Germany. These lectures, which appeared in print in 1894, served as the prototype for what would become the *American Mathematical Society Colloquium Publications*.[55]

With their organizational and mathematical appetites whetted by the success of both the Congress and the colloquium, Moore and his friends next approached the New York Mathematical Society for money toward the publication of the papers read at the Congress. Writing almost fifty years later, Raymond C. Archibald viewed this as a "... major publication enterprise, transcending local considerations and sentiment [which] quickened the desire of the Society for a name indicative of its national or continental character."[56] Owing largely to the promptings of E. H. Moore and his colleagues, the New York Mathematical Society met as the *American* Mathematical Society on July 1, 1894.

Despite its nominal nationalization, though, the Society continued to meet monthly in New York to the virtual exclusion of all but those living in the Northeast. By 1896, Moore and his associates at Chicago had figured out a

[54]On the history of the Chicago's World's Fair, see Reid Badger, *The Great American Fair: The World's Columbian Exposition and American Culture* (Chicago: University of Chicago Press, 1979).

[55]See footnote 48 above.

[56]Archibald, ed., *Semicentennial of the AMS*, **1**: 7.

way to insure the mathematical vitality of the Midwest region as well. In December of that year, Moore mailed an invitation to mathematicians as far west as Kansas and Nebraska and as far east as Ohio to come to Chicago on December 31, 1896 to discuss the possible formation of a "Chicago Section" of the Society. As conceived by Moore, a formally sanctioned Chicago Section would provide not only a vehicle for the official and regular involvement of Midwesterners in the activities of the Society but also an alternative power base in Chicago for the organization. Taking the enthusiastic response to his call to Chicago before the Society early in 1897, Moore succeeded in winning approval for his idea, and the Chicago Section convened for the first time on April 24, 1897.[57]

With this goal achieved, Moore next turned his attentions to the improvement of the printed dissemination of mathematics. Like Sylvester before him, he became involved in the movement to found a new mathematics journal. Prior to 1899, the American mathematical community already supported the *American Journal of Mathematics*, the *Annals of Mathematics* (founded by the astronomer, Ormond Stone at the University of Virginia in 1884), and the *Bulletin of the American Mathematical Society* (begun in 1891 as the *Bulletin of the New York Mathematical Society*). Yet Moore and others sensed the need for a periodical which stressed not only research at a high level but also the work of American contributors. In short, they wanted a journal which showcased *American* mathematics.[58] In 1899, this goal also became a reality when the American Mathematical Society founded its *Transactions* and appointed Moore as the editor-in-chief.[59]

Moore's ascension to the editorship of the *Transactions* underscored his growing political influence within American mathematics. In 1899, he was already serving out a two-year term as vice president of the Society, and in 1900, the membership elected him to its presidency.[60] Moore used his national post to champion the cause of mathematics education at all levels of the curriculum. Like his colleague, John Dewey, he argued for a more active, hands-on approach to mathematics teaching and tried to implement

---

[57]For the history of the Chicago Section, see Arnold Dresden, "A Report on the Scientific Work of the Chicago Section, 1897-1922," *Bulletin of the American Mathematical Society* **28** (1922): 303-307.

[58]According to Moore and many of his contemporaries, the *American Journal*, under the editorship first of Sylvester and then of Simon Newcomb, favored contributions from mathematicians abroad to the exclusion of papers by Americans. The *Annals of Mathematics* had too much of a popular, non-research-oriented flavor, and the *Bulletin* targeted expository and historical work as opposed to research-level mathematics.

[59]On the controversy surrounding the establishment of the *Transactions*, see Archibald, ed., *Semicentennial of the AMS*, 1: 56-59.

[60]Of the first six presidents of the Society, Moore was the only one based in the Midwest and not in the Northeast.

such ideas in his own department at Chicago. One manifestation of this educational progressivism was the Mathematical Club founded in 1892.[61]

Unlike Sylvester's "Mathematical Seminarium," the Mathematical Club functioned as a forum for the presentation of completed research. Graduate students and faculty alike lectured on their current work before the group and answered both questions and criticisms. As Gilbert A. Bliss, one of the early students at Chicago, described it:

> Those of us who were students in those early years remember well the tensely alert interest of these three men [Moore, Bolza, and Maschke] in the papers which they themselves and others read before the Club. They were enthusiasts devoted to the study of mathematics, and aggressively acquainted with the activities of the mathematicians in a wide variety of domains. The speaker before the Club knew well that the excellence of his paper would be fully appreciated, but also that its weaknesses would be discovered and thoroughly discussed. Mathematics, as accurate as our powers of logic permit us to make it, came first in the minds of these leaders in the youthful department at Chicago,....[62]

With its goal of encouraging and promoting the highest standards of research and exposition, the club served as the training and proving ground of a second generation of American mathematicians.

Among this second generation, thirty students earned their Ph.D.'s under Moore's guidance. During Chicago's first fifteen years, Moore's mathematical interests ranged from group theory to the foundations of geometry to the foundations of analysis, and his students' work reflected not only this diversity but also their mentor's insights. Between 1896 and 1907, in fact, the list of Moore's students reads like a *Who's Who* in early twentieth-century mathematics.[63] The algebraist Leonard E. Dickson, the geometer Oswald Veblen, the analyst George D. Birkhoff, and the topologist Robert L. Moore, each grew up on E. H. Moore's brand of mathematical thinking and matured into independent-minded mathematicians who made seminal contributions to their respective fields as well as to the body politic.[64] Together, these four mathematicians published thirty books and over six hundred papers in addition to directing the research of almost two hundred Ph.D.'s. They each

[61]The logbooks of the Mathematical Club from its beginnings through the 1950s are housed in the Department of Special Collections, Joseph Regenstein Library, University of Chicago. In the earlier volumes (prior to 1900), the speaker's name as well as the date and title of his or her talk are accompanied by a short synopsis of the results presented.

[62]Bliss, p. 833. This was also quoted in Parshall, "E. H. Moore and the Founding of a Mathematical Community in America," pp. 329–330.

[63]For a complete list of Moore's students, see Bliss, p. 834.

[64]The statistics which follow were originally presented in Parshall, "E. H. Moore and the Founding of a Mathematical Community in America," pp. 330–332.

also edited major journals, served as Society president, and won election to the National Academy of Sciences. Finally, like their mathematical father, they built or maintained premier departments at their respective institutions with Dickson at Chicago, Veblen at Princeton and later at the Institute for Advanced Study, Birkhoff at Harvard, and R. L. Moore at the University of Texas at Austin.

Doctoral Dissertations Written under Moore at Chicago 1896–1907

1. Leonard Eugene Dickson, "The analytic representation of substitutions on a power of a prime number of letters; with a discussion of the linear group," 1896.
2. Herbert Ellsworth Slaught, "The cross ratio group of 120 quadratic Cremona transformations of the plane," 1898.
3. Derrick Norman Lehmer, "Asymptotic evaluation of certain totient sums," 1900.
4. William Findlay, "The Sylow subgroups of the symmetric group on K letters," 1901.
5. Oswald Veblen, "A system of axioms for geometry," 1903.
6. Thomas Emory McKinney, "Concerning a certain type of continued fractions depending upon a variable parameter," 1905.
7. Robert Lee Moore, "Sets of metrical hypotheses for geometry," 1905 (under the direction of E. H. Moore and O. Veblen).
8. George David Birkhoff, "Asymptotic properties of certain ordinary differential equations with applications to boundary value and expansion problems," 1907.
9. Nels J. Lennes, "Curves in non-metrical analysis situs, with applications to the calculus of variations and differential equations," 1907.

Why did Moore's students succeed where Sylvester's students had failed? While Sylvester proved that American students had the talent to extend the frontiers of at least certain areas of mathematical research, his idiosyncratic teaching style forced them into narrowly focused topics which soon ran dry mathematically. Furthermore, there was no well-established mathematical community in the America of the early 1880s to support their continued development. Without both this broader community and the strong personality of Sylvester to sustain it, Sylvester's school collapsed. Unable to go out and set up graduate-level programs, his students failed to maintain a tradition of training American mathematicians on American soil.

With no viable options for them at home, Americans turned to Europe, and particularly to Felix Klein in Germany, for their mathematical inspiration between 1884 and 1894. During these ten years, Klein willingly accepted the responsibility for the mathematical future of the United States but came to sense that he was playing only an interim role.

By the nineteenth century's close, American universities had made definite, serious, and long-term commitments to graduate education and to the fostering of basic research. With mathematics as the case in point, there were jobs for new Ph.D.'s at institutions which encouraged and nurtured their further growth as mathematicians. Furthermore, through the organizational efforts of Moore, Klein's students, and others, the small enclaves of mathematical research growing in scattered locations like Chicago, New York, Boston, Princeton, Baltimore, Berkeley, and Austin, were unified under the aegis of an even broader support system, the American Mathematical Society. To a large extent, the spectacular developments which took place from Sylvester's arrival in Baltimore in 1876, through Klein's tutelage in the 1880s and 1890s, to Moore's dominance at Chicago by 1900, paved the way for the mathematical preeminence America would come to enjoy in the twentieth century.

Roger L. Cooke was a student of S. Bochner at Princeton University, where he received his Ph.D. in 1966. Since 1968, he has held a position at the University of Vermont. He did research in Fourier analysis before turning to the history of mathematics. During a recent visit to the Soviet Union, he did research on the history of the Moscow School of Analysis associated with N. N. Luzin.
V. Frederick Rickey received his Ph.D. in 1968 from the University of Notre Dame, working under the direction of B. Sobociński. Since then, he has held a position at Bowling Green State University. He did research in logic until his conversion to the history of mathematics. He received a Pólya Award from the MAA for a recent paper on Isaac Newton.

W. E. Story of Hopkins and Clark

ROGER COOKE AND V. FREDERICK RICKEY

INTRODUCTION

The career of W. E. Story (1850–1930) is intimately bound up with the first period (1875–1920) of institutionalized American mathematical research. Until after the Civil War, professors of mathematics in America generally attempted only to understand and transmit to their students the mathematics of previous generations. They rarely engaged in mathematical research, partly because their universities did not foster such activity. It was only during the general cultural expansion immediately following the Civil War that a few Americans began to study mathematics at European universities and some American universities began to offer graduate degrees in mathematics. The establishment of graduate programs at Hopkins, Clark, and Chicago is the clearest sign of a mathematical awakening in America. Although the program at Clark is the least known of these three, it was the leading light of institutionalized American mathematical research in the early 1890s. It also formed a transition between the program at Hopkins, which blossomed during J. J. Sylvester's tenure from 1876 to 1883, and that at Chicago, which developed rapidly in the mid-1890s.

An important figure in America's late-nineteenth-century emergence from the mathematical backwaters was William Edward Story. He graduated from Harvard, earned a Ph.D. in Germany, conducted mathematical research as a faculty member at Hopkins, and developed the graduate program at Clark. Thus not only was Story a central actor in the development of American mathematics, but also his career was a microcosm of the new mathematical activity. These are some of the reasons his biography provides an ideal basis for discussing the mathematical climate of the time. To emphasize the changes in that climate, it is appropriate to begin with his intellectual forbearers, who represent an earlier, less institutionalized phase of mathematical activity.[1]

1. Story's Intellectual Background

Benjamin Peirce (1809–1880), the first great American mathematician, was professor of mathematics at Harvard for nearly fifty years, from 1831 until his death in 1880, but there were only two periods when he had many advanced mathematical students. The first was during the 1850s and early 1860s when the *American Ephemeris and Nautical Almanac* office was located in Cambridge (1849–1867). One member of this group was Charles W. Eliot (1834–1926), who earned his A.B. in 1856 and A.M. in 1858 and then stayed on for three years as tutor in mathematics. In this capacity he helped Peirce introduce written final exams. One of the objections to this reasonable sounding proposal was faculty concern about the students: "more than half of them can barely write; of course they can't pass written examinations" [Flexner 1930a, p. 86]. Eliot also taught chemistry at Harvard (1858–1863) and MIT (1865–1869) before becoming president of Harvard in 1869. Up to this time, most colleges had a lock-step curriculum, but Eliot instituted the free elective system. This system allowed weak students to avoid mathematics and strong students to take as much as they wanted. Peirce was a strong advocate of this system, for it allowed him to devote his energy to the good students. This policy brought Peirce another group of advanced students in the last decade of his life [Anonymous 1911a, p. 7].

The Harvard class of 1871 consisted of 158 graduates, three of whom became mathematicians. Henry Nathan Wheeler (1850–1905) was the author

[1]The authors would like to express their sincere thanks to Dr. Stuart Campbell, University Archivist at Clark University, for his most gracious help with using the archives. Both he and University Historian, Dr. William A. Koelsch, were generous in sharing their knowledge of the early history of Clark University. We would also like to thank Mrs. Cynthia Requardt of Johns Hopkins University for archival assistance. We thank the Milton S. Eisenhower Library at The Johns Hopkins University for permission to publish the letters quoted in §§2–3 below from the Daniel Coit Gilman papers, MS. 1, Special Collections, and cited as "Gilman papers." Finally, we thank the Clark University Archives, Clark University, Worcester, MA, for permission to publish the letters and documents quoted in §§5–10 and for permission to publish the photographs which appear herein.

of half a dozen elementary mathematics texts. He served as proctor and instructor at Harvard until 1882 when he took charge of the educational department at Houghton Mifflin Publishing Company [Anonymous 1921a, pp. 187–188]. William Elwood Byerly (1849–1935) continued his education at Harvard, earning his Ph.D. in 1873 (Harvard's first two Ph.D.s were in this year) with a dissertation on "The Heat of the Sun." After teaching at Cornell for three years, he returned to Harvard where he taught until his retirement in 1913. Byerly was an exceptional teacher and administrator and an early advocate of higher education for women. Of his six textbooks, those on the calculus are noteworthy for initiating the long lists of exercises that are so common today. From 1899 until 1911, Byerly was one of the editors of the *Annals of Mathematics*, which had been founded by Ormond Stone (1847–1933) at the University of Virginia in 1884.

The third mathematician from the class of 1871 was William Edward Story (1850–1930), the main character of our story. Born in Boston on 29 April 1850, Story was the eldest son of Isaac and Elizabeth Bowen Woodberry Story and a descendant of Elisha Story, who came from England about 1700 and settled in Boston. His ancestors included Dr. Elisha Story of Bunker Hill, one of the "Indians" at the Boston Tea Party, and a great uncle, Joseph Story, who was a Supreme Court justice. He was "fitted for College" at the high school in Somerville Massachusetts, where his father was a lawyer [Anonymous 1921a, p. 163].

In the fall of 1867, he entered Harvard College, where he took advantage of Eliot's new elective program and "took all the courses in mathematics then given" including one on elliptic functions and another on the *Theoria Motus* of Gauss (Story to Gilman, 29 July 1876; Gilman papers). Story graduated "with Honors in Mathematics, (being the only graduate who has as yet complied with the requisites for those honors since their establishment in 1870–71)" (J. M. Peirce to Gilman, 4 July 1876; Gilman papers).

In contrast to his classmate Byerly, Story chose to go to Germany for further study. Although a thin stream of students had been going abroad for sixty years, he was ahead of the flood. For the next two and one-half years (September 1871–January 1874), Story studied mathematics and physics at Berlin and Leipzig. He returned home for the spring and summer of 1874 before going back to Germany on a Parker fellowship in October 1874. These fellowships, offered only to Harvard graduates, were designed to encourage study abroad.

Story was one of the first American mathematicians to take a degree at a German university, receiving his Ph.D. at Leipzig on 31 July 1875 for a dissertation entitled *On the Algebraic Relations Existing Between the Polars of a Binary Quantic.* James Mills Peirce (1834–1906) called this "a most masterly treatment, involving considerable originality, of a very abstruse & important subject of the modern 'Higher Algebra' or Theory of Quantics" (to

William E. Story at his desk, 1892–1893. (Clark University Archives)

Gilman, 4 July 1876; Gilman papers). Remembering that J. M. Peirce was a great teacher but not a creative scholar, one might not put much stock in this evaluation; however, Peirce remarked that his father, Benjamin, concurred in the judgment.

According to the Vita in his Ph.D. dissertation, Story attended the lectures of Weierstrass, Kummer, Helmholtz, and Dove in Berlin, and Neumann, Bruhns, Mayer, Von der Mühll, and Engelmann in Leipzig. We have been unable to determine who directed his dissertation, but suspect it was Karl Neumann (1832–1925). It was not Felix Klein as Reid [1978a, p. 21] surmises; we have no evidence of any contact between Story and Klein during Story's student days.

After earning his degree, Story returned to Harvard, where he served as tutor from September 1875 until July 1876.

2. The Johns Hopkins University

On the day before Christmas in 1873, the Baltimore bachelor financier Johns Hopkins died, leaving his entire fortune of seven million dollars to found a university and hospital. While planning the university, the trustees sought, and received, considerable guidance from three university presidents, all of whom had been trained as scientists: Charles William Eliot, president of Harvard from 1869 to 1909, James Burrill Angell, president of Michigan from 1871 to 1909, and Andrew Dixon White, president of Cornell from 1866 to 1885 [Hawkins 1960a, p. 9]. It was decided to interview Daniel Coit Gilman (1831–1908), who was then serving as the first president of the University of California. Gilman quickly made it clear that he wanted to found a university of national scope which promoted advanced scholarship and the training of graduate students [Hawkins 1960a, p. 22]. Since the trustees were already inclined in this direction, they agreed with him and offered him the position. Gilman quickly accepted the invitation to be the first president of Johns Hopkins University.

Gilman told the trustees that if they could hire a great classicist and an outstanding mathematician, everything else would take care of itself [Flexner 1946a, p. 29]. He hired classicist Basil Gildersleeve (1831–1924) and mathematician James Joseph Sylvester (1814–1897), and things did take care of themselves. Sylvester did not come easily or cheaply, but once the complicated negotiations were completed, he was most enthusiastic about "our university." It took $6,000 in gold to get him, a handsome salary considering that Yale's highest salary was then $3,500, Harvard's $4,000, and these were unusually high [Hawkins 1960a, pp. 42–43].

Sylvester arrived in Baltimore in May of 1876, but left again almost immediately for New York City to look for "his most precious box—the one containing his life's work in manuscripts" [Hawkins 1960a, p. 44], which had been lost in transit. As he had found the heat unbearable, he continued North to Harvard to visit his old friend Benjamin Peirce, who had been his host in 1842 and 1843 after Sylvester spent a few months at the University of Virginia. Perhaps it should be added—to quell persistent rumors—that Sylvester did not quit that post because he killed a student [Feuer 1984a].

The Peirces—both Benjamin and his son James Mills—independently recommended that Sylvester hire Story as an assistant professor. Story's dissertation impressed Sylvester, and so he promised to try to meet him. Sylvester had also asked about Story's classmate, Byerly, so J. M. Peirce wrote to President Gilman of Hopkins that Byerly

> is a man of great ability & character, a good mathematician, an assiduous worker, & would be an accession to any university in the country. I told Mr. Sylvester however that I thought Dr. Story would be an even better man for you ... [4 July 1876; Gilman papers]

After describing Story's background, calling him a "mathematician of great promise," indicating that they would hate to lose him yet felt they could not hold him back, and singing his praises for several pages, J. M. Peirce adds:

> My Father wishes me to say that he fully concurs in my opinion of Dr. Story.... We both think him the most promising mathematician that has been produced here for many years, & likely to hold a distinguished position among the Scientific men of America. He is by no means a mere teacher. [Gilman papers]

But Sylvester continued to complain of the heat and "depression"—it plagued him every summer in America—and so decided to return to England for a holiday before classes began at Hopkins. Sylvester's departure left it up to Gilman to negotiate with Story for the position as Sylvester's assistant. Gilman telegraphed an offer to Story, adding:

> If you desire light work and a good place in which to study I think you will find the place of an Associate ... honorable and advantageous. [Hawkins 1960a, p. 44]

Not surprisingly, Story found this a bit condescending, and so Gilman quickly learned that younger mathematicians can be difficult to deal with too. Story replied "as distinctly as possible" that he wished

> to devote my leisure time to original work as a mathematician, not merely as a student. I do not therefore lay so much stress upon having much leisure, but the high character of the work which seems to be demanded at Baltimore is a greater object with me.... I know what work is, and have no objection to it. [Story to Gilman, 29 July 1876; Gilman papers]

Story ended the letter by expressing the desire for an interview. He wanted to explain to Gilman his plans for a mathematical journal and a student mathematical society. He also tried for a better position at Harvard—remember he was only a tutor—but when nothing materialized he accepted the position at Hopkins [Hawkins 1960a, p. 45].

In the fall of 1876, William Story became "associate" at Hopkins. He was the only other faculty member in mathematics at Hopkins besides Sylvester. The title was equivalent to that of assistant professor elsewhere although the rank was not created at Hopkins until 1945. In 1883, when the new rank of "Associate Professor" was created at Hopkins, Story was promoted to that rank.

There is evidence that Story succeeded in founding his student mathematical society. *The Johns Hopkins University Circulars*, which are a rich source of information about the university, contain titles and reports of the talks given at the monthly meetings of the "Mathematical Society." From one of these we learn that when Lord Kelvin lectured at Hopkins in 1884, he spoke to a group of mathematicians who called themselves "the coefficients" [Gilman 1906a, p. 75].

3. The American Journal of Mathematics

On 3 November 1876, only a few weeks after classes began at Hopkins, President Gilman held a dinner in honor of Sylvester. Probably Gilman saw to it that Story's idea of a mathematics journal "emerged," for on 8 November 1876 a crudely duplicated letter was sent out proposing "The American Journal of Pure and Applied Mathematics" (see [French 1946a, pp. 51–52] for the text). The proposed title was doubtless influenced by the British *Quarterly Journal of Pure and Applied Mathematics*, which Sylvester had edited since he and Ferrers founded it in 1855 to replace the *Cambridge and Dublin Mathematical Journal*. The letter was signed by Sylvester, Story, Rowland, and Newcomb.[2] It elicited more than forty responses, all but one favorable. Most promised to subscribe and many offered suggestions. The suggestion of Joseph Henry that the journal be an instrument for education as well as research was, fortunately, ignored. The proposed new journal also aroused interest in the popular press.

[2]The physicist Henry A. Rowland (1848–1901) was the first faculty member and full professor hired by Gilman, whose interest had been piqued when he learned that the *American Journal of Science* had thrice rejected Rowland's papers because of his youth. Today Rowland is remembered for work he did in the 1880s: the invention and ruling of concave spectral gratings to accurately measure wavelengths of light. The mathematical astronomer Simon Newcomb (1835–1909) was associated with Hopkins from its beginnings, first as a visiting lecturer and later as Sylvester's replacement. Although essentially self-educated, Newcomb did study with Benjamin Peirce, getting a degree at the Lawrence Scientific School at Harvard in 1858.

As might be expected of any new journal, the *American Journal of Mathematics* had its initial difficulties. Gilman could not find a publisher to assume ownership, and the trustees of Hopkins refused to take on the burden, although they did provide $500 per volume, or about a fifth of the cost [Hawkins 1960a, p. 75]. This explains why the title page of the new journal proclaimed that it was "issued under the auspices of Johns Hopkins." Sylvester was wise enough to realize that the financial and managerial details of the journal were not his forte and would take time away from his research, so Story was appointed "associate editor in charge."

On 17 March 1878, Sylvester invited Benjamin Peirce to Baltimore "to dine with us and some of the supporters of the Mathematical Journal to celebrate its birth which is now daily expected and which you have done much to promote" [Archibald 1936a, p. 139]. Although the first issue, dated "January 1878" did not appear until at least March of that year [Archibald 1936a, p. 136], Sylvester realized that "Story is a most careful managing editor and a most valuable man to the University in all respects and an honor to the University and its teachers from whom he received his initiation" [Archibald 1936a, p. 139]. Publication deadlines are the scourge of all editors, and Sylvester was no exception. Two years later, on 25 March 1880, he wrote Mrs. Benjamin Peirce that "Our December number of the Journal [vol. 2, no. 4] still tarries in coming out," but, rather than being disturbed by the delay, he is delighted with the issue itself. He continues:

> It will be a glorious number and two contributions from [your son] Charles [Sanders Peirce (1839–1914)] ... will form not the least interesting part of its contents. It opens with Tables of Invariants and concludes with two dissertations on the 15 puzzle [of Kirkman]. So you see we take a wide range. But I tell Dr. Story that the 15 puzzle will be the gem of the number and help to make the other matter go down. [Archibald 1936a, pp. 144–145]

These papers, by W. W. Johnson and Story, were the first to show the impossibility of certain arrangements of the sliding blocks in this puzzle which "was engrossing the minds of millions of people" and is still familiar today to Macintosh users. This paper became part of Story's popular fame. As his obituary states:

> Dr. Story was deeply interested in all kinds of puzzles. His mathematical mind and profound knowledge combined with practice to make him a great expert. Few problems of this description baffled him, no matter how difficult they might be. [*Worcester Evening Gazette*, 10 April 1930]

While there had been some early editorial disagreements between Sylvester and Story [Archibald 1936a, p. 137], matters came to a head with the January

1880 number of Volume 3. Sylvester sailed for England in the late spring, as was his custom throughout his stay at Hopkins, and as he had done the previous year, he left Story in charge, with instructions about how he wanted the issue put together.

In early June, Sylvester wrote Gilman inquiring why he had not received an acknowledgment of a paper he sent Story [Fisch and Cope 1952a, p. 358]. Then, on 22 July 1880, Sylvester sent Gilman an eight-page letter in which his indignation is clearly manifested by his heavy, nearly illegible penstrokes:

> I have sent off a telegram to you this morning requesting to be informed when "Journal did or will appear." A telegram sent to Story a week or two ago has met with no response. His answer by letter to my message through you was utterly unsatisfactory.
>
> He gave no explanation worthy of the name why I had to wait for 8 or 9 weeks before receiving an acknowledgement which I had requested of a communication for the Journal sent from [illegible] on my arrival there. If he treats me in this way how is he likely to act towards other contributors?
>
> He informs me that he has allowed Rowland to exceed the limits of the Journal by 20 pages in flat disobedience to my directions and without referring the matter to me for my opinion and in the face of the fact known to him that I had risked giving offense [to] C. S. Peirce by requesting him (which he complied with) to abridge his most valuable memoir in order that the proper limits might not be exceeded and above all that the publication of the number that was due might not be delayed.
>
> It ought to have appeared (as all the matter had been sent in before my departure) during the month of May or very early in June at latest. It is now the end of July and I am kept by Story on this as on all other matters connected with the Journal (since I left) completely in the dark and am unable to give any reply if asked when it will appear. It is 7 months after time. Every one (persons of the highest position that I can name) says that this delay and irregularity are doing immense injury to the Journal.
>
> When I consider Story's conduct since my absence this year and couple it with the fact of his disobeying my directives when I was absent last year and the inexcusable want of right [?] feelings not to say mala fidés exhibited by him in his [illegible] of Mr. Kempe's valuable memoir, I have come to the conclusion that it is inexpedient that we should continue to act together in carrying on the Journal and as I am primarily responsible to the Public, to the Trustees, and the World of Science for its success, I formally request that arrangements may be made for dissolving the present

> connexion of Story with the Journal and myself as I can no longer work satisfactorily with or feel any confidence in him—for I consider that his conduct has proved him to be wanting in loyalty and trustworthiness—I shall be willing to return to America at any moment when requested and shall be prepared to take upon myself in future any additional amount of labor in connexion with the Journal and will undertake unaided to carry it on satisfactorily and in a businesslike manner. I could and of course would, take means to provide myself with some useful subordinate in whom I could place confidence and would undertake that under no circumstances should the funds of the University be called upon for assistance beyond that stipulated for under the existing arrangements. I feel the deepest and (as mature reflexion and consultation with others who are dispassionate enable me to affirm) well founded displeasure with Dr. Story and no explanation that he might assume to offer can remove this feeling or ever again induce me to place confidence in him—I do not write this under any seal of confidence.
>
> He is at liberty to know of my opinion of his conduct and the wish I have expressed to be released from all further connexion with him in the conduct of the Journal on the ground that I can no longer place any confidence in him.
>
> I am willing to return at the shortest possible notice if in your opinion the interests of the Journal render it desirable that I should do so. [Gilman papers]

We have no information about what blunder Story made in handling A. B. Kempe's now famous paper "On the Geographical Problem of the Four Colours" [*American Journal of Mathematics* **2** (1879), 193–200]. He did follow it with his own "Note on the preceding paper" [*American Journal of Mathematics* **2** (1879), 201–204]. Perhaps Sylvester did not consider this appropriate.

Before continuing the discussion of the contents of Sylvester's letter, we should let Story tell his side of it, as he did in a letter to Gilman of 26 July 1880. It should be kept in mind, however, that he is reacting not to the above letter of Sylvester but to the telegram and letters he received from Sylvester as well as to a note from Gilman of July 24. It was not until 7 August 1880 that Gilman could write Story that he had received Sylvester's letters of July 22 and 24. He did not show them to Story as Sylvester allowed, but notes

> I think a frank explanation to him of the serious difficulties you have encountered and an apology for any delay on your part to answer his telegram and letter would not be amiss... I should be truly sorry to have you lose his confidence and good will. I think they are possessions which you will not lightly forfeit. [Gilman papers]

Here is Story's letter to Gilman in full (from the Gilman papers). He obviously had anticipated the request for a frank explanation.

Catonsville, Baltimore Co., Md.
26. July. 1880.

My Dear Sir:

The first number of vol. III of the "Journal of Mathematics" is not yet out, although all the articles are nearly or quite ready for the press. It has been a very hard number for me. Every page of Stringham's and [C. S.] Peirce's articles has been worked over by me, and I have read Sylvester's and Rowland's as carefully as I could without working all the formulae out. Franklin also read Sylvester's, and he took no little time about it, during which I had to wait. Just before he left Sylvester gave orders to replace Craig's article of 14 pages (the first 14 in the number) by Stringham's: "On regular figures in n-dimensional space", which Stringham had not then in any kind of form. I worked this paper out very carefully with Stringham, giving him constantly suggestions and criticisms, and it was only the day before he sailed that he put the finishing touches to it. This paper was a great cause of delay. There is now no particular reason why the number should not appear as soon as the sheets can be worked off. I shall explain all this to Sylvester in a letter of same date as this. He is very hard to satisfy, especially when away from the field of operations. He writes me that he greatly disapproves of my course in inserting the whole of Rowland's article in this number, thereby causing the number to run over the regular limit by 16 pages. But Rowland insisted on the insertion of his paper entire, although I explained to him that Sylvester expressly desired that the number should not exceed the regular limits. R. said "Cut down Sylvester!" However it is not too late now to change and I shall cut off R.'s paper at the usual end of a number, running on the latter half in the next number. So there are nearly or quite ready for the press 27 pages of Number 2. I cannot please all parties. I have not yet found time for any original work this vacation, but in the necessary pauses in this unremunerative [?] editing have been rusticating a little.

I understand that much dissatisfaction is felt at Harvard on account of the appointment of L. & others over men who have been there some time and who thought they had a right to some consideration.

Very Truly Yours
William E. Story

The closing comment about "L" being appointed at Harvard is enigmatic, but the rest is straightforward. Sylvester ordered Story to replace Thomas Craig's "Orthomorphic projections of an ellipsoid upon a sphere" by a not yet finished paper of W. I. Stringham. As both of them were Hopkins Ph.D.s under Sylvester (1878 and 1880, respectively) and both were teaching at Hopkins, we must presume that Sylvester felt Stringham's enumeration of *n*-dimensional polyhedra was of more interest—remember that higher dimensional geometry was then very much in vogue—than Craig's continuation of Gauss's work on the projection of an ellipsoid on a sphere. Nowadays we would consider Craig's paper more interesting; Sylvester's editorial decision is probably only a reflection of his interest in pure mathematics. Story must have had to work hard to force Stringham's paper into the same fourteen pages that Craig's was to occupy. Probably the two plates that were sewn in took extra time in printing. His only reward was Stringham's "grateful acknowledgement ... especially to Dr. Story, for valuable suggestion[s]" [*American Journal of Mathematics* **3** (1880), 14].

It is understandable that Rowland, one of only four professors at Hopkins, insisted that his paper not be cut into two parts. He did get his way, but it was not by cutting down Sylvester. Story must have decided—or perhaps he was told—to keep the issue to the prescribed size by putting off Rowland's paper to the second number of Volume 3. Thus Number 1 must have consisted of three papers occupying eighty-eight pages: Stringham; C. S. Peirce, "On the algebra of logic;" and Sylvester, "On certain ternary cubic-form equations." Number 2 began with Rowland's paper and was followed by Craig's.

Although Sylvester wanted to lay all the blame for the delay on Story, it appears that a good deal of it was caused by the changes Sylvester initiated. We have already noted the substitution of Stringham's paper for Craig's, and a note that the printer sent Gilman (received 27 July 1880) says that

> We have just learned that Prof. Sylvester's article was only returned to us on Saturday last, and that it was *dreadfully cut up*, and that another proof of it has to go out. [Fisch and Cope 1952a, p. 358]

Perhaps this proof was the one read by Fabian Franklin, but more likely it was Sylvester's own. We do not know when this issue finally appeared, or whether Sylvester persisted in his demand that Story be fired immediately, but we do know that on 7 August 1880 C. S. Peirce wrote Gilman:

> I have received from Sylvester an account of his difficulty with Story. I have written what I could of a mollifying kind, but it really seems to me that Sylvester's complaint is just. I don't think

> Story appreciates the greatness of Sylvester, and I think he has undertaken to get the *Journal* into his own control in an unjustifiable degree. I think that we all in Baltimore owe so much to Sylvester that he should be supported in any reasonable position with energy; I hope the matter may not go to the length of displacing Story because I think he is admirably fitted for it in other respects than those complained of. But Sylvester ought to be the judge of that. It is no pleasure to me to intermeddle in any dispute but I feel bound to say that Sylvester has done so much for the University that no one ought to dispute his authority in the management of his department. [Fisch and Cope 1952a, p. 297]

This attempt at mediation did not succeed for long, if at all, for Story's name last appears as "Associate Editor in Charge" on the title page of Volume 3 which is nominally 1880. The title page, contents and errata would have been the last part of the volume printed, sometime before the spring of 1882 (for C. S. Peirce was then working on his father's celebrated paper on linear associative algebras which appeared in Volume 4 ("1881"), Number 2 [Fisch and Cope 1952a, p. 299]). We suspect that Volume 3 was printed in early 1882, for that is consistent with the fact that throughout the records at Clark, Story listed his term as 1878–1882 (see, e.g., [*Decennial*, p. 546]). It would be interesting to know precisely which issues of the journal Story edited. We also do not know whether he was fired or resigned under duress. Fisch and Cope claim that Story took a "quasi-proprietary interest" in the journal [1952a, p. 358]. We find this too strong a judgment, though perhaps his outlook played a role in his demise as editor.

Story was replaced by Thomas Craig (1855–1900), who was, like C. S. Peirce, dividing his time between the U. S. Coast Survey (Craig was assistant in the Tidal Division) and half-time teaching at Hopkins. Craig, who was Sylvester's first Ph.D. (1878) at Hopkins, had been teaching there since he arrived when the university opened in 1876 (except for the spring of 1878). Whether Sylvester wanted Craig full time at Hopkins so that he could replace Story as associate editor-in-charge we do not know, but from a letter that he wrote to Gilman on 28 March 1881, we learn that he did get him:

> Allow me to express the great satisfaction I feel in the interest of the University at the measures adopted by the Trustees to secure the continuance of Craig and Peirce. We now form a corps of no less than eight working mathematicians—actual producers and investigators—real working men [sic]: Story, Craig, Sylvester, Franklin, Mitchell, [Christine] Ladd [Franklin], Rowland, Peirce; which I think all the world must admit to be a pretty strong team. [Fisch and Cope 1952a, p. 297]

While we are not sure when Craig replaced Story as associate editor of the journal, his name did not appear on the title page until Volume 6, dated 1884, where he is listed as "Thomas Craig, Ph.D., Assistant Editor." Story's name last appears on Volume 3 (1880). Sylvester's alone appears on Volumes 4 and 5. Newcomb becomes the chief editor beginning with Volume 7, with Craig being "associated." This state of affairs continues until Volume 16 (1894), when the journal is "Edited by Thomas Craig with the Co-operation of Simon Newcomb." The same is true the next year, but Craig's name does not appear at all on Volume 21 (1899), when Craig had to resign due to poor health. He died in 1900.

Before concluding this section, we want to go back and consider the question of who founded the journal. On 20 December 1883 at a banquet in honor of Sylvester, who was about to leave Hopkins to take up the position of Savilian Professor of Geometry at Oxford, Sylvester was explicitly given credit by Gilman for founding the journal. Sylvester's reply was as follows:

> You have spoken about our *Mathematical Journal.* Who is the founder? Mr. Gilman is continually telling people that I founded it. That is one of my claims to recognition which I strongly deny. I assert that he is the founder. Almost the first day that I landed in Baltimore, when I dined with him in the presence of Reverend Johnson and Judge Brown, I think, from the first moment he began to plague me to found a *Mathematical Journal* on this side of the water something similar to the *Quarterly Journal of Pure and Applied Mathematics* with which my name was connected as nominal editor. I said it was useless, there were no materials for it. Again and again he returned to the charge, and again and again I threw all the cold water I could on the scheme, and nothing but the most obstinate persistence and perseverance brought his views to prevail. To him and to him alone, therefore, is really due whatever importance attaches to the foundation of the *American Journal of Mathematics* ... [Cordasco 1960a, p. 107]

Sylvester's reluctance because of lack of material had already been countered by Rowland in an article decrying the state of American science and the need for scholarly journals. Regardless of how modest Sylvester might have been on his departure, there is no doubt that President Gilman deserves a very large share of the credit for introducing scholarly journals in this country and especially the series of American journals that he began at Hopkins. This is also indicated in his reminiscence titled *The Launching of a University*:

> When Sylvester agreed to come to Baltimore, he was requested to bring along with him the *Mathematical Journal* of which he had been one of the editors, but this was not practicable. His American colleague, Dr. W. E. Story, independently proposed the

> establishment of an *American Journal of Mathematics*, and, after a good deal of correspondence, it was decided to begin such a journal, in a quarterly form, and to ask the concurrent editorial aid of professors in other universities. It was intended that the *Journal* should be open freely to contributors in any part of the country. [Gilman 1906a, pp. 116–117]

Thus we see that Story independently had the idea of founding the journal. Story's involvement is also evident from his entry in Poggendorff: "Gründete 1878 & edirte bis 1881 Vol. 1–3 mit Sylvester d. 'Amer. J. of Math'." [Vol. 3, p. 1303], although we do not know whether this was written by Story himself or by Poggendorff.

We have provided all of this detail partly because of its inherent interest, but also to contradict the common myth that Sylvester founded the *American Journal of Mathematics*. This appraisal is too simplistic. There is no doubt that his international reputation and connections played a vital role in the development of the journal. But Gilman deserves credit for seeing that the publication of scholarly journals was absolutely vital to the development of his university, and William Story deserves credit for independently seeing the need for and conceiving of a journal of mathematics. Story most likely was also instrumental in encouraging Gilman to get Sylvester involved. But however it began, Story did not get his own journal at Hopkins.

4. Story's Best Student at Hopkins, Henry Taber

While Sylvester was at Hopkins, Story taught a variety of subjects ranging from quaternions, elliptic functions, and invariant theory to mathematical astronomy and the mathematical theory of elasticity. But he seemed to favor higher plane curves and solid analytic geometry, subjects which for him included the general theory of curves and surfaces. In the fall of 1884—Sylvester had left the previous January—Story began giving an "Introductory Course for Graduates" which consisted of short sequences of lectures on the leading branches of mathematics and which was designed to give the beginning graduate student an overview of mathematics [Cajori 1890a, p. 276]. Story's care in the redesign of the curriculum is alluded to by Fabian Franklin (1853–1939), writing after Sylvester's death:

> It would never in the world have done to have a whole faculty of Sylvesters; anything like a systematic programme would have been out of the question, ... the presence of *one* Sylvester was of absolutely incalculable value. Not only did he fire the zeal of the young men who came for mathematics, but the contagion of his intellectual ardor was felt in every department of the university, and did more than any one thing to quicken that spirit of idealistic

devotion to the pursuit of truth and the enlargement of knowledge which is, after all, the very soul of a university. [*The Nation*, 22 October 1908, p. 381]

An essential part of the student's education was the "mathematical seminary," to use their quaint sounding phrase. Sylvester presided during his tenure, but when Newcomb replaced him there were three such seminaries. One was run by Craig, one by Newcomb, and one by Story. The purpose of these seminaries was to get students involved in research [Cajori 1890a, p. 276].

Space does not permit an excursion into the details of the Hopkins curriculum. Instead, we shall trace the studies of one student, whose future career forms an important part of the story we are telling. Fortunately, the *Johns Hopkins University Circulars* make it possible to trace each student in minute detail, since the *Circulars* provide a list of those enrolled in every course, as well as a wealth of other information about the academic program. The student we are interested in is perhaps not typical, since he started in philosophy and switched to mathematics. Also his subsequent career was more distinguished than that of most Ph.D.s, despite the chronic health problems which retarded somewhat his academic progress. Nevertheless, his biography gives insight into the state of American mathematics at both Hopkins and Clark during the period of the current study.

Henry Taber (1860–1936) was born at Staten Island, New York, on 10 June 1860. He entered Yale's Sheffield Scientific School to study mechanical engineering in 1877, but had to leave temporarily because of illness. When he found himself unsuited for engineering, he was allowed to substitute a special course in mathematics for part of his work. Taber finally graduated with a Ph.B. in 1882.

Taber went to Hopkins in the fall of 1882. From the *Circulars*, we know that he attended Story's higher plane curves (three hours), as well as Thomas Craig's elliptic functions (three hours) and calculus of variations (two hours). He also took Charles Sanders Peirce's elementary logic course (four hours each semester), which seemed to attract his interest. In the spring, he took Story's conic sections (three hours). In 1883–1884, Taber took Peirce's advanced logic course (two hours) and his probabilities course in the spring (two hours). Incidentally, Story also was an auditor in the probabilities course, since he is listed among the students in the *Circulars*. Taber took no courses from Sylvester in the three semesters when they overlapped. Since Peirce was not reappointed for the 1884–1885 academic year (for unknown reasons, cf. [Hawkins 1960a, p. 195]), Taber switched fields and began to take more mathematics courses. In the fall of 1884, Taber took three of the five courses (thirteen hours per week) taught by Story: the introductory course for graduates (five hours), number theory (two hours), and modern synthetic geometry (three hours). For the next year and a half Taber took no courses; we conjecture that he was ill.

In the fall of 1886, Taber took two of Story's four courses: quaternions (three hours) and advanced analytic geometry (two hours). In addition, he took Story's seminar (Story again taught thirteen hours). In the spring of 1887, he continued in these three courses and picked up the second half of Story's introductory course for graduates.

In his sixth year (actually the second half of his fourth year, taking account of the eighteen-month hiatus in his enrollment), Taber took Story's linear associative algebras (two hours) and advanced analytic geometry (three hours). Finally, the theory of functions course, taught by Craig using the books of Briot and Bouquet, and Hermite, attracted Taber's interest. In his final semester at Hopkins, Taber took only Story's seminar. It is interesting to note that in this year—1888—Story gave a course in "symbolic logic" which may well have been the first such course in a *mathematics* department; Taber, however, did not take this course.

On 14 June 1888, Henry Taber received his Ph.D. in "Mathematics and Logic" for a thesis entitled "On Clifford's *n*-fold algebras" [*American Journal of Mathematics* **12** (1890), pp. 337–396]. No director is listed for his thesis in *Circular* #67, but undoubtedly it was Story. The next year (1888–1889), Taber was "Assistant in Mathematics" at Hopkins, teaching analytic geometry (two hours) and trigonometry (one and one-half hours) both semesters. Alas, the only thing that has changed in the intervening century is that we now take more hours per week to do this!

In the spring of his year as assistant Taber attended Craig's abelian functions (two hours) and then a very famous course on the "Theory of Substitutions" which met five hours per week for four weeks. The latter was taught by Oskar Bolza, an 1886 Ph.D. of Felix Klein, from whom he had taken a similar course in Germany in the summer after he received his degree. This course represents the first discussion of Galois theory in this country. Ten people attended the course including Craig, Franklin, and Story, i.e., all of the faculty except Simon Newcomb, who never taught courses unrelated to astronomy, and C. Smith, who taught only solid analytic geometry [*Circular* #71]. The absence of Newcomb is rather odd, since it was he who had encouraged Bolza to lecture on the theory of substitutions and its application to algebraic equations [Bolza 1936a, p. 20].

Under ordinary circumstances, it is likely that Story and Taber would both have spent their entire careers at Hopkins, contributing a respectable amount to mathematical research, but not having great impact on the direction in which it developed. However, in 1889, an opportunity arose for Story to mold a mathematics department in his own image. That opportunity changed

Henry Taber at his desk, 1892–1893. (Clark University Archives)

the careers of both Story and Taber and had a significant impact on the development of mathematical research in America.

5. The Founding of Clark University

Jonas Gilman Clark (1815–1900) was a New England farm boy with little schooling whose mother taught him to love books and reading. He learned the wheelwright's trade and then went into business selling manufactured goods, first in New England and then in California. Through sagacious strategy, he captured a large share of the hardware and furniture trade in California for several years, but then, because of health problems, Clark was forced to sell his business. He invested his large profit conservatively and wisely and eventually became an extremely wealthy man. He traveled widely throughout Europe, acquired a large library, and took a deep interest in higher education. The founding of a university by his old California friend, Leland Stanford, and the approach of his seventy-first birthday seem to be the impetus for implementing plans to endow a university of his own. Clark wanted to begin with an undergraduate college and then develop graduate programs later. See [Koelsch 1987a] for further details.

The board of trustees that Mr. Clark appointed chose G. Stanley Hall (1846–1924), a prominent psychologist of unusual intellectual breadth and achievement, as the first president of Clark University. Hall had spent several years studying in Europe, had earned a Ph.D. at Harvard in 1878, and had been a professor of psychology at Johns Hopkins since 1881. He was reluctant to leave Hopkins until he formed the opinion that he could create a purely graduate university. He made this clear in his letter of acceptance, stating that he had no interest in "organizing another College of the old New England type, or even the attempt to duplicate those that are best among established institutions old or new" [Atwood 1937a, p. 4].

Hall's first act as president of Clark University was to take a year-long "pedagogic tour" to study European educational methods and facilities and to hire distinguished faculty if possible. On this tour, he tried "to get a clear idea from the expressed opinion of their colleagues, of the relative merits of each of the best German professors, in each of the departments we contemplate" (Hall to Clark, 22 November 1888; [Rush 1948a, p. 24]). What was "contemplated" by Hall was a "purely graduate institution," with work originally in only five areas: mathematics, physics, chemistry, biology, and psychology. It is clear that Hall set his sights on what he believed was the best. He wrote Jonas Clark on 14 November 1888:

> I have learned on all sides that Professor Klein, of whom we have often spoken as about the very best mathematician in Europe, is widely so considered here by those experts most competent to

> judge. I lately spent several hours with him talking about the possibility of his joining us at Worcester. He is inclined to come if he could have $5000 per year which was offered him at Baltimore. [Rush 1948a, p. 21]

Earlier, Hopkins had attempted to replace Sylvester with Felix Klein (1849–1925) [Reid 1978a], and now Hall was going to attempt to hire him at Clark since he was "a great man enough ... to keep our American mathematical students from going abroad to study higher mathematics" [Rush 1948a, p. 25]. One consideration that prevented Klein from going to Hopkins was still an issue, namely the question of sick pay and pensions, which were universal in Germany, but nonexistent at Clark and elsewhere in the U.S. A new issue was that Klein wished to come for only six months a year for several years, so that he could retain his position in Germany. But the most formidable obstacle was the German *Kultusministerium.* As Hall wrote in the letter quoted above,

> This ministry is very reluctant to lose its best men, and, if there is any talk of their going to America diffuses the sentiment that they love money more than science and are not patriotic. Thus they are discredited among the universities.

While there is no evidence that this was ever done, leaving a prestigious position in Germany would certainly have the same effect. In Klein's case, he was decorated by the government for not going.

Klein was interested, said Hall, since "he told me he was chiefly attracted by the opportunity of doing only very advanced work for a very few men, with whom he could carry on his researches." Mr. Clark liked Hall's plan of hiring Klein and approved of "the policy of securing several of the best men that can be obtained" (Clark to Hall, 4 December 1888). However, less than a month after his first letter, Hall wrote that Klein,

> has at length decided (after my going several times to Göttingen to see him) that his wife is so opposed to going to America (and I fancy that the certainty of his speedy call to the first chair of mathematics in Europe soon to be vacated in Berlin by Weierstrass so very well assured) that even if called he could not leave Germany. [12 December 1888; Rush 1948a, pp. 27–28]

At this point, it is tempting to consider "what if." However, we shall avoid the biographical subjunctive and leave the story as it is.

For a variety of reasons, Hall was unable to hire a European for any of the five departments he contemplated. Thus he was forced to go with native talent. The question then arose: Who was the best American mathematician? Hall made the logical and correct choice: He hired William Edward Story.

At the time, Story was forty years old and the possessor of a Ph.D. from Leipzig, which Hall regarded as the best university in Europe. He was an established and respected mathematician with eleven research papers to his credit. He was a member of the London Mathematical Society (elected March 1879), Corresponding Member of the Natural Society of the Natural and Mathematical Sciences of Cherbourg, and a Resident Fellow of the American Academy of Arts and Sciences (elected May 1876). At Hopkins, he had a reputation as a good teacher and was the senior pure mathematician (not counting the hybrid Newcomb). He was also well known and trusted by Hall, having associated with him at the best and essentially only Ph.D. granting institution on the North American continent. In short, Story was the natural choice.

It is not known whether Hall considered anyone else for the position of chairman of the mathematics department, but a glance at the 80 "starred" names in the first edition of J. McKeen Cattell's *American Men of Science. A Biographical Dictionary* (1906) shows that there was really little choice among mathematicians between the ages of thirty and fifty. In 1903, Cattell had a group of ten mathematicians rank all American mathematicians, the top eighty of which are starred in the 1906 edition. The numerical rankings were published in the 1933 edition of *American Men of Science*, p. 1269. Many of those listed had taken classes from Story at Hopkins, and the best of those ranked (E. H. Moore) had received his Ph.D. only in 1885 (at Yale). An examination of Cattell's list makes it clear that Story was the best-qualified person for the job. We are not claiming that Story was a great mathematician, for he was not, but only that he was the best available at the time.

There were many reasons why Story might have wanted to leave Hopkins. He was not a full professor there, though he had been there thirteen years.[3] He was not the editor of the *American Journal of Mathematics*, which had been one of his youthful ideas. Finally, he had come to feel that Hopkins was not the wonderful place intellectually that he thought it might and should be: "a peculiar organization of the Mathematical department of the Johns Hopkins made me feel that I was not as free to carry out my own ideas as I wished" (Story to Hall, 12 December 1912). On the positive side, there would be a lighter teaching load and that would leave more time for research. But perhaps most importantly of all, he would have the opportunity to develop a department that focused on graduate education and on research. And he could do it the way that he thought best. For all these reasons, it is likely that the opportunity to move to Clark would have attracted Story.

---

[3]Story was passed over in favor of Simon Newcomb. This decision is unreasonable if one feels that the department head should be a pure mathematician, rather than an astronomer. Of course many consider Newcomb a mathematician, for he did serve as president of the AMS. Today, a recreational problem posed by Newcomb is of interest in combinatorics [MR, 58 #10473].

Story was originally hired as acting head of the mathematics department. Why he did not receive the title of head is not known; perhaps Hall still hoped to hire a distinguished European. More likely, simple titles were the style of the day. Hall himself was only "temporary professor of Psychology" [*First Announcement*, May 1889, p. 8].

But once Hall had hired Story, the rest of the faculty was easy to fill in. Taber, of course, was happy to follow his mentor. Curiously though, his resignation from Hopkins was announced in the *Circulars* before that of Story. Craig suggested to Bolza that Clark would be a good place for him [Bolza 1936a]. Since Bolza had been introduced to Hall by Klein when Hall was on his pedagogical tour of Europe, he was inclined to accept also. Thus when Clark opened its doors to mathematicians in the fall of 1889, it offered Story as professor, Bolza as associate, and Taber as docent.

Oskar Bolza (1857–1942) had entered the University of Berlin in 1875. His family hoped that he would enter the family business of manufacturing printing presses, but his scholarly bent won out. His first interest was linguistics, then he studied physics under Kirchhoff and Helmholtz, but experimental work did not attract him, so he decided on mathematics in 1878. The years 1878–1881 were spent studying under Elwin B. Christoffel and Theodor Reye at Strasbourg, Hermann A. Schwarz at Göttingen, and particularly Karl Weierstrass in Berlin. "Undoubtedly, the fact that he was a student in the famous 1879 course of Weierstrass on the calculus of variations exerted a strong influence on the formation of Bolza's mathematical interests, although some twenty years elapsed before he began active research in this field, for which he was to gain world renown" [Dictionary of American Biography]. Bolza received his Ph.D. under Klein in 1886 and the following year he, and his good friend Heinrich Maschke (1853–1908), were in a private seminar with Klein. This had the curious effect of undermining Bolza's confidence. He was awed by Klein's quickness, and felt that Maschke was a better mathematician than he was. Bolza had done some practice teaching at a Gymnasium, and he found the experience too physically demanding; there was no energy left for research. His friends Maschke and Franz Schulze-Berg formed the same opinion and left for the United States in 1891. Bolza followed soon thereafter, for he realized there was little hope of obtaining a university position in Germany. Soon after his arrival, he went to Hopkins, where he presented the famous lectures on Galois theory mentioned above.

6. Three Golden Years, 1889–1892

Clark University's *First Official Announcement* in May 1889 contained almost no information about the Department of Mathematics: "Appliances for this department are also liberally ordered; the names of instructors will soon be announced" (p. 18). The *Second Register and Announcement*, which appeared in May 1890, reported that President Hall had hired three Hopkins mathematicians, William Story, Oskar Bolza, and Henry Taber to serve as faculty for the Department of Mathematics and recorded what they had taught in 1889–1890 . The first year their audience consisted of one scholar, L. P. Cravens, whose previous position was Superintendent of Schools in Carthage, Illinois, and the following five fellows in mathematics: Rollin A. Harris, an 1885 Ph.D. at Cornell; Henry Benner, an 1889 M.S. at the University of Michigan; Joseph F. McCulloch, an 1889 M.A. from Adrian (Michigan) College; William H. Metzler (1863–1943), an 1888 A.B. from the University of Toronto; and Jacob William Albert Young (1865–1948), an 1887 A.B. from Bucknell, who had studied at the University of Berlin in 1888–1889.

Fortunately, we are able to get a very detailed picture of the activities of the faculty and the types and level of mathematics studied from the *Registers*. From the *Second Register and Announcement* (pp. 27–28), we learn that Story "directed courses of reading in the following subjects, supplementing the text books by lectures five times weekly:" (1) modern higher algebra, (2) higher plane curves, (3) general theory of surfaces and twisted curves, (4) theory of numbers, (5) calculus of finite differences, (6) calculus of probabilities, (7) quaternions, and (8) modern synthetic geometry. He also gave a course of lectures twice weekly on (9) analytic mechanics. There is no information on how much time was devoted to each of these topics, but probably Story lectured on these topics sequentially. We do know that he taught seven hours per week.

Of the nine courses described in the *Register*, four no longer make up part of any curriculum (numbers (2), (3), (7), and (8)), at least in anything like the form described, though part of their subject matter is subsumed in courses that students do take nowadays. The others are more or less completely taught in the standard undergraduate curriculum of the present. Number theory dealt with what we now call elementary number theory, through quadratic reciprocity. Probability theory was elementary discrete probability, through Bernoulli trials and the study of errors of observation. The course in quaternions indicates the influence of the British school, the influence of Sylvester on Story. The lectures on mechanics must have helped to expand the rather limited offerings in physics, a department which at that time had only one fellow and one faculty member. But the latter was none other than Albert A. Michelson (1852–1931), who attained permanent glory for his experiments on the velocity of light.

Bolza, the Göttingen Ph.D. who came to Clark by way of Hopkins, represented the closest Hall was able to come to importing a German mathematician. He was just at the beginning of his career, but he was well versed in the mathematics of Göttingen and Berlin. His topics for the year were: (1) definite integrals, (2) calculus of variations, (3) elliptic functions, and (4) the theory of functions. In addition, he gave a special course twice a week on "Weierstrass' theory of elliptic functions and Riemann's theory of hyperelliptic integrals." The first topic, which included line integrals and Fourier series, seems rather elementary, but the others were sophisticated even by present day standards. All of these strongly reflect the ideas of Riemann and Weierstrass. Students who heard Bolza lecture were hearing the latest mathematics that could be said to have attained anything like a definitive form.

Taber's second year of teaching was much more exciting than the analytical geometry and trigonometry that he had taught at Hopkins the previous year and is akin to what postdoctorates do today. His course on the theory of matrices was an exposition of topics related to his dissertation, extending the ideas of Cayley, Sylvester, and Clifford. It was therefore fully in the British school, except that the ideas of Benjamin Peirce on linear associative algebras were discussed.

Harris, who was the author of three papers in the *Annals of Mathematics* and the only Ph.D. among the five fellows, gave lectures on the use of analytic function theory in the construction of maps.

The *Register* also indicates which courses were to be given in Clark's second year (1890–1891), says a bit about the facilities at Clark, lists the publications of Story (eleven), Bolza (four), Taber (one to appear), and Harris (three), and indicates the current research topics of the faculty. Story is investigating non-Euclidean geometry, and Taber is applying matrices to nonions and developing Clifford's geometrical algebras and their applications to non-Euclidean geometry.

In summary, the first year of operation at Clark produced an admirable amount of both research and instruction in the very latest topics. Although direct contact with the established European masters was lacking, the mathematicians who were present had studied with these masters and carried with them some of the zeal and ability in research which characterized this vigorous period. Of the three great centers of research, Britain, Germany, and France, the first two showed a fairly direct influence on the work at Clark. The names of Cayley, Clifford, and Sylvester show beyond any doubt the strong influence of the British school on the direction of research, while the frequent mention of Weierstrass and Riemann in Bolza's course injected a significant German influence. At this stage, only the French influence seemed to be missing; no mention was made in the *Register* of the work of Hermite and Picard, even though these two mathematicians had made enormous contributions to the subjects of linear algebra (as we now regard it) and analytic function theory, which were being taught by Story, Bolza, and Taber. This work, however, was closely related to the work of Weierstrass, and may have been mentioned at least in passing in the lectures of Bolza. Unless some lecture notes are discovered, we shall never know.

The *Third Register and Announcement* of May 1891 revealed that two new mathematicians and one new physicist had been added to the faculty for the second academic year, 1890–1891. Henry S. White joined the department as assistant, and Joseph de Perott as docent. Mathematics was also represented in the physics department by the heavily mathematical physicist Arthur Gordon Webster.

Joseph de Perott (1854–1924) was appointed docent in mathematics. He was born in St. Petersburg, raised in Thumiac, France, studied in Paris and Berlin 1887–1880 but received no degree, and was a close friend of Sonja Kovalevskaya. His interest was number theory. For more information on this most colorful of the Clark mathematicians see [Cooke and Rickey 199?a].

Henry S. White (1851–1943) was appointed assistant in mathematics in the fall of 1890. He was born in Cazenovia, NY, graduated from Wesleyan University in 1882, and then taught for several years. On the advice of close friends on the Wesleyan faculty, including Van Vleck, he decided to go to Germany for advanced study. He first went to Leipzig where he studied with Sophus Lie and Eduard Study for a year and then to Göttingen to seek out Felix Klein. Oskar Bolza, Wilhelm Maschke, and Frank Nelson Cole had left the year before White arrived. He took courses from Schwarz and Schönflies and wrote a dissertation on abelian integrals under Klein's direction.

White returned to the U.S. in March 1890 to a position in the "preparatory department" at Northwestern University. However, G. Stanley Hall, to whom Klein had introduced him during Hall's tour of Europe, offered him a position as assistant in mathematics at Clark. White accepted "Though the salary was hardly adequate for subsistence, I accepted it eagerly in spite of kind offers from Evanston and Middletown. My teaching was mainly algebraic and projective geometry, and the invariant-theory of linear transformations" [White 1946a, p. 24]. White had a productive year at Clark, writing two papers, one on ternary and quarternary [sic] linear transformations, and one giving a symbolic proof of Hilbert's method for deriving invariants and covariants of ternary forms.

Arthur G. Webster (1863–1923) was appointed docent in mathematical physics. After graduating at the head of his class with an A.B. from Harvard in 1885 with honors in mathematics and physics, Webster spent a year at Harvard as instructor in mathematics, before leaving for Europe on a Parker Fellowship. He studied at the Universities of Berlin, Paris, and Stockholm, earning a Ph.D. at Berlin in 1890. Webster became a full professor at Clark in 1900 and was elected to the National Academy of Sciences in 1903 at the early age of thirty-nine. He was noted for the heavy use of mathematics in his physics textbooks, including his *Partial Differential Equations of Mathematical Physics* (1927, which was translated into German in 1930 by none other than Gabor Szegö.

Arthur Gordon Webster lecturing on mathematical physics, 1892–1893. (Clark University Archives)

The university had taken great care to define the position of docent, and their final description had even been reported in the *New York Times*. By the charter of the university,

> The highest annual appointment is that of docent. These positions are primarily honors, and are reserved for a few men whose work has already marked a distinct advance beyond the Doctorate and who wish to engage in research.

During his first academic year of 1890–1891, Perott used the two hours per week allotted to a docent to discuss "the most elementary parts of the theory of numbers," but carried the subject far beyond what Story had done the previous year. Perott even included a sketch of Kummer's theory of ideal numbers.

White took over some of the courses given by Story the previous year and lectured on (1) higher algebra, (2) higher plane curves, (3) plain [sic] cubics and quartics, (4) abelian integrals, (5) algebraic surfaces and twisted curves, and gave an introductory course on (6) modern synthetic geometry.

For his second year at Clark, Henry Taber chose to lecture on (1) quaternions, (2) multiple algebra, and (3) logic. The first part of the logic course dealt with symbolic logic as based on the work of DeMorgan, Mitchell, and C. S. Peirce. The second part dealt with the theory of induction and especially the work of John Stuart Mill and Peirce.

The most exciting new development in the second year was that Story began a "seminary." White emphasizes in his 1946a autobiography how the Hilbert basis theorem excited the participants in Story's seminar.

Clearly, there was much more activity the second year. Not surprisingly, what occurred the first year was at too high a level for some, and only two of the six students returned, namely Metzler and Young, both of whom earned Ph.D.s from Clark. They were joined by five more auditors. Alfred T. DeLury (an 1889 B.A. from Toronto University) and Thomas F. Holgate (an 1889 M.A. from Victoria University in Canada) were the new fellows. The three new scholars were Levi L. Conant (an 1887 Dartmouth M.A.), John J. Hutchinson (an 1889 A.B. at Bates College), and Frank H. Loud (an 1873 A.B. at Amherst College).

The academic year 1891–1892 appears, in retrospect, to have been the brightest in the history of mathematics at Clark University. Story, Bolza, Perott, Taber, and White gave lectures which one would greatly wish to have heard. The catalog descriptions alone, from the *Register and Fourth Official Announcement* (April 1892), are exciting to anyone who has looked at these topics. A few samples will suffice to give the flavor. Story lectured on (1) the history of arithmetic and algebra, (2) some topics of analysis situs,

> 3. Modern Algebra; an advanced course on the covariants and invariants of systems of quantics involving any number of variables, their conditions, numbers, and syzygies. The writings of Cayley, Sylvester, and Hilbert formed the basis of this course, from which the symbolic methods of Aronhold, Gordan, and Clebsch were necessarily excluded. The lecturer presented also the results of his own recent investigations. Twice a week from January to March and weekly during the rest of the year.

and (4) algebraic plane curves of the fourth and higher orders. In addition, "Story has also conducted weekly two-hour meetings of the mathematical department." These seminar topics included "Cantor's hyperinfinite number-system" and "Models illustrating rotation in 4-fold space."

Bolza lectured on (1) definite integrals, (2) elliptic functions, (3) calculus of variations, (4) theory of functions, and

> 5. Klein's Icosahedron-Theory, finite groups of rotations, the corresponding groups of linear substitutions, rational automorphic functions. Twice a week until March 1.

Perott discussed (1) theory of numbers (advanced course), and (2) numerical computations. Taber lectured on (1) modern algebra, (2) applications of the theory of matrices to bi-partite quadratic functions, and (3) symbolic logic. White discussed (1) modern synthetic geometry, and

> 2. Higher Plane Curves (Introductory Course); use of homogeneous coordinates, ordinary singularities of algebraic curves, projection and reciprocal figures, rational curves, Pluecker's relations, envelopes, tactinvariants, configuration and reality of inflexional points on the general cubic, conjugate points on the cubic, quadric transformation and general Cremona-transformations.

He also discussed (3) algebraic surfaces and twisted cubics and (4) theta-functions of three and four variables.

Since this is a report on what was done in 1891–1892 rather than what was planned, it is hard to dispute the claim in the *Register* that "The facilities for the study of the higher mathematics offered by this University are unsurpassed in this country." With this high level of activity, it is amazing how modest the requirements for admission were:

> Differential and Integral Calculus, Plane Analytic Geometry, through Conic Sections, Solid Analytic Geometry, through Quadric

Surfaces, Elements of the Theory of Algebraic Equations. A knowledge of the theory of Determinants and their application to the solution of linear equations, and of Differential Equations is desirable.

The published intentions of Clark University were being admirably fulfilled at this point. The very latest research from Europe was being studied and extended in a new American university. Moreover, the graduate students were being intimately involved in collaboration with the faculty.

This was a superb faculty. Both the German Bolza and the American White were students of Klein, and Bolza had a strong Weierstrassian component as well. Combined with the Berlin-Paris training of Perott, these mathematicians gave Clark a strong continental influence. This was something which was never present at Hopkins. The British and American influences were provided partly by Story and even more strongly by Taber, who had absorbed the spirit of Sylvester's work. Thus the Clark mathematicians were prepared to work in the best traditions of both the British and continental schools, and to continue the American work in logic and associative algebras begun by the Peirces.

It is no exaggeration to say that in 1892 Clark had the strongest mathematics department in the New World. Cattell's 1903 survey in *American Men of Science* is one way to evaluate the department. All five of the Clark faculty except Perott are listed among the top twenty: Bolza is fifth, White eighth, Story fifteenth, and Taber nineteenth. Also, Webster, the physicist, is listed twenty-fifth among the mathematicians, and fifth among physicists. To be sure, Chicago makes an even stronger showing in 1903, when the survey was taken, but the building of Chicago under E. H. Moore had just begun in 1892, the year we are discussing. The list also points up the relative decline of Hopkins during this period, though some account must be taken of the fact that Thomas Craig had died in 1900, just before the survey was conducted.

There is no doubt that for the years 1889–1892, Clark University was the preeminent school of mathematics in the Americas.

7. Revolt and Retrenchment

The early achievements and promise of Clark University were blighted by an unfortunate faculty rebellion that culminated during the 1891–1892 school year. Members of the faculty in several departments became disillusioned with the course of events and many of them left. The faculty members blamed President Hall, whom they felt had not kept his promises. Hall portrayed himself as caught between the faculty and the founder. It now seems that Mr. Clark also wanted an undergraduate school, and so the real disagreement was between him and President Hall. For present purposes, it is unnecessary to analyze the exact causes of the rebellion or attempt to fix the blame,[4] for the mathematicians were not active participants even though two of them, White and Bolza, did leave after that year, Bolza to go to Chicago, White to Northwestern.

In April of 1892, President William Rainey Harper (1856–1906) of the newly formed University of Chicago showed up in Worcester, having heard of the unhappiness among the faculty at Clark. Backed by the wealth of John D. Rockefeller, Harper was able to offer $7,000 to department heads that Clark had paid only $4,000. Although not the only factor, money certainly played a role. In 1891–1892, the five Clark mathematicians had a combined salary of $7,200, the same amount that Harper was offering his new department heads at Chicago. He was able to hire two of the four full professors at Clark; only Story and Hall remained. The greatest loss was the physicist A. A. Michelson. Great as this disaster was, it does show that the Clark faculty had a reputation for quality.

Refusing an offer from Johns Hopkins in 1891, White accepted one from Northwestern in 1892 for reasons which apparently had nothing to do with the general exodus of scholars from Clark in that year. "The inducements were, first a better salary with assured permanency, and second, proximity to the new University of Chicago and my highly valued friend E. Hastings Moore, its new head professor of mathematics. He indeed tried to bring me into his department but could not secure sufficient appropriation" [White 1946a, p. 24]. The higher salary was an understandable motive, since on 28 October 1890 White had been married. Apparently, White left Clark without bitterness, for he corresponded with Hall in September 1893 about Klein's itinerary in the U.S., and in 1903 he asked Hall for a letter of recommendation.

Another serious loss to the mathematics department was Oskar Bolza. One might surmise that Bolza left Clark because Hall promised to hire Maschke, and then backed out, but that does not seem to be the case. Although Bolza had no personal battle with Hall, the rebels had persuaded him to make some commitments to them, which he felt obliged to live up to [Bolza 1936a, p. 23]. In addition, like White, he was attracted by E. H. Moore.

The loss of two of its distinguished faculty reduced the department at Clark to a loyal core of three—Story, Taber, and Perott. They remained as the only faculty in mathematics until their retirements in 1921. In addition to the loss of faculty, the university was impoverished. From 1892 until his death in 1900, Mr. Clark gave no more money to the university. During this period there was only $32,000 per year to support the entire institution.

---

[4]As might be expected, different participants perceive these events differently. For Hall's view, see his *Life and Confessions of a Psychologist*, New York, 1923. While both Atwood [1937a] and Barnes [1925a] are Clark people, they have other views. By far the most balanced presentation is Koelsch [1987a].

Nonetheless, Story was given a salary increase for staying, and Taber was promoted to assistant professor.

One would think that the split in the department would cause lasting animosities, but there is no evidence of that. On the contrary, there are some signs of cooperation and good will. The three University of Chicago mathematicians, Moore, Bolza, and Maschke, worked with White to organize the International Mathematical Congress of 1893, which brought Felix Klein as head of the German universities exhibit. During this visit, Klein resided with his former student White in Evanston, commuting the twelve miles to Chicago every day. Klein's seminars, which were originally to have been divided between Northwestern and the University of Chicago, had to be given at Northwestern because of flooding in Chicago.

Story was elected president of this "zeroth" International Congress of Mathematicians, that is, the Columbian Exposition in Chicago in 1893. This shows that he was held in high regard by the American mathematical community. At this meeting, there were four representatives present from Clark—Story, Taber, Webster, and Keppel. Of the other American universities, only Chicago had as many, and two of those had recently been at Clark. There were only thirteen American residents who presented papers at the meeting, and two of them were from Clark. Taber gave a talk "On orthogonal substitutions," which definitely shows that Bolza had an influence on his work. In absentia, Perott contributed "A construction of Galois' group of 660 elements."

At the World's Columbian Exposition in Chicago, Clark University had 150 square feet of exhibition space wherein "each department will be represented by photographs, descriptive pamphlets, publications of the university and otherwise" [*New York Tribune*, 6 Feb. 1893]. More than 170 of these photographs survive in the Clark Archives, some 25 of which pertain to the mathematics department. Several of these are reproduced here. Most of the mathematical photographs deal with the "set of Brill's admirable models ... and Björling's thread-models of developable surfaces" which Story considered so vital to the teaching of higher geometry. A list of these models occupies eighteen pages in the *Third Annual Report of the President*, April 1893. These were not the only photographs taken, in addition:

> A graduate student at Clark University, Mr. H. G. Keppel, is taking a series of photographs of the mathematical models and portraits of mathematicians to which he has access. It will include stereoscopic views of about one hundred different models. [*Bulletin of the American Mathematical Society* **1**(1894–1895), 127]

There follows a list of thirty-five "portraits already photographed." These photographs of Herbert Govert Keppel (1866–1918) are not in the Clark Archives, and their whereabouts are a mystery.

The mathematics and physics group, 1892–1893. Seated, left to right: (2) A. G. Webster, (3) H. Taber. Standing: (4) T. F. Holgate?, (5) H. G. Keppel?, (7) W. E. Story. Other mathematics students at Clark in 1892–1893 were L. W. Dowling, T. F. Nichols, F. E. Stinson, and W. J. Waggener. Perott is not in the photograph. (Clark University Archives)

Through a grant of $500 from Senator George F. Hoar, a member of the board of trustees, Story was finally able to get his mathematical journal. The first number of the *The Mathematical Review. A Bi-Monthly Journal of Mathematics in all its Forms* was published in July of 1896. The other number in this volume was published in April 1987. It was followed by part of another volume in 1897 and then quietly ceased publication. Although no records survive, it undoubtedly ceased publication because of lack of funds and competition from other journals. It primarily consists of dissertations presented at Clark (recollect that the *Transactions of the American Mathematical Society* were founded primarily to publish dissertations). See §11 for a list of the individuals who received degrees in mathematics from Clark.

Clark University was very proud of its accomplishments, and so to celebrate its tenth anniversary, Story and Hall's right-hand man, L. N. Wilson, prepared a large (vi + 566 pp.) volume entitled *Clark University, 1889–1899. Decennial Celebration*, which is a gold mine of information about the university. We learn, for example, that the "mathematical department was not modelled after that of any other institution, but was determined by the conception of what would constitute perfection in such a department" [p. 68] and that in making appointments to fellowships and scholarships, "We are on the lookout for geniuses" [p. 65]. Before the *Decennial* volume was published, a public celebration was held. As part of this, Émile Picard of the University of Paris was invited to give a series of lectures on mathematics, and Ludwig Botzmann of the University of Vienna lectured on physics. They, along with three other individuals, were granted honorary degrees from Clark University on 10 July 1899. Previously, the only degree was the earned Ph.D. The *Decennial* volume contains a long description of the individual departments and, most importantly, a list of over 500 publications by people who had been associated with Clark in its first decade.

On 10 September 1909, the twentieth anniversary of Clark was celebrated. This time, honorary degrees were given to five mathematicians: E. H. Moore of Chicago, William Fogg Osgood of Harvard, James Pierpont of Yale, Edward Burr Van Vleck of Wisconsin, and Vito Volterra of Rome. Fortunately, we have a picture of this gathering. This was perhaps the most famous meeting ever held at Clark. On the same day, Sigmund Freud of Vienna and Carl Jung of Zürich were given honorary degrees. This was the only such honor that Freud ever received.

When Mr. Clark died in 1900, the university faculty were hoping that he would rescue them from their financial plight. Instead, Clark continued with his original plan, leaving one-fourth of his estate to the University, another one-fourth to the library, and with the remainder, he did what he had wanted all along. He founded an undergraduate college, Clark College.

Photograph taken for the twentieth anniversary celebration at Clark, 1909. Front row, left to right: (1) Robert William Wood, (2) E. H. Moore, (3) Vito Volterra, (4) A. A. Michelson, (5) John Monroe Van Vleck, (6) Edwin Herbert Hall, (7) James Edmund Ives (Clark Ph.D., Physics, 1901), (8) W. E. Story, (9) Norton A. Kent. Second row: (1) Stephen Elmer Slocum (Clark Ph.D., 1900), (2) A. G. Webster, (3) Ernest Rutherford (between rows two and three), (4) unknown, (5) Edward Burr Van Vleck, (6) J. W. A. Young (Clark Ph.D., 1892), (7) Norman E. Gilbert, (8) Ernest Fox Nichols, (9) Guy G. Becknell?, (10) Henry Sedgwick?. Third row: (1) Thomas Lansing Porter, (2) Theodore William Richards, [Rutherford], (3) George D. Olds, (4) Carl Barns, (5) unknown, (6) H. Taber, (7) Albert Potter Wills (Clark Ph.D., Physics, 1897), (8) unknown, (9) Elmer Adna Harrington. Fourth row: (1) Rocket pioneer Robert Hutchings Goddard (Clark Ph.D., Physics, 1911), (2) Joseph George Coffin (Clark Ph.D., Physics, 1903), (3) Arthur W. Ewell, (4) Frank B. Williams (Clark Ph.D., 1900), (5) unknown, (6) Chester Arthur Butman, (7) unknown. William Fogg Osgood, James Pierpont, and the astronomer Percival Lowell were present, but have not been located in the photograph. Perott is not in the photograph. (Clark University Archives)

8. Story's Best Student at Clark, Solomon Lefschetz

One way of determining the quality of the mathematics department at Clark University and its role in the development of graduate mathematical education is to examine the careers of the graduate students there, especially the twenty-four students who received Ph.D.s from 1892 to 1917. We have included some information about all of them in §11 below, but we shall concentrate our attention on the most famous of the group.

In the fall of 1910, two new graduate students arrived at Clark to join the three who were already there. One was Alice Berg Hayes, the first woman to receive a degree in mathematics at Clark. Women were not allowed to be graduate students at Clark until 1900, although Leona Mae Peirce (Ph.D. Yale, 1899) studied informally with Story in the 1890s. Hayes received a master's degree in June 1911 for a thesis entitled "Reduction of Certain Power Determinants" which she wrote under Story's direction. The other new student was Solomon Lefschetz, whom Hayes married in 1913.

Solomon Lefschetz was born of Turkish parents in Moscow on 3 September 1884 and was reared in France. He was a student at the École Centrale in Paris from 1902 to 1905 when he received a degree as "ingénieur des arts et manufactures." He then came to the U.S. where he worked for a few months with the Baldwin Locomotive works, and was then on the engineering staff of Westinghouse Electric and Manufacturing Company in Pittsburgh until 1910.

> He lost both of his hands in 1907; the heroic spirit which later enabled him to overcome all but insurmountable obstacles, and to attain to his present position of eminence, must be unique in the annals of the mathematical brotherhood. [Archibald 1938a, p. 237]

Because of this accident, he soon realized that his "true path was not engineering but mathematics" [Lefschetz 1970a, p. 344]. Going back to his French roots, he read the three-volume treatises of Émile Picard (*Analysis*) and Paul Appell (*Analytical Mechanics*), both of whom were professors at the École Centrale. "I plunged into these and gave myself a self-taught graduate course. What with a strong French training in the equivalent of an undergraduate course, I was all set" [Lefschetz 1970a, p. 344].

In May of 1910, Lefschetz accepted an appointment as junior fellow at Clark, which waived fees and paid $100 in ten monthly installments. When he accepted, he added a postscript to his letter: "I ask, as a special favor, that you should forward me the catalogue of the University for 1910, with programms [sic] for 1910–1911 as I intend to do some hard digging during the summer" (Lefschetz to Hall, 9 May 1910). From this *Register*, Lefschetz learned the philosophy of the department, a philosophy which is well worth emulating today:

> The chief aim of the department is to make independent investigators of such students as have mathematical taste and ability; these naturally look forward to careers as teachers of the higher mathematics in colleges and universities, and we believe that the course of training best adapted to the development of investigators is also that which is most suitable for all who would be efficient college professors, even if they are not ambitious to engage in research. The first essential of success in either of these lines is the habit of mathematical thought, and the direct object of our instruction is the acquisition of this habit by each of our students. With this end in view, we expect every student to make himself familiar with the general methods and most salient results of a large number of different branches of mathematics, conversant with the detailed results and the literature of a few branches, and thorough master of at least one special topic to the extent of making a real contribution to our knowledge of that subject.

Since Lefschetz played such an important part in the rise of mathematics in the Americas, at the Universities of Nebraska (1911–1913), Kansas (1913–1923), Princeton (1923–1953), and Mexico (1944–1966), we shall describe the course work that was announced in the *Register* that he requested. Story planned to teach (1) analytic geometry of higher plane curves, higher surfaces, and twisted curves, five hours, (2) finite differences, two hours, (3) history of mathematics, two hours in the fall, and (4) a seminary for advanced students. Taber intended (1) theory of functions of real and imaginary variables, elliptic functions, and definite integrals, five hours, (2) theory of bilinear forms, two hours in the fall, (3) theory of integral equations, two hours in the spring, and (4) a seminary. Perott was to offer (1) theory of numbers, two hours in the fall, and (2) abelian integrals, two hours in the spring.

Unfortunately, no records survive as to which courses Lefschetz actually took in 1910–1911. We can make some conjectures from the annual report submitted by Story on 10 October 1911 dealing with what was actually taught during 1910–1911. Story did teach higher plane curves, but only three days a week. Since there were two students in this course, and since it had also been offered the previous year, Lefschetz was undoubtedly one of them. We also suspect that he was one of the three students in Story's calculus of operations, including the calculus of finite differences. Story offered no seminar that year and Lefschetz certainly did not attend Story's course in mathematics for practical purposes. Taber's theory of functions was undoubtedly familiar to Lefschetz, but he probably attended the two-hour supplementary course.

The other offerings were half-year courses: Taber's bilinear forms and Perott's advanced number theory. Undoubtedly, most of Lefschetz's time was devoted to research under Story's guidance.

Story assigned Lefschetz the problem of investigating "the largest number of cusps that a plane curve of given degree may possess" [Lefschetz 1970a, p. 344]. This resulted in his dissertation "On the Existence of Loci with Given Singularities" which was published in the *Transactions of the American Mathematical Society* **14** (1913), 23–41. There is no doubt that Lefschetz appreciated the education that he received at Clark. A few years later, when he planned a return visit, he wrote ahead asking to give a series of lectures on his recent research, adding "I know of no other place where I may expect to get an audience as surely as at Clark Univ. & none where I'd care more to have one than there" (Lefschetz to Hall, 16 May 1913).

9. The Library

Yet another measure of the quality of Clark University was its library. Lefschetz wrote:

> At Clark there was fortunately a first rate librarian, Dr. L. N. Wilson, and a well-kept mathematical library. Just two of us enjoyed it—my fellow graduate student in mathematics and future wife, and myself. I took advantage of the library to learn about a number of highly interesting new fields, notably about the superb Italian school of algebraic geometry. [Lefschetz 1970a, p. 344]

This high opinion of the library by a distinguished mathematician can be complemented by information describing the contents and quality of the library.

Jonas Clark was a self-educated man who read widely, collected books and manuscripts, and understood the importance of a good library. Consequently, he donated his personal library of some 3,200 volumes, and set up a separate endowment consisting of $100,000, the income of which was to be used for the purchase of books and the maintenance of the library. The faculty were invited to contribute lists of books that they wanted, and the library purchased whatever they requested. A few of these early lists survive.

Just before Clark University opened, Florian Cajori conducted a survey of American mathematical education. Of the 168 schools responding, 117 subscribed to no mathematics journals, 11 subscribed to only the *American Journal of Mathematics*, 12 to only the *Annals of Mathematics*, and 28 subscribed to several mathematical periodicals [1890a, p. 302]. There is no explanation of what several means, but there is no doubt that Clark was soon to be near the top.

In 1893, Clark subscribed to sixteen mathematical periodicals and to thirteen others which contained articles on mathematics. Of these sixteen, all but two consisted of complete runs. The importance of complete sets of serial publications was well understood by the librarian [*Decennial*, p. 196]. In 1900, Bryn Mawr, by contrast, subscribed to twenty-two mathematical journals. There are some fifty listed in Robert Gascoigne's *A Historical Catalogue of Scientific Periodicals, 1665–1900* (Garland Press, 1985), though, of course, a good many of those were no longer being published in 1890. By the time Lefschetz was at Clark, there were about sixty mathematical periodicals in the collection.

In the *Second Register and Announcement* we read:

> The facilities to be found here for the study of mathematics in its various branches are unexcelled in this country. The library is provided with complete sets of all the more important current mathematical periodicals and the publications of the scientific societies of the world, with the standard treatises on the subjects now particularly engaging the attention of mathematicians, the collected works of the great mathematicians, and many books illustrating and discussing the history of mathematics; to which will be added from time to time such other works as may be needed or appear desirable.

The total library holdings in 1900 of 18,000 volumes may seem meager compared to the half million at Harvard and the 90,000 at Johns Hopkins, but remember that this number represented only four fields (chemistry disbanded after the exodus), and then only with graduate-level works that some faculty member requested.

Still in the library at Clark is a *List of Books in Mathematics in the Clark University Library. Worcester Massachusetts, December 1, 1908* (Z 733 C5). This typescript of seventy-seven pages lists the books according to the classification scheme devised at Clark and gives a real indication of the riches of the library. Section "C 21 Works, complete and select." lists the collected works of more than sixty mathematicians. Our experience has been that this remains a very nice small library to work in if your interests are in late-nineteenth-century mathematics.

In 1900, when Jonas Clark died, he left one-fourth of his estate to the library (received on the death of Mrs. Clark in 1903). Thus they had $32,000 per year for the library, a sum they were never able to spend. He also left $150,000 for the construction of a library building. This was built in 1904 at a cost of only $125,000. An addition was started in 1909 at a cost of $100,000. In 1921, the college and university libraries were merged.

The library also had the admirable habit of trying to obtain a copy of every publication of everyone who had ever been associated with the university.

The records of this collection still exist in a separate card catalog in the Archives. Unfortunately, many of the unbound reprints have been destroyed. Even the collection of master's and doctor's dissertations is not complete.

In addition to the library, Story compiled his own bibliography of the mathematical literature. In the early 1900s, it consisted of some 100,000 cards. He aimed to get it published, but unfortunately that never happened. "In 1931 through the generosity of Clark U. and the initiative of Prof. F. B. Williams the Library [of the AMS] acquired the mathematical Bibliography (156 drawers and 35 boxes of cards) of the late Prof. W. E. Story (1850–1930)" [Archibald 1938a, p. 93]. It is not known if this catalogue still exists.

10. Denouement

The penultimate student to receive a master's degree in mathematics at Clark University was Ida Louise Bullard (Pearson). She graduated from Mount Holyoke College with thirty hours of mathematics courses, although nine of them were precalculus. She received "testimonials" from Anna J. Pell (Wheeler) and Sara Effie Smith, and was appointed "Senior Scholar in Mathematics" in 1918–1919. The faculty was enthusiastic about having her, and they kept her busy. She took fifteen hours of classes in the fall and eighteen in the spring. The classes were taught by the lecture method, with her at one end of a very large table in the mathematics classroom, taking notes as fast as she could. Two boxes of her class notes survive in the Clark University Archives. In addition, she wrote a master's thesis, "Report on the Literature of Fractional Derivatives" (1919).

On 10 August 1973, Louise Pearson was interviewed by University Historian William A. Koelsch. From his notes after this interview, we learn her impressions of the faculty. Story was "a very nice, dignified, grey-headed gentleman" who encouraged her to continue for a Ph.D., but she declined because "women could only find positions at women's colleges." Unfortunately, this is all she had to stay about Story, but we do know more. On 20 June 1878, he married Mary D. Harrison of Baltimore, and they had one son, William E. Story, Jr., who was an undergraduate at Harvard and then earned a Ph.D. in physics from Clark in 1907. Fabian Franklin wrote that the elder Story "was happy in his marriage as in his work" [*American Academy of Arts and Sciences. Proceedings* **70** (1935–1936), 580]. Story was "noted for his skill as a raconteur and his force in discussing scientific matters" [Worcester *Gazette*, 27 April 1921]. In addition, he was an excellent teacher, but with a fiery disposition:

> In the mathematics department the most picturesque figure was Story, who could be daily observed lecturing with the enthusiasm of a Bryan delivering his "Cross of Gold and Crown of Thorns"

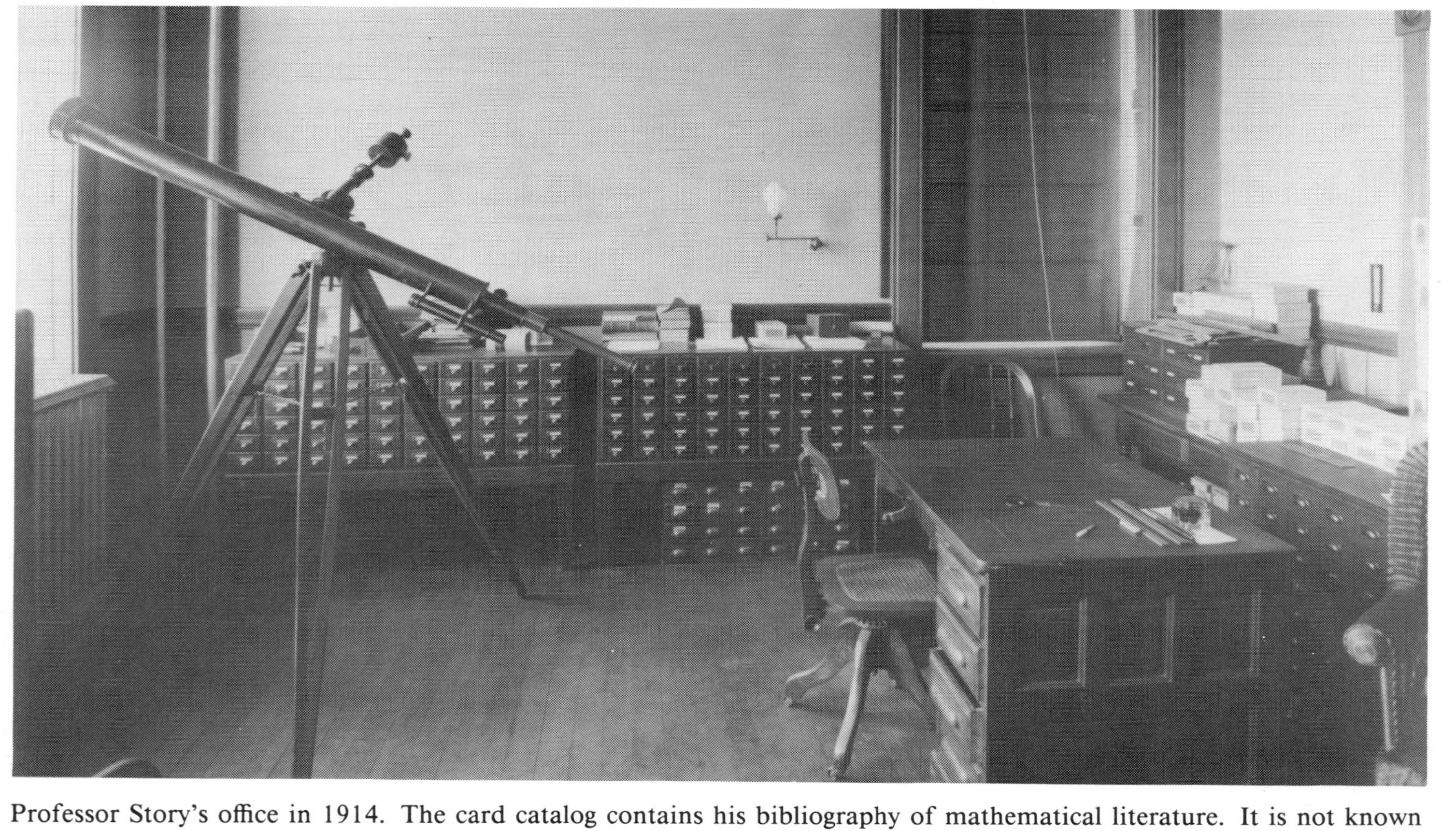

Professor Story's office in 1914. The card catalog contains his bibliography of mathematical literature. It is not known whether it survives. (Clark University Archives)

> speech, to one student in infinitesmal [sic] geometry or the theory of hyperspace, and whose expostulations announced to passers-by in Main street trolley cars that a faculty meeting was being held in the opposite side of the university building. [Barnes 1925a, p. 275]

Besides the scientific honors mentioned earlier, Story was a member of the National Academy of Sciences (elected 1908), fellow and former vice president of the American Association for the Advancement of Science, and a member of the American Mathematical Society. After his retirement in 1921, Story served as president of the Omar Khayyam Club of America from 1924 to 1927. Earlier, he had written an interesting little pamphlet, *Omar Khayyam as a Mathematician* (1918), that reflects his long held interest in the history of mathematics. He died of pneumonia, after a very brief illness, on 10 April 1930.

> The ablest member of the department was Henry Taber—than whom no finer type of American scholar and gentleman has yet been produced, who lectured in polished and dignified English upon the theory of functions, and read the *Nation*, the *New Republic* and *Freeman* unabashed. [Barnes 1925a, p. 275]

In 1891, Taber was elected to the American Academy of Arts and Sciences. In a biographical memoir to their *Proceedings*, Archibald concurred that Taber was "ever ready to champion the cause of one whom he felt wronged" [Vol. 75, p. 176]. He belonged to the Worcester boat club and was an excellent tennis player. Taber had a wide range of interests, including chemistry, history, literature, music, and dancing. Pearson commented on his teaching:

> Tall, thin, reddish or sandy haired, and a vigorous lecturer. Also a classically absent-minded professor, illustrated by two stories: (1) One day Taber walked into the mathematics classroom and began lecturing, and lectured for twenty minutes before noticing that no one was there, and discovering that he was an hour late. (2) The Tabers lived on the second floor of their house, and one day Dr. Taber discovered he was locked out. So he borrowed a ladder from a neighbor, climbed through a second story window, came back down, and returned the ladder, subsequently discovering that he was still locked out.

Taber married Fanny Lawrence of New York in 1886, and they had three daughters. Sadly, his wife died in 1892, so he had to raise the girls alone. Henry Taber died 6 January 1936.

Naturally, it was Joseph de Perott who consumed the bulk of the interview, for "There were always numerous stories circulating about him." Because

Kovalevskaya would not divorce her husband to marry him, so the story goes, he

> vowed he would never again attempt to make himself attractive to women. This accounted for the mass of tangled hair down to his shoulders, which he never combed, keeping it in place with a derby hat jammed tightly over the top of his head.

This report contrasts with his obituary in the Worcester *Telegram*, on 23 May 1924 which described him as a friendly and happy person with a love of nature, a vast knowledge of languages, a knowledge of Shakespeare that Harvard coveted, and that "His long flowing gray hair and his neat but somewhat threadbare clothing, seemed to attract rather than repel the children" (23 May 1924). It was his mane of hair that earned him the nickname "Johnny the Lion" [Koelsch 1987a, p. 62]. For additional information see [Cooke and Rickey 199?a].

In 1919, President Hall, who had served for thirty-two years, asked to be relieved of his responsibilities at Clark University. Simultaneously, President Stanford, of Clark College, resigned so that a common successor could be found who would merge the two units. The Trustees chose the geographer Wallace W. Atwood as successor. Although the graduate departments of psychology and education had achieved an international reputation, there had been almost no new money in twenty years, and so the other departments were stagnating. The trustees decided that Clark could compete with the now larger graduate institutions only if it had something distinctive to offer, and so founded a department of geography. "The Department of Mathematics, which had very few students, discontinued graduate work, and the members of the staff, who had been in the University practically from the beginning, retired on pensions" [Atwood 1937a, p. 16]. It is sad to realize that such a glorious department had come to an end.

The various decisions that Atwood made sparked a report by the American Association of University Professors, their first comprehensive report of administrative practices. In it we read that Story "retired on account of age in 1921" and Taber "on account of health." In fact, the retirements were forced. Story expressed an interest in continuing in active service (Atwood to Story, 10 March 1921) and Taber was working with a master's student at the time (Taber to Atwood, 12 March 1921). At the time, Story was 71, Taber 61, and Perott 67. We have not forgotten them.

In conclusion, it is probably worthwhile to reflect on why mathematics at Clark University was only a brief success. The most important ingredient was there: well-trained mathematicians, with considerable research potential, and a will to excel. There was a good mix of experience and youth, a diversity of backgrounds, yet many shared interests. The library was excellent and, at first, the salaries were adequate. But it was a lack of money and poor administration that led to the internal strife and subsequent loss of faculty. William Story, and his colleagues Henry Taber and Joseph Perott, contributed to the development of mathematics by being carriers of our mathematical culture. Their careers illustrate the importance of dedicated "minor" mathematicians without whose work—learning, teaching, and doing mathematics—the community of research mathematicians would not grow.

11. Graduate Degrees in Mathematics, Clark University, 1889–1921

This is a complete list of the students in mathematics at Clark University who received either a master's degree or a doctor's degree between 1889 and 1921 (but not those after the Ph.D. program was reinstated in 1965). If known, we have given the call number, accession number, and date of accession of each dissertation in the Clark University Archives.

Allen, Reginald Bryant (1872–1938). At Clark 1901–1903 and 1904–1905. Ph.D. under Taber defended May 25, 1905: "On hypercomplex number systems belonging to an arbitrary domain of rationality," *Transactions of the American Mathematical Society* **9** (1908), 203–218. Clark Library: 49105, May 1909.

[**Boyce, James W.** Fellow at Clark 1896–1899. *Science* **10**, 132 lists him as receiving a Ph.D. at Clark in 1899 for a dissertation entitled "On the Steinerian Curve," but there is no reference to this in the Clark records.]

Bullard, Ida Louise (married Charles W. Pearson). M.A. thesis under Story: "Report on the literature of fractional derivatives." Clark Library: B935, 91162, November 1919. Degree received June 23, 1919.

Bullard, James Atkins (1887–1959). At Clark 1911–1914. Ph.D. under Taber received June 18, 1914: "On the structure of continuous groups," *American Journal of Mathematics* **39** (1917), 430–450. Clark Library: 84338, February 1918.

Bullard, Warren Gardner (1867–1927). At Clark 1893–1896. Ph.D. under Story defended June 17, 1896: "On the general classification of plane quartic curves," *The Mathematical Review* **1** (1899), 193–208 + three plates. Clark Library: 49099, May 1909.

Dowling, Linnaeus Wayland (1867–1928). At Clark 1892–1895. Ph.D. under Story defended June 19, 1895: "On the forms of plane quintic curves," *The Mathematical Review* **1** (1897), 97–119 + two plates. Clark Library: D747, 49102, May 1909. Starred in *American Men of Science*, edition 2.

Dustheimer, Oscar Lee (1889–1963). Fellow at Clark 1913–1914. M.A. under Story received June 18, 1914: "The historical order of development and some applications of symmetric determinants." Clark Library: D974, 69705, September 1914.

Ferry, Frederick Carlos (1868–1956). At Clark 1895–1898; Fellow 1895–1896. Ph.D. under Story defended June 15, 1898: "Geometry on the cubic scroll of the first kind," Bd. 21 (1899), Nr. 3, pp. 1–57. Clark Library: F399, 49098, May 1909.

French, John Shaw (1873–??). At Clark 1895–1898; Fellow 1896–1898. Secretary and chairman of the board of admissions, Clark University, 1918–1921. Ph.D. defended March 28, 1899: "On the theory of the pertingents to a plane curve." Dissertation director unknown but probably Story. No copy of dissertation in Clark Library.

Gates, Jesse Nevin (??–1936). At Clark 1900–1904. Ph.D. under Story defended July 1, 1904: "Cubic and quartic surfaces in 4-fold space." Clark Library: 151051, May 1939 (sic), 44 page typescript. Unpublished.

Goodrich, Merton Taylor (b. 1887). At Clark 1911–1912. M.A. under Story received June 18, 1912: "On the forms of plane curves of the fourth class." Clark Library: G655, 61663, July 1912.

Hayes, Alice Berg (Mrs. Solomon Lefschetz). At Clark 1910–1911. M.A. under Story received June 15, 1911: "Reduction of certain power determinants." Clark Library: H417, 57939, Sept. 1911.

Hill, John Ethan (1865–1941). Fellow at Clark 1892–1895. Ph.D. defended June 17, 1895: "On quintic surfaces," *The Mathematical Review* **1** (1896), 1–59. Director unknown but probably Story. Clark Library: H646, 49100, May 1909.

Holgate, Thomas Franklin (1859–1945). At Clark 1890–1893. Ph.D. under Story defended May 9, 1893: "On certain ruled surfaces of the fourth order," *American Journal of Mathematics* **15** (1893), 344–386. Clark Library: H731, 50335, November 1909. He was number 32 on Cattell's 1903 *American Men of Science* list.

Keppel, Herbert Govert (1866–1918). At Clark 1892–1895 and 1900–1901. Ph.D. defended June 13, 1901: "The cubic three-spread ruled with planes in fourfold space." Unpublished. Story was probably director. No copy of dissertation in Clark Library.

Lefschetz, Solomon (1884–1972). At Clark 1910–1911. Ph.D. under Story received June 15, 1911: "On the existence of loci with given singularities," *Transactions of the American Mathematical Society* **14** (1913), 23–41. Listed with star in *American Men of Science*, edition 3.

Leyzerah, Peysah (??–1976). Name later changed to Philip Lazarus. At Clark 1912–1916. Fellow in mathematics in 1915. Ph.D. under Story received June 15, 1916: "On the indeterminate linear inequality with irrational coefficients." Clark Library: L855, 112463, November 1925 (sic), 38 pages. Unpublished.

Lie, Olaf Kristofer (d. 1914). At Clark 1905–1908. M.A. under Story received June 18, 1908: "On the reduction to the canonical forms of the equations of transformation groups with continuous parameters." Holographic copy in Clark Library: L716, 46112, September 1908.

[**Lieber, Lillian Rosanoff** (1886–??). At Clark 1912–1914 as a fellow in chemistry. Ph.D. in chemistry, 1914, under Martin A. Rosanoff, her brother. She is listed here because of her many delightful books on mathematics.]

McCormick, Clarence (1888–??). At Clark 1913–1916. Fellow in 1915. M.A. under Taber received June 18, 1914: "On the theory of finite continuous groups." Clark Library: M131, 70489, December 1914. Ph.D. Columbia 1928.

Metzler, William H. (1863–1943). At Clark 1889–1892. Ph.D. under Story defended January 4, 1893: "On the roots of matrices," *American Journal of Mathematics* **14** (1892), 326–377. Clark Library: M5960, 50336, November 1909. Story is the official advisor, but the paper says: "For valuable suggestions in the working of this paper I am indebted to Dr. Henry Taber." Number 47 on Cattell's 1903 *American Men of Science* list.

Michalopoulos, Aristotle D. (1899–1953). He later abbreviated his surname to Michal. A.B. 1920, assistant in mathematics and physics 1918–1920, fellow 1920–1921, A.M. 1921, all at Clark. M.A. under Taber: "Theory of analytic functions of a single complex variable." He was the first student of Clark College to enter the graduate mathematics program (letter of Story, December 15, 1920). Ph.D. at Rice, 1924. Has star in *American Men of Science*, edition 6. Classnotes in Clark Archives.

Montgomery, William John (d. 1915). At Clark 1907–1911. M.A. received June 18, 1909: "1. On the smallest number of inflexions on a nonsingular odd branch of an algebraic plane curve. 2. On the solution of differential equations of the second order invariant to an infinitesimal transformation." Holographic copy in Clark Library: M788, 46092, August 1908. Ph.D. directed by Story received June 15, 1911: "Singularites of twisted quintic curves." Typescript of thirty-three pages in Clark Library. No accession number. Unpublished. He is the only mathematics student to receive both an M.A. and a Ph.D. from Clark.

Moreno, Halcott Cadwalader (1873–1948). At Clark 1897–1901; assistant 1900–1901. Ph.D. under Story defended June 8, 1900: "On ruled loci in n-fold space," *Proceedings of the American Academy of Arts and Sciences* **37** (1901), 121–157; presented by Story May 8, 1901, received June 1, 1901. Clark Library: 49097, May 1909.

Morley, Raymond Kurtz (1882–1965). At Clark 1907–1910. Ph.D. under Story received June 16, 1910: "On the fundamental postulate of tamisage," *American Journal of Mathematics* **34** (1912), 47–68. Clark Library: 61145, May 1912.

Nichols, Thomas Flint (1870–??). At Clark 1892–1895; scholar 1892–1893, fellow 1893–1895. Ph.D. under Story defended June 20, 1895: "On some special Jacobians," *The Mathematical Review* **1** (1896), 60–80. Clark Library: N622, 49103, May 1909.

Peabody, Leroy Elden (1894–1956). Scholar at Clark 1915–1916; Honorary Scholar 1917–1918. Scholar in mathematics. M.A. under Story received June 15, 1916: "Continued fractions of the second order, with bibliography of continued fractions." Clark Library: P353, 78022, August 1916.

Rettger, Ernest William (1871–1938). At Clark 1895–1898. Ph.D. under Taber defended June 16, 1898: "On Lie's theory of continuous groups," *American Journal of Mathematics* **22** (1900), 60–95. No copy at Clark.

Slobin, Hermon Lester (1883–1951). A.B., Clark 1905. Fellow at Clark 1905–1908. Ph.D. under Story received June 18, 1908: "On plane quintic curves." Privately printed, 25 pp. Clark Library 55556, February 1911.

Slocum, Stephen Elmer (1875–1960). At Clark 1897–1900. Ph.D. under Taber defended June 6, 1900: "On the continuity of groups generated by infinitesimal transformations," *Proceedings of the American Academy of Arts and Sciences* **36** (1900), 85–109. Clark Library S634, 49096, May 1909.

Van der Vries, John Nicholas (1876–1936). At Clark 1897–1901. Ph.D. under Story defended June 14, 1901: "On the multiple points of twisted curves," *Proceedings of the American Academy of Arts and Sciences* **38** (1903), 473–532. He thanks Taber "for his careful supervision of my work in the first two years, and Professor A. G. Webster and M. Joseph Perott for frequent assistance throughout this work."

Waits, Benjamin Lewis (??–1967). Scholar in 1915. M.A. under Story received June 15, 1916: "Fourier's method for the separation of the roots of an algebraic equation." Clark Library: W1456, 78027, August 1916.

Williams, Frank Blair (1871–1933). At Clark 1897–December 1901 as a student; scholar 1897–1898, fellow 1898–December 1901. Became instructor in mathematics at Clark University in 1910; assistant professor, Clark College 1907, professor 1920. Ph.D. under Story defended June 4, 1900: "Geometry of ruled quartic surfaces," *Proceedings of the American Academy of Arts and Sciences* **36** (1900), 85–109.

Young, Jacob William Albert (1865–1948). At Clark 1889–1892. Ph.D. under Story defended September 16, 1892: "On the determination of groups whose order is a power of a prime," *American Journal of Mathematics* **15** (1893), 124–178. He was the first Clark Ph.D. in mathematics. Clark Library: Y735, 49107, May 1904.

Zeldin, Samuel Demitry (1894–1965). At Clark 1913–1917. Fellow in 1915. Ph.D. under Taber received June 19, 1917: "On the structure of finite continuous groups with a single exceptional infinitesimal transformation," *American Journal of Mathematics* **44** (1922), 204–216. Clark Library: 104717, Dec 1923.

Bibliography

Anonymous

1911a *Graduate Work in Mathematics in Universities and Other Institutions of Like Grade in the United States. International Commission on the Teaching of Mathematics. The American Report. Committee No. XII.* United States Bureau of Education Bulletin, 1911, no. 6, whole number 452. Washington, Government Printing Office.

1921a *Eleventh Report of the Class of 1871 of Harvard College*, Cambridge.

Archibald, R. C.

1936a "Unpublished letters of James Joseph Sylvester and other new information concerning his life and work," *Osiris* **1**, 85–154.

1938a "A Semicentennial History of the American Mathematical Society, 1888–1938," in *American Mathematical Society: Semicentennial Publications*, vol. 1, History, Amer. Math. Soc., Providence, RI.

Atwood, Wallace W.

1937a *The First Fifty Years. An Administrative Report*, Clark University, Worcester, MA.

Barnes, Harry Elmer

1925a "Clark University: An Adventure in American Educational History," *American Review* **3**, 271–288.

Bolza, Oskar

1936a *Aus meinem Leben*, München, E. Reinhardt, 45p.

Cajori, Florian

1890a *The Teaching and History of Mathematics in the United States*, Washington.

Cooke, Roger and Rickey, V. Frederick

199?a "Joseph Perott: Mathematical Immigrant," *The Mathematical Intelligencer* (to appear).

Cordasco, Francesco

1960a *Daniel Coit Gilman and the Protean Ph.D. The Shaping of American Graduate Education*, Leiden, E. J. Brill.

Decennial

1899a *Clark University, 1889–1899. Decennial Celebration*, Worcester, MA.

Feuer, Lewis S.

1984a "America's First Jewish Professor: James Joseph Sylvester at the University of Virginia," *American Jewish Archives* **36**, 152–201.

Fisch, Max H. and Cope, Jackson I.

1952a "Peirce at the Johns Hopkins University," *Studies in the Philosophy of Charles Sanders Peirce*, edited by Philip P. Wiener and Frederic H. Young, Cambridge.

Flexner, Abraham

1930a *Universities, American, English, German*, Oxford University Press, New York.

1946a *Daniel Coit Gilman*, New York.

French, John Galvin

1946a *A History of the University Founded by Johns Hopkins*, Johns Hopkins University Press, Baltimore.

Gilman, Daniel Coit

1906a *The Launching of a University*, Dodd, Mead and Co., New York.

Hawkins, Hugh

1960a *Pioneer: A History of the Johns Hopkins University, 1874–1889*, Cornell University Press, Ithaca, NY.

Koelsch, William A.

1987a *Clark University 1887–1987. A Narrative History*, Clark University Press.

Lefschetz, Solomon

1970a "Reminiscences of a Mathematical Immigrant in the United States," *American Mathematical Monthly* **77**, 344–350. Reprinted in this collection, *A Century of Mathematics in America* Vol. 1, Amer. Math. Soc., Providence, RI, pp. 201–207.

Reid, Constance

1978a "The Road Not Taken. A Footnote in the History of Mathematics," *The Mathematical Intelligencer*, Vol. **1**, No. 1, pp. 21–23.

Rush, N. Orwin, editor.

1948a *Letters of G. Stanley Hall to Jonas Gilman Clark*, Clark University Library, Worcester, MA.

Smith, David Eugene and Ginsburg, Jekuthiel

1934a *A History of Mathematics in America before 1900*, MAA Carus Mathematical Monograph, #5. Reprinted 1980, Arno Press, New York.

White, Henry S.

1946a "Autobiographical memoir of Henry Seely White (1861–1943)," *National Academy of Sciences of the United States of America, Biographical Memoirs*, Vol. 25, second memoir, pp. 16–33.

Wilson, Louis N.

1920a "List of Degrees Granted at Clark University and Clark College, 1889–1920," *Publications of the Clark University Library*, Vol. 6, No. 3.

George M. Rosenstein, Jr. received his Ph.D. at Duke University in 1963 with a dissertation in topology under the direction of John H. Roberts. He taught at Western Reserve University, then in 1967 took a position at Franklin and Marshall College. His conversion from topologist to historian of mathematics was aided by a year at the Smithsonian Institution under the tutelage of Uta Merzbach. He is currently the Chief Reader of the calculus examination for the national Advanced Placement program.

The Best Method. American Calculus Textbooks of the Nineteenth Century

GEORGE M. ROSENSTEIN, JR.

INTRODUCTION

The need for calculus textbooks in the United States was met by American authors from the 1840s onward and by 1870 the industry was well established. Between 1828 and 1920, editions of calculus books by about seventy different authors or sets of authors were published in the United States. A bibliography of these texts is appended to this paper. The books range from vanity pieces, privately printed, to the long-lived books of Davies (1836–1901), Loomis (1851–1902) and Granville (1904–1946).

I will examine books which appeared before 1910, which appeared in more than one edition, and which were not vanity pieces.[1] I refer to those books as "commercial texts." By examining these books, I can show that the last quarter of the nineteenth century was a period of experimentation at the end of which various features of contemporary texts became standard.

The first calculus texts published in the United States were American editions of British books and, notably, Farrar's translation of Bézout.[2] However, in 1828 James Ryan published the first calculus text written by an American. By 1850, several "native" books were available. By 1875, nearly a dozen

[1]Only one author, William Batchelder Greene, met the first two criteria but not the last.

[2]See Cajori, pp. 395 ff. A list of references will be found at the end of the paper. American texts to which I refer are in the bibliography.

books had appeared. These books and those that followed up to the end of the century display a remarkable diversity.

Most notable is the diversity of approaches to the derivative. Texts used limits, infinitesimals or rates (fluxions in a new dress) as fundamental notions. During this same period, European authors were also experimenting with appropriate pedagogical schemes for introducing the derivative. For example, the second edition of Serret's text, published in 1879,[3] displays an eclectic approach using limits and infinitesimals that American contemporaries would immediately recognize. In the United States, limits would not become the preferred approach to the derivative until late in the century. As we shall see, there were other differences also.

The Beginning of the Line

When American calculus teachers were presenting fluxions to their students, British texts could serve their needs. An American edition of Vince's *The Principles of Fluxions* appeared in 1812 and editions of Charles Hutton's *Course of Mathematics* appeared between 1812 and 1831. Both of these books were used in American colleges. Texts were not so easy to find, however, when teachers turned to the style of the French.[4]

Once American teachers became convinced that the continental style, represented by the French in the early nineteenth century, was preferable, they were without texts to help them. Few students could read French although Greek, Latin, and often Hebrew were standard parts of their education.[5] These teachers solved their problem by translating, perhaps with modifications, texts that they found exemplary.

The first translation was John Farrar's of a text by Bézout. Farrar was a graduate of Harvard who, after studying theology at Andover, returned to his alma mater in 1805 as tutor in Greek before assuming the chair in mathematics and natural philosophy in 1807. Farrar translated and edited for use at Harvard a large number of classic French texts. These texts formed the series known as Farrar's Cambridge Mathematics. Included were algebras taken from the works of both Euler and Lacroix, Legendre's geometry, Lacroix's trigonometry and, in 1824, *First Principles of the Differential and Integral Calculus, or The Doctrine of Fluxions, intended as an introduction to the physico-mathematical sciences*; *taken chiefly from the mathematics of*

[3]Serret, Joseph Alfred, *Cours de calcul différentiel et intégral*, Paris: Gauthier-Villars, 1879. The first edition was in 1868. In his advertisement, Serret says that the book covers the "substance of the lessons" he teaches each year at the Sorbonne. He begins by introducing limits in a fairly careful way, but quickly shifts to infinitesimals a page later.

[4]Cajori believes (p. 82) that fluxions or calculus was offered in "the better colleges" in the early part of the century. Also see Cajori's "Bibliography of fluxions and the calculus" (p. 395ff) for information on early texts.

[5]Rudolph, p. 25; see also Chapters 3–5.

Bézout, and translated from the French for the use of the students of the university at Cambridge, New England.[6]

Farrar had pedagogical reasons for choosing Bézout. In the Advertisement to the first edition, he writes

> [Bézout's book] was selected on account of the plain and perspicuous manner for which the author is so well known, as also on account of its brevity and adaptation in other respects to the wants of those who have but little time to devote to such studies. The easier and more important parts are distinguished from those which are more difficult or of less frequent use, by being printed in a larger character.[7]

As an introduction to the book, Farrar appended an essay by Carnot[8] that explains the "truth of the infinitesimal method."

Notice Farrar's desire to choose a book accessible to "those who have but little time to devote to such studies." Although calculus was part of the curriculum in a number of American colleges during the first third of the nineteenth century, very little time was devoted to it. For example, at Harvard in 1830, sophomores studied trigonometry and its applications, topography and calculus. Furthermore, this third of a year was the only calculus they studied.[9]

Also notice that Farrar has distinguished the "easier and more important parts" typographically from the more difficult. Although Harvard was one of the first colleges to experiment with electives, most of the curriculum in most American colleges was required until after the civil war.[10] Consequently, Farrar also needed a text that was accessible to those with little talent for mathematics.

The book opens by explaining that the object of calculus is "to decompose quantities into the elements of which they are composed, and to ascend or go back again from the elements to the quantities themselves. This is, strictly speaking, rather an application of the methods, and even a simplification of the rules of the former branches of analysis, than a new branch" (p. 7). Bézout–Farrar develops calculus in a manner similar to that used by Benjamin Peirce sixteen years later and by Peck and Bowser after that.

Farrar and those who followed him approached calculus from the perspective of infinitesimals. Despite the foundational improvements of Cauchy,

---

[6]See Cajori, pp. 127–130.

[7]John Farrar, *First principles of the differential and integral calculus...*, Boston: Hilliard, Gray, & Co., 1836. I have considered this book a translation, as apparently Farrar did, and have not included it in my bibliography. Page references are from this edition.

[8]Carnot, Lazare N. M., *Réflexions sur la métaphysique du calcul infinitésimal*, Paris, 1797.

[9]Cajori, p. 132.

[10]Rudolph, Chapter 5.

this approach persisted into the twentieth century. Its best features, however, were adopted by the authors who favored limits. In this way, infinitesimal language outlived the line of texts that championed it. At the beginning of the line of limits authors is Charles Davies of the United States Military Academy at West Point.

The Military Academy rose to prominence after the arrival of Sylvanus Thayer in 1817. Thayer had been sent to Europe to study the systems military education used there. When he returned, he not only reorganized West Point along the lines of the French system, he also introduced French texts. In 1823, the chair of mathematics was assumed by Davies, a member of the class of 1815.[11]

Davies published over his career a series of books so widely used in the United States that Cajori refers to him thirteen years after his death as "one whose name is known to nearly every schoolboy in our land."[12] In addition to serving for 21 years at West Point, Davies spent four years as professor of mathematics at Trinity College in Hartford, another year at the University of New York and eight years as Professor of Higher Mathematics at Columbia College.[13]

The Davies series in mathematics eventually ran from arithmetic through calculus and included books on surveying and navigation, descriptive geometry, and "Shades, Shadows, and Perspective." Some of those books were translations, but others were more original. We, of course, are interested in his calculus.

Davies' Calculus

Davies' calculus books appeared between 1836 and 1901, with new editions every year or two between 1836 and 1860.[14] In the preface to the "Improved Edition" of 1843, he asserts that he is not writing an exhaustive book on the calculus, but only an "elementary treatise" as a textbook. He also acknowledges his sources:

> The works of Boucharlat and Lacroix have been freely used, although the general method of arranging the subjects is quite different from that adopted by either of those distinguished authors.[15]

[11]For background on West Point, see Ambrose.

[12]Cajori, p. 118.

[13]For biographical data on Davies, see Cullum. For assessments of the importance of the Military Academy in the development of American mathematics, see Cajori, p. 114ff, and Grabiner, "Mathematics".

[14]Although many editions appeared, I do not know how many copies were printed and/or sold of each one. Thus we can only conclude that there was a continuing market of an unknown size in this period.

[15]This and other references come from the 1843 edition of *Elements*.

Both Boucharlat and Lacroix were authors of popular French texts. Boucharlat's was published between 1815 and 1858, with a ninth edition appearing in 1926. (Boucharlat died in 1848, between the fifth and sixth editions.) His text was translated into English by Blakelock in 1828.[16] Lacroix was certainly the better mathematician and the better known.[17] His treatises on calculus appeared in long and short forms and in many editions between 1797 and 1881. His calculus books were extremely influential in Europe, as well as in the United States. In particular, his elementary treatise was translated into English by Babbage, Peacock and Herschel as part of their campaign to bring continental mathematics to England.[18] As we have seen, a number of his other texts were translated and used here.

Davies' book follows Boucharlat quite closely, both in the words he chooses and the examples he uses. Davies begins by noting that "if two variable quantities are so connected to each other that any change in the value of one necessarily produces a change in the value of the other, they are said to be functions of each other." This symmetric view of the functional relationship will prove very handy, for the text emphasizes the calculus of curves, as opposed to that of functions. He next examines, using the specific examples $u = ax^2$ and $u = x^3$, what happens when the independent variable is incremented by h. Looking at the quotient $(u' - u)/h$ where u' is the incremented value of the function, he declares:

> If we examine the second members of these equations, we find a term in each which does not contain the increment h.... If now, we suppose h to diminish, it is evident that the terms $2ax$ and $3x^2$, which do not contain h, will remain unchanged, while all the terms which contain h will diminish. Hence, the ratio
>
> $$\frac{u' - u}{h}$$
>
> in either equation, will change with h, so long as h remains in the second number of the equation; but of all the ratios which can subsist between
>
> $$\frac{u' - u}{h}$$

[16]For biographical data on Boucharlat, see *Nouvelle Biographie Générale...*, Paris, 1862, vol. 6, p. 855f.

[17]See Kline for references to Lacroix's work.

[18]See references in Grabiner and Kline.

> is there one which does not depend on the value of h? We have seen that as h diminishes, the ratio in the first equation approaches $2ax$, and in the second to $3x^2$; hence, $2ax$ and $3x^2$ are the limits toward which the ratios approach in proportion $a[s]$ h is diminished; and hence, each expresses that particular ratio which is independent of the value of h. This ratio is called the limiting ratio of the increment of the variable to the corresponding increment of the function (pp. 17, 18).

Davies is a teacher, and not an extensively educated one. His concern is pedagogical, not mathematical. Yet even Davies is concerned that his readers may not understand this explanation. He tries another.

Davies goes on to say that "the limiting ratio of the increment of the variable to that of the function ... is called the differential coefficient of u regarded as a function of x" (p. 19). He immediately introduces an infinitesimal argument for defining the differential of x, telling students to "represent by dx the last value of h, that is, the value of h, which cannot be diminished, according to the law of change to which h or x is subjected, without becoming zero ...."

After explaining that du is the "corresponding difference between u' and u," Davies attempts once more to help his struggling students:

> It may be difficult to understand why the value which h assumes in passing to the limiting ratio, is represented by dx in the first member and made equal to 0 in the second. We have represented by dx the *last* value of h, and this value forms no appreciable part of h or x. For, if it did, it might be diminished without becoming 0, and therefore would not be the *last* value of h. By designating this last value by dx, we preserve a trace of the letter x, and express at the same time the last change which takes place in h, as it becomes equal to 0 (p. 18).

Notice also that Davies has not established any notation for calculating the derivative. Thus, when he wants to prove a theorem, he must go back to first principles. Davies gets around this stumbling block by introducing a property of the derivative used by Lagrange,[19] namely that

$$u' - u = Ph + P'h^2$$

where P is the differential coefficient and P' will in general be a function of h, as well as of x. He explicitly assumes this result on the basis of his previous examples (p. 21).

---

[19] See Grabiner, *Origins*, p. 118ff, for a discussion of the importance of this result.

Using this tool, Davies goes on to derive the rules of differentiation, including the chain rule. He also develops Taylor's Theorem and Maclaurin's and pays some attention to "cases in which [they do] not apply;" that is, to cases in which one of the derivatives is undefined. Again, this latter material shows the influence of Lagrange. All of the above is achieved in under fifty pages.

Davies' text is "Lagrangian" throughout. For example, he finds the derivative of an exponential function by using the binomial theorem to expand $(1 + b)^h$. Less than eighty pages into the book, he is developing the Taylor series for functions of two variables. His proofs of l'Hospital's rule and of what we sometimes call the first and second derivative tests for extrema are based on the Taylor series expansions.

The only application of the derivative in Davies is curve sketching. However, this subject is treated exhaustively. Indeed, as much space is devoted to this topic as to all of the "theory" of the derivative, including Taylor series. Davies discusses cusps, multiple points, involutes and evolutes, osculating curves, and transcendental curves, such as spirals and the cycloid. Then he turns to integrals.

For Davies, the integral is the antiderivative, "the method of finding the function which corresponds to a given differential." He does note that the integral sign denotes a sum and "was employed by those who first used the differential and integral calculus, and who regarded the integral of $x^m\,dx$ as the *sum* of all products which arise by multiplying the mth power of x, for all values of x, by the constant dx" (pp. 189, 190). Now Davies spends fifty pages on techniques of integration organized in several broad categories, including "Integration by series" (pp. 201–206). The book closes with 40 pages of geometric applications of integration: Rectification of curves, quadrature of curves and curved surfaces and cubature of solids, including double integrals.

Davies' text was the first commercially successful calculus text written by an American. Of course, we must be careful, for, as we have noted, it was largely derived from French work. As we shall see, authors continued to acknowledge sources of inspiration for many years.

The Antebellum Books and Their Authors

Before the civil war, American colleges generally had a fixed curriculum which included mathematics at an elementary level and a smattering of science. As new colleges spread through the West and South, they replicated the style and form of the earlier colleges from which their founders came. By the war, there were about 200 colleges in the United States, the majority of them founded after 1840 and many of them on the frontier.[20] The philosophy of the frontier, as well as other changes in the mood of the country, would affect the nature of higher education in the decades following the war. Until that time, with the exception of West Point and Rensselaer Polytechnic Institute, America needed only brief calculus books to strengthen and decorate the mind.

Beginning with Davies, these books were supplied by eight commercial authors who began publication before 1870. Six of them used limits as the foundation for the derivative; the other two, Peirce and Smyth, used infinitesimals. One (Loomis) had studied abroad for a year, but the remainder had received domestic educations, half of them at the Military Academy.

TABLE 1: Commercial Authors Who Began Publication Before 1860

Name	Dates	Edit	Education	Positions
Davies	1836–1868	many	USMA*	USMA, Columbia
Peirce	1841–1862	3	Harvard	Harvard
Church	1842–1872	many	USMA	USMA
M'Cartney	1844–1848	2	Jefferson	Lafayette
Loomis	1851–1902	many	Yale, Paris	Yale, Western Reserve, U. City of NY
Smyth	1854–1859	2	Bowdoin	Bowdoin
Courtenay**	1855–1876	8	USMA	USMA, U.PA, U.VA
Quinby	1856–1879	6	USMA	USMA, U. Rochester

Dates = span of frequent publication;
Edit = number of editions
*United States Military Academy
**Courtenay died in 1853, leaving a manuscript.

That West Point was so well represented is not surprising. This was the paragon of scientific education in the United States during that period. In addition to supplying engineers for the country's expansion, it was providing educators. Cajori (p. 127) reports that the Academy had provided 192 educators to American colleges, including 119 teachers, numbers that the Board of Visitors of the Academy, at least, found praiseworthy.

The only author on the list who might be called a professional mathematician is Benjamin Peirce, perhaps the outstanding American mathematician of his time and a founding member of the National Academy of Sciences.

[20]See Hofstadter, pp. 11–13, and Tewksbury, Chapter 1.

Peirce was a true nineteenth-century mathematician. His *Linear Associative Algebra* was the first major American contribution to pure mathematics.[21]

Peirce's book, apparently based on Farrar's, is mathematically intriguing but pedagogically painful. Before beginning his discussion of calculus, Peirce devotes a chapter to theorems on infinitesimals, proving for instance that "any power of an infinitesimal is infinitely smaller than any inferior power of the same infinitesimal." Although he doesn't provide a definition for an infinitesimal, he carefully lays out a program of definitions, theorems and corollaries which would delight the mathematician but horrify the sophomore.

The standard defense of the infinitesimal approach, however, is that it is more accessible to students and easier to apply to problems. Peirce's contemporary, William Smyth, justifies his choice of the "method of Leibnitz" on these grounds.

> The recent textbooks, both English and French, are in general based on the method of Newton [i.e., limits]. The expediency of this may well be questioned. The artifice which lies at the basis of the Calculus, consists in the employment of certain special auxiliary quantities adapted to facilitate the formation of the equations of a problem. The limit, or differential coefficient, the auxiliary employed in the method of Newton, is not easily represented to the mind, and being composed of two parts which cannot be separately considered [*dx* and *dy*?], it is with more difficulty applied to the solution of problems. On the other hand the differential ... is simple in itself, is very readily conceived, and adapts itself with wonderful facility to all the different classes of questions which require for their solution the aid of the calculus.[22]

As if to prove his point, Smyth includes in his text sections on applications to mechanics and astronomy in addition to those included in Davies. He considers the problem of a body falling through a hollow tube to the center of the earth, as well as center of gravity and fluid pressure problems. His book runs 240 pages and concludes with a section on the theory of limits.

Growth of the University: 1870–1895

After the Civil War, publication of calculus books from both the limit and infinitesimal lines continued. A new line, the method of rates, appeared, flourished briefly, and failed with the growing awareness in the United States

[21] For biographical information on Peirce, see Carolyn Eisele, "Benjamin Peirce," *Dictionary of Scientific Biography*, vol. 10, pp. 478–481. See also J. Grabiner, "Mathematics," p. 18, Cajori, p. 136, and Thwing, p. 304, for comments on Peirce as a teacher.

[22] Smyth, preface to the first edition.

of Cauchy's foundational work. Until the end of the century, however, authors continued to discuss the best method of presentation.

The educational environment in which these authors worked was different in several ways from that of their predecessors. The elective system replaced the required curriculum as the standard mode of education. The public began to demand a more practical education, a desire that was given support by the Morrill Act. Finally, American industrialists were funding their visions of higher education. All of these changes had begun in the decades before the war, but their impact came later.

Colleges had experimented with elective systems since 1824, when the University of Virginia adopted a completely elective curriculum. That experiment ended in 1831. Another attempt at Harvard about the same time also failed. However, the elective system established itself when the demands of the public for a more practical education and the intellectual demand of the sciences for a larger piece of the curricular pie had to be met.

Science, and mathematics with it, bloomed in the new land-grant colleges designed to encourage the study of agriculture and the mechanic arts, and authorized by Congress in The Morrill Act of 1862. It also flourished in the "Scientific Schools" formed at established colleges. The Lawrence School, established at Harvard in 1847 and the Sheffield School, established at Yale in 1854, both enriched by the gifts of wealthy patrons, are two examples. Finally, science and mathematics benefited through the creation of universities, such as Cornell, in 1869, and Johns Hopkins, in 1874, both named to honor their wealthy industrialist benefactors. In them, research and graduate education assumed a greater role than they had played in the colleges. Thus in the final quarter of the century, the German-style university began to replace the classical college as the model of American higher education.[23]

Scientific education for more, but better motivated, students demanded more advanced mathematics texts. Students needed calculus to study modern science. Ready to meet the demand were not only the earlier texts of Church and Loomis, for example, but also those of a more modern set of authors.

The Authors

Many of these authors (see chart) followed much the same career path as their predecessors. Others went different ways. Buckingham, a Military Academy graduate, was the president of the Chicago Steel Works when his books appeared. Byerly, the first Ph.D. on our list, was one of Harvard's first as well, with a dissertation on the heat of the sun. Along with Byerly, Johnson was one of the early members (the first from outside the New York

[23]See Rudolph on the rise of the American university; see also Grabiner, "Mathematics," pp. 17–23.

area) of the New York Mathematical Society, which soon became the American Mathematical Society. He wrote the first article in the first issue of the *Bulletin*.[24] One of the authors, Newcomb, deserves special mention.

Simon Newcomb was the fourth president of the American Mathematical Society. In Archibald's *Semicentennial History* ..., the biographical sketches of the first three presidents take a total of fourteen pages; Newcomb's takes fifteen. Newcomb was America's foremost astronomer and was recognized internationally for his work. He was also a "scientific statesman," as his membership in many academies of science, his honorary degrees and his editorship of the *American Journal of Mathematics* show.[25]

TABLE 2: Commercial Authors who Began Publication 1870–1895

Name	**Dates**	**Edit**	**Education**	**Positions**
Olney	1870–1885	4	no formal	Kalamazoo, U. Mich
Peck	1870–1877	5**	USMA*	USMA, U. Mich, Columbia
Johnson##	1873–1909	many	Yale	USNA#, Kenyon, St. John's
Buckingham	1875–1885	3	USMA	Kenyon
Byerly	1879–1902	many	Ph.D. Harvard	Cornell, Harvard
Bowser	1880–1907	many	Santa Clara, Rutgers	Rutgers
Osborne	1889–1910	many	Harvard	USNA#, MIT
Taylor	1884–1902	9	Colgate	Colgate
Bass	1887–1905	6	USMA	USMA*
Newcomb	1887–1889	2	Harvard	Naval Observatory, Nautical Almanac, Johns Hopkins U.

Dates = span of frequent publication;
Edit = number of editions;
*United States Military Academy;
#United States Naval Academy;
**includes an edition published after 1877;
##includes the books written jointly with John Minot Rice.
Authors in *italics* used infinitesimals; in **boldface**, rates.

[24]For further information on Byerly and Johnson, see Archibald.
[25]Archibald, p. 124ff.

During this period, some authors cite other works in their prefaces, but some of the cited works are American. Some authors thank professional colleagues. Some do neither. Among the foreign works cited, Bertrand appears on the lists of Bowser, Byerly and Johnson. Bertrand's *Traité de calcul*[26] appeared in 1864 and nominally uses limits as its approach to calculus. However, very early in the book, Cauchy's definition of an infinitesimal, which we discuss below, appears and an extended discussion of orders of infinitesimals of the sort that Peck gives (see below) follows. Other European texts cited include the British books of Price, Todhunter and Williamson and Duhamel's book from France.[27] Generally, outside sources receive less attention than they do in the earlier period: Books are allowed to stand on their own.

Of the ten commercial authors who began publication in this period, only four based their books on limits. Three based their presentation on infinitesimals and three based theirs on the method of rates. (See Table 2.) In this period, we begin to see the merger of the infinitesimal approach into that of limits.

Limits

In the preface to his text, Osborne asserts that he has based his text "on the method of limits, as the most rigorous and most intelligible form of presenting the first principles of the subject." He goes on to state that many students have been introduced to limits in earlier courses and "may be assumed to be fully conversant with it on beginning the Differential Calculus."[28] Byerly says that a feature of his book is "the rigorous use of the Doctrine of Limits as a foundation of the subject," but adds that it's "preliminary to the adoption of the more direct and practically convenient infinitesimal notation and nomenclature ...."[29] Bass, writing later, believes that "the more rigorous and comprehensive method of infinitesimals is suitable only for a treatise, and not for a textbook intended for beginners."[30]

Whatever their belief, they all define and use infinitesimals. Bass introduces them on the page on which he defines the limit of a variable (p. 25). Byerly's introduction is much later (Chapter X). Osborne in his earlier work

[26]Joseph Louis Bertrand, *Traité de calcul différentiel et de calcul intégral*, Paris, 1864–1870. His *Cours...* appeared in 1875 and was republished until the end of the century.

[27]Bartholomew Price, *A treatise on the differential calculus...*, 1848 and later, Oxford. Price used the method of infinitesimals. Isaac Todhunter, *A treatise on the differential calculus and the elements of the integral calculus...*, 1852. Editions appeared until 1923. Benjamin Williamson, *An elementary treatise on the Differential Calculus...*, 1872, London. Jean Duhamel, *Cours d'analyse...*, 1840, Paris.

[28]Osborne, *Elementary treatise*, 1903, p. iii. The preface is dated 1891; the copyright date is 1891.

[29]Byerly, *Differential calculus*, 1879, p. iii. All Byerly references are to this edition.

[30]Bass, *Elements*, 1901, p. iii. The quotation is from the preface to the 1895 edition.

describes dx as an infinitely small Δx and lets it go at that, but in 1908 he talks about infinitesimals in a standard way.

The "standard way" is Cauchy's formulation: An infinitesimal is a variable with limit zero. With this convention, limits authors are able to utilize the advantages of infinitesimal techniques without becoming mathematically suspect, and its use continued well into the twentieth century.[31]

What is gained formally is the legerdemain of replacing limit talk with algebra, as the following proof from Bass illustrates. We wish to prove that if U and V are variables which under their laws of change are always equal, then their limits are equal. If C is the limit of U, then $C = U + e$ where e is an infinitesimal, or $U = C - e$. Since $U = V$, we have $V = C - e$ or $C = V + e$. Hence C is also the limit of V.[32]

We, of course, use this theorem regularly in our calculus courses. If $U = [(x + h)^2 - x^2]/h$ and $V = 2x + h$, then U and V, "under their laws of change," namely h is not zero, are always equal. Since V differs from $2x$ by an infinitesimal, its limit is $2x$ and, by the theorem, so is that of U. Interestingly, Bass gives roughly this example before he states his theorem. Indeed, the role of theory in Bass' book is uncertain.

Although much of the talk in these books seems to be about variables, in fact the authors clearly have functions in mind. For example, Bass defines the limit of a variable as

> a fixed quantity or expression which the variable, in accordance with a law of change, continually approaches but never equals; and from which it may be made to differ by a quantity less numerically than any assumed quantity however small (p. 22).

He makes clear, however, in a footnote, that he means the term "variable", to include all functions. For Bass and for Byerly, a function is a quantity the value of which depends upon another quantity. Indeed, when Bass gets around to defining the differential coefficient and differentials, he is quite explicit in his use of functional notation (although his explanation is quite obscure) (pp. 47, 55).

Notice how modern Bass' definitions of function and limit appear when compared to Davies'. A "modern" application also appears. There are related rates problems, including ships sailing on perpendicular courses and men walking to or from lamp posts (p. 84ff). However, there are no extremum problems[33] and, like Davies, he belabors the geometry of curves. Again, he includes all of the topics in Davies, and adds a number of exercises.

[31] See, for example, Granville et al., 1904 and later.

[32] Bass, *Introduction*, p. 25. Other citations from Bass are from this book.

[33] His 1901 book does include extremum problems.

Bass does not treat integration. Both Byerly and Osborne do, and, again, in a modern way. Integrals and definite integrals are identified as different entities. That requires a Fundamental Theorem of Calculus. Quinby was the only earlier author to prove this. Byerly[34] uses an infinitesimal argument (as did Quinby); Osborne provides an example that shows that the limit of the sum yields the difference of the two values of the antiderivative (my term).

In addition to the applications of the integral to the geometric problems of arc length, area and volume, Byerly and Osborne apply it to physical problems. Byerly treats centers of gravity, mean distances and probability. Osborne discusses moments of inertia.

Finally, these authors used series much more carefully than their antebellum colleagues. All of them, at least by their later editions, worried about convergence. All of them stated and proved—more or less—Taylor's theorem with the Lagrange remainder.[35]

INFINITESIMALS

By contrast, the infinitesimal authors all develop series in a manner essentially identical to that of the earlier group. An expansion of $f(x+h)$ in the form $A + Bh + Ch^2 + \cdots$ is assumed and the necessary values of the coefficients are calculated. Some attention is paid to values for which the development fails, as before.

While their treatment of series reflects a dated foundational view, all of the authors are clearly aware of the limit approach and have deliberately rejected it for pedagogical reasons. Olney chooses infinitesimals because it's simpler and because it facilitates the application of calculus to "practical problems." He goes on to criticize the "general use" of limits in textbooks for "preventing the common study" of calculus. His complaint is the same as Smyth's fifteen years earlier.

> This method is not only exceedingly cumbrous, but it has the misfortune that its element is a ratio. The abstract nature of a ratio, and the fact that it is a compound concept, peculiarly unfit it for elementary purposes. The beginner will never use it with satisfaction, for it does not give him simple, direct and clearly defined conceptions.[36]

[34]There are differences between the 1881 and 1889 editions of Byerly's *Integral calculus*. In the former, he speaks of the computation of the definite integral as the limit of a sum; in the latter, of its definition that way.

[35]Byerly, whose text is much more modern than those of his contemporaries, uses Rolle's Theorem in 1879 to get the remainder.

[36]Olney, 1871, p. v. All references to Olney are from this edition.

While not as outspoken as Olney on the disadvantages of limits, both Bowser and Peck agree with him on the advantages of infinitesimals. It's the easiest method to understand and apply.

But just as the limits authors did not neglect infinitesimals, neither do these authors neglect limits. Bowser devotes his third chapter to "limits and derived functions" and, using limit techniques, rederives the basic formulas. Olney claims (p. 5) that, in fairness he will introduce and use the theory of limits when he wants to. Peck's "Note on the Method of Limits" appears as an appendix, is not listed in the table of contents and is credited to E. H. Courtenay whose publisher was also A. S. Barnes.[37]

Peck, son-in-law of Charles Davies, developed infinitesimals in much the same manner as Bowser and in only a slightly different way from Peirce.[38] He defined a quantity to be infinitely small with respect to another if the quotient was "less than any assignable number." A number that was infinitely small when compared to a "finite number," for instance 1, was called an infinitesimal. He followed these definitions with a discussion of orders of infinitesimals that was less formal than Peirce's, but had the same objectives. Peck concluded that "an infinitesimal may be disregarded in comparison with a finite quantity, or with an infinitesimal of lower order" when added or subtracted (pp. 13, 14).

Now Peck is ready to teach the student to find differentials. He has a very simple algorithm.

> In order to find the differential of a function, we give to the independent variable its infinitely small increment, and find the corresponding value of the function; from this we subtract the preceding value and reduce the result to its simplest form; we then suppress all infinitesimals which are added to, or subtracted from, those of a lower order, and the result is the differential required (p. 14).

He then notes wisely that the method is too long for general use and will be employed only to derive some general rules.

Consider Peck's proof of the quotient rule. If s and t are functions of x, we are required to find $d(s/t)$. Augment x by its infinitely small increment dx. Then s is augmented by ds; t by dt; and s/t by $d(s/t)$. Thus we have

$$\frac{s+ds}{t+dt} = \frac{s}{t} + d\left(\frac{s}{t}\right).$$

[37]Courtenay, *Treatise*.

[38]Peck, *Practical treatise*, 1870. Page references that follow are from this edition.

Subtracting s/t from both sides, finding a common denominator on the left and simplifying, we now have

$$\frac{t\,ds - s\,dt}{t^2 + t\,dt} = d\left(\frac{s}{t}\right).$$

But we can suppress the $t\,dt$ term in the denominator because it is of lower order (p. 17).

Infinitesimals work particularly well for applications. For example, since a curve is made up of infinitesimal elements, the slope of a tangent line to a curve is dy/dx practically by definition, or, as Peck says, "an element of the curve ... does not differ from a straight line. Hence, the slope of a curve, at any point, is measured by the first differential coefficient at that point" (p. 53). Similar arguments provide the motivations for arc length and area.

In the same manner, suppose the object of integration to be, as Bowser says, finding "the relations between finite values of variables from given relations between the infinitesimal elements of those variables, or ... the process of finding the function from which any given differential may have been obtained" (p. 238). Then the area under $y = f(x)$ is, naturally, given by the integral of $f(x)\,dx$, the differential of the area, and the length of the curve is the integral of ds, the differential of the arc length. This intuition also leads to direct solutions of many physical problems. Interestingly, only Peck, Professor of Mechanics in the School of Mines as well as Professor of Mathematics and Astronomy at Columbia College, among the infinitesimal authors provides applications other than geometric ones.

The infinitesimal books are clearly dated when compared to the books using limits in this period. The books using rates, however, were truly from another age. They represent a return to fluxions.

The Method of Rates

While infinitesimal techniques remain part of our heuristics in teaching calculus, the method of rates has been completely discarded. In this approach, calculus is regarded directly as the mathematics of change. The fundamental questions of calculus, proponents of rates argue, are not about tangent lines and areas, but about how one quantity changes in response to changes in another. This point of view is, of course, that of Newton's fluxions, but, except for one British book of 1845[39], it had completely disappeared even in England when the books we are considering were written.

Nevertheless, two very important texts used this method into the first decade of this century: Rice and Johnson and James M. Taylor. As I noted above, Johnson was an important member of the American mathematical

[39]Connell, James, *The elements of the differential and integral calculus*, London: Longman & Co., 1845. This was the only edition of the book.

scene. He served as one of only five elected members of the Council of the American Mathematical Society for the 1892–1893 term.[40] Thus, while his choice may have been eccentric, it was not an ignorant one. The last rates text, that of Edward Nichols, appeared in 1900 with a second edition in 1918.[41]

All of the authors eventually introduced limits. Taylor and Nichols did this in early chapters and then proceeded to use whatever seemed handiest. Rice and Johnson gave in somewhat later. Of course, our interest here is not in the limit portion of the texts, but the more unusual part.

The authors begin by describing uniform change of a variable as occurring when "its value changes equal amounts in equal arbitrary portions of time." Now turning to non-uniform change, Taylor says (Rice and Johnson is quite similar),

> If a variable changes non-uniformly with respect to x, the measure of its rate is what its increment corresponding to the increment 1 of x would be if at the value considered its change became uniform (p. 7).

Now the differential of a variable is its rate of change.

In this system, the differential triangle with legs dx and dy and hypotenuse ds is not simply the representation of an infinitesimal figment by a fine Euclidean object, but the realization of what would have happened had the rates become uniform. Similarly, if $A(x)$ is the area under the curve $y = f(x)$, then dA is $y\,dx$ simply from the definition of the differentials involved. This system thus had advantages for certain applications.

However, the method of rates exacted a terrible cost when the authors tried to prove something as basic as the product rule for derivatives. Rice and Johnson derive this from the rule for differentiating x^2, since $xy = (1/2)(x+y)^2 - (1/2)x^2 - (1/2)y^2$. That rule is not easy to derive. First, they set $z = mx$. Then it follows that $dz = m\,dx$ and that $d(z^2) = m^2 d(x^2)$. Now dividing the second of these equations by the first, substituting z/x for m and separating the variables, they get

$$\frac{1}{z}\frac{d(z^2)}{dz} = \frac{1}{x}\frac{d(x^2)}{dx}$$

At this point the authors invoke their Fundamental Theorem: The value of dy/dx does not depend upon dx, but is a function of x alone (p. 17). Thus, denoting x^2 by $f(x)$, the equation becomes $f'(z)/z = f'(x)/x$ and this is true for all values of x and z, since the constant m was arbitrary. Now we

[40] Archibald, p. 97.

[41] Rice and Johnson, *The elements*. The argument used to derive the product rule is from the first part of the 1874 book. Taylor, *Elements*. Citations are from the 1894 edition. Nichols, *Differential*.

can conclude that $f'(x)/x$ is some constant c, or that $d(x^2) = cx\,dx$. The remaining problem is to find c.

To find c, we apply this last result to the identity

$$(x+h)^2 = x^2 + 2xh + h^2.$$

Since we know that the differential of a constant is zero and that the differential of a sum is the sum of the differentials, we have

$$c(x+h)\,dx = cx\,dx + 2h\,dx \text{ or } (c-2)h\,dx = 0.$$

"Since h and dx are arbitrary quantities, we have $c = 2$, which gives ... $d(x^2) = 2x\,dx$" (pp. 21–23).

Despite some painful developments such as this, the method of rates authors believed that their approach was, if not the best, a satisfactory one. Rice and Johnson were blunt about their choice.

> The difficulties usually encountered on beginning the study of the Differential Calculus, when the fundamental idea employed is that of infinitesimals or that of limits, together with the objectionable use of infinite series involved in Lagrange's method of derived functions, have induced several writers to return to the employment of Newton's conception of rates or fluxions (1877, p. iii).

Taylor was more circumspect by 1898.

> ... [A]n attempt has been made to present in their unity the three methods commonly used in the Calculus. The concept of Rates is essential to a statement of the problems of the Calculus; the principles of Limits make possible general solutions of these problems, and the laws of Infinitesimals greatly abridge these solutions (1898, p. iii).

Taylor goes to considerable pains to defend rates against the charge that it invokes a "foreign element," namely time. However, these late nineteenth century books would be the last gasp of fluxions in America.

THE PROFESSIONAL CLIMATE, 1870–1895

One of the factors in the coming demise of the nineteenth century texts was the growth of a mathematical community. In 1888, the New York Mathematical Society was founded and by 1895 (as the American Mathematical Society), it had 268 members. It also had a new journal that published reviews of books, including calculus texts, and discussions on the teaching of calculus.

Reviews of calculus texts had appeared in earlier journals, but they tended to be complimentary rather than analytic. Issues of the *Mathematical Visitor* regularly contained notices of new books. In January 1880, the editors describe Rice and Johnson's revised edition as "the most extensive work on the Differential Calculus yet published in this country" and "heartily commend it to all who want a good textbook on the subject." In the same issue, they describe Byerly's book as "a good work" and a year later call Bowser's "a work of rare excellence." The *Analyst* published similar reviews.[42]

By contrast, in the first volume (1891–1892) of the *Bulletin* of the New York Mathematical Society, Charlotte Scott[43] wrote a scathing seven-page review of a British text by Joseph Edwards. After noticing how well written it was and how nicely printed, Scott asserted that the book had "many defects" and she proceeded to point them out. Among other faults, she observed that Edwards was not aware of Weierstrass's example of a continuous, nowhere-differentiable function.

Other articles also reflected the intellectual growth of the mathematics community. The opening paper of the October, 1893 issue of the *Bulletin* was a reprint of Felix Klein's inaugural address to the Chicago congress and the following paper in the same issue was a report by T. H. Safford on "Instruction in Mathematics in the United States." The December issue contained a translation of a circular describing the program in mathematics at Göttingen. A new age in mathematics was beginning.

It is not clear, however, that teachers of calculus were ready for the new day. In his 1889 survey of American colleges and schools[44], Cajori discovered that about half of those teaching calculus favored limits. Almost 30% favored infinitesimals. (140 of the 160 colleges and universities responding to the survey apparently taught calculus.) Since the vast majority of the respondents taught mainly from textbooks, the disappearance of the infinitesimal books near the beginning of the new century should have been the cause of some concern.

A New Century

The "old fashioned" books disappeared as new standards for authors emerged by the beginning of the twentieth century. Among those (see Table 3) who began publication in the period between 1900 and 1910, at least 30% had studied in Europe and 45% had doctoral degrees. About half of the degrees were from U. S. universities. This phenomenon, the ascendancy of the Ph.D., is as marked as any textbook feature we have examined.

---

[42]See Cajori, p. 277ff, for comments on journals of the nineteenth century.

[43]Charlotte A. Scott, "Edwards' Differential Calculus," *Bulletin of the New York Mathematical Society*, 1 (1891–92), p. 217ff.

[44]Cajori, pp. 296–360.

TABLE 3: Commercial Authors who Began Publication 1895–1910

Name	Dates	Edit	Education	Positions
Gould	1896–1907	4	E. des Mines	engineer
Fisher	1897–1909	8[1]	Ph.D. Yale, European study	Yale
Hall	1897–1905	7[2]	Lafayette	Lafayette
Lambert	1898–1907	2	Lehigh, European study	Lehigh
Love	1898–1899	2	U. No. Carolina Harvard, Hopkins	UNC, Harvard
Murray	1898–1908	4	Ph.D. J.Hopkins	Dalhousie, NYU, Cornell, McGill
Hardy	1900–1912	2	Lafayette	Lafayette
Nichols	1900–1918	3	VMI	VMI
Echols	1902–1908	2	U. Virginia	Mo. School of Mines, U. Virginia
Osgood	1902–1938	8[3]	Ph.D. Erlangen	Harvard
Granville	1902–1957	many	Ph.D. Yale	Yale, Gettysburg, insurance
Smith			Ph.D. Yale, European study	Yale
Longley			Ph.D. Chicago	Yale, Colgate
Snyder[4]	1902–1912	2	Ph.D. Göttingen	Cornell
Hutchinson			Ph.D. Chicago, European study	Cornell
Campbell	1904–1919	5	Ph.D. Harvard	Harvard, IIT, actuary
Cain	1905–1911	4	NC Mil.Inst.	U. N. Carolina
Keller	1907–1908	2	*	
Knox			*	
Woods	1907–1954	many	Ph.D. Göttingen	Wesleyan, MIT
Bailey			*	
Townsend	1908–1911	5[5]	Ph.D. Göttingen	U. Ill
Goodenough			Mich. Ag. C.	Mich. Ag. C., U. Ill
Brown	1909–1912	2	Cornell	Naval Academy
Capron			Harvard	Naval Academy
Ransom	1909–1949	5	Tufts, Harvard	Harvard, Tufts

Dates = span of frequent publication; Edit = number of editions *Biographical data is missing

[1] Includes two editions published after 1909.

[2] Includes two editions published after 1905.

[3] Includes one edition after 1938; does not include his *Advanced calculus*.

[4] Snyder also published a book with James McMahon in 1898.

[5] Includes 1925 edition.

Authors in *italics* used infinitesimals; in **boldface**, rates

The Ph.D.'s with their uniform approach to the calculus dominated textbook production despite the fact that most calculus teachers did not have doctorates and did not learn their calculus from books like the new ones. In 1899, fewer than 180 Americans held doctorates in mathematics and most of them were located at the universities.[45] However, their number was growing and their professional organization, the American Mathematical Society, was growing too.

Between 1895 and 1907, the number of members of the AMS doubled to 568 and a single section had become four. Moreover, the presidents of the organization were young. Of the first ten (through 1910), only Van Amringe, McClintock and indefatigable Newcomb were over fifty when they presided. Half of them had studied in Germany.[46] In an age that cherished "progress," traditionalists would have been hard pressed to stop the rush of these enthusiastic students of brilliant German teachers to reform the teaching of the calculus.

The Bulletin of the Society continued to be filled with reports on teaching mathematics at all levels and on teaching calculus in particular. Osgood's presidential address in 1907 was called "The Calculus in Our Colleges and Technical Schools."[47] Importantly, calculus books were reviewed critically in the Bulletin.[48] Old publications were pushed out and new "modern" books took their place.

THE PUBLICATION RECORD

Davies' books were published regularly for 44 years and editions appeared regularly over 65 years. Loomis' books appeared regularly for 36 years and the last edition was published 51 years after the first. Rice and Johnson was published for over 35 years and the staying power of Granville, Smith and Longley is legendary. Davies, Loomis, and Rice and Johnson, however, were separated from Granville, et al. by a barrier between the old and the new. Books published before the barrier did not get far beyond it. On the other side were the new books, the books of the new profession.

Looking at Table 4, we can see the abruptness of this change. No commercial author who began publication before 1897 and only three authors who began publication before 1902 had a calculus book published after 1912. This

[45]See Richardson for data on Ph.D.s. Richardson's data show that, in 1935, less than 30% of those teaching mathematics in colleges and universities had doctorates. Also see Kevles, Table 7 on the distribution of employment for productive Ph.D.s in the period up to 1915. Richardson's data and Kevles' are difficult to reconcile. The orders of magnitudes seem to agree, however.

[46]For the early history of the AMS, see Archibald.

[47]*The Bulletin of the American Mathematical Society*, 2nd series, vol. 13, June 1907, pp. 449–467.

[48]I have found reviews of about fifteen calculus texts in the *Bulletin* between 1900 and 1910.

TABLE 4: Publication Dates of Commercial Texts, 1885–1930, of Authors who Began Publication Before 1910

Author	85	90	00	10	20
Davies			*		
Loomis	**	*	*		
Olney	*				
Peck		*			
Rice & ...	*****	* * *	* * * *		
Buckingham	*				
Byerly	*	*** * **	**		
Bowser	***	*** * *	* * * *		
Osborne		*** * *	*** **** *		
Taylor	**	*** **	*		
Bass	* *	*	* * *		
Newcomb	* *				
Gould		* *	*		
Fisher		*	* * * *		*
Hall		****	*	*	*
Lambert		*	*		
Love		**			
Murray		*	** *		
Hardy		*		*	
Nichols		*	*	*	
Echols			* *		
Osgood			* * *		** * **
Granville, et al.			***	**	*
Snyder & ...			*	*	
Campbell			**	* * *	
Cain			* * *	*	*
Keller/Knox			**		
Woods/Bailey			* *	*	* * *
Townsend & ...			*	**	*
Brown/Capron			*	*	
Ransom			*	*	

Authors are listed in order of first publication date.
Authors in italics used infinitesimals; in boldface, rates.
Authors above first dashed line began publication before 1897; those between the lines, between 1897 and 1902. (See text)

was in spite of the fact that twelve commercial authors began publication between 1880 and 1901. Of the three, one was William Shaffer Hall who had two stray editions in 1915 and 1922; one was the fluxions author, Edward West Nichols; and the last was an economist, Irving Fisher. (Virgil Snyder, a Göttingen Ph.D., published with James McMahon in 1898 and with John Hutchinson in 1902 and 1912.)[49] The standard for calculus texts was changing so markedly and rapidly in the final decade of the nineteenth century

[49] Hall, *Elements*, 1897–1922; Nichols, *Differential*, 1900–1918; Fisher, *A brief*, 1897–1937; McMahon and Snyder, *Elements*, 1898; Snyder and Hutchinson, *Differential*, 1902 and *Elementary*, 1912.

that earlier authors, no matter how virtuous their product might have been, simply lost their market to the newcomers.

A "Modern" Book

Just before the end of the century, the last two infinitesimal books appeared.[50] They were books small enough to fit in a coat pocket and cost less than a dollar each, according to the brief review by Thomas Fiske in the *Bulletin*. Although Fiske's tone is quite mild, he is clearly unhappy with the books. Gould's "gives rules without pretense of demonstration and almost without explanation." Fisher's is quoted to show his inappropriate handling of infinitesimals.[51]

A decade later, by contrast, the initial edition of Granville received an extended and generally positive review from Edward Van Vleck.[52] He asserts at the end of the review that he knows "of no work which has greater promise of success in our college classes." His assessment was accurate. Granville, Smith and Longley, successor to Granville's 1904 book, was the standard calculus text, the book against which others were measured in the United States for nearly five decades. The "modern textbook," as its authors described it, had arrived.[53]

Granville, like the authors of earlier books, has explicit pedagogical goals for his text. In the preface, he describes his book as a "drill book" and certainly one feature of the text is its large number of exercises. For example, compared with Snyder and Hutchinson's 1902 edition, Granville's book has about twice as many exercises on calculating the derivative and three times as many extremum problems. However, Granville has more to offer than exercises.

Granville believes that the results in the text "should be made intuitionally as well as analytically evident to the student." He chooses to introduce ideas and results intuitively first, then supply the analytic argument that proves the result (p. iii). His discussions of extrema dramatically illustrate his approach. In the ninth chapter, Granville produces the now-standard derivative tests for extrema by encouraging his readers to examine the graphs; forty pages and six chapters later, he proves the results using the Mean Value Theorem.

However to our ears, over eighty years later, some of his explanations sound forced and artificial. Partially, this is the result of dated ideas, such as subtangents and subnormals, that appear in the book and of language that

[50]Gould, *A primer*, 1896, and Fisher, *A brief*, 1897.

[51]Thomas S. Fiske, "Recent textbooks in calculus," *Bulletin of the American Mathematical Society*, Series 2, 4 (Feb. 1898), p. 237f.

[52]Edward B. Van Vleck, "Granville's Differential and Integral Calculus," *Bulletin of the American Mathematical Society*, Series 2, 11 (Jan. 1906), p. 181ff.

[53]Granville, *Differential*, 1904, p. iii. All references are to this edition.

we find stilted, but partially this is the result of trying to deal with difficult technical notions without using precise mathematical language.

The book has been denigrated for its lack of rigor, and, from our vantage point, there is plenty of material that is open to question.[54] For example, he explicitly assumes that all functions are continuous and continuously differentiable, except possibly at a finite number of points. Some of his proofs, such as that for the chain rule, are faulty, and he seems to prefer quick, if not quite correct, proofs to more complete and careful ones. Indeed, many of Granville's "lapses", such as the chain rule proof, appear deliberate, as if, knowing better, he has chosen not to write a rigorous book. We should not confuse, however, Granville's intent with that of the more rigorous contemporary texts of, for example, Hardy, or with treatises on the theory of functions of a real variable, such as Jordan.[55] Granville is writing an introductory text for American students. In fact, he does avoid the most egregious errors of his predecessors. However, Granville is still talking about limits of variables and infinitesimals.

Granville invokes a new standard formula for the limit of a variable.

> If a variable v takes on successively a series of values that approach nearer and nearer to a constant value l in such a manner that $|v - l|$ *becomes and remains less* [my emphasis] than any arbitrarily small positive quantity, then v is said to approach the limit l, or to converge to the limit l (p. 19).

The phrase "becomes and remains less," or a similar one that reflects Weierstrassian mathematics, appeared in many of the new books, including Snyder and Hutchinson and Echols, and in later books.

Granville's fourth chapter is devoted to the theory of limits and infinitesimals in Cauchy's sense. He proves the theorems on the algebra of limits by first using a mixture of intuition and epsilons to validate—"prove" is too strong—the corresponding results for infinitesimals. He introduces the derivative as a limit in Chapter V. Finally, in the sixth chapter, using the limit theorems, he derives the algebra of derivatives and rules for differentiating all of the elementary functions. Then he turns to applications.

Some of the old geometry remains. In addition to subtangents and subnormals, radius of curvature, evolutes and envelopes are there. However, related rates, curve sketching and extremum problems appear in modern

[54]See Halmos for a nostalgic look at a later edition. I remember discussions about 1957 in which Granville, Smith and Longley, finally out of print, was still being flailed.

[55]Hardy, Godfrey Harold, *A course of pure mathematics*, Cambridge: The University Press, 1908. Hardy's well-known book went through ten editions plus two American editions and several reprintings of the ninth, which first appeared in 1944. Also, Jordan, Camille, *Cours d'analyse...*, Paris, 1882–87. The third edition of this three volume work appeared between 1909 and 1915.

form. The Mean Value Theorem rates a chapter heading and Van Vleck praises Granville's intuitive handling of the theory there. The section on series comes late, rather than early, in the about 280 pages devoted to the differential calculus.

The first seventy pages on integration are devoted to the indefinite integral, and include techniques of integration. The chapter on the definite integral begins with a section on the differential of an area. This, of course, is exactly the same argument that infinitesimal authors had used many years earlier. The infinitesimal line has now been completely absorbed into the limit books. However, Granville does go on to treat integrals as sums in his next chapter. His applications are to areas and volumes and to moments of inertia.

We have seen the features of this book emerging in the nineteenth century. In Granville, they are collected and refined. Nevertheless, Granville does not look like a text from 1985 or even like one from 1955. We have noted that Granville has not utilized the mathematics of his period fully and that some of his presentations are suspect. However, both the author and the reviewer thought of this as a modern book.

These "modern" authors and their contemporaries thanked their friends for help with their text. Some authors mentioned other books to which they had referred. For example, Murray and Echols each mention several, but Granville and Snyder mention none. There is no question, however, that these authors are presenting their own product, a product for the students of the American university.

Conclusion

During the nineteenth century, American calculus textbooks evolved from being translations and copies of European sources to mathematically "modern" resources. As the mathematical community became better educated, texts that were less acceptable to professional mathematicians began to disappear, to be replaced by ones that reflected more closely the foundations of the subject as developed by Cauchy and then Weierstrass.

In this paper, I have examined this evolution. While I have been primarily concerned with the analysis of the changing contents of the books, I have also tried to tie those changes to changes in the professional and educational contexts in which the authors of the books labored and in which the books found their market.

We know that the mathematical community was maturing during the second half of the century. We have seen that this development affected these texts both directly through the increased mathematical sophistication of the authors and indirectly through more stringent critical standards. By the early part of the twentieth century, books differed little from each other in content and approach, although thirty years earlier variety was more normal than similarity.

But the mathematical community itself was a subset of a larger academic environment which had changed markedly during the post-Civil War period. The classical college that had dominated American higher education from the founding of Harvard to that war was replaced by the "practical" colleges encouraged and supported by the Morrill Act. These, in turn, and others endowed by wealthy industrialists began to evolve by the end of the century into the great research universities.

These colleges and universities provided both an expanding market for texts and an academic home for well-trained mathematicians. Although few in number, these mathematicians, in turn, created a demand for better texts; that is, for texts that met the standards of the emerging mathematics profession.

References

Ambrose, Stephen E., *Duty, Honor, Country: A History of West Point*, Baltimore, 1966.

Archibald, Raymond Clare, *A Semicentennial History of the American Mathematical Society 1888–1938*, Amer. Math. Soc., 1938.

Cajori, Florian, *The Teaching and History of Mathematics in the United States*, U.S. Government Printing Office, 1890.

Cullum, George W., ... *Biographical Register of Officers and Graduates of the U.S. Military Academy...*, New York, 1879.

Grabiner, Judith V., "Mathematics in America, The First Hundred Years," in *Bicentennial Tribute to American Mathematics*, 1776–1976, edited by J. Dalton Tarwater, Math. Assoc. of America, 1976.

Grabiner, Judith V., *The Origins of Cauchy's Rigorous Calculus*, Cambridge, MA, 1981.

Halmos, P. R., "Some Books of Auld Lang Syne," *A Century of Mathematics in America, Part I*, Amer. Math. Soc., 1988, pp. 131–174.

Hofstadter, Richard, "The Development of Higher Education in America," in Richard Hofstadter and C. DeWitt Hardy, *The Development and Scope of Higher Education in the United States*, New York, 1952.

Kevles, Daniel, "The Physics, Mathematics, and Chemistry Communities: A Comparative Analysis," from *The Organization of Knowledge in Modern America, 1860–1920*, edited by Alexandra Oleson and John Voss, Baltimore, 1979.

Kline, Morris, *Mathematical Thought from Ancient to Modern Times*, New York, 1972.

Richardson, R. G. D., "The Ph.D. Degree and Mathematical Research," *American Mathematical Monthly*, vol. 43, 1936, pp. 199–215. Reprinted in *A Century of Mathematics in America, Part II*, Amer. Math. Soc., 1989.

Rudolph, Frederick, *The American College and University: A History*, New York, 1968.

Tewksbury, Donald G., *The Founding of American Colleges and Universities before the Civil War*, New York, 1965.

Thwing, Charles Franklin, *A History of Higher Education in America*, New York, 1906.

APPENDIX: CALCULUS TEXTBOOKS BY AMERICAN AUTHORS, 1828–1920

This list attempts to cite all Calculus texts written by residents of the United States or Canada and published between the first such book (Ryan, 1828) and 1920. The closing date is arbitrary. Books that were published after 1920 are included in the list if the author has other texts that appeared before 1920 (e.g., Granville, Osgood). Furthermore, "American author" has been liberally interpreted to include natives of other regions who were in the United States or Canada when their books were published (e.g., Bonnycastle).

This list was compiled primarily by searching the shelves of the Library of Congress and a variety of other libraries for books from the appropriate period. When a book was discovered, the National Union Catalog, Pre-1956 Imprints was searched for additional information. Although I have examined many of the books listed, I cannot claim to have seen them all. While I am confident that the list is off by probably no more than ten authors, I have no confidence that it is complete or entirely correct. I invite (indeed, beg) additions and corrections.

This list differs from Cajori's "Bibliography of Fluxions and the Calculus" not only in the period covered but in the criteria for inclusion. Cajori was interested in textbooks printed in the United States. He therefore included books written by non-American authors (e.g., Hutton, Vince) but published in the United States. He also included books that were translations by Americans of foreign texts (e.g., Farrar's translation of Bézout). As we have seen, there is a fine line between translation and inspiration in some early texts.

Finally, Cajori has annotated his list extensively, while I have restricted my notes to a few technical details, such as a change of publisher.

The list is arranged alphabetically by primary author. The name of the primary author is followed by the name(s) of any collaborators. Beneath this is a chronological list of all calculus books written by the author with the first-listed book being the earliest. With each book is the date of first publication, followed by the date of last publication. The name of the publisher and the place of publication follow the dates. Occasionally, the publisher and/or the place of publication is missing.

Bass, Edgar Wales, *Introduction to the differential calculus* ... 1887, USMA Press, West Point, NY.

——, *Differential calculus* ..., 1889–1892, USMA Press and Bindery, West Point, NY.

——, *Elements of differential calculus*, 1896–1905, John Wiley & Sons, NY and London.

Bayma, Joseph (SJ), *Elements of infinitesimal calculus*, 1889, A. Waldteufel, San Francisco.

Bonnycastle, Charles, *Syllabus of a course of lectures, upon the differential and integral calculus*, 1838 C. P. M'Kennie, Charlottesville.

Bowser, Edward Albert, *An elementary treatise on the differential and integral calculus with numerous examples*, 1880–1907, D. Van Nostrand, NY.

Brown, Stimson Joseph (Capron, Paul), *The calculus, an elementary treatise on the differential and integral calculus, with applications, prepared for the use of the midshipmen of the United States Naval Academy*, 1909–1912, The Lord Baltimore Press, Baltimore.

Buchanan, Roberdeau, *An introduction to the differential calculus by means of finite differences*, 1905, Washington, DC, Reprinted from Popular Astronomy, vol. XIII, nos. 5, 6.

Buckingham, Catharinus Putnam, *Elements of the differential and integral calculus, by a new method, founded on the true system of Sir Issac Newton, without the use of infinitesimals or limits*, 1875–1885, S. C. Griggs & Co., Chicago.

——, *The method of final ratios commonly called the method of limits*, 1879, S. C. Griggs & Co., Chicago.

Byerly, William Elwood, *Elements of the differential calculus, with examples and applications*, 1879–1901 Ginn & Heath, Boston.

——, *Elements of the integral calculus, with a key to the solution of differential equations*, 1881–1902, Ginn, Heath & Co., Boston.

——, *Problems in differential calculus*, Supplementary to a treatise on differential calculus, 1895–1904, Ginn & Co., Boston.

A short table of integrals by B. O. Peirce, added starting in '89. An 1888 edition(?) was reprinted by G. E. Stechart & Co. (NY) in 1941.

Cain, William, *A brief course in the calculus*, 1905–1911, D. VanNostrand & Co., NY. Third edition apparently reprinted in London in 1930. Also paper, "On the fundamental principles of the differential calculus," J. Elisha Mitchell Scientific Soc., 1892.

Campbell, Donald Francis, *The elements of the differential and integral calculus, with numerous examples*, 1904–1919, The Macmillan Co., NY.

Chandler, George Henry, *Elements of the infinitesimal calculus*, 1907, Wiley, NY. I don't have the early publishing history of this book; the 1907 edition is the "3rd ed; rewritten."

Church, Albert Ensign, *Elements of the differential and integral calculus*, 1842–1872, Wiley and Putnam (notes) NY. Publishers: '55, Barnes: '60, '63, '64, Barnes & Burr.

Clark, James Gregory, *Elements of the infinitesimal calculus, with numerous examples and applications to analysis and geometry*, 1875, Wilson, Hinkle & Co., Cincinnati, NY.

Cook, Hiram, *An elementary treatise on variable quantities*, in two parts, the direct and inverse, 1921, Privately printed, Berkeley.

Book published after Cook's death in 1917. Preface dated 1916.

Courtenay, Edward Henry, *Treatise on the differential and integral calculus and on the calculus of variations*, 1855–1876, A. S. Barnes & Co., NY. Courtenay died in 1853.

Davies, Charles, *Elements of the differential and integral calculus*, 1836–1889, Wiley & Long (see note), NY.

____, *Elements of analytical geometry and of the differential and integral calculus*, 1859–1901, A. S. Barnes & Burr, NY.

____, *Differential and integral calculus, designed for elementary instruction*, 1860, A. S. Barnes & Burr, NY.

____, *Differential and integral calculus on the basis of continuous quantity and consecutive differences, designed for elementary instruction*, 1873–1901, A. S. Barnes & Co. (notes), NY & Chicago, Barnes.

Davies' books changed publishers: A. S. Barnes (or Barnes & Burr) took over "Elements ..." in '38; American Book Co. (Cincinnati) published '01 edition of "... elementary ...".

Davis, Ellery Williams (Wm. Chas. Brenke, E. R. Hedrick), *The calculus*, 1912–1930, The Macmillan Co., NY. This book is "edited by Earle Raymond Hedrick." Brenke is sometimes coauthor, sometimes assistant.

Docharty, Gerardus Beekman, *Elements of analytical geometry and of the differential and integral calculus*, 1865, Harper & Brothers, NY.

Echols, William Harding, *An elementary textbook on the differential and integral calculus*, 1902–1908, Henry Holt & Co., NY.

Fisher, Irving, *A brief introduction to the infinitesimal calculus; designed especially to aid in reading mathematical economics and statistics*, 1897–1937, Macmillan & Co., NY.

Franklin, William Suddards (Barry MacNutt & R. Charles), *An elementary treatise on calculus; a textbook for colleges and technical schools*, 1913, published by the authors, S. Bethleham. Barry MacNutt is coauthor with Franklin of a number of physics/engineering books; Rollin Charles authors no other books.

Gould, E(dward) Sherman, *A primer of the calculus*, 1896–1907, D. Van Nostrand, NY.

Granville, William Anthony (Percey F. Smith, Wm. R. Longley), *Elements of the differential and differential calculus*, 1904–1957, Ginn & Co., Boston.

____, *Elements of calculus*, 1946, Ginn & Co., Boston.

See also entries for Smith, Longley. The roles of Smith and Longley on the title page change over time.

Greene, William Batchelder, *An expository sketch of a new theory of the calculus*, 1859, printed for the author, Paris.

____, *The theory of the calculus*, 1870, Lee & Shepard, Boston.

____, *Explanation of "The Theory of the Calculus"*, 1870, Lee & Shepard, Boston.

Groat, Benjamin Feland, *An introduction to the summation of differences of a function; an elementary exposition of the nature of the algebraic processes replaced by the abbreviations of the infinitesimal calculus*, 1902, H. W. Wilson, Minneapolis.

Hackley, Charles William, *Differential calculus, for the use of the senior class of Columbia College ...*, 1856, Baker & Godwin, printers, NY.

Hall, William Shaffer, *Elements of the differential and integral calculus with applications*, 1897–1922, D. Van Nostrand, NY.

Hardy, Joseph Jonston, *Infinitesimals and limits*, 1900–1912, Chemical Publishing Co., Easton, PA.

Hathaway, Arthur Stafford, *A primer of calculus*, 1901, Macmillan and Co., NY, London.

Hayes, Ellen, *Calculus, with applications; an introduction to the mathematical treatment of science*, 1900, Allyn & Bacon, Boston.

Hayward, Harrison Washburn, *Notes on calculus; for the use of students of the Lowell Institute school for industrial foremen*, Massachusetts Institute of Technology, 1915, The Taylor Press, Boston.

Hedrick, Earle Raymond (O. D. Kellogg), *Applications of the calculus to mechanics*, 1909, Ginn and Co., Boston. Translated Goursat's *Mathematical Analysis*; also see Ellery W. Davis.

Hulburt, Lorrain Sherman, *Differential and integral calculus, an introductory course for colleges and engineering schools*, 1912–1943, Longmans, Green and Co., NY. 1943 edition published by Barnes and Noble, New York.

Johnson, William Woolsey (John Minot Rice), *An elementary treatise on the integral calculus founded on the method of rates or fluxions*, 1881–1909, John Wiley & Sons, NY.

——, *An elementary treatise on the differential calculus, founded on the method of rates*, 1904–1908, John Wiley & Sons, NY.

——, *A treatise on the integral calculus founded on the method of rates*, 1907, John Wiley & Sons, NY.

For earlier versions written with Rice, see Rice. "A treatise on the integral ..." is an enlargement of "An elementary treatise ...."

Keller, Samuel Smith (W. F. Knox), *Mathematics for engineering students; analytical geometry and calculus*, 1907–1908, D. Van Nostrand Co., NY.

Lambert, Preston Albert, *Differential and integral calculus for technical schools and colleges*, 1898–1907, The Macmillan Co., NY, London.

Longley, William Raymond (W. A. Wilson, P. F. Smith, {Granville}), *An introduction to the calculus*, 1924.

——, *Analytic geometry and calculus*, 1951.

Wallace Alvin Wilson was coauthor of the first book; Percey Smith (see) and Wilson were coauthors of the second.

Loomis, Elias, *Elements of analytical geometry and of the differential and integral calculus*, 1851–1872, Harper, NY.

——, *Elements of the differential and integral calculus*, 1874–1902, Harper, NY.

The 1902 edition of the later book was published by the American Book Co., NY. This book is unchanged between 1874 and 1902.

Love, James Lee, *An introductory course in the differential and integral calculus; for students in engineering in the Lawrence Scientific School*, 1898–1899, Harvard University, Cambridge.

M'Cartney, Washington, *The principles of the differential and integral calculus; and their applications to geometry*, 1844–1848, E. C. Biddle, Philadelphia. The 1848 edition was published by E. H. Butler & Co., Philadelphia.

March, Herman William (Henry C. Wolff), *Calculus*, 1917–1937, McGraw Hill, NY.

McMahon, James (Virgil Snyder), *Elements of the differential calculus*, 1898, American Book Co., NY, Cincinnati. See Snyder for additional books.

Murray, Daniel Alexander, *An elementary course in the integral calculus*, 1898, American Book Co., NY, Cincinnati.

——, *A first course in infinitesimal calculus*, 1903–1904, Longmans, Green & Co., NY.

——, *Differential and integral calculus*, 1908, Longmans, Green & Co., NY. Murray taught in Canada.

Newcomb, Simon, *Elements of the differential and integral calculus*, 1887–1889, Holt, NY.

Nichols, Edward West, *Differential and integral calculus with applications; for colleges, universities, and technical schools*, 1900–1918, D. C. Heath & Co., Boston.

Nicholson, James William, *Elements of the differential and integral calculus, with examples and practical applications*, 1896, University Publishing Co., NY & New Orleans. There may have been an 1894 edition of this book also.

Olney, Edward, *A general geometry and calculus.* Including part I of the general geometry, treating loci in a plane; and an elementary course in the differential and integral calculus, 1870–1885, Sheldon & Co., NY. The 1870 version contains just the first three chapters (geometry) of the 1871 book.

Osborne, George Abbott, *Notes on differentiation of functions. With examples....*, 1884, J. S. Cushing & Co., Boston.

——, *The differential calculus applied to plane curves and maxima and minima*, 1889–1890, J. S. Cushing & Co., Boston.

——, *The integral calculus applied to plane curves. Successive integration*, 1889, J. S. Cushing & Co., Boston.

——, *Differential and integral calculus, with examples and applications*, 1891–1910, Heath, Boston.

——, *An elementary treatise on the differential and integral calculus, with examples and applications*, 1891–1906, Leach, Boston & NY.

With the 1899 edition, Heath became the publisher of "An elementary treatise ...." "Notes ..." is a 40 page booklet, perhaps originally paperback, marked "Printed, not Published" on title page.

Osgood, William Fogg, *A modern English calculus*, 1902, The Macmillan Co., NY.

——, *A first course in the differential and integral calculus*, 1907–1929, The Macmillan Co., NY.

——, *Elementary calculus*, 1921, The Macmillon Co., NY.

——, *Introduction to the calculus*, 1922–1954, The Macmillan Co., NY.

——, *Advanced calculus*, 1945, The Macmillan Co., NY.

"Introduction ..." is called "a revision of ... 'A first Course...'."

Peck, William Guy, *Practical treatise on the differential and integral calculus, with some applications to mechanics and astronomy*, 1870–1898, A. S. Barnes & Co., NY, Chicago. 1892, 1898 editions are identical to 1870 edition; 1898 edition published by American Book Co., NY, Cincinnati, Chicago.

Peirce, Benjamin, *An elementary treatise on curves, functions, and forces.* 2 volumes, 1841–1862, James Munroe and Co., Boston & Cambridge.

Phillips, Henry Bayard, *Differential calculus ...*, 1916, John Wiley & Sons, NY.

——, *Integral calculus*, 1917, John Wiley & Sons, NY.

——, *Calculus*, 1927–1940, Wiley, NY.

——, *Analytical geometry and calculus*, 1942–1946, Addison Wesley, Cambridge, MA.

The 1940 edition of "Calculus" was published by Cummings, Cambrige, MA; the 1946 edition of "Analytical geometry ..." was published by Wiley.

Ransom, William Richard, *Freshman calculus; a presentation of fundamental conceptions and methods for students of science and engineering*, 1909, lithographed, Boston.

——, *Early calculus*, 1915, Tufts College, Medford, MA.

——, *A working calculus*, 1936, planograph, Boston.

——, *The calculus, according to a new plan*, 1947–1949, Tufts College, Medford, MA. "The calculus ..." is a revision of "A working calculus."

Rice, John Minot (William Woolsey Johnson), *On a new method of obtaining the differentials of functions, with especial reference to the Newtonian conception of rates or velocities*, 1873–1875, John Wiley & Sons, NY.

——, *The elements of the differential calculus founded on the method of rates or fluxions*, 1874, John Wiley & Sons, NY.

——, *An elementary treatise on the differential calculus founded on the method of rates or fluxions*, 1877–1904, John Wiley & Sons, NY. See Johnson for the continuation of this series, authored by Johnson alone.

Robinson, Horation Nelson (Issac Ferdinand Quinby), *Elements of analytical geometry and the differential and integral calculus*, 1856–1859, J. Ernst, Cincinnati.

——, *A new treatise on the elements of the differential and integral calculus*, 1868–1879, Ivison, Phinney, Blakeman, NY. The 1858 and 1859 editions of "Elements ... " were published by Ivison in NY.

Quinby is called the editor of "A new treatise ... ;" all editions appear after Robinson's death in 1867.

Ryan, James, *The differential and integral calculus*, 1828, White, Gallaher & White, NY.

Sestini, Benedict, *Manual of geometrical and infinitesimal analysis*, 1871, John Murphy & Co., Baltimore.

Smith, Percey Franklyn (W.A. Granville, W. A. Longley), *Elementary calculus; a textbook for the use of students in general science*, 1902–1903, American Book Co., NY, Cincinnati.

——, *Elementary analysis*, 1910, Ginn and Co., Boston, NY.

——, *Intermediate calculus*, 1931, Ginn and Co., Boston, NY.

Coauthors: "Elementary ... " with Granville, "Intermediate ... " with Longley. See also Granville, Longley. The latter's independent publications began after 1920.

Smith, William Benjamin, *Infinitesimal analysis ...*, 1898, Macmillan Co., NY.

Smyth, William, *Elements of the differential and integral calculus*, 1854–1859, Sanborn & Carter, Portland.

——, *Elements of calculus*, 1859, Sanborn & Carter, Portland.

Snyder, Virgil (John I. Hutchinson, J. McMahon), *Differential and integral calculus*, 1902, American Book Co., NY, Cincinnati.

——, *Elementary textbook on the calculus*, 1912, American Book Co., NY, Cincinnati.

John Irwin Hutchinson is coauthor of these books. See also James McMahon.

Spare, John, *The differential calculus: with unusual and particular analysis of its elementary principles and copious illustrations of its practical applications*, 1865, Bradley Dayton & Co., Boston.

Strong, Theodore, *A treatise on the differential and integral calculus*, 1869, C. A. Alvord, NY.

Taylor, James Morford, *Elements of the differential and integral calculus with examples and applications*, 1884–1902, Ginn, Heath, & Co., Boston.

Thomas, Robert Gibbes, *Applied calculus; principles and applications, essentials for students and engineers*, 1919–1924. D. Van Nostrand Co., NY. 1924 edition is "an abridged and rev. ed. of *Applied calculus*, with additional exercises and formulas."

Townsend, Edgar Jerome (George Alfred Goodenough), *First course in calculus*, 1908–1910, Holt, NY.

——, *Essentials of calculus*, 1910–1925, Holt, NY.

Goodenough coauthored both books.

Veblen, Oswald (N. J. Lennes), *Introduction to infinitesimal analysis; functions of one real variable*, 1907–1935, Wiley, NY. The 1935 edition is a reprint by Stechert.

Wilson, Edwin Bidwell, *Advanced calculus; a text upon select parts of differential calculus, differential equations, integral calculus, theory of functions, with numerous examples*, 1911–1912. Ginn and Co., Boston, NY.

Woods, Frederick Shenstone (Frederick H. Bailey), *A course in mathematics, for students of engineering and applied science*, 1907–1909, Ginn and Co., Boston, NY.

——, *Analytic geometry and calculus*, 1917–1944, Ginn and Co., Boston, NY.

——, *Elementary calculus*, 1922–1950, Ginn and Co., Boston, NY.

——, *Advanced calculus*, 1926–1954, Ginn and Co., Boston, NY.

Bailey is coauthor of all books.

Young, Jacob William Albert (C. E. Linebarger), *The elements of the differential and integral calculus*, based on Kurzgefasstes Lehrbuch der Differential-und Integral-rechnung, von W. Nernst ... und A. Schoenflies, 1900, D. Appleton & Co., NY.

Editors' note: The following review of a popular calculus textbook is reprinted as an indication of American mathematicians' sophistication and growing insistence on precision and rigor during the late nineteenth century. It illustrates a point made by George Rosenstein in the preceding article.

EDWARDS' DIFFERENTIAL CALCULUS.

An Elementary Treatise on the Differential Calculus, with applications and numerous examples. By JOSEPH EDWARDS, M.A., formerly Fellow of Sidney Sussex College, Cambridge. Second edition, revised and enlarged. London and New York, Macmillan & Co. 1892. 8vo, pp. xiii + 521.

WHEN a mathematical text book reaches a second edition, so much enlarged as this, we know at once that the book has been received with some favour, and we are prepared to find that it has many merits. We are at once struck by Mr. Edwards' lucid and incisive style; his expositions are singularly clear, his words well chosen, his sentences well balanced. In the text of the book we meet with various useful results, notably in the chapter on "some well known curves," and moreover the arrangement is such that these results are easy to find; and in addition to these, numbers of theorems are given among the examples, and, this being a feature for which we are specially grateful, in nearly every case the authority is cited. Recognizing these merits, however, we notice that the book has many defects, some proper to itself, some characteristic of its species; and just because it is so attractive in appearance, it seems worth while examining it in detail, and pointing out certain specially vicious features.

A book of this size may fairly be required to serve as a preparation for the function theory; at all events, the influence of recent Continental researches should be evident to the eyes of the discerning. Mr. Edwards' preface strengthens this reasonable expectation, for he promises us "as succinct an account as possible of the most important results and methods which are up to the present time known." But we soon find that the "important results and methods" are those of the Mathematical Tripos; and in our disappointment we utter a fervent wish that instead of the "large number of university and college examination papers, set in Oxford, Cambridge, London, and elsewhere," Mr. Edwards had consulted an equally large number of mathematical memoirs published, principally, elsewhere. The Mathematical Tripos for any given year is not intended for a *Jahrbuch* of the progress of mathematics during the past year; and as long as so many will insist on regarding it in that light, text books of this type will continue to be published.

Nothing in this book indicates that Mr. Edwards is familiar with such works as Stolz's *Allgemeine Arithmetik*, Dini's *Fondamenti per la teorica delle funzioni di variabili reali*, or Tannery's *Théorie des fonctions d'une variable.* In support of our contention we may instance the definitions of function,

Reprinted from the *Bulletin of the New York Mathematical Society* 1 (1892), 217–223.

limit, continuity, etc. On page 2, Lejeune Dirichlet's definition of a function is adopted. According to this very general definition, there need be no analytical connection between y and x; for y is a function of x even when the values of y are arbitrarily assigned, as in a table. That Mr. Edwards does not adhere to this definition is evident from his tacit assumption that *every* function $\varphi(x)$ can be represented by a succession of continuous arcs of curves. Whatever definition is adopted for a continuous function y of x, it is evident that to small increments of x must correspond small increments of y; but Weierstrass has proved that there exist functions which have this property, but which have nowhere differential coefficients. The well known example of such a function is

$$f(x) = \sum_{n=0}^{\infty} b^n \cos (a^n x\pi),$$

where a is an odd integer, b a positive constant less than 1, and ab greater than $1 + 3\pi/2$. According to the accepted definition, this function of x is continuous; according to Mr. Edwards' definition, it is not continuous, inasmuch as it cannot be represented by a curve $y = f(x)$ with a tangent at every point.

We acknowledge that Mr. Edwards displays a considerable degree of consistency in his view of the meaning of a continuous function, but we insist that after the adoption of the curve definition he should have been at some pains to prove that the numerous series of the type $\sum_1^{\infty} f_n(x)$ scattered throughout the book give rise to curves with tangents, whereas he never even takes the trouble to prove that they are continuous functions of x in any sense of the term. No more damaging charge can be brought against any treatise laying claim to thoroughness than that of recklessness in the use of infinite series; and yet Mr. Edwards has everywhere laid himself open to this charge. One of the most difficult things to teach the beginner in mathematics is to give proper attention to the convergency of the series dealt with. All the more need, then, that a text book of this nature should set an example of consistent, even *aggressive* carefulness in this respect. We do, it is true, find an occasional mention of convergence (pp. 9, 81, 454, etc.), but as a rule it is ignored. Mr. Edwards rearranges the terms of infinite and doubly infinite series, applying the law of commutation without pointing out that his series are unconditionally convergent; he differentiates $f(x) = \sum_1^{\infty} f_n(x)$ term by term, and gets $f'(x) = \sum_1^{\infty} f'_n(x)$, implying that the process is universally valid (*e.g.* p. 84); or, at all events, giving no hint that there are cases in which the differential coefficient of the sum of a convergent series is different from the sum of the differential coefficients of the individual terms. We find no formal recognition of the importance of uniform convergence in modern analysis, nothing even to suggest that he has ever heard of the distinction between uniform and non-uniform convergence. We begin to suspect that he has never looked into Chrystal's Algebra.

The unreasoning mechanical facility thus acquired in performing operations unhampered by any doubts as to their legitimacy, naturally leads Mr. Edwards to view with favour "the analytical house of cards, composed of complicated and curious formulæ, which the academic tyro builds with such zest upon a slippery foundation," *—and to build up many a one. A curious and interesting specimen is

$$f(x) = x^{x^{x^{\cdot^{\cdot^{\cdot}}}}}$$

to be continued to infinity. This expression has been examined by Seidel,† who points out that Eisenstein's paper in *Crelle*, vol. 28, requires correction. Before such an expression can be differentiated, a definite meaning must be assigned to it; but Seidel's conclusion is that, denoting x^x by x_1, x^{x_1} by x_2, x^{x_2} by x_3, and so on, then as x varies from 0 to $1/e^e$, $\underset{n=\infty}{L} x_{2n}$ increases from 0 to $1/e$, while $\underset{n=\infty}{L} x_{2n+1}$ decreases from 1 to $1/e$; beyond these limits for x, the case is different. In particular when $x > e^{1/e}$, the expression diverges. Our objection is not to the non-acceptance of Seidel's conclusions, but to the unnecessary use of a function of this doubtful character. Examples can be found to illustrate every point that ought to be brought up in an elementary treatise on the differential calculus without ranging over examination papers in search of striking novelties.

Feeling now somewhat familiar with Mr. Edwards' point of view, we examine his proofs of the ordinary expansions with a tolerably clear idea of what we are to expect. We find, of course, "the time-honoured short proof of the existence of the exponential limit, which proof is half the real proof plus a *suggestio falsi*"; we find in the chapter on expansions a general disregard of convergency considerations; we find throughout the book the assumption that

* Professor Chrystal, in *Nature*, June 25, 1891.
† *Abhandlungen der k. Ak. d. Wiss.* Bd. xi.

$\varphi(a) = \underset{x=a}{L}\, \varphi(x)$, and that $\varphi(0, 0) = \underset{x=0,\, y=0}{L}\, \varphi(x, y)$ *; we find the usual assumptions as to expansibility in series proceeding by integral powers, with disastrous results further on. We find the usual dread of the complex variable, though Mr. Edwards has given one or two examples involving it, without however explaining what is meant by $f(x + iy)$. We can hardly regard these examples, even with § 190, as a sufficient recognition of the complex variable in a treatise of this size. We must notice also the thoroughly faulty treatment of the inverse functions. For example, no explanation is given of the signs in $\frac{dy}{dx}$, when $y = \cos^{-1} x$ or $\sin^{-1} x$. Mr. Edwards' attitude towards many valued functions is simple enough; as a rule, he ignores the inconvenient superfluity of values. He does, it is true, give in § 54 a note, clear and correct, on this point; but he is very careful to confine this within the limits of the single section, and to indicate, by choice of type, that it is quite unimportant.

We pass on now to the second part, applications to plane curves; and here we must object emphatically to the introduction of so many detached and disconnected propositions relating to the theory of higher plane curves. From Mr. Edwards' point of view this is doubtless justified; we are quite ready to acknowledge that we know of no book that would enable a candidate to answer more questions on subjects of whose theory he is totally ignorant. The deficiency of a curve, *e.g.*, is a conception entirely independent of the differential calculus; but probably this single page will obtain many marks for candidates in the Mathematical Tripos; these we should not grudge if we thought an equivalent would be lost by a reproduction of Mr. Edwards' treatment of cusps. Our spirits rose when we remarked the italicised phrase on p. 224, that there is "*in general a cusp*" when the tangents are coincident. But three pages further on we find that the exception here indicated is simply our old friend, the conjugate point, whose special exclusion from the class in which it appears must be a perpetual puzzle to a thoughtful student with no better guidance than a book of this kind. Such a student, probably already familiar with projection, knows that the real can be projected into the imaginary, and the imaginary into the real. If then the acnode, appearing as a cusp, has to be specially excluded, why not the crunode? But here Mr. Edwards reproduces the now well established

* See *e.g.* p. 122; and on this page note also the assumption that the relation between h, k, while $x + h$, $y + k$, tend to the limits x, y exerts no influence on the result.

error, calling tacnodes, formed by the contact of real branches, double cusps of the first and second species, and excluding those formed by the contact of imaginary branches; he even goes further astray, introducing Cramer's osculinflexion as a cusp that changes its species.

This matter of double cusps is a fundamentally serious one, and not a mere question of nomenclature. This persistent misnaming effectually disguises the essential characteristic of the cusp. It is *not the coincidence of the tangents* that makes a cusp. From the geometrical point of view it is the turning back of the (real) tracing point, expressed by the French and German names, {*point de rebroussement, Rückkehrpunkt*}; from the point of view of algebraical expansions (of y in terms of x, $y = 0$ being the tangent) the essential characteristic of a single cusp is that at some stage in the expansion there shall be a fractional exponent with an even denominator, so that the branch changes from real to imaginary *along its tangent;* from the point of view of the function theory, which is really equivalent to the last, the simple cusp is characterised by the presence of a *Verzweigungspunkt* combined with a double point. The simple cusp, that is, presents itself as an evanescent loop. A double cusp, then, in the sense in which Mr. Edwards uses the term, does not exist. There cannot be two consecutive cusps, vertex to vertex; for the branch if supposed continued through the cusp, changes from real to imaginary; and two *distinct* cusps, brought together to give a point of this appearance, produce a quadruple point.

While on this subject, we must mention Mr. Edwards' rule for finding the nature of a cusp. Find the two values of $\frac{d^2y}{dx^2}$; these by their signs determine the direction of convexity (§ 296). How does this apply *e.g.* to $y^2 = x^3$?

This confusion regarding cusps is made worse by the assumption already noticed that when $f(x, y) = 0$ is the equation of the curve, y can be expanded in a series of integral powers of x. This error is repeated on p. 258, where to obtain the branches at the origin, this being a double point, we are told to expand y by means of the assumption $y = px + \frac{qx^2}{2!} +$ etc. The whole exposition of this theory of expansion is most inadequate. In § 382 there is no hint that the terms obtained are the beginning of an infinite series, giving the expansion of (say) y in powers, not necessarily integral, of x; there is no hint what to do when the first terms of the expansion are found; there is no suggestion of the interpretation of the result when two expansions begin with the same terms. A thoughtful student *may* by a happy comparison of scattered examples (p. 200, and ex. 3, p. 230) arrive at the correct theory ; but he surely deserves better guidance.

One or two more points must be noticed. The theory of asymptotes, when two directions to infinity coincide, cannot be satisfactorily developed without assuming a knowledge of double points ; and the only way of giving the true geometrical significance is to introduce the conception of the line infinity, and to consider the nature of the intersections of the curve by this line. A tangent lying entirely at infinity does *not* "count as one of the n theoretical asymptotes" ; if counted among the asymptotes at all, it has to be counted as the equivalent of two out of the n. This is one of the strongest arguments against including the line infinity in enumerating the asymptotes. The various expressions for the radius of curvature involve an ambiguity in sign ; what is the meaning of this ? The omission of this explanation causes obscurity, notably in § 330. The equation of a curve, referred to oblique axes, being $\varphi(x, y) = 0$, what is the condition for an inflexion ? As a matter of fact it is the same as in the case of rectangular axes, given on p. 264 ; but as this is obtained from a formula for the radius of curvature, the investigation is not applicable. Throughout Mr. Edwards displays an almost exclusive preference for rectangular axes, and seems to regard the metric properties so obtained as of equal importance with descriptive properties. For instance, in the case of an ordinary double point (p. 224) instead of the *three* cases usually distinguished, we have *four*, the additional one being that of perpendicular tangents.

In the third part we notice that in the chapter on "undetermined forms" there is no discussion of the case of two variables, though it is on this that we have to rely for a rigorous proof of the theorem $\frac{\delta^2\varphi}{\delta x \delta y} = \frac{\delta^2\varphi}{\delta y \delta x}$. We recognize an old friend, the discussion of the limit of ∞/∞, in which it is first assumed, and then proved, that the limit exists. The statement of ex. 17, p. 457, is somewhat misleading ; the formula there given for the expansion of $(x + a)^m$ is true when m is a positive integer ; but when $m = -1$, it is evidently not true for $x = -b, -2b$, etc.* The treatment of maxima and minima of functions of two variables (§§ 497–501) is incomplete and incorrect. The geometrical illustration, as given on p. 424, omits the case of a section with a cusp, which is the simplest case that can occur when $rt = s^2$; of the more complicated cases Mr. Edwards attempts no discrimination ; he does not even state correctly the principles that must guide us in this discrimination. The inexactness of the ordinary

* LAURENT, *Traité d'Analyse*, iii., 386.

criteria (given in § 498) appears at once from the example $u = (y^2 - 2px)(y^2 - 2qx)$ [Peano]. The origin is a point satisfying the preliminary conditions; taking then for x, y, small quantities h, k, the terms of the second degree are positive for all values except $h = 0$; when $h = 0$, the terms of the third degree vanish, and the terms of the fourth degree are positive; nevertheless the point does not give a minimum, which it should do by the test of § 498. For we can travel away from O in between the two parabolas, so coming to an adjacent point at which u has a small negative value, while for points inside or outside both parabolas the value of u is positive. The truth is, the nature of the value a of the function u at a point (x_0, y_0) at which $\frac{\delta\varphi}{\delta x}$ and $\frac{\delta\varphi}{\delta y}$ vanish, depends on the nature of the singularity of the curve $u = a$ at this point. If this curve has at (x_0, y_0) an isolated point of any degree of multiplicity, we have a true maximum or minimum of u; but if through (x_0, y_0) pass any number of real non-repeated branches of the curve, we have not a maximum or minimum; in Peano's example the branches coincide in the immediate neighbourhood of the origin, but then they separate, and therefore we have not a minimum value for u.

We object, then, to Mr. Edwards' treatise on the Differential Calculus because in it, notwithstanding a specious show of rigour, he repeats old errors and faulty methods of proof, and introduces new errors; and because its tendency is to encourage the practice of cramming "short proofs" and detached propositions for examination purposes.

CHARLOTTE ANGAS SCOTT.

BRYN MAWR, PA., *May* 18, 1892.

Born in Switzerland, Armand Borel did his undergraduate work at the Federal School of Technology (ETH) in Zürich. He obtained his doctorate degree at the University of Paris in 1952 and then spent two years at the Institute for Advanced Study in Princeton. He has been professor there since 1957.

The School of Mathematics at the Institute for Advanced Study

ARMAND BOREL

In the late twenties, Abraham Flexner, a prominent figure in higher education, had made an extensive study of universities in the U.S. and Europe and was extremely critical of many features of American universities. In particular, he deplored the lack of favorable conditions for carrying out research. In January 1930, while preparing for publication an expanded version of three lectures he had given in 1928 at Oxford on universities, he saw in the New York Times an article on a meeting of the American Mathematical Society (AMS), in which Oswald Veblen, professor at Princeton University, was quoted as having stated that America still lacks a genuine seat of learning and that American academic work is inferior in quality to the best abroad. He immediately wrote to Veblen, saying there was not the slightest doubt in his mind that both statements were true and hoping that Veblen had been correctly quoted. In his answer, Veblen confirmed these views, described the context of his remarks and wrote in conclusion:

> Here in Princeton the scientific fund which we owe largely to you and your colleagues on the General Education Board, is having an influence in the right direction, and I think our new mathematical building which is going to be devoted entirely to research and advanced instruction will also help considerably. I think my mathematical institute which has not yet found favor may turn out to be one of the next steps. Anyhow it seems to me to fit in with the concept of a seat of learning.

The first Faculty of the School of Mathematics (minus J. von Neumann) with the second Director. From left to right: J. Alexander, M. Morse, A. Einstein, F. Aydelotte, Director, H. Weyl and O. Veblen.

(Photograph courtesy of the Institute for Advanced Study.)

Here Veblen was alluding first to the efforts, initiated by Fine and pursued with the help of Eisenhart and Veblen, to improve research conditions in his department and to the construction of what became Fine Hall; second to a plan for an "Institute for Mathematical Research" he had outlined and presented (without success) around 1925 to the National Research Council and to the General Education Board of the Rockefeller Foundation. It was to consist of four or five senior mathematicians who would devote themselves entirely to research, their own and that of some younger men, and of some younger mathematicians. Members would be free to give occasional courses for advanced students. It could operate within a university or be entirely independent of any institution.[1]

Shortly before, Flexner had been approached by two gentlemen who were surveying medical education on behalf of two persons who wanted to use part of their fortune to establish and endow a medical college in Newark. Since Flexner was an authority on medical education in the U.S., it was only natural to seek his counsel. He advised against it, explaining why in his opinion there was no real need for a new institution of the type they had in mind. Instead, he showed them the proofs of his book on universities and outlined his plan for an institution of higher learning, where scholars would pursue their researches and interests freely and independently. They were so fascinated by it that they swayed the potential donors, namely Louis Bamberger and his sister, Mrs. Felix Fuld, born Caroline Bamberger, convinced them to look into this possibility and soon introduced them to Flexner. This initiated a series of discussions and a correspondence extending over several months, at the end of which the Bambergers agreed enthusiastically to back up Flexner's plan, on condition that he would be the first director. A certificate of incorporation for a corporation to be known by law as the "Institute for Advanced Study – Louis Bamberger and Mrs. Felix Fuld Foundation" was filed with the state of New Jersey in May 1930 and the New York Times announced in June the creation of an Institute for Advanced Study, to be located in or near Newark, on a gift of $5 million from Louis Bamberger and his sister, Mrs. Felix Fuld. Veblen learned about it for the first time through that press release, although there had been a little further correspondence between the two about the idea of an Institute, but carried out *in abstracto*, at any rate on Veblen's side. He wrote immediately to Flexner that he was greatly pleased and he expressed the wish that this Institute would be located in the Borough or Township of Princeton "*so that you could use some of the facilities of the University and we could have the benefit of your presence.*" This heralded an increasing involvement of Veblen with this project, first as a consultant, then

[1] For this and the development of mathematics in Princeton until WW II, see William Aspray's article in *A Century of Mathematics in America, Part II* (editor, P. Duren, with assistance of R. A. Askey and U. C. Merzbach), Amer. Math. Soc., Providence, R.I., 1989, pp. 195–215.

as a professor having the primary responsibility for the building up of the School of Mathematics.

The Institute was eventually to consist of a few schools, but Flexner decided early on to start first with one in mathematics, because "mathematics is fundamental, requires the least investment in plant or books and he could secure greater agreement upon personnel than in any other field".[2] He began to make extensive inquiries in the U.S. and in Europe as to who would be the best choices for a faculty in mathematics. Among American mathematicians, the two most prominent names were those of George D. Birkhoff and Veblen. Flexner started with the former, on the theory that Veblen was already in Princeton anyhow. An offer was made, at an extremely high salary and accepted in March 1932, but Birkhoff asked to be released eight days later. After further inquiries, Flexner came to the conclusion that: "*If the Princeton authorities agreed willingly and unreservedly, we could not do better than to select Veblen.*" They did so quickly, and Eisenhart telegraphed to Veblen in June:

> Have talked with those concerned and they approve. Congratulate you heartily. Look forward to big things.

1932 was marked by extensive travelling, wide ranging consultations, and discussions, correspondence and negotiations with Veblen, Einstein and Weyl. (Of course, no outside advice was needed in the case of Einstein, and Flexner forged ahead as soon as he understood that he might be interested.) In October two faculty nominations were announced, that of Veblen, already effective October 1st, 1932 and that of Einstein, effective October 1st, 1933 (as well as the nomination of Walther M. Mayer, the then collaborator of Einstein, as an "associate"). It was also announced that the new Institute would be located in or near Princeton (a shift formally proposed in April 1932) and would be housed temporarily at Fine Hall. The school would officially begin its activities in Fall 1933, but in fact, during the academic year 1932–1933, Veblen already conducted a seminar in "Modern Differential Geometry."

It is well-known that Einstein was enthusiastic from the beginning ("Ich bin Feuer und Flamme dafür," he had stated to Flexner) and excessively modest in his financial requirements, but the negotiations were not all that smooth. In 1933 Flexner learned that Einstein had also accepted a professorship in Madrid and one at the Collège de France. Since their residence requirements were minimal (in the former case, nonexistent in the latter), while those of the Institute were for him only from October to April 15, Einstein did not see any incompatibility; on the other hand, if Flexner felt otherwise, he would agree to terminate the arrangement with the Institute.... The Madrid offer also included the right to name a professor and Einstein tried to use it as

[2]A. Flexner, *I remember*, Simon and Schuster, New York, 1940, pp. 359–360.

a leverage to secure a professorship at the Institute for W. Mayer (without success). In summer of 1933, Flexner had asked whether Einstein could arrive soon enough to participate in a general organizational meeting of the members of the school on October 2nd. Einstein felt he could not because this would entail spending one month away from W. Mayer, which would be too detrimental to his work. He arrived on October 17. He was reminded of that when he complained later that he had not been consulted about invitations and stipends. The collaboration with Mayer was over within a few months.

In Europe, the two names of mathematicians mentioned to Flexner above all others were those of G. H. Hardy and H. Weyl. While in Cambridge, Flexner got readily convinced that there was no way to lure Hardy away from Cambridge and he turned his attention to H. Weyl. (Hardy and Einstein, as well as J. Hadamard, had singled out Weyl as the most important appointment to be made from Europe.) Both he and Veblen, who had received an offer in June and was in Europe at the time, began discussing the matter with Weyl. He was interested from the start, in spite of strong misgivings about leaving Germany, and immediately expressed some desiderata about the school. First he thought it was absolutely necessary to add to Einstein, Veblen and himself a younger mathematician, preferably an algebraist. Weyl commented (in a letter to A. Flexner, dated July 30, 1932):

> The reason lies with the plans for filling the three main positions. By his personality, Veblen is certainly the most qualified American one can wish as the guiding spirit in an institution such as the one you have founded. But he is not a mathematician of as much depth and strength as say Hardy. The participation of Einstein is of course invaluable. But he pursues long-range speculative ideas, the success of which no one can vouch for. He comes less under consideration as a guide for young people to problems which have necessarily to be of shorter range. I am of a similar nature, at any rate I am also one who prefers to think by himself rather than with a group and who communicates with others only for general ideas or for a final well-rounded presentation. Therefore I put so much value on having a man of the type of Artin or v. Neumann.[3]

---

[3]Der Grund liegt mit in der Art der in Aussicht genommenen Besetzung der drei Hauptstellen. Veblen ist zufolge seiner menschlichen Qualitäten sicher der geeignetste Amerikaner, den man sich als führenden Geist in einer solchen Institution wie der von Ihnen gegründeten wünschen kann. Aber er ist doch nicht ein Mathematiker von ähnlicher Tiefe und Stärke wie etwa Hardy. Einsteins Mitwirkung ist natürlich unbezahlbar. Aber er verfolgt spekulative Ideen auf lange Sicht, deren Erfolg niemand verbürgen kann. Als Führer junger Leute zu eigenen, notwendig auf näher gesteckte Ziele gerichteten Problemen kommt er weniger in Betracht. Ich bin von ähnlicher Natur, jedenfalls auch Einer, der lieber einsam als mit einer Gruppe gemeinsam denkt und mitteilsam nur in bezug auf die allgemeinen Ideen oder in der fertigen gerundeten Darstellung. Mit darum lege ich so viel Wert auf einen Mann vom Typus Artin oder v. Neumann.

In fact, this was important enough to Weyl that Flexner included in his official proposal to him: "*the understanding that when the right person has been found, an algebraist of high promise and capacity will be appointed*". Later Weyl also pointed out the necessity for him to be allowed to give now and then regular courses. He was of course assured he would be welcome to do so, and he accepted in principle the offer in December 1932. But then, in three successive telegrams on January 3, 4, and 12, 1933 he withdrew, then accepted "irrevocably" ("unwiderruflich") and withdrew again. Later on he apologized profusely, explaining he had not realized he was suffering from nervous exhaustion. In his last telegram, he had given as his reason that he felt his effectiveness was tied to the possibility of operating in his mother tongue (a worry still faintly echoed in the foreword to his *Classical Groups*). But the deterioration of the conditions in Germany, in particular the passing of laws not only against Jews, but also against Aryans married to Jews (his case) made his leaving Germany all but unavoidable and in the course of the year he accepted a renewed Institute offer and began his activities at the Institute in January 1934.

The year 1933 also saw the addition to the school faculty of James Alexander and John von Neumann. It had been agreed between Eisenhart, Flexner, and Veblen that an offer would be made to either Lefschetz or Alexander, who both wanted the appointment. The choice fell on the latter, for reasons I have not seen stated anywhere. I have heard indirectly that Eisenhart had said he could more easily spare Alexander than Lefschetz. In view of the much greater involvement of the latter in all the activities of the department, this seems rather plausible. It is also well-known that later Lefschetz was not stingy with critical remarks about Veblen or the Institute. (In 1931, Flexner had asked his views first on the desirability, nature and location of an Institute and second on whom he would choose in mathematics, were he asked to do so. His answer to the second question was Veblen, Alexander and himself from Princeton, Morse and Birkhoff from Harvard; from Europe, he would add above all Weyl, but, since he was holding the most prestigious chair in mathematics in the world, there was no chance to attract him.) J. von Neumann had been half-time professor at the University for some time and the University was trying to make other arrangements. Veblen had suggested to offer him a position at the Institute but at first Flexner was reluctant to take a third mathematician from Fine Hall. However, after Weyl redeclined and after a further conference between von Neumann, Eisenhart, Veblen, and Flexner, an offer was made and quickly accepted. It was also agreed that the two institutions would, henceforth, jointly publish (and share the financial responsibility for) the *Annals of Mathematics*, with managing editors Lefschetz (who had been one since 1928) and von Neumann.

The appointment of Marston Morse in 1934, effective January 1st, 1935, brought to six the school faculty, which was to remain unchanged for the next ten years. To have assembled within three years such an outstanding faculty was an extraordinary success by any standard. In a report to the trustees of the Institute in January 1938, Flexner credited for this achievement Veblen and the help received from the University, in particular from L. P. Eisenhart, then dean of the faculty.

It was, of course, a tremendous boost for the development of the school that it could function in the framework of an outstanding department, strongly committed to research, and make full use of its facilities, vastly superior to those of any other mathematics department in the country. President Hibben and Eisenhart felt that the development of the Institute would be mutually beneficial, although the Institute was offering unique conditions for work, superior salaries, and therefore might again be successful in attracting faculty members besides Veblen. But others in the university community apparently had different opinions, so that, after the third appointment from the university faculty, Flexner and some trustees, in particular L. Bamberger, felt they had to assure the university authorities they would not in the future offer positions to Princeton University professors. As far as I can gather from the record available to me, they did so early in 1933 in one conversation with Acting President Duffield, (Hibben was retired by then). Whether this was meant for a limited time or forever, I do not know. I also have no knowledge of an official written statement by the Institute to that effect, nor of one by the University taking cognizance of such a commitment. On the contrary, the only university document of an official character on this matter I know of (prior to 1963, see below) takes a completely different position. To be more precise, L. P. Eisenhart had written to A. Flexner on November 26, 1932:

> I agree with you that the relationship of the Institute and our Department of Mathematics must be thought of as a matter of policy extending over the years. Accordingly I am of the opinion that any of its members should be considered for appointment to the Institute on his merits alone and not with reference to whether for the time being his possible withdrawal from the Department would give the impression that such withdrawal would weaken the Department. For, if this were not the policy, we should be at a disadvantage in recruiting our personnel from time to time. If our Trustees and alumni were disturbed by such a withdrawal, as you suggest, they should meet it by giving us at least as full opportunity to make replacements intended to maintain our distinction. The only disadvantage to us of such withdrawals would arise, if we were hampered in any way in continuing the policy which has brought us to the position which we now occupy. This policy has been to watch the field carefully and try out men of promise at every possible opportunity. If it is to be the policy of the Institute to have

> young men here on temporary appointment, this would enable us to be in much better position to watch the field.
>
> In my opinion the ideas set forth are so important for the future of our Department that it is my intention to present them to the Curriculum Committee of our Board of Trustees at its meeting next month, after I have had an opportunity to discuss them further with you next week.

Accordingly, Eisenhart presented on December 17 to the Curriculum Committee of the Board of Trustees a statement "on certain matters of policy in connection with the relation of Princeton University to the Institute", a copy of which was kindly given to me by A. W. Tucker. One paragraph reproduces in substance, even partly in wording, the first one quoted above. In conclusion, Eisenhart states that he is presenting this statement "with the expectation that you will approve of the position which I have taken...". It was indeed "approved in principle" by the committee. Obviously the latter was empowered to do so and to speak in the name of the Board of Trustees. Had it been solely advisory, Eisenhart could only have asked the committee to recommend to the board that it approve of his position. I am not aware of any other statement by university authorities addressing this question, again prior to 1963.

As already mentioned, Eisenhart was at the time dean of the faculty. Tucker pointed out to me that, in the organization of the University, this position was next in line to the presidency and that there was in fact no president in charge at that time: Hibben had retired in June 1932 and Dodds would be nominated and become president in late spring 1933. During the academic year 1932–1933, there was only an acting president, namely the Chairman of the Board of Trustees, E. D. Duffield, living in Newark, who mainly took care of off-campus, external affairs. Under those circumstances, Eisenhart was in fact addressing the Curriculum Committee as the chief academic officer of the University.

Although Flexner had not mentioned it in his formal report, he was of course acutely aware of another powerful factor for the rapid growth of the Institute, namely the anti-Semitic policies of the Nazi regime, without which the Institute could hardly have attracted Einstein, Weyl, and von Neumann. This was in fact only the beginning of the Institute's involvement with the migration of European scholars to the U.S. It is a well-known fact that Veblen played a prominent role in helping European mathematicians who had to leave Europe to relocate in the United States.[4] He, Einstein, and Weyl,

[4]See in particular the articles by L. Bers, D. Montgomery and N. Reingold in *A Century of Mathematics in America, Part I* (editor, P. Duren, with assistance of R. A. Askey and U. C. Merzbach), Amer. Math. Soc., Providence, R.I., pp. 231–243, pp. 118–129, pp. 175–200, respectively.

through a network of informants, were well aware of many such cases and often aided in a crucial way by offering first a membership, sometimes with a grant from the Rockefeller Foundation.

At the official Institute opening on October 1, 1933, the school already had over twenty visitors. The level of activities was high from the beginning. While emphasizing the importance of the freedom to carry on one's own research, and the opportunity of making informal contacts and arrangements, the early yearly *Bulletins* issued by the IAS list an impressive collection of lectures, courses and seminars. Among those given in the first four years, let me mention: A two-year joint seminar on topology by Alexander and Lefschetz, followed by a two-year joint course on topology, a joint seminar (extended over several years) by Veblen and von Neumann on various topics in quantum theory and geometry, a course and a seminar by H. Weyl on continuous groups (the subject matter of the famous *Lecture Notes* written by N. Jacobson and R. Brauer), followed by a course on invariant theory, courses and seminars by M. Morse in analysis in the large, a two-year course by von Neumann on operator theory, lectures on quantum theory of electrodynamics by Dirac, on class field by E. Noether, on quadratic forms by C. L. Siegel, and on the theory of the positron by Pauli. In 1935 H. Weyl started and for a number of years led a seminar on current literature. There was also of course a weekly joint mathematical club. The membership steadily increased and Veblen could state around 1937 that in Fine Hall there were altogether approximately seventy research mathematicians and an intense activity. This figure included the members and visitors of the University, too. There was no physical separation in Fine Hall between the two groups, which intermingled freely.[5] Many faced the familiar dilemma of having to choose between attending lectures or minding one's own work. There were also some grumblings that all this was too distracting for the graduate students. The trustees, mindful of the financial aspect, were asking for some limitation and even a reduction of the number of members; Veblen apparently was not too receptive. Almost from the start, Princeton had become a world center for mathematics, the place to go to after the demise of Göttingen.

That the Institute had in this way a considerable impact on mathematical research in Europe and in the United States needs hardly any elaboration. Less evident, and maybe less easy to imagine nowadays, is its role in the improvement of the conditions in American universities by the sheer force of the example of an institution providing such exceptional conditions and opportunities to faculty and visitors. In 1938 Flexner was pleased to quote to the trustees from a letter written to him on another matter by the secretary of the AMS, Dean R. G. D. Richardson of Brown University: "... *The Institute*

[5]For many recollections about Fine Hall at this time, see *The Princeton Mathematics Community in the* 1930*s. An Oral History Project*, administered by C. C. Gillespie edited by F. Nebeker, 1985, Princeton University (unpublished, but available for consultation).

has had a very considerable share in the building up of the mathematics to its present level.... Not only has the Institute given ideal conditions for work to a large number of men, but it has influenced profoundly the attitude of other universities."

The School of Mathematics developed along lines certainly consonant with the vision of the founders, as outlined in the first documents, but not identical with it. Underlying the original concept was a somewhat romantic vision of a few truly outstanding scholars, surrounded by a few carefully selected associates and students, pursuing their research free from all outside disturbances, and pouring out one deep thought after another. Einstein, Weyl, and Veblen soon decided they were not quite up to that lofty ideal and that the justification for the Institute would not be just their own work but, even to a much greater extent, to exert an impact on mathematics, in particular mathematics in the United States, chiefly through a vigorous visitors program. The visitors (called "workers" initially, "members" from 1936 on) were to be mathematicians having carried out independent research at least to the level of a Ph.D. and to be considered on the strength of their research and promise, regardless of whether or not they were assured of a position after their stay at the Institute. Furthermore, their interests did not have to be closely connected to those of one of the faculty members. Originally it was intended that the Institute would also have a few graduate students (but no undergraduates) and would grant degrees. It was officially accredited to do so in 1934. But already then, Flexner stated that it had been done because this seemed a wise thing to do, but it would not be a policy of the Institute to grant degrees, earned or honorary. Indeed, it has so far never done so. This view was confirmed in the 1938 issue of the yearly *Bulletin*, which stated that the Institute had discarded undergraduate and graduate departments on the ground that these already existed in abundance.

In short, the School of Mathematics had very early taken in many ways the shape it still has now, albeit on a different scale, at any rate for the visitors program. It was called School of Mathematics, although its most famous member was not a mathematician. In fact, when asked which title he would want to have, Einstein chose Professor of Theoretical Physics. However, it had been understood from the start that the school would also include theoretical physics. Internally, it was sometimes referred to as School of Mathematics and Theoretical Physics and there were always some visitors specifically in theoretical physics. The faculty had contemplated early on the addition of theoretical physicists; in particular Schrödinger was suggested by Weyl in 1934 and then also by Einstein. Dirac was also mentioned. But the director felt that he could not increase the faculty in the school: He was at the time starting two other schools, in economics and politics and in humanistic studies. Moreover, the financial situation caused some worry and he and the trustees felt some caution was called for. Still, Dirac was a visiting professor in 1934–1935 and Pauli the following year. Later, Pauli spent the war years at the Institute and was offered a professorship in 1945. He was interested but felt he could not commit himself before he had gone back at least for a while to Zürich, where his position had been kept open for him. He stayed at the Institute for one more year with the official title of Visiting Professor, but functioning as a professor and chose later to go back definitely to Zürich. The first real expansion in theoretical physics took place under the first half of Oppenheimer's directorship. As theoretical physics grew at the Institute, the two groups operated more and more independently from one another until it was decided, in 1965, to separate them officially by setting up a School of Natural Sciences. In the sequel, "School of Mathematics" will be meant in the narrow sense it has today.

The Institute developed first very informally. As already stated, Flexner relied for mathematics largely on outside advice, mainly that of Veblen. He had to: "*Mathematicians, like cows in the dark, all look alike to me*", he had said to the trustees at the January 1938 meeting. But this was to be an exception. He had already much more input in the setting up of the School in Economics and Politics and he expected fully it would be so in most aspects of the governance of the Institute. The correspondence with Veblen had shown already some differences of opinion on the eventual shape and running of the Institute, but they were not urgent matters at the time and could be overlooked while dealing with the tasks at hand, on which Flexner and Veblen were usually fully and warmly in agreement. However, as the Institute grew, differences of opinion between the director and some trustees on one hand, and the faculty on the other, became more apparent and relevant. The former liked to view the Institute as consisting of three essentially autonomous schools. They were willing to let each one run its own academic affairs; but there was a rather widespread feeling that professors were often conservative, parochial, not really able to see the Institute globally. Besides it was wrong for them to get involved in administrative matters (after all, Flexner had so often heard professors complain about those duties, which take so much precious time away from research and there he was offering them the possibility of having none...). On the other hand, the faculties of the three schools, which had been chosen quite independently and did not know one another, began to meet, to discuss matters of common interest, to compare views and problems and as a consequence to develop some feeling of being parts of one larger body. Understandably, they wanted to have at least a strong consultative voice in important academic matters. This came to a head when Flexner appointed two professors in economics without any faculty consultation. Added to earlier grievances, it led to such an uproar that Flexner had to resign. But, at a more basic level, there was no attempt to reconcile these two rather antagonistic attitudes in order to arrive at a *modus vivendi* offering a better framework to resolve any conflict that might arise again. None did arise under the next director, Frank Aydelotte (1939–1947), who earned the confidence of the faculty by his way of handling Institute matters (but, as a counterpart, less than unanimous approval from the trustees). Some conflict did surface, not to say erupt, under the next two directors, J. Robert Oppenheimer (1947–1966) and Carl Kaysen (1966–1976). Fortunately, except in one case to which I shall have to come back, these disputes had comparatively little visible impact on the workings of the School of Mathematics, as unpleasant and distracting as they were to its faculty, so that with relief I may pronounce these matters as outside the scope of this account and ignore them altogether. To conclude this long digression, let me add that a prolonged, in my opinion largely successful, effort was made over several years and concluded in 1974 to set up some Rules of Governance for handling in an orderly way between trustees, faculty and the director all aspects of the academic business of the Institute. There has been no such crisis under the present director, Marvin L. Goldberger (1986–), nor under the previous one, Harry Woolf (1976–1986).

In the fall 1939, a new chapter in the life of the Institute began with the moving of the Institute into the newly built Fuld Hall, on its own grounds. In preparation for this change, the school had begun to build up a library, aided in this first of all by Alfred Brauer, whom Weyl had taken as his assistant for this purpose. (Brauer did the same later on, on a bigger scale, for the Mathematics Department of the University of North Carolina at Chapel Hill.) In spite of the war, the Institute operated normally, although some professors were engaged in war work, albeit on a somewhat reduced scale. The influx from Europe increased and, again this had a direct bearing on the school: Siegel was given permanent membership, converted to a professorship in 1945. Kurt Gödel, after having been a member for about ten years, became a permanent member in 1946 and a professor in 1953. Why it took so long for Gödel is a matter of some puzzlement. There was of course unanimous admiration for his achievements and some faculty members had long favored giving him a professorship. The reluctance of others reflected doubts not on his scientific eminence, but rather on his effectiveness as a colleague in dealing with school or faculty matters (Siegel has been quoted to me as having said that one crazy man (namely himself) in the school faculty was enough) or on whether they would not be too much of an imposition on him. As a colleague of his in later years, I would say I found that, his remoteness not withstanding, he would acquit himself well of some of the school business, hence that those fears were not all well founded. On the other hand, I have to confess that I found the logic of Aristotle's successor in more difficult affairs sometimes quite baffling.

After the war, the activities of the school and its membership increased gradually. There was a conscious effort to have members from Eastern Europe or East Asia, in particular Poland, China, India. 1946 was also the beginning of the first (and so far only) venture of the Institute outside the realm of purely theoretical work, namely the construction of a computer under von Neumann's leadership. This has been described in considerable detail by H. Goldstine in his book,[6] to which I refer for details. The computer was used for a few years by a group working on meteorology and von Neumann wanted this to become a permanent feature at the Institute. But the faculty did not follow him. Even the faculty members who had a high regard for this endeavour in itself felt that it was out of place at the Institute, especially in view of the fact that there was no related work done at the University. The computer was given to the University in the late fifties.

Of the first faculty, Alexander resigned in 1947, remaining for some time as a member, Einstein became Professor Emeritus in 1946, Veblen in 1950 and Weyl in 1951. Siegel resigned in 1951 to return to Germany. Added to the faculty in 1951 were Deane Montgomery and Atle Selberg, who had been permanent members since 1948 and 1949 respectively, followed in 1952 by Hassler Whitney.

I came to the Institute in the fall of 1952, not knowing really what to expect. The only recommendation I can remember having received was to appear now and then at tea. This may have been prompted by memories of more formal days, but I soon realized that they were not counting heads. Instead, I found a most stimulating atmosphere, many people to talk to, and suggestions came from many sides. Let me indulge in some reminiscences of those good old days, with the tenuous justification that it is not out of order to describe in this paper some of the experiences and impressions of one visiting member.

F. Hirzebruch, whom I had known in 1948 when he spent some time in Zürich, came once to my office to describe the Chern polynomial of the tangent bundle for a complex Grassmannian. It was a product of linear factors and the roots were formally written as differences of certain indeterminates; Hirzebruch proceeded to tell me how to interpret them but he could not finish: they looked to me like roots in the sense of Lie algebra theory and this was just too intriguing for me to listen to any explanation. An extension to generalized flag manifolds suggested itself, but it was not clear at the moment whether this was more than a coincidence and wishful thinking. A few days later however, it became clear it was not and that marked the start of our joint work on characteristic classes of homogeneous spaces, to which we came back off and on over several years. Conversations with D. Montgomery and H. Samelson led to a paper on the ends of homogeneous spaces. A Chinese member, the topologist S. D. Liao, lectured on a theorem on periodic homeomorphisms of homology spheres he had proved using Smith theory. Having the tools of "French topology" at my finger tips, I tried to establish it in that

[6] H. H. Goldstine, *The computer, Part III*, Princeton University Press, Princeton, N.J., 1972.

framework, succeeded and then, by continuation, obtained new proofs of the Smith theorems themselves. This was the beginning of an involvement with the homology of transformation groups. Of much interest to me also was the seminar on groups, let by D. Montgomery, including his lectures on the fifth Hilbert problem, solved shortly before by him, L. Zippin and A. Gleason, and the contacts with H. Yamabe, his assistant that year.

At the University, Kodaira was lecturing on harmonic forms ("a silent movie" as someone had put it. The lectures were perfectly well organized, with everything beautifully written on the blackboard, but given with a very soft, low-pitched voice which was not so easy to understand.) Tate was lecturing on his thesis in Artin's seminar. The topology at the University gravitated around N. Steenrod, and his seminar was the meeting ground of all topologists. Among those was J. C. Moore, whom I had looked for immediately after my arrival with a message from Serre. This was the beginning of extensive discussions, and a friendship which even moved him to put his life and car at stake by volunteering to teach me how to drive.

My discussions with Hirzebruch went beyond our joint project. He was at the time developing the formalism of multiplicative sequences or functors, genera and experimenting with reduced powers, the Todd genus and the signature. In the latter case, this was soon brought to a first completion after Thom's results on cobordism were announced. Sheaf theory, in particular cohomology with respect to coherent sheaves, had been spectacularly applied to Stein manifolds by H. Cartan and J.-P. Serre; Kodaira, Spencer, Hirzebruch were naturally looking for ways to apply such techniques to algebraic geometry. So was Serre, of course. Being in steady correspondence with him, I was in a privileged position to watch the developments on both sides, as well as to serve as an occasional channel of communication. The breakthroughs came at about the same time in spring 1953 (I shall not attempt an exact chronology) and overlapped in part. Serre's first results were outlined in a letter to me, to be found in his *Collected Papers* (I, 243–250, Springer-Verlag, Berlin and New York, 1986); included were the analytic duality and a first general formulation of a Riemann–Roch theorem for n-dimensional algebraic manifolds. It was soon followed by the analogue for projective manifolds of the Theorems A and B on Stein manifolds. Spencer and Kodaira gave in particular a new proof of the Lefschetz theorem characterizing the cohomology classes of divisors. Soon came a vanishing theorem, established by Kodaira via differential geometric methods and by Cartan and Serre via functional analysis. Attention focussed more and more on the Riemann–Roch theorem, whose formulation became more precise, still with no proof. During the summer, we parted, I to go to the first AMS Summer Institute, devoted to Lie algebras and Lie groups (6 weeks, about thirty participants, roughly two lectures a day, a leisurely pace unthinkable nowadays) and then to Mexico (where I lectured sometimes in front of an audience of one, but not less than one, as Siegel is rumored to have done once in Göttingen, a rumor which unfortunately I could not have confirmed).[7]

Back at the Institute for a second year, I found again Hirzebruch, whose membership had also been renewed. The relationship between roots in the Lie algebra sense and characteristic classes had been made secure, but this whole project had been left in abeyance, there being so much else to do. Now we began to make more systematic computations, using or proving facts of Lie algebra theory and translating them into geometric properties of homogeneous spaces. Quite striking was the equality of the dimension of the linear system on a flag variety associated to a line bundle defined by a dominant weight and of the dimension of the irreducible representation with that given weight as highest weight. Shortly after, I went to Chicago, described this "coincidence" to André Weil, and out of this came shortly what nowadays goes by the name of the Borel–Weil theorem. After I came back, Hirzebruch was not to be seen much for a while, until he emerged with the great news that he thought he had a proof of the Riemann–Roch theorem. This was first scrutinized in private seminars and found convincing. I also provided a spectral sequence to prove a lemma useful to extend the theorem from line bundles, the case treated by Hirzebruch, to vector bundles. A bit later, Kodaira proved that Hodge varieties are projective. All this, and the work of Atiyah and Hodge giving a new treatment of integrals on algebraic curves, completed a sweeping transformation of complex algebraic geometry. Until then, it had been rather foreign to me, with its special techniques and language (generic points and the like). It was quite an experience to see all of a sudden its main concepts, theorems and their proofs all expressed in a more general and much more familiar framework and to witness these dramatic advances. This led me more and more to think about linear algebraic groups globally, in terms of algebraic geometry rather than Lie algebras, an approach on which I would work intensively the following year in Chicago, benefitting also from the presence of A. Weil.

During that second year, I also gave a systematic exposition of Cartan's theory of Riemannian symmetric spaces and got personally acquainted with O. Veblen, on the occasion of a seminar on holonomy groups he was holding in his office. I had of course no idea of his role in the development of the Institute, nor did I know about Flexner and his avowed ambition to create a "paradise for scholars". But I surely had felt it was one, or a very close approximation, so when I was offered a professorship in 1956, I was strongly inclined to accept it. It raised serious questions of course. I realized that, viewed from the inside, with the responsibilities of a faculty member, paradise might not always feel so heavenly. I had also to weigh a very good

[7](Added in proof) B. Devine just drew my attention to the interview of Merrill Flood by A. Tucker in the collection referred to in footnote five above, according to which such an incident did indeed take place once in Fine Hall.

university position (at the ETH in Zürich) with the usual mix of teaching and research against one entailing a "total, almost monastic, commitment to research", (as someone wrote to us much later, while declining a professorship). In fact, the offer had hit me (not too strong a word) while I was visiting Oxford and in a conversation the day before, J. H. C. Whitehead had made some rather desultory remarks about this "mausoleum". To him it was obviously essential to be surrounded by collaborators and students at various levels. I also had to gauge the impact on my family of such a move. But, after some deliberation and discussions with my wife, who left the decision entirely to me, I felt I just could not miss this opportunity.

My professorship started officially on July 1st, 1957, but I was already here in the spring. I found Raoul Bott, with whom I had many common interests. Sometime before, Hirzebruch and I had made some computations on low-dimensional homotopy groups of some Lie groups and, to our surprise, some of our results were contradicting a few of those contained in a table published by H. Toda. There ensued a spirited controversy, in which the homotopists felt at first quite safe. Bott was very interested; he and Arnold Shapiro, also at the Institute at the time, thought first they had another proof of Toda's result on $\pi_{10}(G_2)$, one of the bones of contention, but a bull session disposed of that. Later, Bott and Samelson confirmed our result. Eventually, the homotopists conceded. At the time, I had not understood why Raoul was so interested in those very special results, but I did a few months later when he announced the periodicity to which his name is now attached: Our corrections to Toda's table had removed a few impurities which stood in the way of even conjecturing the periodicity.

There was also a very active group on transformation groups around D. Montgomery who, with the Hilbert fifth problem behind him, had gone back fully to his major interest. My involvement with this topic increased, culminating in a seminar held in 1958–1959.

But I was now a faculty member in mathematics (together with K. Gödel, D. Montgomery, M. Morse, A. Selberg, H. Whitney, as already mentioned, Arne Beurling, who had joined in 1954, and A. Weil from fall 1958 on) and had to have some concerns going beyond my immediate research interests. Foremost were two, the membership and the seminars. As regards the former, it was not just to sit and wait for applicants and select among them, but also of course to seek them out. Weil and I felt that in the fields somewhat familiar to us, a number of interesting people had not come here and I remember that for a few years, in the fall we would make lists at the blackboard of potential nominees and plan various proposals to the group. In this way, in particular, we contributed not insignificantly to the growth of the Japanese contingent of visitors, which soon reached such a size that the housing project was sometimes referred to as "Little Tokyo" and that a teacher at the nursery school found it handy to learn a few (mostly disciplinary) Japanese words.

After a few years however, there was no significant "backlog" anymore and no need to be so systematic. As to the seminars, there were first some standard ones, like the members' seminar and the seminar in groups and topology, led by D. Montgomery. Others arose spontaneously, reflecting the interests of the members or faculty. We felt that the Princeton community owed it to itself also to supply information about recent developments and that beyond the graduate courses offered by the University and the research seminars, there should be now and then some systematic presentations of recent or even not-so-recent developments. In that respect, J.-P. Serre, a frequent fall term visitor during those years, and I organized in fall 1957 two presentations, one on complex multiplication and a much more informal one where we wrestled with Grothendieck's version of the Riemann–Roch theorem. As soon as he arrived, Weil set up a joint University–Institute seminar on current literature, thus reviving the tradition of the H. Weyl seminar, which he had known while visiting the Institute in the late thirties, and had also kept up in Chicago. The rule was that X was supposed to report on the work of Y, Z, with $X \neq Y, Z$. Later on, the responsibility for this seminar was shared with others. It was quite successful for a number of years, but was eventually dropped for apparent lack of interest. As I remember it, it became more and more difficult to find people willing to make a serious effort to report on someone else's work to a relatively broad non-specialized audience. Maybe the increase in the overall number of seminars at the University and the Institute, at times somewhat overwhelming, was responsible for that, I don't know.

During those years, algebraic and differential topology were in high gear in Princeton. In 1957–1958 J. F. Adams was here, at the time he had proved the nonexistence of maps of Hopf invariant one (except in the three known cases). Also Kervaire, while here, proved the non-parallelizability of the n-sphere ($n \neq 1, 3, 7$) and began his joint work with J. Milnor. In fall 1959 Atiyah and Hirzebruch developed here (topological) K-theory as an extraordinary homology theory, after having established the differentiable Riemann–Roch theorem; Serre organized a seminar on the first four chapters of Grothendieck's EGA. During that year, Kervaire, then at NYU came once to me to outline, as a first check, the construction of a ten-dimensional manifold not admitting any differentiable structure! M. Hirsch and S. Smale were spending the years 1958–1960 here, except that Smale went to Brazil in 1960. Soon Hirsch was receiving letters announcing marvelous results, so wonderful that we were mildly wondering to what extent they were due to the exhilarating atmosphere of the Copacabana beach, but they held out. (At the Bonn Tagung in June, as the program was being set up from suggestions from the floor, as usual, the first three topics proposed were the proofs of the Poincaré conjecture in high dimension by Smale and by Stallings and the construction of a nondifferentiable manifold by Kervaire; Bott, freshly arrived and apparently totally unaware of these developments, asked whether this was a joke!)

During these first years at the Institute, my active research interests shifted gradually from algebraic topology and transformation groups to algebraic and arithmetic groups, as well as automorphic forms. That last topic was already strongly represented here by Selberg, and had been before by Siegel. This general area was also one of active interest for Weil, and it soon became a major feature in the school's activities. Without any attempt at a precise history, let me mention a few items, just to give an idea of the rather exciting atmosphere. I first started with two projects on algebraic groups, one with an eye towards reduction theory, on the structure of their rational points over non-algebraically closed fields, the other on the nature of their automorphisms as abstract groups. Some years later, I realized that Tits had proceeded along rather similar lines and we decided to make two joint endeavours out of that. But I was more and more drawn to discrete subgroups, especially arithmetic ones. Rigidity theorems for compact hermitian symmetric spaces, hyperbolic spaces and discrete subgroups were proved by Calabi, Vesentini, while here, Selberg and then Weil. It is also at that time that I proved the Zariski density of discrete subgroups of finite covolume of semisimple groups. Weil was developing the study of classical groups over adeles and of what he christened Tamagawa numbers. I. Satake, while here, constructed compactifications of symmetric or locally symmetric spaces. It became more and more imperative to set up a reduction theory for general arithmetic groups. The Godement conjecture and the construction of some fundamental domain of finite area became prime targets. The first breakthroughs came from Harish-Chandra. I then proved some results of my own; he suggested that we join forces and we soon concluded the work published later in our joint *Annals* paper. This was in summer 1960. The next year and a half I tried alternatively to prove or disprove a conjecture describing a more precise fundamental domain and finally succeeded in establishing it. Combined with the other activities here and at the University, this all made up for a decidedly upbeat atmosphere. But in 1962 rumors began to spread that it was not matched by equally fruitful and harmonious dealings within the faculty. Harish-Chandra, who was spending the year 1961–1962 here, asked me one day, What about those rumors of tremors shaking the Institute to its very foundations? We were indeed embroiled in a bitter controversy, sparked by the school's proposal to offer a professorship to John Milnor, then on the Princeton faculty.

Before we presented this nomination officially, the director had indeed warned us, without being very precise, that there might be some difficulty due to the fact that Milnor was at the University, and we could hardly anticipate the uproar that was to follow. The general principle of offers from one institution to the other and the special case under consideration were heatedly debated in (and outside) two very long meetings (for which I had to produce minutes, being by bad luck the faculty secretary that year). A number of colleagues in physics and historical studies stated that it had always been their understanding that there was some agreement prohibiting the Institute to offer a professorship to a Princeton University colleague. In fact, the historians extended this principle even to temporary memberships. Fear was expressed that such a move would strain our relations with the University, which some already viewed as far from optimal. In between the two meetings, the director produced a letter from the chairman of the Board of Trustees, S. Leidesdorf, referring to a conversation he had participated in between Flexner and the president of the University, in which it had been promised not to make such offers. He viewed it as a pledge, which could be abrogated only by the University.

Those views were diametrically opposite to those of the mathematicians here and at the University, which were in fact quite similar to those of Eisenhart in the letter quoted earlier or in his statement to the curriculum committee, both naturally and unfortunately not known to us at the time. He really had said it all. First of all, the school used to give sometimes temporary memberships to Princeton faculty. This was on a case-by-case basis, not automatic, and it had never occurred to us to rule it out *a priori*. We also felt that our relations with Fine Hall were excellent and would not be impaired by our proposal. In fact D. Spencer had told us right away we should feel free to act. D. Montgomery stated that Veblen had repeatedly told him, in conversations between 1948 and 1960, that there had never been such an agreement. J. Alexander, asked for his opinion, wrote to Montgomery that he had never known of such an agreement (whether gentlemanly or ungentlemanly). He also remembered certain conversations in which an offer to a university professor was contemplated, or feared by some university colleague, conversations which would have been inconceivable, had such an agreement been known. Finally he had "no knowledge of deals that may have been consummated in 'smoke-filled rooms' or of 'secret covenants secretly arrived at.' All this sort of stuff is over my depth." A. W. Tucker, chairman of the University Mathematics Department, consulted his senior colleagues and wrote to A. Selberg, our executive officer, that in their opinion (unanimous, as he confirmed to me recently) the Institute should be free to extend an offer to Milnor. Of course, were he to accept it, this would be a great loss, but any such "restraint of trade" was distasteful to them and could well prove damaging in the long run. It would be much better, they felt, if the University would answer with a counteroffer attractive enough to keep Milnor. The point was repeatedly made that, when two institutions want the services of a given scholar, it is up to the individual to choose, not up to administrators or colleagues to tell him what to do; also, as Eisenhart had already pointed out, that such a blanket prohibition might be damaging to the recruiting efforts of the University.

In the course of the second faculty meeting a colleague in the School of Historical Studies, the art historian Millard Meiss, stated it had indeed been his understanding there was such an agreement; he noted that the mathematicians and his school acted differently with regard to temporary memberships; he felt the rule had been a wise one in the earlier days of the Institute, but was very doubtful it had the same usefulness today. Accordingly, he proposed a motion, to the effect that the faculty should be free to extend professorial appointments to faculty members of Princeton University, with due regards to the interests of science and scholarship, and to the welfare of both institutions. He also insisted that this should occur only rarely. This motion was viewed as so important ("the most important motion I have voted on in the history of the Institute", commented M. Morse) that it was agreed to have the votes recorded by name, with added comments if desired. It was passed by fourteen *yes* against four *no*, with two abstentions.

After this, it would have seemed most logical to take up the matter with the president of the University, R. Goheen, but nothing of the kind was done at the time and the tension just mounted until the trustees meeting in April. There, as we were told shortly afterwards by the director, the Milnor nomination did not even come to the board: The trustees had first reviewed the matter of invitations to Princeton University faculty, with regard to the Meiss motion, and had voted a resolution to the effect that the agreement with Princeton University to refrain from such a practice was still binding.

In this affair we had worked under a further handicap: In those days, it was viewed as improper to talk about a possible appointment with the nominee before he had received the official offer (nowadays, the other way around is the generally accepted custom). Consequently, none of us had ever even hinted at this in conversations with Milnor. But he had heard about it from other sources and it became known that he would have been seriously interested in considering such an offer. The director and the trustees may not have felt so fully comfortable with their ruling after all. At any rate, they soon proposed to offer some long-term arrangement to Milnor, whereby he could spend a term or a year at the Institute during any of the next ten years. This was of course very pleasant for Milnor, and we gave this proposal our blessing, but it fell short of what we had asked for. Finally, eighteen months later, in October 1963, we were informed that, following instructions from the trustees, the director had taken up the matter of general policy with President Goheen in January 1963 and we received a copy of a letter written on January 21, 1963 by President Goheen to the director, outlining one. Although cautious in tone, it allowed one institution to extend an offer to a faculty member of the other, after close consultation "*to the end of matching the interests of the individual with the common interests of the two institutions to the fullest extent possible.*" In conclusion, he urged that "*this agreement supplant any specific or absolute prohibition that we may have inherited from our predecessors.*"

Right after the next trustees meeting the director wrote to Goheen on April 22, in part: "*The Trustees asked me to tell you that they welcome your letter, and that they have asked me to let it be a guide to future policy of the Institute.*" As far as I know, the matter was never reconsidered and this agreement is still in force. At the time we were apprised of this (October 1963), it would have therefore been "legally" possible for us to present again our proposal, although Milnor was still a Princeton faculty member.

But we could not! During 1962–1963, we had asked for two additions to our group; they had been granted and no chair was available to us anymore. How had this come about?

This experience had left strong marks. It was not just the decision of the trustees, but the way the matter had been handled and the breakdown in relations within the faculty (also contributed to by conflicting views on some nominations in the School of Historical Studies), the ruling from on high by the board, without bothering to have a meaningful discussion with us, bluntly disregarding our wishes, as well as those of the faculty as expressed by the Meiss motion, all this chiefly on the basis of a rather flimsy recollection of the chairman of the board, promoted to the status of an irrevocable pledge. Some of us were wondering whether to withdraw entirely into one's own work or to resign, and were sounded out as to their availability. One Chairman, who had for some time wanted to set up a mathematics institute within his own institution, toyed with the idea of making an offer to all of us. We still had the option of making another nomination and there were indeed two or three names foremost on our minds. But just choosing one and presenting it would not suffice to restore our morale. Something more was needed to help us rebound. It was Weil who suggested that we present two nominations instead of just one, as was expected from us. After some discussions, we agreed to do so and nominated Lars Hörmander and Harish-Chandra.

This took the rest of the faculty and the director completely by surprise. The latter did not raise any objection on budgetary grounds. He also made it clear at some point that if granted, this request would have no bearing on faculty size for the other groups. Since our nominations were readily agreed to be scientifically unassailable, it would seem that our proposal would go through reasonably smoothly, but not at all. Our request had been addressed by A. Selberg, still our executive officer, directly to the director and the trustees, bypassing several steps of the standard procedure for faculty nominations, which seemed unpracticable in the climate at the time, and also not fulfilling one requirement in the by-laws. And it is indeed on grounds of procedure that the director and some colleagues raised various objections. There was overwhelming agreement on the necessity of major changes in our procedure for faculty appointments. The question was whether this review should precede or follow the handling of our two nominations. Again, this grew into a full-size debate and we did not know how our proposal would fare at the

April trustees meeting. There, as we were told at the time, the director recommended to postpone the whole matter, but the trustees, after having heard Selberg present our case, voted to grant our request under one condition, namely that a faculty meeting be held to discuss our nominations. This was really only to restore some semblance of formal compliance with the by-laws, and they were anxious that this matter be brought with utmost dispatch to a happy conclusion, so that the Institute would soon regain its strength and some measure of serenity. This meeting was held within a week and the offers were soon extended.

Harish-Chandra accepted quickly, Hörmander after a few months. Finally, this sad episode was behind us. We felt and were stronger than before and could devote ourselves again fully to the business of the school. In fall 1963 there were the usual seminars on members and faculty research interests. Harish-Chandra started a series of lectures, which became an almost yearly feature: every week two hours in a row, most of the time on his own work, i.e., harmonic analysis on reductive groups (real, later also p-adic), documenting in particular his march towards the Plancherel formula. He was not inclined to lecture on other people's work. One year however he did so, he "took off", as he said, viewing it as some sort of sabbatical, and lectured on the first six chapters of Langlands' work on Eisenstein series (then only in preprint form). There were also some seminars on research carried out outside Princeton: I launched one on the Atiyah–Singer index theorem, for non-analysts familiar with all the background in topology. Eventually, R. Palais took the greater load and wrote the bulk of the *Notes* (published in the *Annals of Math. Studies* under his editorship). The following year, there was similarly a "mutual instruction" seminar on Smale's proof of the Poincaré conjecture in dimensions ≥ 5. Still, we felt some imbalance in the composition of the membership and the activities of the school. Of course, there is no statutory obligation for the school membership to represent all the main active fields of mathematics. In any case, in view of the growth of mathematics and of the number of mathematicians, as compared to the practically constant size of the school (the membership size hovering around 50–60 and that of the faculty around 7–8), such a goal was not attainable anymore. Nevertheless, it has always been (and still is) our conviction that the school will fulfill the various needs of its membership best if it offers a wide variety of research interests, and that this is a goal always to keep in mind and worth striving for, even if not fully reachable. For this and other reasons we decided in 1965 to have more direct input in part of the work and composition of the school by setting up a special program now and then. This idea was of course not to have the school fully organized all of a sudden, rather to add a new feature to the mathematical life here, without supplanting any of the others. Such a program was to involve as a rule about a quarter, at most a third, of the membership, with a mix of invited experts and of younger people. It would often be centered on an area not well represented on the faculty, but not obligatorily so. We did not want to refrain from organizing a program in one of our fields of expertise, if it seemed timely to gather a group of people working in it to spend a year here. It was of course expected that such a program would include a number of seminars for experts to foster further progress, but we also hoped it would feature some surveys and introductory lectures aimed at people with peripheral interests, and would also facilitate to newcomers access to the current research and problems. Pushing this "instructional" aspect a bit further, we also decided to have occasionally two related topics, hoping this would increase contacts between them.

The first such program took place in 1966–1967 and was devoted to analysis, with emphasis on harmonic analysis and differential equations. In agreement with the last guideline stated above, the second one (1968–1969) involved two related topics, namely algebraic groups and finite groups. As a focus of interaction, we had in mind first of all the finite Chevalley groups and their variants (Ree and Suzuki groups). They played that role indeed, but so did the Weyl groups and their representations, as can be seen from the *Notes* which arose from this. The third program (1970–1971) centered on analytic number theory.

In 1971, again with an eye to increasing breadth and exposure to recent developments, another activity was initiated here, namely an ongoing series of survey lectures. In the sixties and before, the dearth of expository or survey papers had often been lamented. The *AMS Bulletin* was a natural outlet for such, first of all because the invited speakers for one-hour addresses are all asked to write one. But this did not seem to elicit as many as one could wish and various incentives were tried, with limited success. It had always seemed to me that most of us are cold to the idea of just sitting down to write an expository paper, unless there is an oral presentation first. But the example just mentioned showed that this condition was not always sufficient. Already in my graduate student days, I had been struck by some beautiful surveys in the *Abhandlungen des Math. Sem. Hamburg.* They were usually the outgrowth of a few lectures given there. This suggested to me that one might have a better chance of getting a paper if the prospective author were invited to give some comprehensive exposition in a few lectures, not just one. However I had done nothing to implement such a scheme, just talking about it occasionally, until the 1970 International Congress in Nice. There K. Chandrasekharan, then president-elect of the IMU, told me he wanted to set up a framework for an ongoing series of lectures sponsored by the IMU, to be given at various locations, with the express purpose to engender survey papers. Would I help to organize it? Our ideas were so similar that we quickly agreed on the general format: A broad survey, for non-specialists, given in four to six one-hour lectures, within a week or two. Expenses would be covered, but the real fee would be paid only upon receipt of a manuscript suitable for inclusion in this series. A bit later, I suggested as an outlet for publication the *Enseignement Mathématique*, mainly for two reasons: First, it is in some way affiliated to the IMU, being the official organ of the International Commission for Mathematical Education. Second, it has the rare, if not unique, capacity to publish as a separate monograph, sold independently, any article or collection of articles published in that journal.

The first two such sets of lectures were given at the Institute in the first quarter of 1971, by Wolfgang Schmidt and Lars Hörmander (who was a visitor, too, having resigned from the faculty in 1968), both soon written up and indeed published in the *Enseignement Mathématique.* But a difficulty arose with our third proposal, namely to invite Jürgen Moser, then at NYU, to give a survey on some topics in celestial mechanics. From the point of view of the IMU, these lectures were meant to promote international cooperation. Accordingly, the lecturer was to be from a geographically distant institution, so that the invitation would also foster personal contacts. They felt that we did not need an IMU sponsorship to bring Moser from NYU to the Institute. They certainly had a point. On the other hand, it was also a sensible idea to have such a set of lectures from Moser. In the school, we were really after timely surveys, whether or not they were contributing to international cooperation, while this latter aspect was essential for the IMU. Also, they wanted of course to have such lecture series be given at various places and their budget was limited. Since we planned to have about one or two per year, our requests might well exceed it, so that some difficulties might be foreseen also on that score. We therefore decided to start a series of similar lectures of our own, and to call them the Hermann Weyl Lectures, an ideal label, in view of Weyl's universality: It was a nice touch to be able on many occasions to trace so much of the work described in those lectures to some of his. We planned to publish them as a rule, though not obligatorily, in the *Annals of Mathematics Studies.* Otherwise, the conditions and format of the lectures were to be the same. Our series started indeed with J. Moser's lectures, resulting in an impressive two-hundred page monograph. For a number of years, the H. Weyl lectures were a regular feature here, at the rate of one to two sets per year. As to their original purpose, namely to bring out survey papers, I must regretfully acknowledge that our record is a mixed one, and that the list of speakers who did not contribute any is about as distinguished as that of those who did. Maybe Moser's contribution was a bit daunting, although F. Adams and D. Vogan rose to the challenge, even topping its number of pages (slightly in the former case, largely in the latter). Overall, the high quality of the monographs growing out of the H. Weyl lectures has made the series very worthwhile. Their frequency has declined in recent years. Since we started this, "distinguished" lecture series have sprung up at many places. Also, symposia, conferences and workshops on specific topics have proliferated, often leading to publications containing many surveys or introductory papers. There is indeed nowadays quite a steady flow of papers of this type so maybe the need for our particular series has decreased. One of the nice features of the Institute is that we need not pursue a given activity if we do not feel it fulfills a useful function in the mathematical community. So we may well leave this one in abeyance and revive it whenever we see a good opportunity.

In 1966 C. Kaysen had taken up the directorship and found the school faculty in good shape. He thought that, at least with our group, he would not face requests for new appointments. But we pointed out to him that our age distribution was a bit unfortunate and would later create some problems, with retirements expected in 1975, 1976, 1977, and 1979. Therefore it might be desirable to consider some advance replacements; also that some minimal expansion might be to the good. He agreed. In 1969 Michael Atiyah joined the faculty. Originally, this appointment had been meant to be an expansion, but it was not anymore, after Hörmander had resigned in 1968. Later, we made offers successively to John Milnor and Robert P. Langlands, who came to the faculty in 1970 and 1972 respectively.

In the sixties, considerable progress was made in the general area I had already singled out as a very strong one here: Algebraic groups, arithmetic groups and automorphic forms, number theory, harmonic analysis on reductive groups. Much of it was done here, but also at the University by G. Shimura, and by R. P. Langlands who was there for three years. It continued unabated, or even at an increased pace, after Langlands joined us. This whole general field had become such an active and important part of "core mathematics" that it was all to the good. However, that was not matched by activities of similar scope in other areas and created some imbalance, accentuated by Atiyah's resignation in 1972. For reasons already explained, in our view it was not in the best interest of the school in the long run and to correct it by increasing activities in other areas became a concern. There were two obvious means to try to remedy this: the special programs and new faculty appointments. But they were not available to us during the energy crisis and the immediately following years. The financial situation of the Institute was worrisome and we had not even been authorized to replace Atiyah. Also, we had not been able to take care completely within our ordinary budget of the special programs, which entailed invitations to well-established people. We always had had to get some outside support, besides our standing NSF contract, and that was hard to come by in those years. But we resumed both as soon as it became possible: Enrico Bombieri came to the faculty in 1977 and Shing-Tung Yau in 1980, broadening greatly its coverage. We had also to wait until 1977 for the programs but have had one almost every year from then on.

In 1977–1978, our program was devoted to Fourier integral operators and microlocal analysis with the participation in particular of L. Hörmander and M. Kashiwara. This was again an attempt to increase contacts between two rather different points of view, in this case the classical approach and the more recent developments of the Japanese school around M. Sato. It led to a collection of papers providing a mix of both. The next one was on finite simple groups and brought here a number of the main participants to the collective enterprise to classify the finite simple groups. 1979–1980 was the year of the biggest program to date, on differential geometry and analysis, in particular nonlinear PDE. The number of seminars was somewhat overwhelming. Several were concentrated at the end of the week, so as to make it easier for people in neighboring (in a rather wide sense including New York and Philadelphia) institutions to participate. Roughly speaking, the main activities were subdivided in three parts: differential geometry, minimal submanifolds, and mathematical physics, with seminar coordinators L. Simon for the second one, S. T. Yau for the other two. A remarkable feature of the third one (devoted to relativity, the positive mass conjecture, gauge theories, quantum gravity) was the cooperation between mathematicians and physicists, probably a first here since the early days. Two volumes of *Notes* resulted from this program.

There was none the following year but then, in 1981–1982, we had one on algebraic geometry, at least as big as the previous one. Again, seminars were also attended by visitors from outside, two even coming from Cambridge, Massachusetts: D. Mumford and P. Griffiths would visit every second or third week for two to three days, each to lead one of the main seminars. We had decided to concentrate on the more geometric (as opposed to arithmetic) aspects of algebraic geometry, since we intended to have in 1983–1984 a program on automorphic forms and L-functions. But even with that limitation, it was of considerable scope (Hodge theory, moduli spaces, K-theory, crystalline cohomology, low-dimensional varieties, etc.). Griffiths' seminar also led to a set of *Notes*. This was again very successful but the evolution of these seminars betrayed a natural tendency, namely to try each time to improve upon the previous one, leading not unnaturally to bigger and bigger programs. As already stated, our original intention had been to add an activity, not to suppress any, and we began to wonder whether these programs, carried out at such a scale, might not hamper somewhat other important aspects of the mathematical life here, such as variety, informality, the opportunity for spontaneous activities and unplanned contacts, quiet work, etc. So we decided to scale them down a bit. Again, this was not meant as a straightjacket; rather, that the initial planning would usually be on a more modest scale. But, if outside interest would lead to a growth beyond our original expectations (as is the case with the present program on dynamical systems), we would of course do our best to accommodate it. We were aided in fact in our general resolve by the emergence of the Mathematical Sciences Research Institute at Berkeley: Big programs are an essential feature there and they have more financial means than we to carry them out. There is no need to compete for size.

S. T. Yau had resigned in 1984 and was soon replaced by Pierre Deligne. The retirements we had warned C. Kaysen about had caught up with us for some time and our group was reduced to six, two fewer than the size we were entitled to at the time, so that we had the possibility of making two appointments. We were anxious to seize this opportunity to catch up with some new major trends in mathematics. There had been some very interesting shifts in the overall balance of research interests, partly influenced by the development of computers, notably towards nonlinear PDE and their applications (with which we had lost first-hand contact after Yau's resignation), dynamical systems, mathematical physics, as well as an enormous increase of the interaction with physicists, the latter visible notably around string theory and conformal field theory (CFT). These last two topics were very strong at the University, but underrepresented here (not only in the faculty, but also in the membership). As a first attempt to improve this situation, I suggested in fall 1985 to E. Witten to give at the Institute a few lectures on string theory aimed at mathematicians. They were very well attended, so that the next logical move was to think about organizing a program in string theory and to ask Witten whether this seemed to him worth pursuing and, if so, whether he would agree to help, first as a consultant and then as a participant. That same year, we made two successful offers to Luis Caffarelli and Thomas C. Spencer, thus increasing considerably our range of expertise in some of the "most wanted" directions.

The first question put to Witten was not entirely rhetorical, given the abundance at the time of conferences and workshops on these topics. But it was agreed after some thought that a year-long program here would have enough features of its own to make it worth trying. A bit later, an expert to whom I had written about it warned that, in view of the usually rather frantic pace of research in physics, this might be all over and passé at the time of the program (1987–1988); but it seemed to us there was enough new mathematics to chew on for slower witted mathematicians to justify such a program on those grounds (later, that expert volunteered to eat his words). Anyway, we went ahead. The program had originated within the School of Mathematics, but the School of Natural Sciences became gradually more involved and eventually contributed to the invitations. In fact, the borderline between the two schools became somewhat blurred, the physicists D. Friedan, P. Goddard and D. Olive being members in mathematics, while the mathematicians G. Segal and D. Kazhdan were invited by the School of Natural Sciences. A primary goal of this program was to increase the contacts between mathematicans and physicists and to help surmount some of the difficulties in communication due to differences in background, techniques, language and goals. Accordingly, we had invited several mathematically minded physicists and some mathematicians with a strong interest in physics, all rather keen to contribute to the dialogue. The program was very intense, too, with an impressive array of seminars, notably many lectures on various versions of CFT, and many discussions in and outside the lecture rooms.

Our last two appointments, succeeded by that of E. Witten in the School of Natural Sciences, have quickly made the Institute a major center of interaction between physics and mathematics and also increased significantly the membership in analysis. Altogether, the school faculty seems to me to be about as broad as can be expected from seven people. I hope it is not just wishful thinking on my part to believe that by its concern for the school and its own work, it is well on its way to maintain a tradition worthy of the vision of the first faculty.

The reader will have noticed that, from the time I came to the Institute, this account is largely based on personal recollections and falls partly under the label of "oral history", with, as a corollary, an emphasis or maybe even an overemphasis on the events or activities I have been involved with or witnessed from close quarters. Even with those, I have not been even-handed at all and this paper makes no claim to offer a balanced and complete record of the school history and of all the work done there.[8] Such an undertaking would have brought this essay to a length neither the editors nor the author would have liked to contemplate. Also absent is any effort to evaluate the impact of the school on mathematics in the U.S. and beyond: How much benefit did visitors gain? How influential has their stay here been on their short-range and long-range activities? What mathematical research was carried out or has originated here? How important has been the presence and work of the faculty? These are some of the questions which come to mind. To try to answer them would again have had an unfortunate effect on the length of this paper. Besides, an evaluation of this sort is more credible if it emanates from the outside, at any rate not solely from an interested party of one. Moreover, as a further inducement for me to refrain from attempting one, two evaluations of relatively recent vintage do exist. First, a report by a 1976 trustee–faculty committee, whose charges were to review the past, evaluate the Institute and provide some guidelines for the future. Its assessment was based in part on the letters of a number of scholars and on the answers (over five hundred from mathematicians) to a questionnaire sent to all past and present members on behalf of that committee. Second, one by a 1986 visiting committee, chaired by G. D. Mostow. Both, though not exempt from

[8]In that connection, let me mention that *A Community of Scholars. The Institute for Advanced Study, Faculty and Members 1930–1980*, published by the Institute for Advanced Study on the occasion of its fiftieth year, contains in particular a list of faculty and members up to 1980 and, for most, of work related to IAS residence.

criticisms, conclude that the School of Mathematics has been successful in many ways. As a brief justification for this claim and without further elaboration, let me finish by quoting from a letter written in 1976 by I. M. Singer to the chairman of the review committee, Martin Segal, who was happy to share it with the committee:

> Their [the members'] stay at the Institute under the guidance of the permanent staff affects their mathematical careers enormously. Their contacts with their peers continue for decades. They leave the Institute, disperse to their universities, and carry with them a deeper understanding of mathematics, higher standards for research, and a sophistication hard to attain elsewhere.
>
> Such was the case when I was here twenty years ago. Last fall when I signed the Visitors' Book I turned the pages to see who was here in 1955–1956. Many are world famous and they are all close professional friends. I notice the same thing happening now with the younger group. Before I came in 1955, the Institute was described to me as I am describing it to you. It remains true now as it has been for the last thirty years.

In preparing this article I benefitted from the use of some archival material. I thank E. Shore and M. Darby at the Institute for their help in dealing with the Institute archives and R. Coleman at the University for having kindly sent me copies of some documents in the University archives. I am also grateful to A. Selberg and A. W. Tucker for having shared with me some of their recollections, and especially to D. Montgomery for having done so in the course of many years of close friendship.

During most of his career, Edgar R. Lorch has been connected with Columbia University. In 1924, he entered Columbia as an undergraduate. He received his B.A. in 1928 and his Ph.D. in 1933, writing a dissertation under the direction of J. F. Ritt. He was appointed to the Columbia faculty in 1935, and he served there until his retirement in 1977. He was Chairman of the Department at Barnard College in 1948–1963 and at Columbia in 1968–1972. His research has focused on operators in linear spaces, normed rings, and topology. Among his many publications is a research monograph entitled Spectral Theory.

Mathematics at Columbia during Adolescence

EDGAR R. LORCH

"Now, really, these French are going too far. They have already given us a dozen independent proofs that Nicolas Bourbaki is a flesh and blood human being. He writes papers, sends telegrams, has birthdays, suffers from colds, sends greetings. And now they want us to take part in their canard. They want him to become a member of the American Mathematical Society (AMS). My answer is 'No'." That was the reaction of J. R. Kline, the AMS secretary, as he strode out of the Society's office on the third floor of Low Library. Kline was a charming person, especially warm with us younger colleagues. It was always a pleasure to be in his company and as we walked from Miss Hull's office to the Faculty Club for lunch or to the afternoon session of the Society in Pupin Hall he would unfailingly tell me some anecdote on one of our flamboyant members. One of them, concerning Norbert Wiener, deserves retelling here.

It seems that the Klines and the Wieners had adjacent summer cottages on a lake in New Hampshire. It was Norbert's habit every summer to swim from his dock to a small island not too far away in the middle of the lake. Thus he would convince himself that his physical capacity did not lag behind his mental sharpness. On these swims, JRK would keep company in a rowboat carrying on a conversation with the convex body which was slowly progressing to the goal. Trying, as usual, to keep the initiative within his own hands, especially since, as he approached the island, he was becoming quite winded, Norbert puffed out his trump card: "Kline, who are the five greatest living mathematicians?" And JRK quietly: "That is an interesting question. Let's see." And he mentioned without delay or difficulty four names. Then, full stop. "Yes, yes, go on," burbled NW, not having heard the name of his favorite candidate. But JRK, with delicate humor, never revealed the identity of Mr. Quintus.

There was a sequel to the Bourbaki episode. About that time the leading mathematical societies signed reciprocity agreements allowing any member of one contracting society to become on demand a member of another. Entrance into the Society was attempted for NB under reciprocity and led to an astonishingly large correspondence (See Everett Pitcher, *A History of the Second Fifty Years, American Mathematical Society, 1939–1988*, AMS Centennial Publications, Vol. I, pp. 159–162.) Today's younger mathematicians cannot easily imagine the heat produced by the episode.

Up to the fifties, Columbia was a Times Square for mathematics, a meeting place for the entire Northeast corridor. This was natural since so many meetings took place on our campus. The beautiful Society office, presided over by the beautiful Miss Hull, was here; the treasurers of the Society seem to have been Columbia people (were we really more honest than the others?), the Society library was here, more or less mixed up with the Columbia collection. We were really privileged at Columbia, and it broke our heart when the Society at the tender age of 60 or so decided to leave the nest and start life on its own in Providence. That was a bit before our department decided that enough was enough and it was time to become modern.

Among the very first of the visitors I remember at Columbia was G. D. Birkhoff who came in the summer of 1929 to teach in our summer session. How many of today's mathematicians have ever taught during the summer? GDB radiated power and good will, and being in his company was a privileged way of starting a career. I was a first year graduate student at the time, a very critical period for a young person. In his lectures, GDB had an unconscious knack for associating himself with substantial stage props, both "im grossen" and "im kleinen." In his course entitled "Mathematical Elements of Art," we navigated from Greek and Japanese vase forms (a rather obvious and easy subject) to the writing of poetry via formulas in which the listener could test and grade himself against Keats and Shelley. Then at the end of the course, there was music, in which I was particularly interested. Due to the tightness of the program, there was only one day left over for this subject. Full of expectation, I went to 202 Hamilton Hall in 88° temperature (plus humidity), and there I found a magnificent Steinway grand piano, all eleven feet of it, in its imperial ebonized glory. What was in store for us now? Well, precisely nothing. GDB spoke in a general way about a variety of things but never, I mean never, was the piano touched.

I remember another episode some years later when Birkhoff gave one of the inaugural lectures for the founding of the Institute for Advanced Study (IAS).

He had already mobilized on the board an astonishing quantity of symbols when he stopped short, looked up and down, and said with surprise, "But there is no colored chalk here." After an inaudible gasp of consternation on the part of his Princeton hosts, a young local professor got up and raced out of the room. GDB proceeded. After a brief pause, the young man reappeared, a bit breathless but also sheepishly jubilant, carrying a lovely box which bounded an 8×10 matrix of a rainbow assortment of chalk. Birkhoff looked at him over his spectacles and said, "That's all right. I don't plan to use it," and went on with his exposition.

As a young student, I was fully aware of the exceptional role played by Columbia in the first years of the Society. Indeed, the original name could well have been the Columbia Mathematical Society. Four of the first seven presidents of the AMS were associated with Columbia. J. H. Van Amringe held a professorship at Columbia over the period 1863–1910 and was Dean of the College from 1896 onward. "Van AM" was a popular teacher who inspired the creation of some old Columbia student songs [Archibald, *A Semicentennial History of the American Mathematical Society 1888–1938*, Amer. Math. Soc., Providence, RI, 1938, pp. 110–112]. He was the first president of the New York Mathematical Society (now the AMS) in 1888. G. W. Hill had close ties with Columbia but worked for much of his career at his home in West Nyack, New York. He lectured on celestial mechanics at Columbia in 1893–1895 and in 1898–1900. He was president of the AMS in 1895–1896. His fundamental contributions to the theory of the lunar orbit earned him an international reputation. Hill's differential equation is now well known in celestial mechanics. R. S. Woodward, the fifth president of the AMS (1899–1900), taught mechanics and mathematical physics at Columbia during the years 1893–1904. He was an astronomer and geographer of first rank who later served as President of the Carnegie Institution of Washington (1904–1921). Thomas S. Fiske was educated at Columbia and was on the faculty from 1888 to 1936. After founding the AMS (NYMS) as a graduate student in 1888, he became its seventh president in 1903–1904.

The offices of the Society at the very beginning must have been the desk of Professor Fiske. When I came to Columbia some thirty-six years later, the Society had its quarters in space provided by the University. Still later, when I was book review editor for the *Bulletin*, I remember making frequent visits to its beautiful sunny quarters on the third floor of Low Memorial Library, which, until the Butler Library was built, housed the main university collection.

An older professor of stature at Columbia in the twenties and thirties was Edward Kasner, a delightful, kind man who had done distinguished work in differential geometry. We used to share an office together, and in my mind's eye, I still see him so well coming in at 10:50 on a chilly fall day, peeling off his topcoat, jacket, and sweater, then putting the jacket back on in preparation for his lecture in fundamental concepts. As a last step in the preparation, he would turn his back to me, pull an envelope from his jacket containing his false teeth, and snap them on audibly. Then forth to the fray.

Kasner's course for M.A. candidates was very popular yet very elementary. He spent a great deal of time working with large numbers. I do not know how many class days were spent estimating the number of grains of sand on the earth. His favorite large number was 10^{100}, well beyond any number arising in the physical universe. He asked his two-year-old nephew what name to give this monster, and the little boy gurgled "google." The name stuck.

As the reader may guess, Kasner was not without his idiosyncrasies. He loved nature and hiking and would regularly walk up Riverside Drive to the New Jersey ferry, cross the river (cost, one nickel) and climb the Palisades, the top of which was covered by a respectable "wild" forest. On each of these walks, and this he recounted to me at least ten times, he would dig a hole at the base of some tree and bury a nickel. Why? So that he would never find himself *depourvu* of ferry fare on his return. Come now, you younger mathematicians who are ostentatious about your pecularities, let's see you match that.

Mathematics departments have their ups and downs, and during the twenties, honesty requires one to admit, Columbia was much on the down side. The administration was keenly aware of the situation and was just as keenly proceeding to do something about it. The rule of action here at our University based on the rule of thumb "il ne faut pas se prendre pour de la merde" is to start at the top creating an ordered list of the world's greatest mathematicians, to make offers starting at number one, and to see what happens. Well, here is one thing that happened as it was told to me some years later by one of the more talkative members of our department. Hermann Weyl received a princely offer. It was discussed, and special conditions were made and agreed upon. One of them was that his assistant, a young woman named Lulu Hoffman, was also to come to Columbia. This raised a problem immediately because Columbia had only male professors. However, the problem was easily solved. Dr. Hoffman was to teach in Barnard. This actually took place, and at Barnard she was the first woman mathematics teacher. The Columbia-Weyl bargaining went on, and finally Weyl decided not to accept. As he put it, and this is the part about which my talkative colleague insisted, Weyl pointed out that Göttingen was the center of the mathematical universe, that he was very happy there, and that he did not wish to change things by accepting Columbia's offer. We fellow instructors used to laugh wholeheartedly picturing Hermann Weyl, on a deck chair of the *SS Bremen* or *Hamburg* crossing the ocean to New York with the center of gravity of world mathematics following obediently some one hundred yards behind the propeller's wash.

There were other attempts by the authorities to obtain the services of a very distinguished man, but these were equally unfruitful. (However, Columbia was nowhere near winning the university sweepstakes for the greatest number of successive turndowns.) It was then decided to engage brilliant promising younger scientists. In this way, the department was enriched by the presence of Bernard Koopman from Harvard and Paul Smith from Princeton. That was an astute move on the part of the university, which paid off handsomely.

Paul was a topologist to the marrow. He was very quiet and very concentrated. His compass always pointed towards Princeton with its solar system of topologists. He was not the one who said, "Whenever I see a derivative it gives me nausea," but he probably thought it. He held some beliefs with a strange intensity. One was love and reverence for Vermont and all it stood for. He had a summer home there. One of his great regrets was that he had not been born there instead of New Hampshire. It was in the late forties that Paul was instrumental in bringing Sammy Eilenberg into the department with consequences for its development and emphasis that lasted decades.

Koop, or Bernie, as we called him, had a completely different personality. In society, he was lively, wide-ranging, playful, and mordant. He loved to open a conversation, size up the strength and weakness in his fellows, and needle them on. My close contacts with him were of the greatest value to me in opening up new horizons, in encouraging me, and in planning some steps of my future.

During the thirties, the mathematicians and the physicists would eat lunch together at the Faculty Club every day at a round table with a normal capacity of six but with as many as eleven trying to reach their plates. The physicists included Rabi, Quimby, Kusch, Fermi, Lamb, and Townes, also on occasion Szilard or Teller; the mathematicians were Ritt, Koopman, Smith, A. C. Berry and myself. There were also Schilt and Eckert from astronomy and Selig Hecht from biophysics. At these lunches, no holds were barred, no subject was taboo. The only rule was no shop talk. The game was to produce the most froth. In this, Rabi and Koopman were the leaders. Alas, WWII put an end to our daily intercourse, and all concerned were the losers.

Koopman was heavily involved in questions of statistical mechanics and kept in constant touch with both G. D. Birkhoff and John von Neumann, who were both super specialists in the subject. On the occasion of one visit to Princeton in the fall of 1931, Koop learned that von Neumann, using one of Koop's ideas, had given a proof of the mean ergodic theorem, based entirely on the theory of unitary transformations in Hilbert space. Tremendously excited, Koopman passed on this bit of news to Birkhoff, indicating proofs. Presumably, Birkhoff did not comment in detail, but, harnessing all of his powers, succeeded during the next weeks in proving a theorem giving convergence almost everywhere as against von Neumann's weaker convergence in the mean. He immediately set about sending in his proof to the *Proceedings*

Paul Smith

Joseph F. Ritt

(Photograph of Paul Smith courtesy of David Plowden/Columbia College Today. Photograph of Joseph F. Ritt reprinted from *Biographical Memoirs*, Vol. 29, 1956, with permission from the National Academy Press, Washington, DC.)

of the NAS where it was published one year before von Neumann's corresponding result. Let us be more precise. GDB's results were communicated on November 27 and December 1 and appeared in the December 1931 *Proceedings.* Von Neumann's proof came out in the January 1932 issue. This brought on a near collision of our two meteors, and Koopman had to work hard to extricate himself and them—which he did in an explanatory article in the *Proceedings* written by himself and von Neumann (May 1932).

I was puzzled and irritated by Koop's attitude towards Bourbakism. Here was a movement which in my mind had been of such inestimable value in uprooting the stuffy leftovers of nineteenth century mathematics, and he, for his part, was persistently deriding it. I think there was a Dedekind cut in time on who became a Bourbakist and who on the other side was doomed to wander about in the once flourishing oases of the previous century. And I, for one, seemed to fall right at the cut or just to the right of it. As we enthusiasts grew up on our side of the cut, we collected some fifteen to twenty "fascicules" of the great man, read him, and cursed him roundly for his style (to read Bourbaki is like chewing hay), and were grateful. Naturally, the movement was overdone. The second generation of Bourbakists included some educationalists who promptly put the "new math" into the grade schools where there was an overkill. I am reminded of a cocktail party in Rome at which a mother of a fourth grade hopeful came to me and proudly announced, "My son has started studying "insiemistica." I was at first puzzled by what she meant, but pulling the word apart, it became all clear: insieme + mistica, that is, the mystique of sets (oder so etwas)!

On August 1, 1914, my father, who most of his life had been a loyal subject of King George V of England, discovered that he had made a serious mistake in finding himself and his family in Frankfurt, Germany. Within forty-eight hours of the declaration of war (WWI), he was arrested and marched off to a concentration camp in Berlin, called "Ruhleben" (life of peace), where he met hundreds of fellow Britishers who were destined to be his stablemates for the coming months. One of these camp mates was James Chadwick who discovered the neutron in 1932. (Ruhleben was the Berlin racecourse. When war was declared, racing was stopped, the stables were emptied, and the empty race course which was surrounded by its high fence to keep out the nonpaying public was adjudged an ideal place to keep the unlucky Englishmen.)

The fortunes of war determined that in 1918 I found myself in Englewood, N. J. where I was duly enrolled in the excellent public schools. It was there, sometime later, that I came to know from a distance an older upper classman who stood out from his peers. His name was Marshall Harvey Stone. Upon finishing high school, I was admitted next door to Columbia in 1924 as a pre-engineer. It was the dean, Herbert Hawkes, a student of J. W. Gibbs, who called me in one day after my advanced calculus course and pointed out that to him I looked more like a future mathematician than an engineer. That was close to the first time that I realized that one could make a livelihood following our Muse. The next semester, I met my first mathematician full on: J. F. Ritt, in differential equations, and it was a revelation. My decision had by now been made and in my senior year I entered graduate study by taking theory of functions (real and complex variables) with Thomas Scott Fiske.

There were at the time some 150 students registered in a more or less loose way in the graduate program. First year graduate courses had a population of sixty or so and for the first time after four years of living in the all male desert of Columbia College, there were women in the class. This cohort of 150 or so students sifted itself out over the years. Some went into the secondary school system or "ended up" at the Bureau of Standards. At the time, Columbia was producing one or maybe two Ph.D.'s a year. I have been given to understand that in the thirty preceding years there had been circa five woman doctorates. I recount here with reluctance and embarrassment an incident which was communicated to me without intermediaries. F. N. Cole, in giving advice to his successor at Barnard College, told him, "Don't ever employ any woman in your department. They'll give you only trouble".

T. S. Fiske was a kind, courteous, and distinguished person. Extremely handsome with his full mane of silver-white hair, his very ruddy complexion crowned by a sharp nose, and dressed always like the governor of the state rather than as a college professor, he imposed his personality on his class, which followed in awe. However, he had long ago given up his research activities, and it was an open secret that if one was to learn function theory, one had to do it on one's own. I don't remember many ε's appearing on the board and I am ready to swear that he never divided ε by 2 or by n in order to accommodate many clients in a proof. On the complex level, he made us read what he affectionately called "my little book" (*Functions of a Complex Variable*, 97 pp., John S. Wiley, 1907), but it was clear that to learn the subject one had to read Konrad Knopp or Osgood. Some years later, after his retirement, three of us younger instructors were assigned to his office. There we found two very heavy dumbbells (evidently hefting the fledgling AMS was not demanding enough for his young muscles) and, unless I am dreaming, a mounted head of a moose, presumably culled on a hunting expedition in the woods of Maine.

The basis of the graduate program consisted of three courses: real and complex variables, algebra, and projective geometry. Algebra was given by W. Benjamin Fite, a group theorist who right to the end contributed papers on his subject. The text used was Dickson's *Modern Algebra*. Inflicting such a book on students was most certainly not an act of kindness. It was awful. Fite taught the class as if we were reading Xenophon's *Anabasis*. Two pages every lesson, during which he reproduced the proofs on the board line by line as they appeared in the book. If Dickson used i and j as subscripts, the professor never made the mistake of using p and q instead. Fite was an exceptionally kind and sweet man. One almost forgave him his pedagogical deficiencies. My algebraic horizons were opened three years later when I read van der Waerden. I am proud to say that I gave the first course in "modern algebra" at Columbia in the spring of 1938 using this marvelous book. We also used to call it "abstract algebra." In fact, one of my students, Robert Schatten, raised questions with me on the first day as to whether the course was abstract enough for him, who evidently was anxious to get to the heart of the matter without foreplay. Schatten had a very disconcerting habit of calling his shots, sometimes years in advance. He was seldom wrong.

The course in projective geometry was given by a younger man, George Pfeiffer. It was based on Veblen and Young and was a good course. As we all know, that kind of course disappeared from the graduate curriculum of most universities. I gave the course at Columbia the last time it was offered. There was a spirit in the department which encouraged the younger members to broaden themselves by giving courses away from their main track. I took much advantage of this attitude over the years. I remember, in particular, giving the only course ever given in our department on mathematical logic. It was the summer of 1950. The heat was unbearable. All doors and windows were open. Next door, the great Jean Dieudonné was lecturing on group theory. Not lecturing but thundering. Since my class had heard his entire exposition in addition to mine, I offered to let my students take his final examination as well as my own.

The most vibrant mathematician at Columbia, and nationally recognized during the thirties and forties was Joseph F. Ritt. Here was a highly original and introspective thinker who developed his ideas and obtained his problems by reading the opera of the past great: Jacobi, Abel, Liouville. A tremendous worker, beset by poor health, he labored in solitude seldom "rubbing elbows" with contemporaries. His work was in a highly classical spirit, and since he did not need the recent mathematics of the twentieth century, he did not learn it. On many occasions, he questioned me on the theory of measure and integration, but although he seemed interested, he was evidently satisfied with the Riemann integral and more recent advances were nice but not too important. In some cases, he was contemptuous of recent trends. Thus, as a longtime worker using only real or complex numbers, he referred to finite fields as monkey fields.

When I was a young instructor (in the post depression one could remain at this level for six to eight years), I came to be quite intimate with Ritt. In fact, I was for many years his closest colleague. He had forgiven me for having dropped the earlier classical interests to which he had introduced me and to have turned my attention to linear spaces. Around 1941, I showed him the proof I had devised that the only complex normed algebra which is a field is the set of complex numbers. He was thrilled. (Gelfand's paper "Normierte

Ringe" containing this theorem did not reach our library until 1942 due to the German invasion of Russia. Of course, Mazur's earlier announcement of a proof was unknown to me.) This result helped to reconcile him to the power of modern methods. Ritt was a proud man and was much upset as years went on that no prize was awarded to him. In laughing about this misfortune, he would recite to us the epitath that he had composed for his tombstone:

Here at your feet J. F. Ritt lies;
He never won the Bôcher prize.

A principle at Columbia was that after receiving the Ph.D. one had to go off on a fellowship for a year or two before coming back to Columbia to become an instructor at $2700 teaching twelve hours a week, including trigonometry. The standard places to go to receive this coat of varnish were Harvard and Princeton. I applied and obtained a National Research Council Fellowship and was soon on my way to Harvard to study under M. H. Stone. There I met a fellow Fellow, Deane Montgomery, and we used to break up our life of continuous daytime study by meeting in his furniture-free apartment at night sipping beer cross-legged on the floor. The following year, I received an offer from the IAS to be von Neumann's assistant.

One of the perks for being a professor at the Institute was to have an assistant. The work load placed on this person's shoulders varied from ε to $1/\varepsilon$ depending on the professor involved. I went to Oswald Veblen for an indication of what would be expected of me. Veblen quickly, and with a modicum of annoyance, described four categories of activity:

1) Follow JvN's lectures, take notes, complete proofs, prepare mimeograph sheets of them, distribute them to the auditors.

2) Assist in the editing of the *Annals* of which JvN was leading editor. Prepare all accepted manuscripts for the printer. (Give all instructions: Greek, boldface, German, etc. Indicate displayed formulas.)

3) The *Annals* were being printed in the USA for the first time and no longer by Lütke and Wolf in Nazi Germany. The assistant was to go to Baltimore two afternoons a week to teach the printers how to set up subscripts, superscripts, etc.

4) JvN was at the time still writing up his many 100-page papers in German. The assistant was to translate, type up, and prepare these many papers for publication.

Veblen added with firmness that the above were the normal duties of the assistant but it would be fair game to add other duties which could not at the moment be foreseen. (I myself questioned the need of a translator at the time. Von Neumann had been lecturing in most fluent English (modulo some idiosyncrasies: "infinite serious") and seemed more than at ease. I was present at an after-lecture party in Harvard in 1934 where someone mentioned

Lewis Carroll's *The Hunting of the Snark.* Von Neumann and Wiener who stood nearby were set on fire by this spark and began to recite at "il più presto possibile" some 150 lines of the poem. So far as I could tell, the race was a dead heat.)

I went back home in a rather downcast mood. Upon arrival in New York, I found that Columbia had awarded me a Cutting Traveling Fellowship, worth $1800, which allowed me to travel freely to any and all countries and to devote all of my time to my studies. I reluctantly turned down the offer of the Institute and eagerly accepted the traveling fellowship which allowed me to spend nine months in the intimate company of Frederic Riesz in Szeged, Hungary. Here I had lunch (2 hours) and dinner (over 2 1/2 hours) with this kind great genius five times a week. In addition, during Carnival we would meet three nights a week at a hotel where I danced with the local talent until 3 A.M. I was told later that the work load originating from von Neumann's stellar position at the Institute was parceled out to four distinct people. I felt certain that each of the four young people who filled these positions were reasonably tired at the end of the day from their paramathematical activities.

During the critical years shortly after 1950, Columbia was the home of a distinguished group of stars including Claude Chevalley (who was said to have refused admission into his linear algebra course to anyone who had previously studied matrix theory), and Harish-Chandra, who stayed briefly before going to the IAS. Then there were at various times the French visitors: Hadamard, Denjoy, and Brillouin in physics. I remember an evening in the nine-room apartment of Leon Brillouin at Columbus Circle where he had on view ten to fifteen of the most spectacular Modiglianis (three full-sized canvasses per room) that one can imagine. An anecdote on Denjoy is in order. He was giving a series of about six lectures to an audience that started with a substantial number and plunged to a bare three graduate students after four lectures. And these three decided to go on strike claiming that their situation was untenable. Consternation in the department. Finally, the strikers, after much urging, agreed to go back to the lecture hall but on one condition: that Denjoy should cease lecturing in English and switch over to French.

Even earlier, there were several younger colleagues starting brilliant careers and contributing much newer-generation strength: Francis Murray, Ellis Kolchin, and Walter Strodt. It was said of Murray that any course that he taught became, in short order, a course on linear operators in Hilbert space. Kolchin and Strodt developed many ideas launched by J. F. Ritt in his ground-breaking work in algebraic differential equations.

An account of the "early" years at Columbia would not be complete without mentioning those outstanding mathematicians in the New York area who should have been members of our faculty and whose distinction earned them the title of Corresponding Members of our department. I am thinking principally of Jesse Douglas and Emil Post. Douglas' work towards solving the

Plateau problem led to his receiving the most distinguished of the many minimum wage prizes that society dangles before our profession: the Fields Medal. Its presentation at the Oslo Congress in 1936 is still sharp in my mind. Douglas himself was not present (maybe, like Bourbaki, he was too busy working at home on his problems) and Norbert Wiener stood in his stead, radiating personality, as he was listening to the glowing citation and as he was photographed by bevies of Norwegian newsmen. That afternoon a few local newspapers, not quite understanding the last minute change of cast, printed the story of Professor Jesse Douglas accepting the Fields Medal and showed with it the glowing photo of Norbert.

Emil Post was another one of us, although his manner was so soft-spoken and his subject so distant from our interests that no one paid much attention to him. Little did we know that we were in the company of a great person of mathematical logic.

Like other older American institutions of higher learning, Columbia changed from being a mere college to a university at the end of the nineteenth century. The Faculties of Political Science, of Philosophy, and of Pure Science were founded in 1880, 1890, and 1892, respectively. It is not a coincidence that the American Mathematical Society was founded during this same period. The fundamental underlying impulse was much the same. It is of some interest to note that, from the point of organizational structure, both the Society and the University had very much in common during the early years: a direct simplicity, a lack of superstructure, a type of growth that was not induced but to a large extent just happened. Yet each was taking care of those things that mattered. The two organizations were like siblings growing through a glorious adolescence and each one leading a protected existence. The cooperation between the two was close. Meetings of the Society were held in classrooms, members slept in dormitories, dinners were held on campus. The University, for its part, encouraged young students who heard the call of the Muse to take the critical step. Mathematics was a calling. The large broth of graduate students was allowed to simmer on its own. The chosen few surfaced by virtue of their gritty perseverance. The Society, on its side, had its six or seven officers. Miss Hull took care of the office. There were few publications, and three young ladies read proof for these. A library developed by accident through exchanges and was housed on an upper floor, where it would not be in anyone's way. The younger people got to know their brilliant elders, who seemed to enjoy their company.

A few years after the termination of WWII, this relaxed and slow moving laissez faire came to an end for both the Society and for the University, represented in our case by the Mathematics Department. The two siblings put behind their adolescence and became energetic, forward looking, and also aggressive institutions. The Society moved out and settled in Providence on its own real estate. Meetings were transferred from classrooms to hotel grand ballrooms seating a thousand or more. Lucky the person who knew five percent of the attendees. Members slept in four-star hotels at $70 per night. Publications multiplied. All the ills of society were fair game for discussion at Council meetings. The services of an Executor Director to act as a chief executive officer of a large corporation were obtained.

The Columbia department, for its part, underwent a parallel transformation. In the first place, mathematics became a profession, like law or dentistry. The department was awarded its own building, thus protecting its members from being contaminated by a stray philosopher or professor of English. Then graduate admission was strictly supervised. Each year some fifteen or more students were admitted to the Ph.D. program with the expectation that eighty percent of them would get a degree in four or five years. These students received free tuition and a stipend to "live." The teaching of calculus was revolutionized. Instead of having sections of twenty freshmen taught by impoverished graduate students who had "been around" for several years, the young were herded into large classrooms of eighty or one hundred and were lectured at by an expert in automorphic functions who had a platoon of graduate students as assistants. The professors applied to the National Science Foundation for grants which allowed them to lighten their teaching load and exempted them from the drudgery of teaching in summer session. Professors freely boasted of their contract appeal. The administration of the department was being carried out by a staff of five secretaries, some of whom would even type in TeX.

"Run like a country store," you could say of both the Society and the Columbia department some fifty years ago, whereas now they resemble more closely a highly efficient mail-order house. However, it is not necessary for us either to sink into nostalgia for the good old times or to swear by the leading edge of progress toward the future.

Both systems allow the greatest freedom in grappling with mathematics, in following the Muse. And what counts more for us than to consecrate ourselves to that Goddess of which Schiller, had he been a mathematician, would have sung:

$M\alpha\theta\eta$ schöner Götterfunken, Tochter aus Elysium

Dirk Jan Struik was born in Rotterdam and graduated from Leiden University. From 1917 *to* 1924, *he was assistant at the Technical University of Delft and collaborated with J. A. Schouten in his work on tensor analysis. This led to his doctoral thesis,* Grundzüge der mehrdimensionalen Differentialgeometrie, *at Leiden in* 1922 *under W. van der Woude. From* 1924 *to* 1926, *he visited the Universities of Rome and Göttingen with a Rockefeller Fellowship, and from* 1927 *to his retirement in* 1960, *he taught at M.I.T. His main scientific interests have been in differential geometry and the history of mathematics. Among his books are* Einführung in die neueren Methoden der Differentialgeometrie *(with J. A. Schouten),* Yankee Science in the Making, A Concise History of Mathematics, Lectures on Classical Differential Geometry, The Land of Stevin and Huygens, *and* A Source Book in Mathematics 1200–1800.

The MIT Department of Mathematics During Its First Seventy-Five Years: Some Recollections

DIRK J. STRUIK

The Massachusetts Institute of Technology was chartered in 1861 and opened its doors in 1865. At this Boston engineering school the teaching of mathematics, for many years, was directed by John Daniel Runkle, pupil and protegé of Benjamin Peirce of Harvard, first at its Lawrence Scientific School, where he graduated in 1851, then for many years at the Nautical Almanac office in Cambridge. He was the right-hand man of William Barton Rogers, the founder and first president of the Institute, and both men set their stamp on its whole educational policy. When Rogers had to take leave of absence, between 1870 and 1878, Runkle was president, in which function he was able to weather the severe financial crisis of 1873. He introduced several laboratory courses, had women admitted as students, and after 1878 devoted much of his energy to the teaching of mathematics. In this he was first assisted by Dr. William Watson, in charge of descriptive geometry (in the accepted tradition of the French Polytechnique), later by George A. Osborne and after 1884 by Harry Walter Tyler, an MIT graduate in chemistry

Harry W. Tyler

Clarence L. E. Moore

Henry B. Phillips
1941

Frederick S. Woods

(Photographs courtesy of the MIT Museum.)

Philip Franklin
1949

Norman Levinson

Jesse Douglas

Eric Reissner

(Photographs courtesy of the MIT Museum.)

who turned to mathematics and passed through the ranks from assistant to a full professorship in 1893.

Runkle saw the mathematics department strictly as a service department for the instruction of budding engineers, on a par with the language instruction. When he died in 1902, Tyler succeeded him as head of the department, a position he held until 1930. This department was section III of course IX, General Studies, when I joined it in December 1926. I remember Tyler as a greying, very correct, gentleman of middle size, with short beard and mustache, kind but disciplined, with a keen eye for administrative and educational efficiency. He belonged to a newer generation than that of Runkle, had learned some of the modern mathematics obtainable in Europe, having listened to Felix Klein in Göttingen and to Paul Gordan and Max Noether in Erlangen, where in 1889 he received his Ph.D. (his thesis dealt with certain types of determinants). Back at MIT he applied himself mainly to administrative tasks. Known for years as "Secretary of the Faculty", he was active in a number of leading positions, in the American Academy of Arts and Sciences, in the American Association of University Professors (AAUP), even in the Appalachian Mountain Club. But, having tasted a bit of modern mathematics, he was no longer satisfied in keeping his department purely as a service establishment for the teaching of undergraduates. Supported by the energetic president Richard McLaurin, Tyler saw to it that the mathematics department was considerably enlarged and creative scientific work encouraged by judicious appointments, like those of Moore, Phillips, Woods and Hitchcock. He taught for many years a course in the history of science together with his colleague, W. T. Sedgwick, the biologist and public health authority. The *Short History of Science* (1917) by Tyler and Sedgwick was one of the first such books in the English language, republished in a revised edition of 1939. Because Sedgwick had died, Tyler found as co-author another biologist-colleague, Robert P. Bigelow.[1]

Geometry, in its many forms from projective and differential geometry to quaternions and tensors, was popular with this first generation of Tech men engaged in research. First of all, there was C. L. E. Moore, "research advisor for mathematics of course IX." Clarence Lemuel Elisha Moore, Ohio born, with a Ph.D. from Cornell (1904), had traveled for a year in Europe, where he was profoundly influenced by E. Study in Bonn and by C. Segre in Turin—as was Julian Lowell Coolidge at Harvard. From 1904 on he had been on the teaching staff at MIT and had published a number of papers on projective and differential geometry, some in collaboration with others. A tall, lumbering, heavily built man, with poor eyesight, always willing to listen to others and to encourage younger men, he enjoyed with them the results of their studies. He was of particular support to young assistant professor Norbert Wiener,

[1]Incidentally, I had not, as the preface claims, "read the complete manuscript and made suggestions." I only offered suggestions on the mathematics.

who, in the days I came to Tech, had already done fundamental work on Brownian motion and harmonic analysis, Wiener being one of the first in this country to understand the importance of Lebesgue integration also for fields of applied mathematics. Despite these achievements Wiener remained uncertain of himself, being a man of many moods and disturbed by the fact that so far little attention had been paid to his work, especially in the USA. Wiener himself, in his autobiography,[2] remembers Moore as a "tall, slightly awkward, humorous and kindly man, with the human gift of affection and love of mathematics." Moore could not always follow Wiener—for that matter, who could?—after all, he was no expert in the more subtle forms of modern analysis. In my own case he could see exactly what I was doing, had even applied tensors in his research; his admiration for Ricci was such that he had Miss Richardson, his secretary, type out the whole of Ricci's *Lezioni sulla teoria delle superficie*, a rare, lithographed book of 1898—those were the days before Xerox.

A paper Moore wrote, in collaboration with his colleague Phillips, was on linear distance in projective geometry (1912), a paper I liked because it ties in, as Moore showed, with those cases in (imaginary) developables where ds^2 is the square of a linear form. He also published on surfaces in more dimensional space with E. B. Wilson, for a while head of the physics department before he went, in 1922, to the Harvard School of Public Health as statistician.

Moore died in 1931. We lost in him a mentor not easily replaced. His memory is kept alive by an instructorship in his name.

Wiener also pays his respect to Henry Bayard Phillips, a North Carolinian with a Ph.D. from Johns Hopkins (1905), who came to MIT in 1907. A widely read man, productive both in pure mathematics and in its applications, he drew Wiener's attention to the statistical mechanics of Willard Gibbs, which led Wiener to the discovery that the Lebesgue integral can play a role in matters of statistics, such as in Brownian motion. We saw already Phillips' interest in geometry. He wrote several textbooks, the one that always interested me was that small-sized book on differential equations (course M22), because it contained an abundance of pretty little problems in mechanics and physics. Some were a bit of brain teasers and there were instructors (myself included) who had trouble finding the solution. Moral dilemma: Shall we pick the brain of a colleague, perhaps of Phillps himself? Humiliating. Shall we hope that a clever student finds the solution first? Not quite cricket, as the British say. Let's try once more, OK now, and we can face our class with steady eye....—Wiener calls Phillips an individualist, and he certainly had philosophical ideas of his own, ideas I could not always follow, but that is neither here nor there.

[2]N. Wiener, *I am a Mathematician*, MIT Press, Cambridge, MA, 1964.

Another geometer was Frederick S. Woods, easy going yet dedicated, with a devotion to the Klein tradition. Like Tyler and so many others who would build a strong mathematical climate in the USA, like Osgood (for Harvard), Van Vleck (for Wisconsin), White (for Vassar), and Cole (for Columbia), he had joined the Gideon band of young mathematicians who in the 1880s had crossed the Atlantic in order to find in Europe what was not yet to be found in their homeland. Woods received his Ph.D. in 1895 under Klein himself, on a thesis about minimal surfaces in what we now would call Minkowski space $(++-)$. Most of his further work remained in the Klein tradition. Arriving in 1895 at MIT as an assistant professor (this was at "Boston Tech", not at the present monumental establishment across the river in Cambridge, dating from 1916, when McLaurin was president), he met as a colleague Frederick H. Bailey, a Harvard graduate, and began to collaborate with him on a series of textbooks that had a wide circulation. Among them was the two-volume *Course in mathematics*, published first 1907–1909, in which the calculus was taught didactically interlaced with algebra and analytical geometry, thus discarding the traditional boundary between these fields (going back, probably unconsciously, to the Leibnizian origins). In different modifications and reprints these "Woods and Bailey" books have been used for years all over the USA. There even was a French edition, a *Mathématiques générales* (1926).

Woods also published other books of interest, such as *a Non-euclidean geometry* of 1911 and a *Higher Geometry* of 1922, the latter still a very readable introduction to such Kleinean notions as line and pentaspherical coordinates.

Woods succeeded Tyler as head of the department in 1930, and left it to Phillips in 1934. He stayed on as an honorary lecturer. He died in 1950.

Still another geometer, or better geometer-algebraist was Frank L. Hitchcock, a Harvard graduate of 1910, the year in which he joined the department at MIT. Originally a chemist, much of his work was on the applied side; like Phillips he wrote a text on differential equations with nice little problems, but new for use in applied chemistry, with Clark E. Robinson as co-author. A modest gentleman, almost self-effacing, friendly, very helpful to students, a hard worker (I read that he published 200 papers) he may not have expected that his work would be useful in computer programming, but there is indeed a Bairstow–Hitchcock method of finding complex roots to polynomials (paper of 1944 by Hitchcock).

Hitchcock's thesis was on vector functions, and much of his mathematical work was dedicated to quaternions and their offspring. They were popular with this "older" generation at MIT, and not only here. We meet in this subject an old, let us call it, Anglo-Saxon hobby. Born under the famous Dublin bridge, quaternions were welcomed under the Stars and Stripes by Benjamin

Peirce, where they led him to the composition of the *Linear Associative Algebras* (1870), the first original mathematical book written in the USA. Then Gibbs, at Yale, crippled the poor quaternions and got in vectors a better insight into Maxwell's theory, as did Heaviside in England. Gibbs' method was explained by his pupil, E. B. Wilson (whom we already met at MIT in the 1920s) in a book of 1901 widely known as "Gibbs–Wilson". Quaternions etc. continued to fascinate American mathematicians; in 1910 both president and treasurer of the "International Association for promoting the Study of Quaternions" were Americans, the one in Ontario (A. Macfarlane), the other in Illinois (J. B. Shaw). Gibbs' dyadics led to Ricci's tensors, also at MIT; in the MIT *Journal of Mathematics and Physics* we find papers by several authors on this topic. Vector analysis was taught at MIT from a book of 1909 by a man with the dismal (Nantucket) name of Coffin, later replaced by a book by Phillips (1933).

In 1922 the department felt strong enough in its research efforts to publish, with the assistance of the administration and some members of other departments, this *Journal of Mathematics and Physics.* Here Moore, Wiener, Franklin and others could publish their results. It could show the mathematical world at large that MIT had reached a certain confidence in the field of the exact sciences.

Among the papers in the early issues of the journal we find some by Joseph Lipka. Lipka, Polish born, was a student of Edward Kasner at Columbia University, where he received his Ph.D. in 1912. He continued to work in that Kasner specialty of geometrical considerations related to classical dynamics, in particular the so-called natural families of curves in n-space. Lipka traveled to Italy and Levi Civita, represented MIT at the 700th anniversary of Padua University, but died of an operation soon after his return. This was in 1924, and he was no more than forty years of age. Since I came to MIT in 1926, I never met him. But he was remembered mainly through his conducting the mathematical laboratory they had at MIT (M54), reflected in his *Graphical and Mechanical Computation* (1918), often referred to as Lipka's Tables. This was still the time of the slide rule, and other mechanical computers such as harmonic analyzers. Wiener and several members of the electrical engineering department under Vannevar Bush had plenty of new ideas, which eventually led to the electronic computers. But that came later.

I have still to mention Lepine Hall Rice, with poor eyesight that grew worse, which led to his being pensioned off (or so I hope, pensions in academia were not what they are now, at any rate in the leading universities; some came from private foundations like that established by Carnegie). Rice's specialty was determinants. He left his large collection of reprints to me and they remained for years under my care—the MIT library had little use for reprints. Then Providence led me one day, while walking on Belmont Hill, to a grandson of the Thomas Muir who wrote the four-volume *Theory of Determinants* (1906–1923). He was living on the hill and after a question or two I received permission to have the reprints transferred to the family of that great expert on determinants.

Among the other members of the faculty there were men who concentrated on undergraduate teaching, a task taken seriously at MIT ever since Runkle's days. I remember Frederick H. Bailey, co-author of the Woods and Bailey books, Dana P. Bartlett who also taught least squares and had written a text about them, Nathan P. George, George Rutledge, author of papers relating to numerical calculation, and Leonard Macgruder Passano, the most colorful of the team.

Passano was a Baltimore man, and not only the author of some mathematical textbooks, but also of a school text on the history of Maryland and of several essays and plays ("A Family Affair", "Zimri the Kind", etc.). Tall, immaculately dressed, with neatly trimmed beard and spats, he saw himself as a man of the world, which he showed by having a large reproduction of Manet's *Olympia* (or was it Goya's *Maya*?) above his desk in his office. He had been at MIT since 1902. He could be witty; on one occasion when plans were discussed to strengthen the applied side of the department he opposed it, probably wanting also his Euclid bare: "Mathematics, the queen of the sciences, should not become its quean."

2

Among the younger men, men of my age, I found, apart from Wiener (whom I already had met in Göttingen), Samuel D. Zeldin, Raymond D. Douglass and Philip Franklin. Zeldin, born in Russia, had come to the USA sufficiently prepared to obtain his Ph.D. at Clark in 1915. His specialty was continuous groups, but after he joined the MIT staff in 1919 he concentrated more and more on teaching undergraduates, who appreciated his kindness and warmth. Douglass came from the University of Maine, wrote some papers with Rutledge, who also supervised his doctoral thesis (1931), but also was primarily a teacher, somewhat of a drillmaster, but an excellent and popular one at that—he had been in the Navy himself. Teaching remained, as I said, an important task and good teaching counted much in promotion. On the whole, I believe the students were satisfied; among the complaints about poor teaching heard at MIT during the years I do not remember many directed against the mathematics department.

New courses were added (and some courses were dropped, as those on elementary mathematics), such as the one of Wiener on harmonic analysis and the one of my own on differential geometry and tensors. There remained gaps, of course; even Harvard had them. To fill omissions to a certain extent a reading course (M90) was added, where students could study special subjects under the tutelage of a willing professor. The first Ph.D. was conferred on

James E.Taylor (1925), who went to Pittsburgh; the second to William Fitz Cheney (1927), whose thesis was on tensors, supervised by Moore. Cheney sat in on my lectures in the first year and was helpful in correcting my English ("don't try to make jokes in a language you don't fully control"). He was great on mathematical puzzles and was a popular performer at Open House. For many years he headed the department at the University of Connecticut at Storrs. The third Ph.D. was Carl Muckenhaupt—see Wiener's autobiography.

Philip Franklin was Wiener's brother-in-law, having married Norbert's sister Constance, a mathematician in her own right. The two men had met during the war on the Aberdeen proving grounds. Phil was an even-tempered, mild, humorous man, "of almost Mr. Chips proportions", as Dean Harrison said in 1965 at his funeral service—and a mathematician of many parts. His Princeton Ph.D. thesis of 1922 was in the Veblen topology field and a contribution to the four-color problem. Coming to Harvard first and then to MIT he brought a new field to Cambridge. "Franklin", Marshall Stone writes, "gave us [Harvard] our first systematic introduction to topology." In the MIT Journal of 1933–1934 he extended his studies to the six-color problem for one-sided surfaces. He was well versed in many fields of geometry, algebra and analysis. In 1936 he lectured before the American Mathematical Society on transcendental numbers. In the early 30s he published with Moore a set of papers on algebraic Pfaffians. Since Franklin brought topology to MIT in his "analysis situs" form, and Wiener in its "point-set-Lebesgue" form, we see that it came to the Institute through two brothers-in-law.

I have always had the feeling that living in the shadow, so to speak, of his overwhelming brother-in-law, cramped his style. At any rate, he devoted much of his time to the writing of eight excellent textbooks, such as his *Differential Equations for Electrical Engineers* (1931) and *Methods of Advanced Calculus* (1944).

Wiener's activities need not be discussed here at any length, since you can find them described in often fascinating detail in his autobiographical *I am a Mathematician* (1956). A Harvard Ph.D. of 1913 in mathematical logic (he was not yet twenty), he had joined the teaching staff at MIT in 1919 after a stay in Europe on a traveling fellowship. He was not only the most original thinker of our group with the most extended interests and a quick mastery of new topics, but through his many travels was also personally acquainted with many outstanding mathematicians in America and Europe. In his work, abstract mathematics blended with its applications to physics and engineering (later also to medicine), and in turn received much inspiration from workers in this field. In discussions with his student, Claude Shannon, and members of the electrical engineering department he laid the foundations of communication theory, and in discussion on harmonic analysis and the computers of that time, such as Bush's differential analyzers, he paved the way to the modern electronic types of computers, as well as to what he himself later would baptize cybernetics. All of this is in my—and many other people's—recollection connected with the picture of Norbert rambling along the corridors of the Institute, entering offices and labs, while buttonholing colleagues to test his newest ideas or worrying about Hitler or the latest foolishness of the State Department.

With the physicists, and especially with Manuel Sandoval Vallarta, scion of an ancient Conquistadores family of Mexico, who was remarkably well-informed, Wiener used to talk relativity and quanta. This was the time that Vallarta collaborated with the Abbé Georges Lemaître, a pupil of Eddington, on cosmic rays and the expanding universe. I participated in many of those discussions, a result of which was a paper that Wiener and I published in our MIT Journal, a paper in which we tried, using a theory of invariants by E. Cotton, to construct a partial differential equation embracing, by suitable normalization, both relativity and the Schrödinger equation. Rereading it recently, I was pleased by our referring to the then just-published five-dimensional theory of Kaluza-Klein, a theory recently found attractive to astrophysicists.

Wiener also gave a course on the history of philosophy. I never sat in on it, but we had many talks. I always enjoyed the way he had of grasping those elements in the philosophies of the past relevant to understanding the science of the period, or of vital importance to the present (and future) state of science. Hence, his admiration for Leibniz.

His teaching was erratic; many students could not follow him. But for the happy few with mathematical enthusiasm, like Levinson, Paley or Shannon, he was a lasting inspiration.

The electrical engineering department, aware of the role that the theory of probability was playing in telephone and related traffic problems, invited Thornton C. Fry of the Bell Telephone Laboratories in New York to lecture on this subject. The content of these lectures, held during 1926–1927, was highly appreciated; their substance can be found in Fry's book of 1928. When Fry returned to Bell, the mathematics department was asked to take over the course. In a reckless moment I volunteered; it was an adventure since all I knew was what I had picked up in a course at Delft. But it was good fun. At first I was only three lectures ahead of my class, but studying the books of Bertrand, Czuber, Fréchet and Coolidge gave me the chance to supplement Fry. My students shared my enthusiasm and the course became an annual event (M76). Eventually other departments organized their own courses and we could replace Fry with more rigorous material, in this case, Uspensky.

That probability came so late as a regular subject may seem strange today. Equally strange is the fact that there were no seminars or colloquia in the mathematics department. Those who had traveled knew them; the Hilbert seminar at Göttingen had a certain fame. After talking it over with Tyler and

Moore, I discussed the plan for a joint Harvard-Tech seminar with Marshall Stone, then a younger man at Harvard, and Stone was sympathetic.

We ran into a snag. Harvard, at that time, had already for many years a strong department with such leading men as Osgood, Bôcher, Huntington, Coolidge and Birkhoff, with promising younger mathematicians, and with some experience in seminars. But there existed an old resentment due to attempts by Harvard to "take over" Tech, and some of the older men had taken it personally. Moreover, the idea of "parity" did not appeal to some Harvard men. A colloquium, yes, but a Harvard one with MIT people graciously invited to attend. I talked it over with Moore. I still see him with his long body slumped in his chair, visor over his eyes, hands behind his head. "Well, Struik," he grinned, "I know these men. For them the MIT is still the vocational school down the river, and on top, there is still the old resentment. Go ahead, accept the offer, and everything will be straightened out in the long run."

And so it was. For years we had seminars at Harvard and MIT with sometimes excellent lectures and members of both institutions equally welcome. After 1948 Brandeis came in, together with an increasing number of specialized seminars.

Among the lectures I remember with great pleasure was J. A. Schouten from Delft. He came in 1931 and talked about the then new subject of spinors. He had a way of starting with a whole series of definitions—"*entia non sint multiplicanda sine necessitatem*", whispered Norbert into my ears—you recognize Occam's razor. The "necessity", of course, came up soon enough. Another lecturer was Felix Bernstein, in a seminar on probability, in which Eberhard Hopf also participated. This was the time of axiomatization of probability in set theory and of the discovery of ergodic theorems. Hadamard also lectured, although he had difficulty in being admitted to the USA because of his communist sympathies; he had just visited Brazil and was excited about the ferns he had seen.

This was a lively time, and a time in which the mathematics department was greatly strengthened, due to new appointments, more than once from the ranks of excellent graduate students. The whole of MIT was changing with the new Compton administration. Karl T. Compton, who became president in 1930, was an outstanding physicist with a long record of achievements mainly in electronic research. He understood fully that a modern engineering school can only be first grade if it is also a leading school of science, which at that time meant mainly chemistry, physics and mathematics. Thus began the transformation of MIT from a still essentially undergraduate college into a research institute of the first rank, but also maintaining or improving its educational facilities. Through the appointment of Vannevar Bush to the vice presidency Compton obtained the cooperation of this aggressive, administrative and engineering genius. Things were moving, and so was the mathematics department.

Just as the electrical engineering department had been emancipated from the physics (already in 1902), the mathematics department obtained its independence from General Studies in 1933 and changed from IX–C to course XVIII. The graduate courses attracted outstanding students. Among those who eventually joined the faculty we find George Wadsworth, Prescott D. Crout, and Norman Levinson, all with doctorates from MIT. The rising reputation of Norbert Wiener—in 1934 he received, with Marston Morse, the Bôcher prize—was a great attraction for students, and also for established mathematicians to accept appointments, as did William Ted Martin and Robert H. Cameron. Crout and Wadsworth did their best work after the period I am dealing with, Crout in computational research in applied fields, Wadsworth in meteorology and oil geology (using ideas of Wiener). Levinson, who started in electrical engineering, became one of Wiener's most brilliant disciples. His work covers many fields of analysis, in nonlinear equations and in prime number theory. Some of his results appeared in his *Gap and Density Theorems* of 1940.

Another follower of Wiener, equally outstanding, was Raymond Paley, a young Englishman fresh from Hardy and Littlewood. He had long sessions with Norbert, showing "a superb mastery of mathematics as a game"; a result of this collaboration was the influential *Fourier Transforms in the Complex Domain* (1934). Paley was as reckless in sport as in mathematics; he found an early death while skiing in the Canadian Rockies. "If he had not come to an untimely end he would be the mainstay of British mathematics at the present time", wrote Wiener in 1956.

A particularly original mind was that of Claude E. Shannon, who started in 1933 as a student, got his Ph.D. in 1940, became a member of the faculty until he left for the Bell Telephone Laboratories, then returned in 1959 as a permanent member of the faculty. Under Wiener's and Bush's inspiration he wrote his *Mathematical Theory of Communication* (1949), which, with his later work, has made him one of the creators—if not the creator—of information theory, developing his ideas from his observation that switching circuits in automatic telephoning can be based on the algebra of logic. Making communication engineering possible, Shannon's work has turned toward the age of the modern computer, as did that of Bush and others at MIT.

Norbert Wiener was, as we see from all this, pretty much the center of research in the department of those days. From outside, as I said, came William Ted Martin, also influenced by Wiener and turning his research into function theory of more variables. Martin left MIT to head the department at Syracuse, but returned to MIT to head the department of mathematics, where he succeeded Phillips in 1947 as efficient head of the department.

No wonder that Phillips, in his report on the department of 1935, could write that it "is now regarded as one of the strongest in the country, both in teaching and research." In 1933 the visiting committee, pointing out that all principal fields of mathematics were covered except perhaps number theory, concluded, perhaps with some exaggeration, that Cambridge had become "the most inspiring mathematical center in America."

The first woman Ph.D. in mathematics at MIT was Dorothy Weeks; her thesis of 1933 was on coherency matrices. For many years she was on the faculty of Goucher College in Maryland.

Among those who came to the faculty and eventually left were Robert H. Cameron, Jesse Douglas, Eberhard Hopf, and Otto Szász. Cameron, a Cornell Ph.D. (1932) like Martin, with whom he co-authored some papers, was influenced by Wiener and was at MIT from 1935 to 1945, when he left for the University of Minnesota. I remember with pleasure the department picnics he and his lady organized, one or two at Walden Pond. Hopf was a Ph.D. from the University of Berlin (1930), had worked in celestial mechanics, and had come to Harvard, I believe, to study with Birkhoff. Through Wiener's influence he became a member of the MIT department, where he stayed from 1932 to 1936. One well-known result of his stay was the Wiener–Hopf equation, expressing Hopf's ideas on cosmic radiation joined to Wiener's insight into prediction theory. Hopf returned to Germany in the Hitler period, attracted by a good professional offer, not because Nazi philosophy appealed to him or his wife. Needless to say, we did not like his choice.

This is the place to remember Jesse Douglas, nervous, emotional, and a remarkably sensitive mathematician. Like Lipka a Kasner graduate of Columbia, he was at MIT from 1930 to 1936. An analyst with a subtle feeling for its geometrical side, he did his best known work on existence theorems in the problem of Plateau, for which he received the Fields medal in 1936 and the Bôcher prize in 1943. We had long discussions, in which he worried about the papers by his rival Tibor Radó. He was fond of anecdotes, some quite funny, mixed with his own special little prejudices; my experience, he said, is that geometers are as a rule nice fellows, and analysts are nasty. He had his own lifestyle which did not include coming to class on a regular schedule, so that Phillips, who stuck to the Runkle discipline of conscientious teaching, had to let him go, to my and others' regret. He lived mostly on fellowships, but spent the last ten years of his life at CCNY. He died in 1965.

With Otto Szász we come to the refugees who came from Central Europe after 1933, often after a sojourn in England or France. It was again Wiener who, through his multifold connections, took much of the initiative in bringing mathematicians over and finding places for them—after all, he was a Jew and knew anti-Semitism from personal experience. Some of us also did our best, with hospitality or by signing affidavits required by the immigration authorities. Placing was not easy; the depression was not over yet and there was still a good deal of anti-Semitism in American academia. Among those who came and became members of the staff, I remember the fine personality of Otto Szász, a Hungarian versed in the methods of his teacher and friend Lipot Fejér in questions pertaining to Fourier analysis in its widest sense. He left in 1936 for Cincinnati. Equally welcome was Witold Hurewicz, topologist and student of Brouwer in Amsterdam, who enriched his field by the influential concept of homotopy groups. He was the second brilliant mathematician at MIT to lose his life in an accident: Paley died in Canada, Hurewicz died in the Yucatán, on a sightseeing trip after attending an international conference in Mexico City (1956).

Hurewicz became a permanent member of the faculty, and so did Eric Reissner, of the Technische Hochschule in Berlin; he obtained his Ph.D. at MIT and remained there, investigating problems in mechanics and elasticity, often questions of bending and buckling of plates. He thus became a force in building the applied side of the department. C. C. Lin came in somewhat later.

There were several others who came and went, like Antoni Zygmund, exponent (with S. Saks) of the Polish approach to analytic function theory. But I think back with particular pleasure to the lectures of Stefan Bergman on functions of several complex variables, not only because of his enthusiasm, but also because of the plastic way he combined his analytical developments with geometrical illustrations of four-dimensional figures. Felix Bernstein came for a while, then became involved in the mathematics of population genetics, on which he lectured at a conference on probability we had at MIT in December 1933; Eberhard Hopf also presented a paper, with some interesting ideas he had on the relationship between causality and probability. These were the days of the foundation in set theory of probability (Kolmogorov's book in the *Ergebnisse* is of 1933) and of the formulation of ergodic theories, in which Birkhoff and Wiener participated; Bernstein's paper was one indication of how stochastic ideas were penetrating all fields of science, as Fry's book had already shown for engineering. The conference took place at the occasion of the yearly meeting of the American Mathematical Society, then held in Cambridge. (The annual dinner was held at the Walker Memorial at MIT with Julian Lowell Coolidge as toastmaster; he was good at it: "Hi, hi, the gang's all here!") Indeed, we met many very bright but economically very unhappy men in those 1930s—men and women. Emmy Noether also visited MIT; the department should have given her an appointment. I do not know why it did not work out.

So much was going on at MIT that I will not go into more details—only mention some names to revive old memories: Sammy Saslaw, Harold Freeman, Nat Coburn and Norman Ball, Shikao Ikehara and Yak Wing Lee. The awarding of Ph.D. degrees was now an annual event; other talented students had to be satisfied with bachelor's or master's degrees, often going on to complete their studies elsewhere.

Among my own students in differential geometry I like to mention Domina Spencer, now well-known as an author of books on mathematics and engineering with her late husband, Parry Moon of the engineering department, Hsin P. Soh and Alfonso Nápoles Gandara. Soh came to us with rather erratic mathematical knowledge but interesting ideas on relativity. But, with a fellowship we got for him, Franklin, Vallarta and I supporting it, he published a paper in the MIT journal of 1932–1933 in which he outlined a field theory based on a complex Riemannian line element, the real part expressing gravitation, the imaginary part electromagnetism. Soh left us for China at the time of Japan's open aggression, which upset him very much. I wonder what happened to him; attempts to find out have failed. If a colleague in China reads this paper, can he tell me something about Soh's later life?

Nápoles was a Mexican with Aztec blood in his veins, not a Castillian like Vallarta. He was a pupil of Sotero Prieto, who against many odds, had been trying to introduce modern mathematics into Mexico. Nápoles, after his return to Mexico, continued, not without success, Sotero's work at the Universidad Nacional where he became head of the department. He invited me to lecture in the summer of 1934—foreign lectures were still quite a novelty at the time, so that my lectures received a remarkable publicity, even my blue eyes were mentioned. My visits, later repeated, may have helped to increase interest in, and respect for mathematics in Mexico. But the honor of creating a school of creative mathematicians goes to Solomon Lefschetz, who spent many months during the years at the Universidad.

After the excursion to Mexico I spent my sabbatical year, 1934–1935, in the Netherlands, where I collaborated with Schouten on our two-volume book on tensors and their application to Riemannian geometry, and with him visited a symposium on tensors in Moscow.

One other mathematical excursion worth mentioning occurred at that time, that of Wiener to Beijing, then called Peiping, also at the invitation of one of his students, in this case Y. W. Lee. You can read about his adventures in his autobiography. When he returned, he could speak Chinese, and tried it out on Chinese students, who told us that his Chinese was very good. Norbert was always good at languages. Curiously enough, he stayed away from Russian, although his father had been a professor of Slavic languages and a translator of Tolstoy.

We now have arrived at the 1940s and the distinguished role MIT mathematicians have played during the war years. But this part of the story I must leave to others.

Wilfred Kaplan has been associated with the University of Michigan since 1940. After completing his Ph.D. at Harvard in 1939 under the guidance of Hassler Whitney, he spent a year at the College of William and Mary before accepting T. H. Hildebrandt's bid to come to Ann Arbor. His research has concerned the topology of curve families, dynamical systems, and complex function theory. He is the author of influential textbooks on mathematics for engineering students, including Advanced Calculus, Ordinary Differential Equations, and Operational Methods for Linear Systems. *For many years he has played an active role in the AAUP, serving for a time on the national Executive Committee.*

Mathematics at the University of Michigan

WILFRED KAPLAN[1]

Introduction

This article is confined to the story of the Mathematics Department in Ann Arbor. For the period up to 1940 an excellent history appeared in [**2**]. This provides much detail about the professors and curriculum. Because of the availability of this source, the early period will be treated rather concisely.

The Period to the End of World War II

The first hundred years. The University of Michigan traces its beginning back to 1817, when a Catholepistemiad of Michigan was created in Detroit [**4**, Chapter 1]. The primitive conditions, however, prevented realization of the plan until 1837, when regents were appointed for an institution in Ann Arbor. It took four more years before buildings could be erected and five professors appointed. On September 25, 1841 instruction began, with seven students in classes taught by two professors: the Reverend George P. Williams for mathematics and science, the Reverend Joseph Whiting for Greek and Latin.

[1]The present article has been prepared, in accordance with advice from the editors, as a shortened version of an article on file at the departmental office in Ann Arbor, including a list of all faculty from 1841 to 1988.

Alexander Ziwet
(about 1920)

George Y. Rainich
(about 1930)

Raymond L. Wilder
(about 1930)

Theophil H. Hildebrandt
Chairman, 1934–1957
(about 1940)

(Photographs of A. Ziwet and R. L. Wilder courtesy of Michigan Historical Collections and photograph of T. H. Hildebrandt courtesy of University of Michigan News and Information Services.)

By 1854 there were sixty-three freshmen and a comparable number in the three higher classes and Professor Williams was being assisted by professors from other disciplines. The curriculum covered algebra, geometry (Legendre), trigonometry, analytic geometry, calculus.

In 1863 Williams became Professor of Physics and Edward Olney was appointed Professor of Mathematics. Until 1872 Olney and one instructor did the teaching. From 1872 to 1877 the staff gradually rose to five. The curriculum expanded slightly, with encouragement to those who wished to pursue topics such as quaternions, calculus of variations and calculus of finite differences. Olney wrote several textbooks for the courses. By 1887 there were courses on projective geometry and the theory of functions, including elliptic functions.

The University's library had started with 3707 volumes (purchased for $5000), covering many fields. It offered little in mathematics and grew very slowly. A major improvement came in 1881 when a complete set of Crelle's Journal was donated.

From 1888 on the department expanded steadily. Notable additions were Alexander Ziwet and Frank N. Cole (Ph.D. Harvard, 1886), the first Ph.D. in mathematics to join the department. Both were much involved with the New York Mathematical Society, which became the American Mathematical Society (hereafter referred to as the Society) in 1894.

Cole was inspired by Felix Klein, whose seminar in Germany he attended in 1883–1885. In 1885–1886, as a graduate student at Harvard, he lectured on the new geometric function theory. He came to Michigan in 1888 and remained until 1895, when he went to Columbia. His years at Michigan were especially productive, yielding eight papers on group theory and a translation of E. Netto's *Theory of Substitutions.* This work stimulated further important work in the field. While in Ann Arbor, Cole had as student and colleague G. A. Miller, who later had an active role in the Society and a distinguished career in group theory. From 1896 to 1920 Cole was secretary of the Society. Concerning his career one is referred to [**1**], especially pp. 100–103.

Alexander Ziwet was on the publication committee of the Society from 1898 to 1912 and was vice-president in 1903–1904. His career at Michigan lasted from 1888 to 1925. He did much to improve the courses and the library, donating a large personal collection of books. He also left a bequest of about $20,000 to the University. The Ziwet Fund has supported a series of Ziwet lectures by outstanding mathematicians, beginning in 1936. From the obituary in the *American Mathematical Monthly* (vol. 36, 1929, p. 240), we quote: "Professor Ziwet was outstanding as a scholar and teacher. His range of knowledge was not limited to mathematics, especially from the applied point of view, but extended to many sections of pure mathematics, history of mathematics and the humanities. As a linguist he was perhaps unsurpassed by any member of the University faculty.... He was a potent influence in the University, not only for high ideals in connection with engineering education, but also in the promotion of graduate work and research."

Another important figure was James W. Glover (Ph.D. Harvard, 1895), who was in the department from 1895 to 1937. In 1902 he offered the first courses in actuarial mathematics and over the years did much to build a strong program in this area. The interest in insurance mathematics had arisen much earlier: John E. Clark and Charles N. Jones were department members from 1857 to 1859 and 1875 to 1888, respectively, who later had careers with insurance companies.

We also mention Walter B. Ford (Ph.D. Harvard, 1905), who wrote on asymptotic series and summability theory; he was active in the Society, holding various posts; he was in the department from 1900 to 1940, but continued to be active in research until his death in 1971 at the age of 96. He was concerned about the college-level curriculum, wrote several textbooks for it, and was a great supporter of the Mathematical Association of America (MAA), of which he was president in 1927–1928. Early investments in IBM made him very wealthy and he gave generously to many philanthropies, as well as to the Chauvenet Fund of the MAA. In 1973, after his death, his son Clinton B. Ford gave a large sum to the MAA to create the Walter B. Ford Lecture Fund. The obituary (*Amer. Math. Monthly*, vol. 78, 1971, pp. 1094–1097) refers to his high standards for exposition: "A doctoral candidate under his supervision could always expect to prepare at least twenty drafts of his dissertation before its linguistic format would be approved."

Louis C. Karpinski (Ph.D. Strasbourg, 1903) was in the department from 1904 to 1948 and had a distinguished career in history of mathematics. Clyde E. Love joined the department in 1905 and had wide influence through his textbooks.

By 1908 the department had grown to twenty: four professors, two junior professors, five assistant professors, nine instructors. The curriculum included Fourier series and spherical harmonics, ordinary and partial differential equations, theory of substitutions, theory of numbers, theory of invariants, potential theory, courses for teachers.

In 1909 Theophil H. Hildebrandt (Ph.D. Chicago, 1910) joined the department and remained until 1957. A student of E. H. Moore, he did important work in functional analysis and integration theory. For example, in 1923 he gave the first general proof of the principle of uniform boundedness for Banach spaces, before the work of Banach and Steinhaus (1927). In 1928 he published a basic paper on the spectral theory of completely continuous transformations (compact operators) on Banach spaces, completing earlier work of F. Riesz. His pioneering research in these developing areas of analysis is described by Dunford and Schwartz, *Linear Operators, Part I* (*Interscience*, 1958).

Tomlinson Fort was in the department from 1913 to 1917; he was active in the Society for many years. Harry Carver joined the department in 1916 and had a distinguished career in statistics; he was in many ways a pioneer in the development of this field in the U. S., having personally started the *Annals of Mathematical Statistics* and taking a leading part in the founding of the Institute of Mathematical Statistics. He remained until his retirement in 1961.

Beginning in 1901 there was a gradual separation of mathematics instruction for engineering students, with Ziwet in charge. This lasted until 1928, when the engineering department was absorbed in the original Mathematics Department in the college of arts and sciences. It should be remarked that at the University of Michigan, as at many other universities, the question of who should teach mathematics to engineering students has remained a bone of contention over all the years; many Engineering College professors have taught "engineering courses" indistinguishable from mathematics courses.

Around 1920 the curriculum expanded by introduction of courses in applied mathematics: vector analysis, hydrodynamics, elasticity, celestial mechanics; also courses in infinite series and products, divergent series, history of mathematics, graphical methods.

Wooster W. Beman functioned as chairman of the department from 1887 to 1922. He was succeeded by Joseph L. Markley, who held the title for only four years, when Glover became chairman. During Markley's term, there were several important additions to the staff: James A. Shohat, Ruel V. Churchill, Cecil C. Craig, Ben Dushnik. Shohat made important contributions to analysis, including a book on the moment problem written with J. D. Tamarkin; he was in the department from 1924 to 1930. The other three remained in the department until retirement. Churchill did much for the applied mathematics program and had wide influence through his books on applied analysis. Craig did important work in statistics. Dushnik was active in set theory.

Glover also brought in some new talent: George Y. Rainich and Raymond L. Wilder in 1926; Walter O. Menge in 1925; William L. Ayres and Arthur H. Copeland in 1929. Rainich, in relativity theory, and Wilder, in topology, did much to strengthen the teaching program and research, especially through seminars. Ayres was also an outstanding topologist and was active in the Society; he left the department in 1941. Menge strengthened the actuarial program; he remained until 1937. Copeland made important contributions to probability theory.

In many ways Hildebrandt, Rainich and Wilder brought the department to a higher level of breadth and seriousness. Each had appreciation for mathematics far beyond his special field and encouraged students and younger staff in all fields.

Most of Rainich's papers and his greatest achievements were on the theory of relativity. In a series of papers in the 1920s he showed that the mathematics of the general theory which Einstein had made to supply a model for gravitation, also supplied one for electromagnetism. Rainich ran an "orientation seminar" for advanced undergraduate and beginning graduate students, covering a broad spectrum of topics; he had a remarkable talent for building enthusiasm of the young students and encouraging them to make careers in mathematics. He and his wife, Sophie, often entertained new and old members of the department at their home and thereby did much to bring the new ones into the life of the department. As émigrés from Russia they brought cultural breadth to university life.

Wilder was a topologist of first order, a product of the R. L. Moore school and (as seemed to follow axiomatically) a superb teacher, using the Socratic method to let the students do the discovering. He had twenty-five Ph.D. students. Wilder was a pioneer in the development of the topology of manifolds. By methods of algebraic topology, he extended to higher dimensions many of the results of set-theoretic topology in the plane and 3-space. Some of his best achievements are found in his AMS Colloquium Publication *Topology of Manifolds* (vol. 32, 1949). He had very broad interests in topology and hence could recognize new talent of great variety. He also promoted interest in logic and foundations through a very popular course. He was president of both the Society (1955–1956) and the MAA (1965–1966).

Hildebrandt did much to encourage work in the rapidly developing area of functional analysis. He was very active in the Society and in the MAA, serving as president of the Society in 1945–1946. In 1929 he was the first recipient of the Chauvenet Prize, given for a 1926 paper on "The Borel theorem and its generalizations." In recognition of his leadership as chairman over twenty-three years, the T. H. Hildebrandt Research Instructorships (later Assistant Professorships) were introduced in 1962. He also had a great interest in music, acquiring a degree in that field in 1912 with the organ as specialty. A testimonial to him after his death in 1980 stated: "He was more than an outstanding scientist and enthusiastic expositor of mathematics; he was a leader who took a deep interest in the personal as well as the mathematical growth of his students and colleagues."[2]

In 1934 Hildebrandt became chairman and further appointments were made: Edwin W. Miller in 1934, specializing in set theory—he died of a

[2]The testimonial is from the minutes of a meeting in Fall 1981 of the faculty of the College of Literature, Science and the Arts of the University of Michigan. The memorial was drafted by George E. Hay and Cecil J. Nesbitt.

heart attack in 1942; Paul S. Dwyer in 1935, working in statistics; Sumner B. Myers in 1936, in differential geometry; Robert M. Thrall and Cecil J. Nesbitt, algebraists, and Robert C. F. Bartels in applied mathematics, in 1937–1938; Herman H. Goldstine in 1939, in functional analysis. The last five named had noteworthy career changes: Myers turned to functional analysis, Thrall to operations research, Nesbitt to actuarial mathematics, Bartels in 1967 became the first director of the University's computing center, Goldstine went on leave in 1941 but did not return, having been drawn into basic research on digital computing with John von Neumann (see his article "A Brief History of the Computer," which has appeared in Part I of *A Century of Mathematics in America*, American Mathematical Society, 1988, pp. 311–322).

Over the years 1922–1941 courses were steadily added, so that at the close of the period all the main branches of mathematics were covered, with a fair number of courses at the graduate level. The department had its first Ph.D. in 1911: W. O. Mendenhall, who wrote on divergent series under the guidance of Ford. By 1922, eleven doctor's degrees had been granted; by 1941, ninety. Among the recipients were Ralph S. Phillips and Charles E. Rickart, both students of Hildebrandt in functional analysis.

As a fitting climax at the end of the first 100 years, the department sponsored a two-week Topology Congress in June 1940, with Wilder and Ayres as organizers. The speakers included S. Eilenberg (who then joined the department), E. Van Kampen, S. Lefschetz, H. Whitney, S. Mac Lane, C. Chevalley.

The war years. As elsewhere, World War II had a devastating effect on the University and, in particular, on the mathematics program. Enrollments diminished and some faculty took leaves for military research. There were several military training programs on campus, such as the Air Force Meteorology Program and the Navy's V-12 Program.

The department did add a few professors at the time: Edward F. Beckenbach, in analysis, who remained only two years; Wilfred Kaplan, who had come for the Topology Congress; George E. Hay, in applied mathematics, who later became chairman (1957–1967); Erich Rothe, in functional analysis; the topologists Samuel Eilenberg and Norman E. Steenrod.

There were also several Ph.D. students finishing up, who helped to sustain interest in research. Among these were L. J. Savage, who went on to a career in statistics, and S. Kaplan in topology.

One unusual by-product of the war was a seminar on meteorology, bringing together several professors including G. Y. Rainich, the physicist G. Uhlenbeck, the geologist R. L. Belknap, the aeronautical engineer A. Kuethe. A motivation was to work on a topic that might have practical applications and help the war effort.

Those who were on leave for military research and those who joined the department after the war, having done such research, gained breadth by the experience and their subsequent research and teaching showed a better understanding of such topics as control systems, operations research, applied statistics, communication networks.

Immediately following the war the enrollments shot up and soon the returning GIs appeared, bringing a highly motivated group of students.

Research discussion groups. There were many formal and informal gatherings to discuss research. A Mathematics Club meeting monthly has existed for about 100 years; according to Jones [**3**, p. 9], this began prior to 1891 in Prof. Ziwet's parlor. In [**1**, p. 50], Cole in 1891 referred to a "Mathematical Society of the University of Michigan." In [**6**, p. 3], Wilder tells of a secret mathematics club of about twelve members, including professors from physics and philosophy, meeting during the period 1927–1934. The writer also recalls a similar club organized by S. Eilenberg about 1940, called "The Gauss Group."

Eventually the secrecy was abandoned. In the 1940s regular colloquia were held and research seminars arose in ever-increasing numbers.

THE POSTWAR YEARS

The professional staff and research activities. Under Wilder's leadership, the University of Michigan quickly grew to be a major center of topology, with such leaders as Eilenberg,[3] Steenrod, Bott and Samelson. There were many Ph.D.'s in the field, including S. Smale in 1957 (winner of a Fields medal in 1966). In real and functional analysis Hildebrandt and Myers provided strength; they were followed by E. H. Rothe, L. Cesari, P. Halmos and many others. Complex analysis took on great vigor following a two-week international conference in 1953; this was the beginning of an important "Finnish connection," involving many visits of faculty to and from Finnish universities—F. Gehring, who joined the department in 1955, became a leader in this enterprise. Number theory was fostered by W. J. LeVeque, D. J. Lewis and others. Applied mathematics had a strong old tradition in the department. Statistics continued as a strong interest, but a separate department was formed in 1969. Logic was promoted by Wilder through a very popular course and book on foundations; the program was sustained by R. Lyndon and others. A small group, including P. Jones and C. Brumfiel, ran a program for the training of secondary-school teachers of

[3] Major joint work of Eilenberg and Mac Lane was a by-product of Ziwet lectures given by Mac Lane in 1941. During one of these lectures (on group extensions) Eilenberg suddenly left the room! The audience wondered whether something was wrong, but later learned that he had just then realized the important connection between group extensions and topology. In the following days Eilenberg and Mac Lane were often found conferring intensely.

mathematics. In algebra and algebraic geometry strength gradually developed, with R. Brauer in the department from 1948 to 1952. Actuarial mathematics had an ancient tradition, created by Glover; this was continued by Nesbitt, A. L. Mayerson, C. H. Fischer and D. A. Jones. For some years Michigan's program in actuarial mathematics was generally considered the strongest in the country.

The instructional program. The table below gives basic information on the growth of the department and of mathematics instruction up to 1985. In column 2 the count is made at the beginning of the fall term and includes visitors but excludes regular staff on leave. The count is affected by the creation of a Computer Science Department in 1965 (later, in 1983, attached to the Electrical Engineering Department) and of a separate Statistics Department in 1969. For column 5 comparable figures are not available for 1940–1960. Although large lecture sections have been used for some second- and third-year courses, the freshman calculus course has been taught only in small sections, generally by teaching assistants, with some coordination by professors.

Growth of the Department, 1940–1985

1	2	3	4	5	6
Year	Departmental Roll		Under-graduate Majors	Graduate Students	Ph.D.s Granted in Previous 5 years
	Staff	Teaching Fellows			
1940	35	8	53		29
1945	35	11	32		21
1950	48	20	80		24
1955	61	34	56		56
1960	63	71	165		43
1965	84	116	290	303	66
1970	65	112	281	250	105
1975	64	80	114	165	101
1980	64	81	74	123	63
1985	66	92	111	131	61

The scope of the undergraduate program has broadened over the years to reflect increasing use of mathematics in engineering and the sciences, especially the social sciences, and the revolution in computing. The following are typical: courses in advanced mathematics for engineers, linear programming, algorithms, operations research, numerical analysis at various levels, mathematics for the social sciences. The concern for K–12 mathematics education has led to expanded offerings in teaching of mathematics, including summer institutes and in-service programs for teachers.

In order to meet the needs of gifted students, two special sequences of courses were introduced in the post-Sputnik era: one proceeding somewhat more rapidly through calculus and differential equations; the other an honors sequence over the first three years. The enrollments in these sequences in 1970 and in 1985 were: rapid sequence: 1970, 218 and 1985, 196: honors sequence: 1970, 73 and 1985, 20. In 1970, 152 students graduated with majors in mathematics; in 1986, 64 did so.

A general master's degree has been offered for many years. Recently specialized programs were introduced in applied mathematics, secondary mathematics education and scientific computing. The number of degrees granted in 1970 was 101 and in 1985 was 24.

The actuarial program produced about 400 graduates from 1903 to 1940, about 300 from 1940 to 1960 and about 200 in each succeeding decade. Among these is the distinguished mathematician T. N. E. Greville, who received a Ph.D. in 1933. (See a forthcoming history of the actuarial profession in North America, titled *Our Yesterdays*, by Jack Moorhead.)

The Ph.D. program has been a major interest of the department. Since 1940 it has been supervised by a committee which has considered entrance requirements, progress of candidates and evaluation of dissertations.

Among those receiving the Ph.D. at Michigan are many who have gone on to successful careers in mathematics. These include W. Feit, 1955, M. Jerison, 1950, D. J. Lewis, 1950, J. R. Munkres, 1956, R. Phillips, 1939, F. Raymond, 1958, S. Smale, 1957, L. J. Savage, 1941, J. R. Schoenfield, 1953, E. H. Spanier, 1947, F. L. Spitzer, 1957.

Since 1960, the Sumner B. Myers Prize has been awarded to those whose dissertations have been deemed outstanding.

Other aspects of department life. Various endowed lecture series have brought outstanding mathematicians to the department for a week or more. The oldest of these is that named after Alexander Ziwet; the first lecturer was Edouard Čech in 1936 and the series has provided twenty-six visits up to 1986. The series named after George Y. Rainich has had three speakers: Lipman Bers in 1983, Michael H. Freedman in 1986, Richard M. Schoen in 1988.

During the academic year there has long been a tradition of a weekly colloquium speaker, very often a visitor. In addition fifteen or more specialized seminars have flourished, from beginning graduate to advanced research levels.

The Mathematics Club has long preserved special customs. The meetings are held in the evenings and are also social occasions, with refreshments served. The talks are expository, aimed at a broad audience, and are often followed by vigorous discussions. Each meeting opens with a reading of the minutes of the previous meeting. Then come unannounced "three-minute talks," presented by staff or students, on some novel ideas found in research or teaching. Finally (and this may be an hour later) the announced speaker is allowed to take over. The informality and spirit of fun at these evenings have done much to build *esprit de corps.*

In 1952 the *Michigan Mathematical Journal* was initiated, under the leadership of Rainich. He became an early fan of "desktop publishing" when he discovered that the new departmental typewriter could justify margins; what better way to exploit this than in a new mathematics journal? (The machine was very primitive by modern standards, and the result was unattractive. A better printing process was soon found which did not, however, justify margins.) From 1954 to 1975 George Piranian was editor; his high standards and dedicated labor put the journal on firm ground. He was succeeded by Peter Duren (1976–1977) and by James Kister and Carl Pearcy with some alternation in the following years.

In 1964 *Mathematical Reviews* moved from Providence to Ann Arbor. Its presence has brought other mathematicians to the area and provided a lively association with the department.

The department has generally avoided being embroiled in political matters. However, one member of the faculty, H. Chandler Davis, was a victim of the "McCarthy period" of the 1950s. He cited the First Amendment to the Constitution as a basis for refusing to testify to a congressional committee investigating communism. He was dismissed from the University in 1954 by action of the regents. There was widespread faculty criticism of this step. A later investigation by AAUP led to a censuring of the University. Davis has described the events of this period in his article "The Purge", which has appeared in Part I of *A Century of Mathematics in America* (American Mathematical Society, 1988), pp. 413–428.

REFERENCES

1. Raymond C. Archibald, "A Semicentennial History of the American Mathematical Society," 1888–1938, *American Mathematical Society Semicentennial Publications*, Vol. I, American Mathematical Society, New York, 1938.

2. John W. Bradshaw, James W. Glover, and Harry C. Carver, *The Department of Mathematics in The University of Michigan—An Encyclopedic Survey, Part IV*, pp. 644–657, The University of Michigan Press, Ann Arbor, 1944.

3. Philip S. Jones, *The Mathematics Department: 1944–1974*, unpublished manuscript on file in University of Michigan Mathematics Department, Ann Arbor.

4. Howard H. Peckham, *The Making of the University of Michigan*, 1817–1967, The University of Michigan Press, Ann Arbor, 1967.

5. Robert M. Thrall, "Some Recent Developments in Mathematics at the University of Michigan," in *Research: Definitions and Reflections, A Sesquicentennial Publication of the University of Michigan Press*, Ann Arbor, 1967.

6. Raymond L. Wilder, transcript of an oral presentation on July 24, 1976, revised in the summer of 1977, reprinted in this volume, pp. 191–204.

Raymond L. Wilder (1896–1982) received a master's degree in actuarial mathematics from Brown University in 1921, then went to the University of Texas intending to complete his actuarial training. Instead he became a student of R. L. Moore and received a Ph.D. in mathematics in 1923. He served on the faculty of the University of Michigan from 1926 until his retirement in 1967. He was President of both the AMS and the MAA, an AMS Colloquium Lecturer and Gibbs Lecturer, a recipient of the MAA Distinguished Service Award, and a member of the National Academy of Sciences. Among his books are Topology of Manifolds *and* Introduction to the Foundations of Mathematics. *His reminiscences, recorded in 1976, are published here for the first time.*

Reminiscences of Mathematics at Michigan

RAYMOND L. WILDER

This is Raymond L. Wilder, Professor Emeritus of Mathematics, speaking on July 24, 1976. At the request of Professor Phillip S. Jones and also of Professor Allen Shields, Chairman of the Department of Mathematics at the University of Michigan, I am making an informal recording of my impressions of my years of active teaching here at the University. I came here in the fall of 1926. That spring, I believe Professor James W. Glover became chairman of the department and (according to the information which John W. Bradshaw gives in his history of the department) "immediately set himself to a task of revivifying the department". The curriculum at that time was of a fairly classical type. It gave a set of courses through the advanced calculus, and I believe some Fourier series. These courses in advanced analysis were given by Professor W. B. Ford. Applied courses in geometry, projective geometry and synthetic geometry I believe were given by Bradshaw. The history of mathematics was represented by Louis Karpinski. All in all it was a very good curriculum, representative of the time. However, it did need modernization and this is one of the first things that both G. Y. Rainich and I set out to undertake when we came here.

It might be interesting to point out Professor Glover's method of going about getting new members of the department. He evidently made dittoed copies of a flyer that he sent around to those he considered promising young mathematicians, inviting them to respond if they felt they might be interested in a position at Michigan. In my own case this flyer came to me while I was at Ohio State University in Columbus, Ohio. Apparently it was thrown on the porch by the mailman and picked up by my oldest daughter who was around two years of age at the time, and she tore it up into small pieces. Later on my wife came out on the porch, found the pieces and put them together again and when I discovered what it was, I did write to Professor Glover. This is how close I came to never coming to the University of Michigan! I am sure that if I had not responded, Professor Glover would not have taken any further action in my case. The result of his research was to bring here Professors G. Y. Rainich, whom we informally called Yuri, and James Nyswander, as well as myself.

I should have mentioned that two prominent people who were at the University at that time, namely T. H. Hildebrandt and Professor Alexander Ziwet, were on the engineering side. At that time, the mathematics department in the engineering college was separate from the L.S.A. department. Hildebrandt was a student of E. H. Moore and showed great promise in real analysis. Alexander Ziwet was not a research man as I understand it, but he was active in the affairs of the American Mathematical Society and saw to it that the library received the foundation of a good collection of mathematical journals and treatises. I also should have mentioned that the department, under the stimulation of Professor Glover and his assistant Harry C. Carver, built up an actuarial program as well as a statistical program. In 1926–1927 I believe Professor Wicksell from Sweden was a visiting professor here in statistics.

After consulting the early catalogs, I find that Professor Rainich introduced courses in differential geometry and relativity in 1926 and I myself introduced a course in analysis situs (the term, originally introduced by Gauss, by which one indicated the subject of topology). I notice in the 1927–1928 catalog that Rainich gave a course in quadratic forms and quadratic numbers. Nyswander gave a course in algebraic theory, Hopkins was giving a course in celestial mechanics of the classical type, and Karpinski a course in the theory of numbers. In the 1928–1929 catalog I notice that I introduced two courses in the foundations of mathematics and Rainich was giving a course in continuous groups. I apparently was also running a seminar in analysis situs, having at that time acquired enough students to justify holding such a seminar.

So far as I can determine it had not been the policy of the department to hold seminars in addition to the regular courses, with exception that Professor Wicksell evidently gave a seminar in statistics during the year that he spent here. In the year 1928–1929, in addition to the seminar which I was giving, there was a seminar on functions of a complex variable given by Rainich, and one in differential equations presumably given by Nyswander; also a seminar the second semester in differential geometry given by Rainich. From then on, as I recall it, the custom of giving seminars became quite common. I might note that although these first seminars apparently received credit and were obviously, or presumably, counted in a man's teaching load, as time went on the number of seminars increased, no credit was given, and also the time given to such seminars was not normally included in a man's teaching load. The teaching load in those days, I think, was around twelve hours. It might vary from eleven to twelve, depending on the number of hours of credit given to a course.

In 1929, two new research men were brought in, namely, Arthur H. Copeland, a Harvard Ph.D., and William L. Ayres, a Pennsylvania Ph.D. Ayres was a topologist and Copeland had apparently specialized in Boolean algebras and foundations of probability. At the end of his history of the department, Professor Bradshaw notes that the number of doctorates which had been given up to 1922 was only eleven, but that in the following eighteen years there were seventy-four doctorates given. That brings us up to 1940. He makes a statement, "increased interest and activity in mathematical research on the part of members of the staff have naturally accompanied this growth", referring to the growth that had occurred during that period to 1940. I don't want to leave the impression, however, that the interests of the department became solely devoted to research. I think it fair to say that all three of us who came in 1926, as well as the later additions in 1929, were generally good teachers, and Rainich in particular was very much interested in the development of the students here at Michigan and gave an unusually large amount of his time to conferring with students. However, we realized that mathematics was not a static thing; it was a growing thing, and in order for the department to take its place among the foremost departments of the country, it was necessary to build up the number of courses in modern mathematics, as well as to keep up interest in what was going on in the journals and in mathematical research generally.

One thing that I must speak of which is not recorded anywhere (certainly in the department records) and which I think had a great deal to do with building up mathematics here at Michigan, was the formation in 1927, a year after we came here, of a Research Club by Rainich and myself. We felt that the Department Club which met monthly in the evening was not accomplishing very much in the development of interests in research. This small club that we founded came to be called "The Small C" as distinguished from the large club, the one that met monthly. However, because we wanted to include only people who were active in research, we did keep it secret and this was perhaps not a good feature of it. It was our practice to meet at one of the members' homes every Tuesday evening. We had a portable blackboard which was taken care of by Professor Ben Dushnik. We had an hour's scientific paper, normally on research being done by a member of the club, sometimes on a mathematical result of great importance which we felt that the members should know about. I don't recall the exact composition of the Small C when it started; I know Rainich and I and also, I believe, Professors Denton (whom Professor Bradshaw mentions in his writing), Dushnik, Donat Kazarinoff, and Shohat were members. In all there were, I believe, eight members of the mathematics department, and one member of the philosophy department, namely, Professor Harold Langford whose specialty was mathematical logic, and three members of the physics department, Professors Otto Laporte, George Uhlenbeck, and Samuel Goudsmit. Professor Rainich took the responsibility of sending out notices of where the meetings were to be held on Tuesday evening. I have endeavored to find any records which he may have left of these meetings, but so far as I can tell, they were all destroyed.

It was our custom whenever a visiting mathematician or physicist of note came to the University to give a talk, to invite him to talk to the Small C, and he was unofficially made a member at that time. I recall now two doctoral students who were in the Small C in that early period, namely L. W. Cohen, who later became head of the Mathematics Panel of the National Science Foundation, and also Edwin Miller, who was very active in mathematical research until his untimely death during the war period. I recall also that Professor T. H. Hildebrandt was made a member a few years after the formation of the club.

In 1934 Professor Hildebrandt was made chairman of the department. This perhaps created a situation which ultimately we felt was not too healthy for the status of the Small C. Since he was a member of it, and since the existence of the Small C inevitably became known to members of the department who were not engaged in research, this led to a general feeling on the part of the latter that the Small C was a political organization and that department affairs were being settled unofficially in its meetings. Now, it is true that during the refreshment period which followed the paper at a meeting, there was some discussion of possible new members of the department, as well as of things that were going on in the department; but so far as settling anything in regard to the department was concerned, the Small C certainly did not do this. By the time Hildebrandt became a member, the Executive Committee system had been introduced in the department. The Executive Committee was composed of five members, in addition to the chairman, consisting of representatives of the graduate division of the department, the Literary College, the engineering side of the department (which had now been combined with the L.S.A. department), and a member-at-large who had a one-year appointment. It was in the Executive Committee that new appointments were made and policies discussed. The only influence that the Small C could have had on this was that inevitably, in addition to the chairman, there would be members of the Executive Committee in the Small C, and anything that was discussed in the Small C might presumably influence the opinions expressed in official meetings of the Executive Committee. However, I believe it was not until 1947 that we agreed that the Small C should be disbanded. We had become well aware of criticisms being made by non-members, but more important, the department by this time had acquired enough new research people that it was impossible to get them all into the Small C and continue our informal way of meeting at one another's houses. So we felt that we should disband and promote as well as we could the introduction of a weekly colloquium to be held in the afternoon by the mathematics department, it being understood that this colloquium should be devoted to research papers of a current nature. The only one who objected to disbanding the Small C that I remember was one of the founders, namely, Professor Rainich. But even he could understand the impracticality of continuing the activities of the group.

I should say something about the effect of World War II on the mathematics department. Of course there was a greatly increased demand for courses during the war, particularly because of the participation by the University in the meteorological program of the Air Force. I recall that we used to have large mathematics classes in the Law Building, and these big classes were cut up into sections later to be handled by instructors and teaching fellows. Periodically the Air Force sent examinations to be held and these were conducted in the large auditorium in the Rackham Building. The problem of increased staff was met by bringing in people from other departments who were mathematically competent, and in some cases using people such as faculty wives who had received master's degrees in mathematics before they were married. I remember that Professor Langford, whom I mentioned in connection with the Small C, was one of those who taught courses in the department. (I suppose that there wasn't much demand for philosophy courses during that time, so that it was easy to secure his services.) At any rate, the department did manage to go through the war years without too much great suffering on the part of the staff, although the increased teaching load was undoubtedly a factor holding back research to some extent.

However, the effects of the war and its aftermath were not confined to these matters. There was, perhaps, a much greater impact made by the introduction very soon after the war, in the later 1940s, of the system of grants for research by the various government agencies. I believe the Office for Naval Research was one of the first of these, and of course later, in addition to the Army, Navy, and Air Force, the National Science Foundation was formed and a system of grants instituted by this agency. I can recall that on the Executive Committee there was considerable discussion about the effect that these grants were going to have. We were particularly worried that recipients of grants would be taken from their teaching, since faculty members, in addition to sabbatical leaves, would be able to take extra leaves because of their grants. It is not easy to oversee the research of a student who is in one place and whose thesis adviser is somewhere else. However, as the years went by I think it was generally conceded that the system of grants was beneficial, especially as student grants ultimately became available. It took a good deal of adjusting and as of now, 1976, government grants seem to be a fixed feature of the university scene. Basing my philosophy on the old "if you can't lick 'em, join 'em", I myself have had grants and certainly these have sometimes made possible things which I couldn't otherwise have done. In particular, I had a grant early in the era of grant disposals, in the year 1949–1950, when I went out to California Tech and wrote the first version of my book, *Introduction to the Foundations of Mathematics*, as well as doing research in topology. So I am not of the opinion that the grant system was an entirely bad influence on university research and development.

There were also new areas of mathematics which owed their stimulation, possibly their existence, to the effects of the war. I remember that both Professors Thrall and Copeland were interested in the new mathematics that was being created in the theory of games and mathematics for the social sciences and, of course, the introduction of the electronic computers was greatly accelerated by the war. If I had the time to do so I could probably take the catalogs and note the evolution in new courses and so on that went on. In my own field of topology there occurred the introduction of courses in algebraic topology and later in differential topology.

Another factor which I believe had a very beneficial influence on the evolution of the department was the Ziwet lectures. These were founded as a result of a bequest to the college by Professor Ziwet in 1929. The first Ziwet lectures were given in 1936 by Professor Edouard Cech. Professor Cech was a Czechoslovakian topologist who was responsible for the so-called Cech homology theory and was also known for other works in the field. He lectured for a two-week period, setting the pattern for later Ziwet lecturers. The later Ziwet lectures were given by such prominent mathematicians as Professor John von Neumann, Saunders Mac Lane, Claude Chevalley, Henry Whitehead, and others whose names I don't recall at this particular time. I think we had one or two lecturers a year until the war started; and afterwards, at intervals of four or five years. I think these lecturers had a very beneficial influence on the department because the lecturers would mingle both professionally and socially with members of the department during their visits, so that they really had quite an influence over the long range. I might also say something about the emergence of the *Michigan Mathematical Journal*, which is now one of the best mathematical journals publishing research articles. During the late 1920s, a committee was appointed by Professor Glover, consisting of Rainich as chairman, and Harry Carver and myself, to look into the possibility of establishing such a journal. We turned in a report to the chairman, and I believe that the idea of financing such a journal was put in the alumni magazine, along with some other worthwhile projects, as something that might attract some alumnus or other. However, nothing came of this, and I believe that after the war when the journal was really established, we looked for this report that we had gotten out earlier and couldn't find it. (As a matter of fact, at that time we were unable to find any of the department records that accumulated during Glover's administration.) There is no question, however, that the establishing of the journal has enhanced the reputation of the Michigan mathematics department and that it has justified whatever it has cost to run such a journal.

I think I should say a few words about the policy concerning the way in which courses were assigned instructors. When I came here in 1926, I recall that, as I think I mentioned before, Professor Bradshaw was teaching the geometry courses and that Professor Ford was teaching the courses in the classical analysis. The policy seemed to be that whoever represented a field was to teach the courses in his field. Now before I came here I had taught courses in such subjects as Fourier series. I had gone to considerable trouble to set up courses of this type at Ohio State, and I remember that I was rather taken aback when I found that I could not teach such courses here at Michigan. As a matter of fact, I found myself teaching courses in mathematics of finance (because of my previous training in actuarial mathematics), some courses in elementary algebra and trigonometry, and graduate courses and seminars. This went on, as I recall it, for quite a few years. This pattern may have been a hangover from the olden days; I don't know how widespread it was in American universities. Staffs were not large and presumably there might not be more than one man in a given field. I recall that at the University of Texas, R. L. Moore made it a policy not to let anyone teach the courses in his field of point set topology. As a matter of fact, if a student who had earned his degree under Moore didn't go on to another institution he just stayed at Texas and had to teach other kinds of courses. That was a policy that Moore had established for himself there. So the pattern may have been quite general. However, at the University of Michigan there has been clearly a gradual weaning away from this idea, particularly taking advantage of the fact that the staff increased so much in size over the years. It was no longer considered, after a number of years, that a man who belonged to a field which was already represented here could not be hired. For instance, I had been here only three years when W. L. Ayres was given an assistant professorship in 1929. This was at my request. However, it was ten years later, I believe, before I brought in another topologist, namely, Sammy Eilenberg, who came over from Poland just as Hitler was about to strike that country. This was partly a result of wanting to save a life of a person, and at the same time to build up the department here. Eilenberg came here as my student and the Graduate School accepted him on that basis, although there was some opposition from Professor Peter Field who was at the time on the graduate board and felt maybe I was bringing in Eilenberg as a new member of the faculty, which I did not have in mind at that time. However, since the war affected the United States soon after, and, as I've already mentioned before, teachers were in demand, it was only natural that Eilenberg (who knew English very well) was given courses to teach, and then he ultimately became a regular member of the staff. We also brought in a former student of mine, not one of my doctorates, but a man who had done his first research under my direction here at Michigan, namely, Norman Steenrod. I don't recall the year he came, probably around 1947. For a little while, then, we had four topologists in the department; viz., Ayres, Eilenberg, Steenrod, and me. Ayres left in 1941 to accept the mathematics chairmanship at Purdue, leaving three of us. However, there was no question about the teaching of courses. The courses in topology were passed around one to another, according to each individual's desires and what he felt he was competent to teach. Later on we brought in Hans Samelson. Now I am beginning to forget the order of appearance; I think perhaps Moise and Young came next, and then Raoul Bott. The field of topology has been gradually built up here by this policy of bringing in new material in the field and making sure that all aspects of this rapidly growing field (topology had perhaps its greatest growth during this period) were represented, and different individuals had chances to teach the aspects of the subject in which they were most interested. I don't know whether this influenced the department in any way to do this in other fields, although it may have.

I believe that if I were asked to describe the evolution of the mathematics department at Michigan, I would divide it into three periods: in the first period I would place all of the development up to 1926 when Professor Glover became chairman. I think that at that time the bringing in of new material, particularly of the calibre of Rainich, was greatly responsible for the rapid development from that point on. Then the next period, I think I would designate as from 1926 up to and including World War II. I think in the third period I would place everything from the end of World War II up to the present, calling this perhaps the modern period. This way the department would have its early period, a second period of rapid development, and then a modern period. Certainly in the modern period the rapid development has continued; during this period the department has had the benefit of grants from the federal government and other sources, and this has been an accelerating factor. Of course, all designation of *periods* in the development of an institution is bound to be somewhat arbitrary. I have not wanted to imply that during the first period up to Glover's succession of the chairmanship there wasn't any research done. For instance, I do feel, however, that the curriculum at that time was representative chiefly of the mathematics of the nineteenth century. However, I do not know well what the contents of all the courses were then. For example, I should imagine that whenever Hildebrandt taught the course in real functions, he certainly must have taken into account such subjects as Lebesgue integration, since he, being a product of the E. H. Moore School at Chicago, certainly was up-to-date in these subjects. Possibly in statistics many twentieth century ideas were brought in. However, I cannot speak with knowledge of that period, of course, since I didn't come in until 1926 myself. I do think that the curriculum at that time was a good curriculum, a strong curriculum. I have no idea what the standing of the department was; i.e., how it rated nationally. As Bradshaw pointed out in his history, there were doctorates given earlier. I don't know who gave these, but I would guess offhand that they were probably done by such staff members as W. B. Ford, perhaps Louis Hopkins in celestial mechanics, and possibly Hildebrandt.

I am going to look now at the items or questions which were raised in a letter to me under date of February 4, 1976, by Professor Phillip Jones. I think I've already touched upon some of these. In his "Section I", he asks, "What was the status of the department when you arrived? Item a, adequacy and modernity of the course offerings and of the staff." I think I have touched upon this fairly well, certainly as far as I could. I failed to mention Karpinski, who was strong in the history of mathematics, no doubt had a good national standing at the time, and probably had been responsible for some of the doctorates which Bradshaw mentioned. Referring again to Professor Jones' letter, major item 2 asks, "What were major changes over the years and the causes? Item a, hiring and promotion policies." I think I have already touched upon this topic. The policy has always been, as I recall it, to hire people who were both good teachers and capable of advancing the frontiers in their own field by their research. There has been a very liberal policy all along, in my opinion, regarding the fields represented by the new appointments. I haven't said anything about applied mathematics. The development of mathematics generally, in this country, during what I call the first period, was gradually from what was considered "applied" (a practical mathematics) to "pure" mathematics. So that during the second period, the University of Michigan, as in most mathematics departments, established itself in what we call research in pure mathematics. About the time of the war, I believe, there was some agitation for getting in more people in applied mathematics. Applied mathematics up to the time of the war seemed generally oriented towards the needs of the engineers and was not, as I recall it, a very strong representative of what we were coming to think of as applied mathematics in the modern sense. I recall distinctly one instance that might throw light on this, and this concerns Professor Friedrichs, who was a Ziwet lecturer in 1946. In inviting Professor Friedrichs here at the time, we felt that since he was one of the most outstanding and most promising people available in modern applied mathematics we should invite him and consider the possibility of offering him a position here. Now I know that there was a considerable discussion of this on the Executive Committee in the department, but it was finally turned down, and I have felt that this was perhaps a mistake. It is well known that Professor Friedrichs went to New York University and became one of the leading lights in the Courant Institute, and I think that the University of Michigan missed out at that time on a good chance to enhance its reputation in the field of applied mathematics.

Professor Jones' second item, 2b, concerns the development of seminars, who stimulated them and when. I think I've already touched upon this and indicated that Rainich was particularly active in this regard. Educated in both Russia and Germany, he had a very broad knowledge of mathematics and undoubtedly enjoyed more the development of students via seminars, than doing his own research. The department probably went somewhat "overboard" by the time the third period developed, in that we had about twenty seminars going at one time, and I began to feel that maybe the students were spending too much time in seminars and not enough time on their own mathematical research. It was not unusual, I think, for a student to spend more time in seminars and reading in the library than doing his own thesis work. In regard to Professor Jones' third item, 2c, "changes in funding", I believe I already touched on this in my remarks regarding government grants. The funding here was, of course, that of what I've called period three, i.e., postwar period, and is now a permanent, or semipermanent, feature of the mathematical scene.

The next item, 2d, "changes in teaching load, hours, levels". When I came here in 1926, I believe the teaching load was from twelve to sixteen hours per week. Instructors were given sixteen hours, I believe. Possibly those of professorship rank had twelve-hour loads. I recall distinctly what happened in 1932 when I was asked to give the Symposium Lecture at the Chicago Section of the American Mathematical Society. I felt that in order to do an adequate job I ought to have a little more time at my disposal to work in the General Library. These Symposium Lectures are no longer given, but they were a feature of the spring meeting in Chicago of the Midwestern Section of the American Mathematical Society. There were two hour-lectures; they were given in the afternoon, one lecture for an hour, then an intermission, and then one lecture for another hour. One didn't accept the responsibility of giving one of these lectures very lightly. Unfortunately that year was during the period 1930–1932 when Professor Field was acting chairman of the department, during Glover's absence. I asked Field if I could have my teaching load reduced to eight hours while preparing my Symposium Lectures and he said no, it was impossible to give time off for the writing of advanced papers; these, as I recall, were his exact words. I presume this was a general attitude at that time. What one did in his research was something *extra*, something outside the regular academic program. That naturally has changed.

Today I think most of the larger universities have teaching loads of six hours per week and this is general for the whole staff, not just for the professors.

Passing to Professor Jones' fourth item, labeled "miscellaneous, item a, how did we happen to build strength in topology?", I think I have covered that. I was the first topologist here so that I feel as though the topological program here was sort of my baby. Item 2b, "was it true that some of the courses assigned to some topologists turned out to be topology?" Now this is a very interesting question and would not have occurred to me, but it should have occurred to me perhaps, for I recall when I started my foundations course, I found that despite the description of the course in the catalog, many of my colleagues thought that I was giving a course in topology. As a matter of fact, I remember that during Professor Field's incumbency of the chairmanship (this was about three or four years after I started the course), he suggested at one time when we were discussing courses for the following year that I give the foundations course to Professor Ayres to teach. Well it was immediately apparent he thought the foundations course was a topology course, and I explained that it wasn't, and I believed that anyone who taught the course should have had some interest in, or some grounding in mathematical logic, the theory of the infinite, etc. Though this is just a sample, it may be that in the later periods there was some feeling of this type. Particularly, perhaps, when a topologist taught a course in real analysis, he might bring in more topology than would normally be brought in, wherever it was applicable. However, I didn't know of any cases where the course turned out to be topology; I think that would be an exaggeration.

Coming now to Professor Jones' third item, 3c, "when, why, how did a conscious effort to bring in foreigners develop?" He gave examples, Eilenberg, Rothe, Brauer, and so forth. Well, I suppose that when Glover brought Rainich here there was no thinking on his part that he was bringing in a foreigner. This is my firm impression. Certainly when I induced the administration to bring in Eilenberg, I wasn't thinking of him as a foreigner; I was thinking of him as a *mathematician.* I think in general there has been no discrimination in this regard, but possibly I am wrong. I believe that we have been quite fortunate at Michigan in the foreigners that we have brought in, and that they did not feel that it was their sole function to do research and a small amount of lecturing. They generally participated very little, however, in such things as committee work (Rainich was an exception), which is one area certainly where I think I've heard the criticism made that foreigners would not in general be doing their part. It was not so much that they would be unwilling to do so, in most cases, but simply that they were not familiar with our ways in general and they couldn't be expected to serve efficiently on committees. I believe that there may have been a feeling around the country that the University was taking in foreigners in order to make positions available to people who otherwise could not get positions. In other words, it was deemed a sort of charitable gesture. I don't recall ever having this feeling in the case of the University of Michigan. I remember one case where a Japanese mathematician in this country had no university position in prospect; his name was Kodama. Realizing that he was a good mathematician who was in somewhat desperate straits, I spoke to the chairman about getting him here. However, I don't think I did this because he was a foreigner, so much as because I thought he was a good mathematician who was available. Incidentally, we did not keep him as a permanent member of the staff; he may have been here around two years. I recall also that I was involved in one other case, namely, Rubens Lintz, a Brazilian who seemed by his publications to have had considerable ability and who I felt would profit greatly by coming to this country. We brought him here on my contract; I don't recall whether it was an NSF or Air Force contract. Later, however, I believe we did give him some teaching. Again, we did not keep him. Afterwards he went to Canadian universities. Accordingly, my judgment is that generally we did not bring a man in because he was a foreigner.

Coming to Jones' next item, item d, "who stimulated and supported Michigan conferences in topology, complex variables, etc. The University, NSF, donors, University Press?" Well, here I can only speak for the conferences in topology of which I recall two. One of these was the topology conference of 1940, for which I recall talking to Graduate Dean Yoakum and asking for help to bring outstanding topologists here. I remember that he gave me a budget of $1,000. The war made it impossible for foreign topologists to come, although some did send abstracts of papers. We did have a good representation of topologists from the United States, and I remember I turned back around $35 of the $1,000. I don't recall that we gave anyone an honorarium, although we did help with travel expenses. I believe among the present members of our staff who first came to Michigan at the time of this conference were Professor Wilfred Kaplan and Professor Erich Rothe, who later became permanent members of our staff. The University Press later published a volume called *Lectures in Topology* which contained most of the papers, in complete or abstract form, which were given at this conference. Not a large edition was published. I don't know how large it was, maybe 300 or possibly 600 copies. They were all sold out shortly, and later the press felt that perhaps the demand would warrant publishing a new edition, or new printing. The department chairman, whose advice was sought, felt that this was perhaps not warranted, that there would not be enough demand. However, I can recall getting requests in recent years for copies of this volume which, of course, was no longer available. I don't know who financed the printing; I don't think it came out of my $1,000, but probably it was financed by the University Press itself. Then there was a topology conference in 1967 which was conceived of as being in my honor at the time of my retirement, and which I believe was funded by the National Science Foundation. Professor Frank Raymond can tell more about this so far as its funding, etc. was concerned. I don't know about the conference in complex variables, and I presume that Wilfred Kaplan could give information in this regard. Neither do I know about possible other conferences.

Going on to Professor Jones' item e, "how was the *Michigan Mathematical Journal* formed?"; I believe I have really covered that. Item 4, also labeled, "miscellaneous", asks, "what do you regard as interesting and/or significant about the history of the University of Michigan's Mathematics Department?" This is a question that requires some reflection and possibly I haven't given it enough. I have, in thinking of this question, set down what I considered reasons for the growth and reputation of the Michigan mathematics department: First, the policy of hiring people who were good in both teaching and research. I know of several cases where people did not gain tenure because of the fact that their teaching did not measure up to our standards, and, of course, I also know of cases where people were let go that we had considered to be promising in research, but who later did not live up to their promise. Secondly, I have put the building up of a good library. This is something that Michigan is noted for amongst mathematicians the world over, I think. We have here at the University of Michigan a collection of books going way back in history, and which ordinarily could not be found anywhere except in places like the John Crerar Library, Library of Congress, and Harvard University Library, and possibly the Brown University Library, to name some that come to me offhand. I don't think this is due to any one person, but certainly Alexander Ziwet is to be credited very largely for this. Pick at random any book which was published during the first part of the century or prior thereto, and you are likely to find Alexander Ziwet's signature in it, as having donated it to the library, and there is no question that the support of the University in giving funds for the library is to be credited in good part for the library here. Karpinski used to make periodic visits to Europe to buy books, both for himself and for the library. Thirdly, I think that the influence of the Small C, which I have already mentioned, had considerable to do with the building up of the department. I think it was a healthy influence and until the beginning of what I call period three, I think it contributed indirectly to the bringing to the University of outstanding people. Fourthly, I want to mention the policy of inviting eminent visitors. The using of the Ziwet bequest for bringing outstanding lecturers who could spend a period of around two weeks here has certainly had a great influence, and in addition to that, of course, there has been the bringing in of lecturers who have given one or two lectures, possibly paid for by somebody's grant. This kind of thing is stimulating to the department and it enhances the reputation of the University. Fifthly, the expansion in fields such as algebra, analysis, statistics, topology, foundations of mathematics, and so on, contributed much to the department's reputation. I have not gone much into statistics because it is not my field of interest, and I think that it will be found later that Professor Harry Carver would be willing to contribute something in this regard.

Finally, I again want to credit my colleague, Professor Rainich, who gave so much of himself to stimulating the interests of students, suggesting innovations, and giving advice to the chairman. Generally, I think the chairmen, and I think this is particularly true during Professor Hildebrandt's chairmanship, have been anxious to have good advice. I won't say that the chairman always acted on it, but this is not to say that he didn't accept advice generally and his decisions were usually in the best interest of the department.

I think that I have now covered most of the items that I had in mind when I started this oral history, if one can call it that. I realize that I may have made some mistakes here and there. Generally, however, I think what I have said fairly represents my memory and opinions, and if there are any points at which amplification is needed and I am able to do so, I would be very glad to cooperate.

Albert C. Lewis was a student at the University of Texas at Austin from 1963 to 1975, receiving B.A. and M.A. degrees in mathematics and a Ph.D. in history of mathematics. Supported by a Humboldt Fellowship in 1975–1976, he did research in Europe on the work of H. Grassmann. He then helped to found the Archives of American Mathematics, serving as Curator of the History of Science Collections at Texas from 1977 to 1981. For the last five years he has worked with the Bertrand Russell Editorial Project at McMaster University in Hamilton, Ontario. His publications include articles on Grassmann, Russell, and the Texas mathematicians Halsted and R. L. Moore.

The Building of the University of Texas Mathematics Faculty, 1883–1938

ALBERT C. LEWIS

INTRODUCTION

In 1938, when the American Mathematical Society was celebrating its semicentennial, R. L. Moore of the University of Texas at Austin was president of the society. Also in that year, R. H. Bing, a twenty-four-year-old high school teacher in Palestine, Texas, took one of Moore's summer courses for teachers. In 1973, Bing returned to Texas after twenty-eight years at Wisconsin and four years later became the second professor at Texas to be elected president of the American Mathematical Society. These personal honors of Moore and Bing also mark periods of growth and of attention at Texas to achieving or regaining high standing among departments across the country.

R. L. Moore, a student at Texas from 1898 to 1901 and a member of the faculty from 1920 to 1969, has to play a leading role in any complete history of mathematics at Texas as the most famous member of the department.[1] But there were mathematicians, principally Harry Y. Benedict and Milton B. Porter, whose names were relatively unknown outside of Austin but whose associations with Texas were just as long and intimate as Moore's and whose

[1]There is a fairly substantial literature on R. L. Moore and the Moore method of teaching. The principal ones are [Traylor], [Wilder 1976], and [Wilder 1982].

G. B. Halsted
ca. 1896

M. B. Porter
1912

R. L. Moore
ca. 1905

H. S. Vandiver
ca. 1955

(Photographs courtesy of the University of Texas at Austin, Archives.)

contributions to mathematics at the university were, in their way, essential to achieving the success it has had. And then, most important, there was George Bruce Halsted, who started it all.

This account focuses on faculty relationships and recruitment for the first fifty-five years, including some of the influences leading to comings and goings of faculty members. It does not attempt to include the many other aspects that would go into making up a complete history of the department: students, libraries, buildings, curricula, visitors, relationships with other departments at the university, and other external influences. And, the most important caveat, it talks around the mathematics which is at the center of the lives of the people described here. Most of the principals in this account are well represented in standard references or other works cited here. The subject at hand is confined to their institution building.

The years spanned by this account are rather well covered, in both their problems and their triumphs, by materials at the University Archives of The University of Texas at Austin, including the nationally-oriented Archives of American Mathematics. Documents have been preserved dealing with topics which in more modern times might be considered too sensitive or controversial even to put in writing in the first place, let alone be preserved. This is fortunate for those trying to explain the development of a department—or, for that matter, of any social group—since it is often the calamity that marks a turning point, for better or worse. The signal for the historical researcher that he is nearing the end of this revealing archival vein occurs when he finds a letter written in the 1920s to the university president at the bottom of which the president has written "Answered by telephone."

The First School of Mathematics: A False Start

The Texas Constitution of 1876, re-expressing a concern stated as early as the 1827 Constitution under Mexico, called for the establishment "as soon as practicable" of "a University of the first class." It also provided one million acres of West Texas grazing lands, supplemented in 1883 by another million, as an endowment for the University and the Agricultural and Mechanical College of Texas.[2]

A faculty could not be formed, however, until 1883. The leader in getting to that point was the very active president of the first Board of Regents, Ashbel Smith, who was born in Hartford, Connecticut and obtained a medical degree from Yale in 1828. He came to Texas in 1837, served the government of the Republic of Texas in several capacities, and helped prepare the way for the annexation of Texas to the United States in 1845. Among his subsequent activities, he served in the Mexican War, was appointed to the board

[2]Detailed accounts of the legislative history are provided in [Lane], [Benedict], and [Griffin].

of visitors of the U.S. Military Academy at West Point in 1848, and was appointed a juror for the 1876 centennial celebration in Philadelphia. As one of many doctors who joined the Texas militia, he played one of the leading roles in the defense of Texas during the Civil War. A lifelong bachelor, as a member of the Texas legislature, he was credited in large measure with the establishment of the Texas school system.[3] To be charged with the founding of a university modelled after the University of Virginia was probably seen by him as the crowning contribution to his services to Texas at the age of 77—Thomas Jefferson was in the same decade of his life when he proposed and pressed for the Charter for the University of Virginia.

Smith wrote to professors and presidents at the older universities, relying especially on Virginia, soliciting advice and nominations for a faculty. Even at this early stage, the mathematics professorship seems to have given more trouble than the others—at least most other appointments appear to have been handled in a unanimous fashion and not brought to a recorded vote, let alone two votes. Apparently, the Board of Regents initially agreed to seek senior people whose reputations were well established and who would thereby bring immediate prestige to the University. Thus in 1882 the minutes record that agreement was reached on establishing one professor in the "School of Mathematics Pure and Applied" at $3,500.[4]

The vote of the board in November was General LeRoy Broun, 1; Professor Bruce Halsted, 1; General Kirby Smith, 2; Professor Alexander Hogg, 2. Since no choice was made, a second ballot was held with the results of General LeRoy Broun, 4 and General Kirby Smith, 2.[5] The military titles serve as a reminder of the postbellum atmosphere of academia to which many former officers returned after the war. Edmund Kirby Smith, for example, originally from Florida, was a general in the Confederate Army, having graduated from the U.S. Military Academy in 1845 where he also taught mathematics from 1849 to 1852. In 1883, he taught mathematics at the University of The South in Sewanee. Alexander Hogg, from Virginia, received his A.M. degree from Randolph-Macon College and was the first professor of pure mathematics at Texas A&M College (now Texas A&M University). He left Texas A&M in 1879 to become a civil engineer with a railway company.[6]

Halsted, born in Newark, New Jersey, was a promising 29-year-old graduate from Princeton University, and at that time an instructor there. His major publications up to that date included several on logic and geometry and a textbook on mensuration. He later described what he was doing at Princeton in

---

[3][Handbook].

[4]Minutes of the Board of Regents, 17 August 1882. Unless otherwise stated, all original archival sources cited in this paper are either in the University Archives, which also handles the Archives of American Mathematics, or in the Eugene C. Barker Texas History Center. Together these repositories form a part of the General Libraries at The University of Texas at Austin.

[5]Minutes of the Board of Regents, 16 November 1882.

[6]On Smith, see [Wakelyn]. On Hogg and Smith, see [Geisser].

what were probably much the same terms he submitted to Texas. Upon obtaining his Ph.D. at Johns Hopkins University under J. J. Sylvester in 1879, "after further study at the University of Berlin, [he] was called to Princeton in 1881 to plan and inaugurate a system of post-graduate instruction in mathematics, and having established it, he remained to give instruction...." His only connection with the South at this point appears to have been that his mother was from South Carolina. His father was a lawyer who worked in Washington D. C. during the war and was a friend of Abraham Lincoln.[7]

William LeRoy Broun best fit the requirements of the board. Born in 1827 in Virginia, with an M.A. from Virginia in 1850, before the war he had been a professor of mathematics at the University of Georgia and afterwards a professor of physics and astronomy. This was followed by the presidency of the State College in Georgia. He had been Professor of mathematics at Vanderbilt University in Nashville for the past seven years. During the war, he commanded the Richmond arsenal, a part of the Confederate Ordnance Bureau, where he became friends with J. W. Mallett, formerly a professor at the University of Virginia, now the chairman of the new faculty at Texas.[8]

When Broun acknowledged his official notice of election in November 1882, he wrote Smith asking what the date of opening was, what buildings would be completed at that time, what provision had been made for obtaining scientific apparatus, and what would be the available income from the University's endowment. The following month he asked for assurance that the salary would not be reduced after he arrived in Texas and asked, "Is the term of office for 'good behavior and satisfactory performance of duty', or is it for a term of years only?" Finally, in January 1883, Broun accepted the position. On 31 August, he arrived in Austin on the same train from New Orleans that Mallett was on.[9]

Neither of them were to stay in Austin for long. Mallett left after one year to return to the University of Virginia. Broun was elected chairman of the faculty in his place in May 1884 but stayed for only a short period, having already given notice in January that he wished to resign because of his daughter's sickness. Some of the regents wished to take advantage of this opportunity to split the chair of pure and applied mathematics and to have a separate chair of applied mathematics which would include engineering. Regent Simkins, in particular, said that he heartily endorsed any such move towards "practical attainments" and "the sooner we begin, the better for the University."[10] Under Broun, the applied mathematics offering as given in

[7][Halsted 1893]. *The New York Times*, 3 July 1871. There is no collection of Halsted papers at Texas, or any known elsewhere, and thus we have only his letters in other collections. A few brief biographical articles exist: [Tropp], [Lewis 1973], and [Lewis 1976].

[8][Lane, p. 268], [Broun].

[9]Ashbel Smith papers: Broun to Smith, 29 and 28 December 1882, 20 January 1883.

[10][Vandiver F, p. 441]. Smith papers: Regent Thomas D. Wooten to Smith, 26 January 1884; E. J. Simkins to Smith, 8 May 1884.

the catalogue consisted of applications of calculus to mechanics and physics "for those who have completed the course in Pure Mathematics."

Another regent's views seem to have had more immediate effect. Thomas D. Wooten, a medical doctor who moved to Texas from Kentucky in 1865, wrote Smith that he wanted to consider younger people for Broun's position.

> I think their vim and enterprise would suit the genius of our people better and are more likely to stand by the university as a means of their own advancement and success. I am not disposed to favor any more confederate... [*illegible*]. But am in favor of selecting if possible living moving progressive men, men who will accept the situation and the university as it is, accept it as their own and willing to work for it stand by it identify themselves with it and the state and people of the state and if they are not likely to have such a fealty we don't want them. I believe it might be well to require them to take some such oath and thus require them to live on Texas soil the year round. As it now is they can hardly wait until the close of the session to get away, as though they had to escape some pestiferous clime or moral infection. So far as I now feel they may all go to the devil or any where else they may choose to go.... Broun has not for some time come up to the true measure of a great man in my estimation. He suggests in his last conversation that we ought to have a president. I think he felt his inability to head the institution and grapple with the situation....

When Ashbel Smith died in 1886, Wooten became chairman of the board and remained on the board as its head for the record period of nearly eighteen years.[11]

It was in 1884 that William Sydney Porter ("O. Henry") came to Austin and a time, as one of his biographers has put it, when the notation 'Gone to Texas' placed beside a man's name made it suspect that "he was on the verge of bankruptcy, unwanted marriage, tuberculosis or some other disaster."[12] A proper university in the capital city would go far to overcoming this sort of image and Regent Wooten's urging to take a chance with willing younger people who would grow with the university was followed—albeit without the formal oath requirement—in choosing the next head of mathematics.

George Bruce Halsted, 1884–1902: A Faculty of One

At the board meeting in August 1884, Halsted was elected for the professorship at a salary of $3,500 plus $500 for housing. Though Halsted never

[11] Smith papers: Wooten to Smith, 24 January 1884.

[12] [O'Connor, p. 18].

seemed to shy from blowing his own trumpet, he probably did get good letters of recommendation from Princeton and Johns Hopkins. These might well be the letters he had printed up in 1883 in a twenty-five-page pamphlet, *Some Testimonials and Credentials of G. B. Halsted*, which included letters by Sylvester, Simon Newcomb, C. A. Young, Josiah Royce, and a former student, Henry B. Fine. Fine had just received his A.M. degree and was to become a professor of mathematics and dean at Princeton and president of the American Mathematical Society in 1911. Fine wrote,

> ...through [Halsted's] influence I was turned from the Classics to Mathematics, and under his instruction or direction almost all of my mathematical training has been acquired. From personal experience, as well as from what I know of the general opinion of his Princeton pupils, I can testify that Dr. Halstead [sic] has the gift, so rare among teachers, of throwing a charm about the very difficulties of his subject, and of awakening real enthusiasm in all who have the least aptitude for it.[13]

Once established in Texas, Halsted wrote back to Princeton leaving no doubt with his former classmates that he had made the right move:

> My lines have fallen here in pleasant places, and I am actively happy as the official head of pure science in a state larger than the German Empire.... I am thoroughly in love with Texas and have purchased ten thousand dollars worth of its soil. I have not yet married and so am open to engagements. My salary here is four thousand dollars for nine months at two hours a day, and besides I have furnished to me an assistant who is paid two thousand dollars a year; so you see that monetarily my position is better than that of the President of Princeton.[14]

Halsted was to marry and raise three sons in Austin. His salary, however, did not grow and, in fact, was to be reduced fourteen years later in unhappier times.

As for the more "practical" side of mathematics which Regent Simkins urged, a new instructor, Alvin V. Lane, was appointed to handle applied mathematics courses given in parallel with pure mathematics and designed as preparation for engineering: "Engineering, Surveying, Mechanical Drawing, etc." Lane was made an associate professor in 1885, but he left in 1888 to join a Dallas bank and was replaced by T. U. Taylor, a graduate of the University of Virginia, who stayed for forty-eight years and came to embody

[13]Letter dated 25 September 1882 as printed in the copy in Princeton University Library.

[14]*Decennial Record, Class of 1875* (Princeton, 1885), p. 43.

engineering at Texas. The following year, the catalogue listed applied mathematics with Taylor as a division under Halsted's School of Mathematics. The next year, the School of Applied Mathematics was listed in a completely separate fashion, but Halsted's domain still had the more general name of School of Mathematics. Halsted's title, however, changed from Professor of Pure and Applied Mathematics to Professor of Pure Mathematics, while Taylor became Associate Professor of Applied Mathematics. The courses offered were non-overlapping. Halsted made a motion at the November 1894 faculty meeting that the regents be asked to "formally and officially" separate the School of Applied Mathematics from the School of Pure Mathematics and to put Taylor in "complete charge" of the former.[15] Whatever the difference may have been between the official catalogue statement and the actual working arrangement, such discrepancy characterized the relationship between pure and applied mathematics (or, more precisely, between two groups of mathematicians classified under these rubrics) for most of the history of the department(s), even after they were officially merged in 1953. In 1896, however, the official distinction was evident. The School of Pure Mathematics under Halsted was followed in the catalogue by the School of Applied Mathematics under Taylor.

Beginning in 1888, Halsted had a succession of students from Texas schools who specialized in mathematics under him and either became instrumental in shaping the future of mathematics at the university or led distinguished careers elsewhere. In 1888, M. B. Porter, from Sherman, Texas, entered the university after attending various schools, followed the next year by H. Y. Benedict, with little formal education, from land in West Texas settled by his parents when they came from Kentucky. Leonard E. Dickson took his first course under Halsted in 1890 and was in the same sophomore class as George Washington Pierce. Florence P. Lewis obtained her bachelor's degree in 1897. R. L. Moore came from Dallas in 1898 from a good school background and after having studied W. E. Byerly's *Differential Calculus*, the text used at the university.[16]

Benedict, as a fellow (i.e., teaching assistant) and then tutor (a full-time, post-graduate teaching position) in Halsted's department in 1892 and 1893, taught freshman courses, and, after getting his M.A. degree in 1893, worked at the University of Virginia astronomical observatory. He credited T. U. Taylor, under whom he had taken undergraduate and graduate courses in applied mathematics, for guiding him in general and getting him this job at Taylor's alma mater. After two years, he was able to enter Harvard where he

[15]Faculty minutes, 6 November 1894.

[16]R. L. Moore papers: Halsted to Moore, 18 February 1898. [Vandiver H 1961], [Archibald] on Dickson, [Greenwood 1988]. For references to classes and grades, here and following, the source is the records in the Registrar's Office of The University of Texas at Austin.

attended lectures by Maxime Bôcher, Byerly, B. O. Peirce, and Osgood, and obtained his doctorate in 1898.[17]

Porter graduated from the university in 1892 and, after tutoring on a sugar plantation, went to Harvard where he obtained a Ph.D. in 1897 under Bôcher. He then returned to Texas for two years as an instructor before going to Yale.

Dickson and G. W. Pierce obtained their B.S. degrees from Texas in 1893 and their M.A. degrees the following year. Pierce went on to get a Ph.D. at Harvard under J. Trowbridge and W. C. Sabine, and continued as a professor of physics there until his retirement in 1940. Dickson was the first ranked graduate of the "academic departments" at Texas in 1893 and delivered the oration at the commencement: "A plea for pure science."[18] He had been a fellow in Halsted's department in his last year but submitted a letter of protest to Halsted which the latter passed on directly to the regents. (One cannot be certain that Halsted did not request Dickson to make the protest.) Dickson said he was getting paid as a fellow ($300) for doing the teaching work of a tutor. Halsted requested the creation of a tutor for his department "so that classes could go on this year as every other year of the University's existence." Regent Wooten agreed after consultation to try to provide for a tutor "with all the pay at the disposal of the executive committee." The university's treasury had only $200 at the time, but the next $100 to come into their hands would go to make up the total of $600 needed, "the Committee being anxious to accommodate you." Presumably this was done—Dickson continued teaching through the rest of the 1893–1894 year.[19] We shall return to Dickson shortly.

Florence P. Lewis studied for part of 1900 at the University of Zürich, left Texas the following year for a $1,000 position in Mississippi,[20] and then returned to teach at Texas for 1902–1903. In 1913, she received her doctorate from Johns Hopkins and taught for most of her career at Goucher College.

W. L. Prather resigned from the regents in 1899 to become president of the university, as he deemed it his "duty" to do so. In 1895, the university had taken up what W. L. Broun called for when he left in 1884, the establishment of an office of president appointed by the regents, instead of a chairman or president elected by the faculty. The president preceding Prather, Winston, left under a cloud, and rather fierce competition ensued to fill the position. For a time, a major contender for the position was Dudley G. Wooten whose father, T. D. Wooten, was president of the Board of Regents. He withdrew his candidacy in July 1899 at the same time his father left the board.[21]

[17]H. Y. Benedict papers: Benedict to Taylor, 2 June 1933; lecture notes, 1895–1898.

[18]Commencement Program, 21 June 1893.

[19]Benedict papers: Halsted to Regent T. M. Harwood, 21 September 1893.

[20]Halsted, "The School of Pure Mathematics," *The University Record*, June 1901, p. 146.

[21]T. S. Henderson papers: G. Winston to Henderson, 22 June 1899; D. G. Wooten to Henderson, 12 July 1899; Prather to Henderson, 7 October 1899.

After his graduate year at Texas, Dickson went to Chicago where he was one of E. H. Moore's first doctoral students. After getting his degree in 1896 and spending time briefly at the Universities of Paris and Leipzig, he went to the University of California as an assistant professor in 1899. In April of that year, Dickson wrote Judge Clark, who, as proctor at Texas, acted as a useful and widely trusted intermediary between students, faculty, and regents:

> Upon learning that Dr. Porter had been appointed to *Yale* instructorship I dropped a line to Prof. Halsted asking if they could not manage to keep Porter perhaps as Assistant Professor; as the former had so intimated his intention of recommending advance next year.... It has occurred to me that Texas could afford to make an assistant professor at least—and that the many ties binding me to Texas people would make it very congenial for me there. But I write this confidentially to you—as there is no need for me to go begging for a place! Of course I could not accept an instructorship under my present status here and offers east. Last spring I had an offer at Michigan and later an increased offer there—and came *near* going. The Chicago people, with all of whom I have most cordial relations, have corresponded considerably about my going there—but I see no need of changing except for a *better* position, which they could not offer, as the expected *vacancy* there did not occur.
>
> Please say nothing of my willingness to be considered—until (granting, of course, Porter's resignation) the authorities are willing to provide an adjunct [assistant, in modern terminology] professorship. In the latter case I can get the warmest support from the heads of the departments at Columbia, Michigan, Chicago, California, Indiana, etc.[22]

Dickson agreed to a three-year appointment as associate professor at Texas, beginning in the summer of 1899, but then took up an offer in April 1900 to go to Chicago as an assistant professor. This was taken by some regents and faculty members as at least a gross insult if not morally and legally wrong. Regent Cowart, from his Washington, D.C. office, wrote the chairman of the board, T. S. Henderson, that he was "disgusted" with Dickson's resignation: "It seems we are fated to develop men of extraordinary mathematical genius, and then other institutions of learning appropriate them as soon as they show any promise." Cowart also wrote to Proctor Clark: "I am simply disgusted with the infernal way that some of our professors have of coming before us constantly and clamoring for an increase of salary, and when any prize glitters before their eyes in a northern University they quit us without any

[22]Memorabilia of the University of Texas: Clark—Dickson to Clark, 1 April 1899.

excuse. I think Dickson's conduct is simply infamous, and I intend, if I can, to have a set of resolutions adopted characterizing it as it should be."

The regents took their retribution by not paying Dickson for the last two months he taught—the remainder of the spring term. Dickson attempted to get this money with the intervention of a lawyer from his hometown of Cleburne, Texas, but the records do not show if he ever received it. There is no record of Halsted's stance in this affair. T. M. Putnam, who had come to Texas as an instructor from California, apparently with Dickson, left with him for Chicago and was one of his first doctoral students there. During Dickson's last brief tenure in Texas, he was R. L. Moore's calculus instructor.[23]

No official announcement of this unexpected vacancy was made by the university, but Dickson's leaving and its circumstances soon became widely known. Cowart thought that they could get Porter if they could "put him on an equality" with "G. A. L. Halstead," and he tried personally to get him to take Dickson's place at the same rank and salary and thought for a while he had succeeded. "I consider him," Cowart wrote later to Henderson, "fully the equal of Dickson, and in a few years I think we could arrange it so that he could be put in charge of that school instead of the Barnum who now disgraces it."[24]

"Cowart has gone too far," Prather wrote to Henderson, "in asking Dr. Porter to accept Dr. Dickson's place." The university may not have the money to pay this salary, and it may not be the "wisest arrangement" at present. Whether the university backed up the offer or not, Porter did not then come to Texas. It has been conjectured that he was put off from returning because of the Dickson incident. Perhaps it is significant that when he did return, it was to take Halsted's place.[25]

Cowart, in referring to Halsted as "that Barnum," may have had in mind an incident from several years previous on another front. Arthur Lefevre, originally from Baltimore and a graduate of the University of Virginia, received a degree in civil engineering from Texas in 1895, at age thirty-two, and in 1894 started as an instructor in the School of Pure Mathematics. In 1896, he published a book, his only publication in mathematics, *Number and Its Algebra*, which referred to Halsted but which went substantially beyond the latter's work on the subject. Apparently, a symposium based on the book was to be held. C. S. Peirce was invited to attend such an event sometime

[23]Henderson papers: Prather to Regent Henderson, 6 April 1900; Cowart to Henderson, 21 April 1900, 8 May 1900, 10 May 1900; W. J. Ramsey to Henderson, 6 January 1901. Benedict papers: Cowart to J. B. Clark, 31 May 1900.

[24]Henderson papers: Cowart to Henderson, 21 April 1900, 8 May 1900. Evidently, "G. A. L." is Cowart's evocation of a form of insult whereby a man is called a girl. In a letter of 6 November 1901, he says of another displeasing man that he "would have made a fine girl."

[25]Henderson papers: Prather to Henderson, 22 May 1900, [Greenwood 1988, p. 13].

before 1898.[26] In 1897, Lefevre appealed to the Board of Regents to try to counter what he saw as injustices being done to him by Halsted and by a president, Winston, who avoided involvement. Lefevre's appeal provides one person's view of what it was like to work as a colleague with Halsted. Lefevre says that Halsted encouraged him to oppose an order of the president's increasing the number of sections of freshman mathematics. Halsted himself, according to Lefevre, said that he had already lost $500 of his salary over casual oppositions and thus was timid about doing battle over this. But the president claimed to Lefevre that Halsted had advocated the increase all along with him and he was just supporting the head of the school. In a related incident described by Lefevre, Halsted had spent an hour in Porter's sophomore classroom with the main purpose of criticizing Lefevre's preparation of the freshmen. Apart from the inappropriateness of the action, a large proportion of the freshmen had not even taken freshman mathematics at the university, according to Lefevre, but had been admitted with credit from an "affiliated school" accredited by Halsted himself. Also, Lefevre claimed that the whole university knew that Halsted had, in his classes, charged Lefevre with unacknowledged appropriations in his book from Halsted's work.

In May 1902, on Halsted's recommendation, E. H. Moore was willing to accept R. L. Moore as a student at Chicago, especially since Halsted ranked him, with equal training, as superior to Dickson. But there were, E. H. Moore replied, no fellowships available for the coming year.[27] Thus Halsted proposed to hire Moore as a tutor. Moore had been a fellow the previous year but since he would no longer be a student he would have to be appointed tutor. There was such a position open since one of the current tutors, E. P. R. Duval, had resigned to become an instructor at the University of Oklahoma. The president and regents had another candidate in mind, Mary Decherd, a school teacher who had been recommended in one letter to the regents as a "relative of Governor Sayers and...from one of the oldest and best families in Bastrop County." Letters of recommendation also came from the Commissioner of the General Land Office and from a former principal of Austin High School.[28] After Moore had left Texas to teach at a high school, Halsted wrote to him:

> I raised the five hundred dollars to pay for you here with me, and made the proposition to Mr. Brayther, and he rejected it. ...Of course I made a fuss about it.

[26][Eisele, p. 70] where "MS 183" should be "MS 229." Max H. Fisch has provided this reference.

[27]R. L. Moore papers: copy (made by Halsted?) of a letter of E. H. Moore to Halsted, 15 May 1920.

[28]Henderson papers: W. E. Maynard to Henderson, 25 April 1902; Charles Rogan to Henderson, 5 June 1902; C. S. Potts to Henderson, 10 June 1902.

> He sent a letter after me, saying that after the present session my "services would not be required" in the University, and threatening, if I divulged his villainy, to cut off the remainder of this year's salary.
>
> Of course that would put a stopper on my work in getting out my book. So you see there is no hope for you here. What should I do?

And, after several more letters offering suggestions to Moore for a position for the next year, Halsted wrote: "I have also another praise of your work coming out in the next number of the great Educational Review, and your future is assured. I wish I could say as much for my own."[29]

A praise appeared in the *Educational Review* for December[30] but an even more direct one appeared in the October issue of *Science* in an article ostensibly about the Carnegie Institution. Halsted's terms of praise were somewhat self-defeating: "And finally among the sifted [sic] few who have the divine gift and the divine appreciation of their gift, the exquisite bud in its tender incipiency may be cruelly frosted." After citing Moore's work on Hilbert's axioms of geometry, which had gotten E. H. Moore's attention, Halsted pressed the point further:

> This young man of marvelous genius, of richest promise, I recommended for continuance in the department he adorned. He was displaced in favor of a local schoolmarm. Then I raised the money necessary to pay him, only five hundred dollars and offered it to the President here. He would not accept it. ...The bane of the state university is that its regents are the appointees of a politician. If he were even limited by the rule that half of them must be academic graduates, there would be some safety against the prostitution of a university, the broadest of human institutions, to politics and sectionalism, the meanest provincialism.[31]

The regents at their meeting on Saturday, 6 December, unanimously adopted a resolution "that in their judgment the interest of the institution requires that [Halsted] should be removed and that his place be declared vacant from this date, his salary to be paid for the current month." On Sunday, official word was sent to Halsted via the janitor. T. U. Taylor later wrote simply that Halsted's services were terminated "on account of misunderstandings." "He was," Taylor added, "too free in his criticisms of the University authorities...." This was evidently just the last straw in what

[29]Moore papers: Halsted to Moore, 8 September 1902; Halsted to Moore, 13 November 1902.

[30]"The teaching of geometry," *Educational Review* **24**(1902), 456–470.

[31]"The Carnegie Institution," *Science* (n.s.) **16**(1902), 644–646.

had become an increasingly strained relationship between Halsted, on the one side, and the regents and Prather, the former regent, on the other. In themselves, Halsted's remarks were only the public expression—though admittedly in Halsted's usual purple style—of a situation which even Regent Cowart had expressed in private: "I don't see how a Board of Regents in with every administration can keep out of politics."[32]

In April of 1903, Moore, teaching high school in Marshall, Texas, received word that he would have his Chicago fellowship. In a 1972 interview, Moore said that he could not analyze his relationship to Halsted, that his appreciation of Halsted was not something he could explain, but that, nevertheless, he was certain there was "no one at all who I wish had been professor of mathematics [at Texas] instead of Halsted." Halsted himself would probably not be at a loss for words to describe the nature of his value to someone like Moore. Though Halsted did do some original and influential mathematics, especially in the axiomatic treatment of non-Euclidean geometry, he was primarily a prolific writer of expository papers and textbooks, and a teacher. In an essay he published in 1876, while a fellow in mathematics at Johns Hopkins working under Sylvester, Halsted described the rise of three separate men in place of the single classical mathematician: the writer of research papers, the teacher, and the reader—"the last class including the writers of non-original treatises and all textbooks." Halsted probably saw himself already in 1876 as a reader, as one of those who, "wishing to be of most use to their race, carefully read these memoirs, and after long and patient study of them, digested them into connected treatises, supplying the missing links and making them really part of the available mental wealth of the world."[33]

Causes for Halsted's dismissal other than the airing of dirty linen have been indicated in a document prepared in 1951 by the Dean of Arts and Sciences in response to a suggestion that an instructorship be named after Halsted. After consulting with the two most senior members of the university community at the time, Porter and W. J. Battle, the dean reported his findings:

> Both Professor Battle and Professor Porter are in agreement that Professor Halsted fully deserved the dismissal he got. According to Dr. Battle, Halsted was associated with Edwards (Biology), Everhardt (Chemistry), and McFarlane (Physics) in an effort to discredit the services of Messrs. Waggener, Wooldridge, and Wooten of the Board of Regents. Several of these men were discharged for their campaign, but Professor Halsted was continued on the Faculty with his salary reduced by $500 for each of three years. The

[32]Minutes of the Board of Regents for 6 December 1902. [Taylor, p. 87]. Benedict papers: R. E. Cowart to J. B. Clark, 7 May 1900.

[33]Moore papers: Halsted to Moore, 14 April 1903. Moore in an interview with A. C. Lewis on 1 November 1972. A similar statement by Moore is reported in [Traylor, p. 20]. Halsted, "Modern mathematicians as educators," *Nassau Literary Magazine*, November 1876, p. 98.

final act which appeared to have led to his dismissal was stated by Dr. Battle and Dr. Porter to have been Halsted's "stuffing the ballot box" in connection with his candidacy for president of the Texas Academy of Science.

Dr. Battle feels that under no conditions should any fund be named in honor of G. B. Halsted. The record seems clear enough in the matter to support this conclusion.[34]

It should be noted that this memorandum was composed at a time when its author was attempting to diminish Moore's influence at the university and the proposal for the instructorship had been initiated by H. J. Ettlinger who could be regarded as staunchly in Moore's camp. Though Battle was professor of Greek from 1893 to 1949 and had served in administrative positions beginning with Dean of Arts, which included mathematics, in 1908, corroboration for either of these points—discrediting the named regents or stuffing a ballot box—has not been found. There were problems associated with the departures of the professors Battle lists but the precise natures of them seem at least as unclear from the existing record as in Halsted's case. Some facts are not quite right: Alexander P. Wooldridge, who was secretary to the Board of Regents during Halsted's time, and Leslie Waggener, who was chairman of the faculty and president, were never regents. Wooten, as has been mentioned above, was a regent from 1881 to 1899. It is true that Halsted's 1884 salary of $3,500 had become $3,000 by 1902, presumably this happened in 1898, or before, judging from what Lefevre reported in the incident described above. Though Halsted, as the professor with greatest seniority, might have been entitled to more, $3,000 was still the top salary, which five of the eighteen full professors received, and President Prather's salary was $3,333.34. The local newspaper reported at the time that there had been a *general* reduction of salaries "several years ago" and that it was rumored that Halsted would resign then. Furthermore, no mention of either of these two points against Halsted has yet been found in the voluminous archives from that period which are not devoid of documentation of what were taken to be scandalous matters (for example, Porter's supposed liaison, treated below). Thus it does not seem advisable to attach much weight to this late and superficial gathering of evidence by someone not exactly disinterested.[35]

Whatever the reason for his dismissal, by the end of 1902, Halsted could well have agreed with the observation made by Mallet, former chairman of

[34]Moore papers: copy of memorandum to file, C. P. Boner, 24 October 1951.

[35]Ettlinger interview with A. C. Lewis, 26 May 1975. Henderson papers: "Salaries September 1, 1901–August 31, 1902." *Austin American Statesman*, 11 December 1902, also quoted in [Traylor, p. 36] (where "Regent Lomax" should read "Registrar Lomax").

the faculty: "Texas can send a man up higher, and let him down lower, than any other region on the face of the earth."[36]

PORTER AND BENEDICT, 1902–20: AN ERA OF DIPLOMACY

R. L. Moore recalled in 1972 that "there was a big difference between the personalities of Halsted and Dickson" and that Porter and Benedict "did not appreciate Halsted." Of all Halsted's students at Texas, Moore was probably the closest personally. Their correspondence continued through Halsted's succession of unhappy jobs and up to Halsted's death in 1922, and it covered a wide range of subjects, both mathematical and personal. Now, as far as the university was concerned, a new beginning could be made and Prather was quick to try to bring Porter back to Texas once again. He could report, just four days after Halsted's dismissal, that Porter had responded to his offer of an associate professorship but was holding out for professor.

In place of Halsted, the regents and administration were probably ready for a head of mathematics whom they could be well assured in advance would be less cause for concern and now, thanks largely to Halsted, they had qualified people available who were Texas alumni, such as Benedict and Porter. We have an overview of this period from J. W. Calhoun (B.A. 1905, M.A. 1908), originally from Tennessee, who entered the university in 1901. Starting as a tutor in pure mathematics in 1905, he continued as a teacher and administrator at the university until his death in 1947. He described this early transition period in a history written about 1946 in response to "vigorous and somewhat heated" discussions following H. S. Vandiver's transfer from the pure to the applied mathematics department to avoid working with R. L. Moore. According to Calhoun,

> ...President Prather...was very fond of Benedict and had a high opinion of his ability [and] desired to appoint him to the place. T. W. Gregory was at that time a member of the Board of Regents and desired to have M. B. Porter appointed. Benedict, who was a classmate of Porter at the University of Texas and had been his roommate in Divinity Hall at Harvard, also wished Porter to be appointed. Porter [who] was then an Assistant Professor at Yale would not leave for less than a full professorship (and as at that time there could be only one professor in a department) Benedict was willing to accept an associate professorship in order to have Porter come as a Professor.

On the surface, this would appear to be a good way to sow the seeds of more problems. In fact, this was the beginning of a new era of diplomacy

[36][Mallett, p. 17].

in which the mathematics faculty played a more normal role in the university and gradually began to seek new members who were graduates of other institutions.[37]

Benedict had already proved himself useful as a moderating influence in a difficult time and was not to be overlooked. Benedict had been asked by Prather to take over the School of Pure Mathematics immediately after Halsted's dismissal. In 1906, the School of Applied Mathematics, last seen in 1903 when it was effectively absorbed by the engineering department, was revived. It now also included astronomy, and Benedict was made chairman of it, assisted by C. D. Rice, who had also been a student under Halsted. Its offerings overlapped pure mathematics and were in some cases cross-listed with pure mathematics, but the intent—besides providing Benedict a full professorship—was the mathematical training of engineering students. At the same time, an engineering college was established with Taylor as dean. According to Calhoun, Taylor "did not wish engineering students taught Mathematics by women," and it was Taylor who proposed the school be headed by Benedict. Benedict was now earning $2,400 compared with Porter's $2,500 salary. Thus began Benedict's move into administration where he was to rise through the ranks: chairman, director of the extension division, dean, and, finally, president.[38]

Porter, while an instructor at Yale from 1899 to 1902, came to know Edward Lewis Dodd, then working on an M.A. degree in mathematics. Dodd, born in Cleveland, Ohio, received his doctorate from Yale in 1904. In 1907, Dodd had been teaching at the University of Illinois for a year and came to Texas at Porter's invitation as an instructor in pure mathematics at $1,600. Dodd wrote that, in advanced mathematics, "I am perhaps best prepared in function theory, vector analysis and differential equations." "I wish," he added,"to be as useful to the University of Texas as possible, and will gladly prepare myself to teach any course that may be desired." His first publication was in 1905. In 1911, he tried to obtain a position back at Yale but was unsuccessful. He then became more mathematically active. In 1912, he offered for the first time at Texas a course in actuarial mathematics, which, according to the catalogue, was "modelled after Broggi's *Traité des assurances sur la vie avec développements sur le calcul des probabilités*." This followed the establishment the previous year of a School of Business Training. Dodd wrote to President Mezes in the spring of 1913: "Since last spring, I have written

[37]W. J. Battle on Calhoun in [Handbook]. [Calhoun, p. 2].

[38]Benedict papers: Prather to Benedict, 8 December 1902. Presidents' papers/College of Arts and Sciences/Pure Mathematics, 1907–1929 [=PM 1907–1929]: Taylor to President Houston, 9 January 1907; Houston to Benedict, 13 June 1907; Houston to Porter, 13 June 1907. [Calhoun, p. 2].

eight papers for publication, on the general subject of mathematical probability with special attention to the theory of measurement and statistics."[39] Dodd thus began his rise through the ranks to become in 1923 Professor of Actuarial Mathematics. His reputation in the actuarial field was such as to attract the attention of R. L. Wilder who came to Texas in 1921 to study with him.

After completing his doctorate at Chicago in 1905, R. L. Moore taught at the University of Tennessee for one year and then went to Princeton, Halsted's alma mater, as instructor. While there, Moore made enquiries about joining the Texas faculty. Oswald Veblen, with whom Moore had worked at Chicago and who came to Princeton in the same year as Moore, wrote also on Moore's behalf pointing out that Moore had seven young men ahead of him at Princeton (including Veblen himself) and that he would have better prospects at Texas. He added that "in his speciality, the foundations of geometry, he is one of the best men in the country."[40] Moore did leave Princeton in 1908 but for Northwestern instead of Texas. In 1911, he moved to the University of Pennsylvania.

An apparently minor, personal, incident in 1910 was probably the closest thing to a scandal during M. B. Porter's watch as the senior mathematician at Texas. There is evidence that during Porter's absence from the campus a faculty member, not in mathematics, told others that Porter had had some sort of illicit affair with a married woman. The actual accusation, which appears not to have caused any lasting damage, does not have any historical relevance to the present account, but the way in which it is referred to in the existing documents provides a valuable illustration of a combination of gentility and frankness that probably characterized the handling of such potential crises. Porter wrote to President Mezes expressing concern about the damage the "slanderous stories" might have for the reputations of the woman and the university. A colleague from another department also wrote to the president in support of Porter's good character:

> It is a monstrous thing that members of the University Faculty should directly or indirectly traduce the character of their associates and trample in the mud the fair name of an excellent and innocent woman, whose husband was not here to protect it.
>
> In the South ordinarily such things have but one ending, but this must by all means be avoided. The gravity of the situation is such as to cause me much anxiety.[41]

[39]Presidents' papers/[PM 1907–1929]: Dodd to Porter, 8 May 1907; carbon copy Mezes to J. Pierpont, 30 January 1911; Dodd to Mezes, 15 April 1913.

[40]Presidents' papers/[PM 1907–1929]: Moore to Houston, 6 September 1907, Veblen to "Dear Sir," 29 April 1908.

[41]Presidents' papers/[PM 1907–1929]: Porter to Mezes, 13 September 1910; W. B. Phillips to Mezes, 13 September 1910.

On a less personal issue, Arthur Lefevre, in his capacity in 1912 as Secretary for Research of the Organization for the Enlargement by the State of Texas of Its Institutions of Higher Education, published a critique of the present state of affairs at the university, especially calling attention to the fact that though doctoral programs were announced "a few years ago" in the catalogue, no one has completed one. He attributed the root cause to inadequate support:

> The average salary paid the teaching force of the University of Texas thirty years ago was double the present average salary. How could an intelligent man demand of the University of Texas, in its present circumstances, the first-class research and manifold services to the general public which have come to be essential characteristics of the modern university?[42]

Lefevre's study, prepared on behalf of an alumni group, did not make numerical comparisons with other universities and did not go into the fact that the "teaching force" was being increased at the lower end of the salary scale. One sign that the mathematics instructors' salaries, at least, offered by Texas were still competitive, in spite of the absolute decline he noted, is the addition in 1913 of a new instructor from Harvard to the School of Applied Mathematics, Hyman J. Ettlinger, at the usual rate for Texas of $1,200.[43] Raised in St. Louis, Ettlinger had been very active in Jewish affairs from his school days, through Washington University in St. Louis, and at Harvard University where he obtained his M.A. in 1911 and was to get his Ph.D. in 1920 under G. D. Birkhoff. It was about the time of Ettlinger's Ph.D. degree that Harvard instituted its *numerus clausus* which set a limit on the number of Jews admitted and which Norbert Wiener described as killing "the last bonds of my friendship and affection for Harvard." There were thirty Jewish students at Texas when Ettlinger arrived there, and he helped establish a Menorah Society for them. In 1915, a rabbi was appointed to the Board of Regents. If there were any problems caused by Ettlinger's being Jewish, either at Harvard or at Texas, he made nothing of them in his later reminiscences. This is not to say there were no problems, but he claimed there was only one incident in sixty-one years. It occurred around 1917 when Ettlinger was an assistant football coach at Texas. In an argument with a caretaker who refused to unlock a door so that Ettlinger could get a referee's whistle, the caretaker said that he would not do it for anyone "and certainly not for a Jewish...," whereupon the 210-pound Ettlinger floored him. He would have ignored the man's remark, Ettlinger has said, but for the fact

[42][Lefevre, pp. 42–43].

[43]Presidents' papers/College of Arts and Sciences/Applied Mathematics, 1909–1913: copy of telegram from Rice to Ettlinger, 14 June 1913. Salaries at Kansas were comparable to those at Texas but in [Price, pp. 187–188] it is maintained that the national scale was rising and that Kansas suffered as a consequence.

that his football players witnessed it and he felt a lesson was needed on the spot. Ettlinger was accused of using less violent but still physical tactics in departmental controversies of subsequent years, but that goes beyond the scope of this account. Moore, smaller than Ettlinger in stature but a proficient boxer at Princeton who kept himself in shape, was also known on occasion to make aggressive use of his physical capabilities in academic disagreements.[44] In Texas, at least, the successful use of such nonverbal language need not detract from one's reputation. In fact, for an established male scholar, it adds a certain cachet which can probably only help one's reputation outside the scholarly world. Still, it seems that no other department at Texas has had members with quite this reputation.

The first doctoral degree in mathematics was earned by Goldie Prentis Horton in 1916 with a thesis entitled "Functions of limited variation and Lebesgue integrals" done under Porter's supervision. She came from Elms, Texas, as an undergraduate in 1904, received a bachelor's degree in 1908, and then alternated teaching in schools with earning a master's degree at Smith College and studying at Bryn Mawr. She accepted a tutorship at Texas in 1913. In 1917, she was promoted to instructor at $1,000. (Porter's salary was then $3,000, Dodd's $1,900, and, to select another but more senior instructor, Mary Decherd's was $1,200.) Dodd, as chairman of pure mathematics, had appealed in 1916 for more staff to help cope with increases in enrollment. In the primary undergraduate course, the enrollment was 675 split up into 22 classes, and there were seven advanced classes.[45]

This increase in enrollment was soon met with a substantial increase in teaching force, at least of those who were below the rank of associate professor. Besides Porter, Dodd, and Calhoun (transferred to pure mathematics the previous year), in 1915–1916 there were nine others listed in pure mathematics including student assistants. Three years before, there were only three besides Porter and Dodd in pure mathematics. There was no such increase in staff for applied mathematics, and, in fact, its total number stayed fairly constant for the next twenty years.

Continuing in an expansionist direction, Texas hired A. A. Bennett, an instructor at Princeton, as an adjunct professor in pure mathematics in 1916. Veblen, Bennett's doctoral supervisor at Princeton, wrote Porter that Bennett had published in the *Annals of Mathematics* and "has wider knowledge of mathematics than any man of his age whom I know." Princeton would try to keep him, Veblen said, but "I am not sure, however, that they will be able to meet your offer...." Veblen also mentioned others that Texas might

[44][Wiener, pp. 271–272]; [Greenwood 1986]. Ettlinger interview with A. C. Lewis, 6 February 1974. [Traylor, pp. 71–72, 89, 127].

[45]Presidents' papers/[PM 1907–1929]: W. J. Battle to Calhoun, 26 April 1916, Dodd to President Vinson, 16 October 1916.

be interested in, but on the back of this letter Porter wrote, "I vote for Bennett." Once the offer was made, Veblen wrote again to Porter saying that he would recommend Bennett take the offer, unless Princeton could duplicate the position and salary, and volunteered more information about him: "no doubt you will be satisfied with Bennett... the only handicap which he has is a certain rapidity of utterance—and this I should think would be less of a drawback in the less effete atmosphere of Texas."[46]

Applied mathematics also increased its staff at this time by the addition of Paul M. Batchelder from New Hampshire. He received an M.A. from Princeton in 1910 and took courses under G. D. Birkhoff. When Birkhoff moved to Harvard, Batchelder went with him and, after two years teaching at Northwestern University, in 1916 obtained his doctoral degree under him. W. F. Osgood recommended Batchelder to Texas and Harvard sent a biography which included an evaluation Birkhoff wrote in 1913: "While I should not characterize him as a man of unusual powers of original investigation, I feel he is possessed of a clear insight and that he has the unusual gift of clear presentation." Birkhoff also said that he found Batchelder likable and "well worth while as a friend." Ettlinger wrote to the chairman of his department at Texas that Batchelder "would make us a good man": "I knew him fairly well. His temperament is very much like Barrow's [David F. Barrow, Ph.D. Harvard, 1913, instructor at Texas, 1914–1916], quiet and retiring. His health at one time was precarious." Batchelder was promptly offered $1,300, and he joined the faculty as an instructor in both pure and applied mathematics. He retired in 1954 as an associate professor.[47]

R. L. Moore and H. S. Vandiver, 1920–1938: New Beginnings

The global confrontations of World War I seem to have had no appreciable effect on the development of the mathematics faculty, but 1917 saw a political battle in Texas which greatly influenced the future place of the university within state politics. This event also helped to eventually bring more emphasis to rewarding faculty members for scholarly achievement rather than for seniority. Though there were undoubtedly supporters for this notion before on the campus, an attack by the governor on the independence of the university and its Board of Regents provided an impetus for this reform.

James E. Ferguson, a banker, had been elected governor in 1914 as a supporter of business and a rescuer of the small farmers who were major victims of the depression of 1913–1914. The trouble appears to have begun

[46]Presidents' papers/[PM 1907–1929]: Veblen to Porter, 24 April 1916 and 8 May 1916.

[47]Presidents' papers/Dean of Arts and Sciences/Applied Mathematics, 1913–1919: Osgood to Rice, 12 July 1916; Ettlinger to Rice, 22 July 1916; typed sheet from Harvard University [?], July 1916; Graff (Secretary to the President) to Rice, 26 July 1916.

by his insistence on a direct say about the university's budget in all its details. When there was resistance, he declared that the regents of this "autocratic University" were in for "the biggest bear fight in Texas." The university president, R. E. Vinson, wrote from the depths of uncertainty to a potential faculty member in June 1917 that the "Governor of Texas has vetoed the entire University appropriation for the next biennium [1917–1919], because the regents failed to accede to a demand of his that four members of the faculty and I be discharged from the University... feel yourself entirely free to accept other employment, in case we are not able to keep the University open."

A coalition of opposition forces joined the supporters of the university and brought about successful impeachment proceedings against Ferguson in the legislature in August of 1917. New regents were appointed by the acting governor and all but one of the fired professors were reinstated.[48]

Just as at other state universities in the country which went through similar catastrophes, there was the positive effect of delineating and maintaining—at least for a period—the boundaries of authority between the governor, on the one side, and the regents and university administration on the other.[49] Perhaps in the long run, it also caused a more vigorous campaign to take advantage of the increasing appropriations being granted the university to attract high quality faculty from outside. None of the mathematics faculty were among those fired or threatened in 1917 and none of the senior faculty, at least, left. In fact, new members were added during the next three years. The post-war increase in the student population and the consequent demand for mathematics opened up jobs across the country.

Pure mathematics in 1919 had been doing without the services of A. A. Bennett who had been taking leaves of absence each year since 1917 to do government war-related work on computation of ballistics tables. He offered to resign in 1919, but the university was willing to try to retain the connection. The applied mathematics faculty needed to be augmented to meet the new demands, and two instructors were added in 1919. One, A. E. Cooper, was a doctoral student of L. E. Dickson at Chicago and helped with Dickson's three-volume *History of the Theory of Numbers* (Carnegie Institute, Washington, D.C., 1919–1934). When an offer of $1,600 was made to Cooper, he at first said that his present position for a private company paid twice that and he asked if he would be getting more later. Evidently, the university stood firm,

[48]Presidents' papers/[PM 1907–1929]: Vinson to T. M. Simpson of Chicago, 4 June 1917. [Frantz, pp. 72–81], [Fehrenbach, pp. 638–639], [Gould].

[49][Price, pp. 184–186] recounts a similar experience at The University of Kansas in the period 1917 to 1924.

before, was still there, though now as an instructor, and would remain until her retirement in 1944.[54]

Moore's first doctoral student at Texas was Raymond L. Wilder. In the fall of 1921, Wilder came to Texas with an M.S. degree from Brown University. R. G. D. Richardson had suggested Wilder for any vacancy that Porter and Dodd might have, and a Brown University statement about Wilder included the fact that his principal work was in accounting. Porter asked President Vinson to offer an $1,800 instructorship "at once lest we lose him."[55]

Though it was Dodd he came to work with, Wilder has told how he decided to take some additional mathematics and Dodd suggested Moore's course. Wilder introduced himself to Moore during registration:

> I soon realized that he was very negative about my enrolling in his course...I had two counts against me, as I analyzed it later. One was that I was a Yankee. The second was that I was an actuarial student, and what in the world was an actuarial student doing taking a course from Moore? Well, this went on for some time, and I didn't want to give up. He finally made the mistake of asking me, "What is an axiom?" I had pretty good training at Brown, and I knew what an axiom was. His guard was down, and I think he, in utter frustration, said, "OK, go ahead, and take the course." Actually, I wasn't really in the course until I proved what we called Theorem 15 in those days.[56]

Wilder eventually became one of the three Moore students, with Bing and G. T. Whyburn, who were elected presidents of the American Mathematical Society.

Two years later, Wilder had his Ph.D. and an offer of $2,750 to go to Oklahoma A&M (now Oklahoma State University). Moore tried to keep him at Texas. Benedict thought he could be given $2,400, and Moore and Porter went together to argue their cause with Acting President T. U. Taylor. But Taylor—with his many years of experience vis-à-vis pure mathematics and an emphasis on seniority in service to the university rather than current academic market value—made an offer to Wilder of only $2,200. Taylor maintained that pure mathematics was "well-equipped in man-power having such top-notchers as Porter, E. L. Dodd, Moore, Ettlinger, A. A. Bennett,

---

[54]Presidents' papers/[PM 1907–1929]: Moore's letter and Porter to Vinson, 1 July 1920, regarding Mullikin.

[55]Presidents' papers/[PM 1907–1929]: Richardson to Porter and Dodd, 23 February 1921; Porter to Vinson, 12 March 1921; Vinson to Wilder, 14 March 1921; Wilder to Vinson, 17 March 1921.

[56]Remarks at the presentation breakfast of The University of Texas at Austin Mathematics Award honoring the memory of Professors R. L. Moore and H. S. Wall, San Antonio, 24 January 1976, recorded and transcribed by Lucille E. Whyburn. The award (to William T. Eaton) and the event was organized by H. J. Ettlinger. More details of Wilder at Texas are given in [Bing].

and also such Instructors as Miss Goldie Horton." "I recall," he continued, "that Miss Horton has had the Ph.D. degree for many years and is yet an Instructor. I also recall the fact that the Administrative Council has been very kind to the Department of Mathematics ["departments" now instead of "schools"] and gave Professor Moore a jump in salary that broke the Southern record." Taylor pointed to another faculty member offered $3,500 by the same Oklahoma institution and claimed he could not offer him more of a raise though his services to his department were more necessary than Wilder's to pure mathematics. Next September, Wilder tendered his resignation so that he could accept an assistant professorship at Ohio State University.[57]

The higher administration appeared for a while to be unenthusiastic about providing for Moore's further advancement. In September of 1921, Benedict, as Dean of Arts and Sciences, conveyed his balanced recommendations for promotions to President Vinson:

> With some misgivings... I append a list of those persons who seem to me to be superior to some men, to say the least, who are now ranked above them, and who ought, therefore, sooner or later, to be promoted in rank: ... Cooper, Dodd, Calhoun, ... The cases of Bennett and Moore demand special comment. Moore is perhaps the more talented; Bennett the more persistent. Both are starred men in Cattell's authoritative list [*American Men of Science*]. (They and [the zoologist] J. T. Patterson constitute our only three live stars, an almost disgraceful situation.) They have been here but a short time; they would have to be made professors purely on scientific merit; but their promotion, particularly if accompanied by a statement of policy in regards the Ph.D. and its accompaniments, would tone up the situation among the "intellectuals." But what about making three full professors in mathematics at once?[58]

The wavering stance of the letter hints at the potential for discontent over what some might regard as an unseemly haste and indeed it took two more years before any of these promotions in pure mathematics came about.

The university could not complain at this time about salaries, at least as far as averages went. Benedict made a comparative study in 1921 from which he concluded that "the salary scale in the College of Arts and Sciences compares very favorably with that found at other state institutions. Our averages are as good as any in the case of associate professors, adjunct professors, and

[57]Presidents' papers/[PM 1907–1929]: text for telegram from Moore to Porter, 1 September 1923; copy Taylor to Wilder, undated; "Acting President" Taylor to "Acting Acting President" Sutton, 10 September 1923; Wilder to Splawn, 8 September 1924.

[58]Presidents' papers/Dean of College of Arts and Sciences, 1923–1924: Benedict to Vinson, 15 September 1921. Porter had also been starred in the first edition of Cattell.

instructors. We fall behind California and Wisconsin in professors' averages because of a few $6,000 and more salaries in those institutions."[59]

In the summer of 1923, before Wilder went to Ohio State, that institution made inquiries to see if Moore himself could be tempted away. Dodd came close to leaving the same year when he took a leave of absence to teach at Williams College.[60] Finally, both Dodd and Moore were made full professors in 1923. In the spring of that year, the university's endowment of grazing land was enhanced when Santa Rita #1, the university's discovery oil well, blew in.

Approval was given in 1924 by the administration for hiring Harry Schultz Vandiver from Cornell as an adjunct professor in pure mathematics at $2,800 to take the place left temporarily by Batchelder. The latter had just been made adjunct professor, perhaps as a security against permanently losing him to Brown University, where he went to teach for the year 1924–1925. Vandiver had been recommended to Porter by L. E. Dickson with whom Vandiver, like Cooper, had worked on the *History of the Theory of Numbers*. By 1924, Vandiver had twenty-three publications on number theory starting with a collection of problems and problem solutions in the *American Mathematical Monthly* from 1900 to 1904. It was through these problems that he got to know G. D. Birkhoff, and the two co-authored Birkhoff's first paper in 1904. Vandiver came to epitomize pure mathematics at the university to an even greater extent than Moore. Whereas Moore played two of the three roles Halsted had posited in 1876, the teacher and original writer, Vandiver played mainly that of original writer. Though they had about the same number of years in academic positions, Vandiver, retiring in 1966 at age 84, and Moore in 1969 at 86, Vandiver took leaves of absence for research, while Moore had ten times as many doctoral students as Vandiver and devoted himself to regularly teaching undergraduate as well as graduate courses. Vandiver had the further distinction of being almost entirely self-taught and having no degrees, or even a high school diploma, until he was awarded an honorary degree in 1946 by the University of Pennsylvania.[61]

Bennett, who was chairing the Department of Pure Mathematics in 1922–1923, remained an associate professor and did not participate in the promotions of that year. In January of 1925, the three full professors of pure mathematics—Porter, Dodd, and Moore—requested the administration raise the salaries of Bennett and Ettlinger to $3,600. They pointed out that Texas Technological College had offered "one of our staff" $3,750. The requested raise was approved by the regents, but by the time it was to take effect later in

[59]Presidents' papers/College of Arts and Sciences, 1921–1924: report of 10 June 1921.

[60]Presidents' papers/[PM 1907–1929]: copy of R. D. Bohannan to Moore, 4 June 1923; Dodd to Vinson, 8 April and 23 April 1922.

[61]Presidents' papers/[PM 1907–1929]: Acting President Sutton to Vandiver, 16 July 1924. [Greenwood 1983], [Vandiver H 1963].

1925, Bennett had been attracted back to Brown University, his alma mater.[62] He is the only member of pure mathematics of adjunct rank or higher who came to Texas between 1902 and 1938 and ever left for another university. (There were two such in applied mathematics, J. N. Michie and C. A. Rupp.)

Porter had worked towards the establishment of a graduate faculty since at least 1920. In 1925, thanks in good part to his efforts, this was accomplished. That same year, R. G. Lubben, a student since 1916, received his doctorate under Moore and stayed on as a member of the faculty until 1959 when he retired due to illness. In 1927, Gordon T. Whyburn, who had his B.A. and M.A. in chemistry from Texas, became Moore's third doctoral student, and his brother, William M. Whyburn, became Ettlinger's first. A report on the department of pure mathematics stated that as of September 1927 there were 1,068 students in 42 freshman sections, and 135 students in higher classes, not counting those in astronomy or aeronautics, which were taught in pure mathematics in 1927–1928. Also, National Research Council Fellowships had been received by Lubben for study in Göttingen and by W. M. Whyburn for work at Harvard. On his return, W. M. Whyburn went to the University of California at Los Angeles where he served as chairman from 1937 to 1944. In 1927, W. T. Reid received an M.A. degree on a subject suggested to him by Dodd, and went on to get a doctorate with Ettlinger in 1929.[63] In 1925, Lucille Smith entered the university to major in English, and worked as a computer (on the Monroe calculator) for Vandiver. She took a course with Moore, became interested in mathematics and, as Mrs. G. T. Whyburn, in 1936 she obtained an M.A. degree under Moore.

With pure mathematics thriving, the perennial question of its relationship to applied mathematics was raised again in 1926 by T. U. Taylor. "Several years ago," he wrote to President Splawn, "the Engineering Faculty unanimously recommended that the Department of Applied Mathematics be included in the College of Engineering like the Department of Drawing.... The Department was created during the Houston Administration solely for this purpose." His recommendation to fix this "illogical" situation was to appoint Benedict professor of astronomy and transfer him to pure mathematics and then transfer the applied mathematics department to engineering. Benedict's response was that he had long thought the departments should be "fused." Calhoun, still in applied mathematics, requested that the president give the matter careful consideration.[64] The only outcome of the consideration was to change the name to Department of Applied Mathematics and Astronomy.

[62]Presidents' papers/Dean of Arts and Sciences, 1924–1929: note by Benedict [?], undated, "Bennett and Ettlinger...". Presidents' papers/[PM 1907–1929]: Porter, Dodd, Moore to President and Dean [January 1925].

[63][Vandiver H 1961], [Greenwood 1988, p. 14]. Presidents' papers/[PM 1907–1929]: memorandum, 16 July 1928.

[64]Presidents' papers/Dean of Faculty/Applied Mathematics, 1924–1929: Taylor to Splawn, 1 March 1926; Benedict to Splawn, 10 March 1926; Calhoun to Splawn, 13 March 1926.

Taylor died in 1941, and the interdepartmental situation remained essentially unchanged until the shotgun merger of the departments in 1953. The new building, then housing the new group, was named after Benedict.

Benedict's rise through the administrative ranks reached its pinnacle in 1927 when he succeeded Splawn as president of the university. Benedict's fellow student in Halsted days, L. E. Dickson, contributed a brief encomium for the occasion:

> All are familiar with his success as dean, due to his unerring judgment, rare talents as an executive, and deep affection for the University. But I wish to emphasize the fact that the man having all these essential qualities is also a scientist. This is the age of science. Himself an astronomer, Benedict is just the man to make a success of the new astronomy observatory so amply endowed. ...A university is no longer counted as a great one unless it is a center of research in the various sciences. And only then does it serve adequately the needs of modern life. Benedict is the ideal man to steer the University of Texas toward greatness.[65]

The department of applied mathematics began to grow in 1928. Ernst George Keller, a graduate from Chicago, was added as an adjunct professor at $2,800. When he went on a leave of absence for the following year, a replacement, Homer Vincent Craig, was hired at $2,000. The budget for the year 1931 of the Great Depression shows that Benedict was receiving $10,000 (presumably his total salary as president), Calhoun (only part time in mathematics) $4,000, Cooper $3,750, Cleveland $2,800, Keller $2,617, and Craig $2,600. In 1932, R. N. Haskell was appointed an adjunct professor. Haskell had been recommended by Griffith C. Evans at Rice Institute (now Rice University) as an "attractively married" mathematician who had published two papers, including his 1930 thesis, with Evans on potential theory.[66]

A physics professor and Dean of the College of Arts and Sciences evaluated the effect of Craig and Haskell in a report of 1950:

> Both of these men are inspiring teachers. Both of them do an excellent job of teaching the fundamentals of the subject. Both are interested in their students and spend a lot of time with their students at any and all hours. The major swing of science students

[65] *The Alcalde*, 27 November 1927.

[66] Presidents' papers/Dean of Faculty/Applied Mathematics, 1924–1929: Calhoun to Benedict, 25 June 1928; Benedict to Calhoun, 28 August 1929. Presidents' papers/Budget and Departments/College of Arts and Sciences/Applied Mathematics and Astronomy, 1929–1939: Calhoun to Benedict, 12 August 1931; Cooper to Parlin, 8 September 1932; Evans to Cooper, 3 September 1931.

> from Pure Mathematics to Applied Mathematics is attributable in considerable measure to these two men.[67]

On the pure mathematics side, G. T. Whyburn was awarded a John Simon Guggenheim Memorial Foundation Fellowship for 1929–1930 and he and his wife went to Vienna and worked with Hans Hahn. On his return, he went to Johns Hopkins and in 1934 from there to the University of Virginia. Moore gave the Colloquium Lectures of the American Mathematical Society for 1929 (published in 1932, revised in 1962, and reprinted in 1970), became Visiting Lecturer for the American Mathematical Society for 1931–1932, and was elected to the National Academy of Sciences in 1931 after a practically unanimous vote of the mathematics section. Vandiver received a Guggenheim fellowship for 1927–1928, in 1931 was awarded the Cole Prize in the theory of numbers by the American Mathematical Society, and delivered the Colloquium Lectures for 1935 (unpublished). In 1934, he was elected to the National Academy of Sciences.[68]

Porter and Goldie Horton collaborated on a textbook, *Plane and Solid Analytic Geometry* (Edwards Brothers, Ann Arbor), and when it was published in 1934 they were married. They both continued in the department, until he retired in 1945 and she, after teaching part time from 1958, retired in 1966.[69]

People from Benedict's day can vividly recall when they learned of his death in 1937. Apart from being a well-liked president, it happened in a way that seemed in keeping with his hard-working West Texas upbringing. As one who was a student at the time put it, he "dropped dead from a heart attack on the sidewalk in front of the old YMCA building, a center of campus activity for many generations. In another world you would have expected him to have fallen dead behind a plow or in his Fordson tractor seat." Calhoun took over as acting president, and, in spite of a reputation as a tight-fisted comptroller of the university, put through the first faculty pay raise in many years. Just the previous year, the pure mathematics department was having problems holding on to its instructors. Porter, Moore, Dodd, Ettlinger, and Vandiver unsuccessfully petitioned the dean for an extra $900 to enable them to keep C. W. Vickery, a 1932 Moore doctoral student, as a full-time instructor "to help with the excessive size of freshman classes." In the summer of 1937, O. H. Hamilton, who had just received his doctorate with Ettlinger, declined an

[67]Presidents' papers/C. P. Boner, five-page memorandum to file, 23 December 1950. Boner's main purpose was to document his belief that R. L. Moore was no longer a positive influence for mathematics at Texas. In a conversation with the author on 20 March 1975, Boner, who died in 1979 at age 79, shed no light on this matter, or the Halsted instructorship incident cited above, and only recounted some earlier and friendlier memories of Moore.

[68][Whyburn], [Archibald, pp. 21, 39, 73]. G. D. Birkhoff papers (Library of Congress): letter to members of the mathematics section of the academy, 27 April 1930.

[69][Greenwood 1972], [Vandiver H 1961].

offer of $900 for a half-time instructorship so that he could accept an offer from Oklahoma A&M.[70]

For 1935–1936, pure mathematics had as instructors, besides Vickery, several others who had recently received doctorates with Moore: Edmund C. Klipple (1932), Robert E. Basye (1933), and F. Burton Jones (1935). Robert E. Greenwood, who had received his B. A. in 1933, was the only instructor who was not a Moore student. After receiving his M.A. and Ph.D. from Princeton, he was to return in 1938 as an instructor in applied mathematics and was one of those who helped to mediate between the two groups on the occasions when their relations deteriorated. He retired as a full professor in 1981.

Money for the top level of faculty was somewhat more forthcoming in 1937 than for instructors. The Texas legislature approved a bill establishing a category of Distinguished Professor and providing $6,500 salaries for nine months to three "nationally distinguished" faculty members to be nominated and voted upon by the graduate faculty. With 58 voting by ballot for the first recipients of this honor, the historian Eugene C. Barker received 33 and Moore 29. The other nominees, geneticist T. S. Painter with 23 votes, zoologist J. T. Patterson with 11, and Vandiver with 7 did not have majorities. Apparently, these results were handed on by Acting President Calhoun to the Board of Regents who then selected Patterson as the third. Vandiver sent in a late ballot for Moore and for the geologist E. H. Sellards, and Vandiver in turn received votes only from Dodd and Porter in pure mathematics. There is no record of Moore taking part in the vote but, in addition to Vandiver, his other colleagues in pure mathematics, Dodd, Porter, and Ettlinger, also voted for him. Later alignments of this group make it tempting to read more into Moore's nonparticipation and the failure of Ettlinger to vote for Vandiver than the documents support, but it is at least clear that there was no complete reciprocity in supporting each other. On the other hand, relationships had clearly not broken down to the extent they were to do by ten years later when it was unlikely either Vandiver or Moore would support each other for anything favorable.[71]

Whatever the relationships were before the balloting, the outcome did not sit well with Vandiver who appealed to the president in 1939. He maintained that reputations could be damaged by the implication that those not selected as Distinguished Professor were in fact not distinguished. "There is in my opinion," Vandiver wrote, "no individual in the faculty or any group of individuals who are at present in a position to estimate the national or international reputation of any particular member of the faculty." Vandiver

[70][Frantz, p. 139]. Presidents' papers/Dean of Arts and Sciences/Pure Mathematics, 1929–1939: Vickery to Parlin, 8 October 1936; Hamilton to Regents, 10 June 1937.

[71]Presidents' papers/General Administration/Distinguished Professors, 1937–1938; [Greenwood 1983, p. 20].

requested that the president make this determination himself and enclosed biographical information, including lists of research grants and publications. He was appointed a Distinguished Professor in 1947.[72] Benedict's death in 1937 seems to have marked the end of a period of relative harmony in the mathematical community at Texas.

Postscript

The University of Texas had passed its fifty-year milestone in 1933. In 1939, the president's office sent a questionnaire to the faculty which invited them to evaluate to what extent the university was "a University of the first class" as called for by the 1876 constitution. To the key question, "Does Texas have a University of the First Class?", Dodd answered "No," Ettlinger "Yes, but it can be improved," and Porter "No." There is no record of responses from Moore or Vandiver (who was on leave for part of the year). Porter's replies stand out among all the faculty responses because he simply and directly called for just those practical reforms which did, in fact, take place before long. The university "needs more first class professors. Fellowships are also needed." To the question "Is the intellectual atmosphere conducive to having a University of the first class?", Porter replied "No." "There should be more $6,500 professors for people that deserve them." There was "a lack of understanding and appreciation of high grade research." "I believe there should be a senate to decide important questions of policy and that competent committees should be appointed to assist the Deans in the selection of new professors above assistant. There should be an aggressive and well-equipped Dean of the Graduate School...." He noted at the end, "That skill in teaching should receive full recognition goes without saying. I believe this is less apt to be overlooked than other qualifications."[73]

Though Porter participated in the founding of the graduate school, he never held a higher administrative position than departmental chairman. During his most active years, he might have done much for the university at a higher level, but, as it was, the mathematics faculty was the main beneficiary and arguably came closer than any other department at the university to the high standard he expressed.

Acknowledgments

In addition to the cited interviews, interviews with the following have provided background information: Anne Barnes, R.H. Bing, Robert E. Greenwood, James M. Hurt, Richard P. Kelisky, R. G. Lubben, and Lucille E. Whyburn. Help in locating material has been provided by the archivist

[72]Presidents' papers/General Subject/Vandiver, 1939. Vandiver to Rainey, 20 October 1939.

[73]Presidents' papers/General Policy/University of the First Class, 1939.

of the Archives of American Mathematics, Frederic F. Burchsted, and by the reading-room staff of the Eugene C. Barker Texas History Center. Uta Merzbach and R. E. Greenwood provided valuable critiques of early drafts. Professor Greenwood has been the departmental memory for many years and has been principally responsible for the faculty memorial resolutions (listed under his name in the bibliography) which show a personal and sensitive acquaintance with the subjects. The present account seeks to provide an archival complement to these biographies.

References

[Archibald] Raymond Clare Achibald, *A Semicentennial History of the American Mathematical Society*, Amer. Math. Soc., New York, 1938.

[Benedict] H. Y. Benedict, *A Source Book Relating to the History of the University of Texas: Legislative, Legal, Bibliographical and Statistical*, University of Texas Bulletin No. 1757, October 10, 1917, University of Texas, Austin.

[Bing] R. H. Bing, "Award for distinguished service to Professor Raymond L. Wilder," *Amer. Math. Monthly* **80** (1973), 117–119.

[Broun] Thomas L. Broun, compiler, *Dr. William LeRoy Broun*, Neale Publ. Co., New York, 1912.

[Calhoun] J. W. Calhoun, *Mathematics Pure and Applied. University of Texas 1883–1946*, mimeographed typescript, 1946.

[Eisele] Carolyn Eisele, "Peirce's philosophy of education in his unpublished mathematics textbooks" in *Studies in the Philosophy of Charles Sanders Peirce, Second Series* (E. C. Moore and R. S. Robin, eds.), University of Massachusetts, Amherst, 1964, pp. 51–75.

[Fehrenbach] T. R. Fehrenbach, *Lone Star: A History of Texas and the Texans*, American Legacy Press, New York, 1983.

[Frantz] Joe B. Frantz, *The Forty-Acre Follies*, Texas Monthly Press, Austin, 1983.

[Geisser] S. W. Geisser, "Men of science in Texas, 1820–1880," *Field and Laboratory* **27** (1959), 43.

[Gould] Lewis L. Gould, "The University becomes politicized: The war with Jim Ferguson, 1915–1918," *The Southwestern Historical Quarterly* **86** (1982), 255–276.

[Greenwood 1970] Robert E. Greenwood, "Memorial Resolutions for C. M. Cleveland," *Documents and Minutes of the General Faculty*, University of Texas at Austin.

[Greenwood 1972] ——, "In Memoriam, Mrs. Goldie Horton Porter," *Documents and Minutes of the General Faculty*, University of Texas at Austin.

[Greenwood 1974] ——, "In Memoriam, Paul Mason Batchelder," *Documents and Minutes of the General Faculty*, University of Texas at Austin.

[Greenwood 1975] ——, "In Memoriam, Robert Lee Moore," *Documents and Minutes of the General Faculty*, University of Texas at Austin.

[Greenwood 1982] ——, "In Memoriam, Homer Vincent Craig," *Documents and Minutes of the General Faculty*, University of Texas at Austin.

[Greenwood 1983] ——, "Mathematics," *Discovery*, Centennial Issue, pp. 18–22.

[Greenwood 1986] ____, "In Memoriam, H. J. Ettlinger," *Documents and Minutes of the General Faculty*, University of Texas at Austin.

[Greenwood 1988] ____, "History of the various departments of mathematics at The University of Texas at Austin (1883–1983)," unpublished typescript.

[Griffin] Roger A. Griffin, "To establish a university of the first class," *The Southwestern Historical Quarterly* **86** (1982), 135–160.

[Halsted 1893] George Bruce Halsted, Halsted entry in *The National Cyclopaedia of American Biography*, vol. 3, p. 519.

[Handbook] *The Handbook of Texas*, 2 vols., The Texas State Historical Society, Austin, 1952.

[Lane] J. J. Lane, *A History of the University of Texas Based on Facts and Records*, Henry Hutchings State Printer, Austin, 1891.

[Lefevre] Arthur Lefevre, *The Organization and Administration of a State's Institutions of Higher Education: A Study Having Special Reference to the State of Texas*, Von Boeckmann-Jones, Austin, 1912.

[Lewis 1973] Albert C. Lewis, "Halsted's translation of Lobachevskii's *Theory of Parallels*: An historical introduction," *The Texas Quarterly* (1973), 85–91.

[Lewis 1976] ____, "George Bruce Halsted and the development of American mathematics", in *Men and Institutions in American Mathematics*, Graduate Studies, Texas Tech University, No. 13, pp. 123–129.

[Mallett] J. W. Mallett, "Reminiscences of the first year of The University of Texas," *The Alcalde*, April 1913, pp. 14–17.

[O'Connor] Richard O'Connor, *O. Henry: The Legendary Life of William S. Porter*, Doubleday, New York, 1970.

[Price] G. Baley Price, *History of the Department of Mathematics of The University of Kansas, 1866–1970*, The University of Kansas, Lawrence, Kansas, 1976.

[Taylor] T. U. Taylor, *Fifty Years on Forty Acres*, Alec Book Company, Austin, 1938.

[Traylor] D. Reginald Traylor, *Creative Teaching: Heritage of R. L. Moore*, University of Houston, Houston, 1972.

[Tropp] Henry Tropp, "George Bruce Halsted" in *Dictionary of Scientific Biography* (C. C. Gillispie, ed.), 16 vols., Scribners, New York, 1970–1980.

[Vandiver F] Frank E. Vandiver, "John William Mallett and The University of Texas," *The Southwestern Historical Quarterly* **53** (1950), 422–442.

[Vandiver H 1961] H. S. Vandiver with J. A. Burdine and R. A. Law, "In Memoriam, Milton Brockett Porter," *Documents and Minutes of the General Faculty*, University of Texas at Austin.

[Vandiver H 1963] ____, "Some of my recollections of George David Birkhoff," *J. Math. Anal. Appl.* **70**, 271–283.

[Wakelyn] Jon L. Wakelyn, *Biographical Dictionary of the Confederacy*, Greenwood Press, Westport, Connecticut, 1977.

[Whyburn] Lucille E. Whyburn, "An American in Göttingen 1926–1927: Letters from J. R. Kline to R. L. Moore," unpublished talk delivered at the American Mathematical Society Meeting, Atlanta, Georgia, January 1978.

[Wiener] Norbert Wiener, *Ex-prodigy: My Childhood and Youth*, The M.I.T. Press, Cambridge, Massachusetts, 1953.

[Wilder 1976] Raymond L. Wilder, "Robert Lee Moore, 1882–1974," *Bull. Amer. Math. Soc.* **82**, 417–427.

[Wilder 1982] "The mathematical work of R. L. Moore: Its background, nature and influence," *Archive for History of Exact Sciences* **26**, 73–97.

Charlotte Angas Scott (1858–1931)

PATRICIA CLARK KENSCHAFT

Biography

Charlotte Angas Scott was born on June 8, 1858, in Lincoln, England, the second of seven children of Caleb (1831–1919) and Eliza Exley Scott. The only extant information about her mother is references in her father's obituaries. They report that the marriage was "a source of profound happiness" to him and that she died in 1899 when he was on his way home from the United States, where he had attended the International Congregational Council and visited his "eldest daughter at Bryn Mawr" ("Ministers Deceased" 1919).[1]

However, a great deal is known about Caleb and his father, Walter Scott (1779–1858), because they were both ministers of the Congregational Church and presidents of colleges training such ministers. Walter Scott was a hard-driving man who struggled for education of the working classes and against slavery and alcohol consumption. His eighth offspring, Caleb, had had three successful years in business and had obtained two degrees by the age of twenty-three. Since their religion was "Non-conformist," and Cambridge and Oxford Universities required a vow of loyalty to the Church of England, Walter and Caleb developed alternative sources of education for young men of their religion.

Since there were no colleges in England open to women while Charlotte Scott was growing up, and almost no secondary schools either, the support of her family and church was indispensable to her education. In a speech to newly elected deacons, Caleb admonished their wives, "Let the innocent tastes and tendencies of youth not be all repressed and stifled in the iron mould of any conventionalism" (Scott 1865). This was a man who encouraged his family to think and to enjoy life, and Scott's later writing indicates

[1]Patricia Clark Kenschaft, "Charlotte Angas Scott (1858–1931)," in *WOMEN OF MATHEMATICS A Biographic Sourcebook*, Louise S. Grinstein, and Paul J. Campbell, eds. (Greenwood Press, Inc., Westport, CT, 1987), pp. 193–203. Copyright ©1987 by Louise S. Grinstein and Paul J. Campbell. Reprinted with permission.

that mathematical games were part of their home entertainment. In 1865 Caleb became principal of the Lancashire Independent College (now called the Congregational College) and thus was able to provide good tutors for an ambitious daughter.

In 1876, at the age of eighteen, she won a scholarship on the basis of home tutoring to the recently opened Girton College. Most of her classmates had never attended a secondary school either. However, secondary schools for girls were springing up in England, so educated women suddenly had career opportunities as teachers. Thus there were eleven students, an unprecedented number, in Scott's entering class at Girton College, the first college in England for women. Life was austere. "When retiring for study after an extremely simple 'tea' in the Commons, they would pick up three things en route to their rooms...two candles, a bucket of coals, and a chamber pot" (Silver 1981).

Girton College had opened in 1869 with five students at a different location and in 1873 had moved to a modest three miles from Cambridge University, thereby enabling its students to attend the lectures of the twenty-two (out of thirty-four) Cambridge professors who were willing to let women listen to them. Such women had to be carefully chaperoned, because until 1894 Cambridge University maintained the "Spinning House," a special prison for prostitutes and "suspected prostitutes," where any unescorted woman would be summarily sent, her entire future thereby ruined. One student of the 1890s told her son-in-law that women attending lectures sat in the back behind a screen, obviously posing special problems to mathematics students (Silver 1981).

Any further instruction was from idealistic, or at least flexible, young tutors. Since the male Cambridge undergraduates received bachelor's degrees with honors by taking the Tripos examinations, the women wanted to pass these examinations too. Three of the first five students had done so in 1872, and songs in the memory of these "Girton Pioneers" were sung during the long dark winter evenings of Scott's student days.

Women would not receive degrees at Cambridge until 1948, but every year after 1872 women applied to take the Tripos exams, and some were given special permission to do so. On nine bitterly cold days in January 1880, Charlotte Scott spent over fifty hours taking the mathematics Tripos. When word leaked out that she had done as well as the eighth man in the entire university, the news permeated England that a woman had succeeded in a "man's" subject.

Because she was female, she could not be present at the award ceremony, nor could her name be officially mentioned. However, a contemporary report says, "The man read out the names and when he came to 'eighth,' before he could say the name, all the undergraduates called out 'Scott of Girton,' and cheered tremendously, shouting her name over and over again with tremendous cheers and waving of hats." The young men of Cambridge gave honor where it was due, even though their elders followed the established rules. At Girton College there were cheers and clapping at dinner, and a special evening ceremony where she was led up an "avenue of students" while they sang "See the Conquering Hero Comes." She stood on "a sort of dais" while an ode written by a staff member was read to her, and then she was crowned with laurels, "while we clapped and applauded with all our might" (Megson and Lindsay 1961, 31).

In 1922 James Harkness, who was only a schoolboy in 1880, remembered that Scott's achievement impressed even him at the time, its widespread impact marking "the turning point in England from the theoretical feminism of Mill and others to the practical education and political advances of the present time" (Putnam 1922). The publicity resulted in pressure on Cambridge University to admit its resident female students to university examinations as a matter of policy, not just special privilege, and to post their names with those of the male students, an important step toward qualifying for jobs. After a year of controversy, this resolution was passed on February 24, 1881. Its national implications are reflected by the fact that at the newly opened college for women at Oxford, the news was proclaimed loudly in the dining room, "We have won! We have won!" (Bradbrook 1969, 55).

Arthur Cayley, a renowned algebraist, was one of those leading the effort for this recognition of women's education, and for the rest of his life, Scott was the recipient of "his kindness" (Scott 1895). She attended his lectures, did her graduate research under him, and obtained her first and only position outside Girton College on the basis of his recommendation. Meanwhile, she was hired as a resident lecturer by Girton College and taught there until receiving her doctorate in 1885.

Although by Scott's time Cambridge University no longer required an oath of allegiance to the Church of England, it would not grant her a degree, because of her sex. Fortunately, the University of London began granting "external" degrees to women in 1876, so Scott took two entirely different sets of examinations from two universities, one to place her with her peers, and the other to obtain degrees. She thus received a B.Sc. in 1882 and a D.Sc. in 1885 from the University of London, both "First Class," the highest possible rank.

Bryn Mawr College, which opened in Pennsylvania in 1885, was dedicated to providing both undergraduate and graduate education of the highest level to women. Since comparable positions for women were virtually nonexistent in Europe, Scott went to Bryn Mawr, becoming its first mathematics department head and the only mathematician on its founding faculty of eight. There was one other woman, a biologist. There were no better options in the world for a woman mathematician during the next forty years, so Scott remained there.

Occasionally her father or brother Walter visited her at Bryn Mawr. Her older sister, with whom she had grown up, died the spring before Scott left for Girton College, and her youngest sister died as an infant; so in her adulthood she was the oldest of five siblings, with two younger brothers and two younger sisters. Her will also mentions her "beloved" sister-in-law, Walter's widow. Walter, who was in the machinery business, died suddenly in Scott's home on August 7, 1918, a great blow to her. One of her sisters worked for a while in an orphanage and then married. The other remained home and cared for her father, Caleb, in his old age. The family was a close and loving one; surviving relatives remember with affection "Auntie Charley [pronounced 'Sharly']."

The early Girton College community had strictly observed the social mores of the time. The existence of the Spinning House left little margin for experimentation, and the prevailing opinion was that personal conservatism was required to promote women's educational and political equality. Charlotte Scott maintained this view throughout her life, disapproving of smoking and makeup, but her disapproval extended equally to both sexes. She bobbed her hair before arriving at Bryn Mawr in 1885, although short hair for women was still controversial in the 1920s. She had at least one close male friend outside her family, Frank Morley, whose time studying mathematics at Cambridge University overlapped hers. He told his son that the social conventions made it more acceptable for her to visit him and his family in Baltimore than for him to visit her, and she did so often.

Scott's relationship with M. Carey Thomas, the first dean of Bryn Mawr College and its president from 1894 to 1922, was always formal, despite the fact that Scott was only one year younger. Thomas had become the first American woman to earn a doctorate in any field (linguistics), in 1882 at the University of Zurich, and had visited Girton College on her way home. A biographer of Thomas says that Scott was hurt by her initial coldness after Scott's lonely trip across the ocean (Finch 1947, 194). Thomas had the impatience of many dynamic reformers, and her correspondence with Scott also includes a confession that mathematics had always been her most difficult subject, suggesting a special tension because of this. In 1906 Scott wrote a letter apparently in response to Thomas's desire to know when a certain student would finish her Ph.D. Patiently she explained, "If it were simply a matter of surveying the field, collating papers and stating the contents clearly, she could do the thesis before June certainly; but to produce an original piece of work is quite another matter..." (Scott Papers). Thomas's lack of knowledge about mathematics is also reflected in much earlier correspondence about the necessity of mathematics journals for the library; Scott was always fighting for her discipline on her home turf.

During Scott's first three years at Bryn Mawr, there was a total of only four serious mathematics students—three undergraduates and one "graduate" student who had studied nothing higher than differential equations before she came. Scott worked intensely, writing her lecture notes "*after*, not before, the lecture... at the end of a busy day... word perfect... knowing that at nine a.m. tomorrow [she would give another lecture].... But the next delivery showed no lack of spontaneity for changes and improvement were made until the notes could be, and as a matter of fact were, used as text-book material" (Maddison and Lehr 1932). Gradually her classes grew larger, and by her ninth year there were six new mathematics students, two undergraduates and four graduates, including her first two successful doctoral candidates. Indeed, three of the nine American women to earn doctorates in mathematics in the nineteenth century studied with her. Her professional correspondence shows her intense involvement with each student, arguing against the doctoral candidacy of one who demonstrated "everything except that one essential, capacity" for doctoral work, and for one who has been discovered to have tuberculosis but has already published good work. She pleads on behalf of a student who inadvertently left a notebook in an examination room, and against those sitting on a fire escape to eavesdrop on a faculty meeting. Former students remembered her kindness and her ability to help them solve their problems.

Her Girton propriety and calm exterior slipped on January 12, 1898, when she wrote to President Thomas:

> I am most disturbed and disappointed at present to find you taking the position that intellectual pursuits must be "watered down" to make them suitable for women, and that a lower standard must be adopted in a woman's college than in a man's. I do not expect any of the other members of the faculty to feel this way about it; they, like (nearly) all men that I have known, doubtless take an attitude of toleration, half amused and half kindly, on the whole question; for even where men are willing to help in women's education, it is with an inward reserve of condescension. (Scott Papers)

The word "nearly" is inserted in small lettering above the handwritten letter. It is indeed unfortunate that Scott's entire correspondence with her family has apparently been lost.

Thomas wrote to her niece in 1932 that "in my generation marriage and an academic career was impossible" (Dobkin 1979, xv), and this fact was basic to Scott's life too. She wanted to build her own house but was unable to find a suitable plot, so in 1894 she moved from a small apartment on the Bryn Mawr campus to a house rented from the college. Her cousin Eliza Nevins joined her to become her companion and housekeeper until Nevin's death in 1928. Others, including an early doctoral student, lived with them occasionally. Scott was a leader among the tenants in campaigning for such mundane matters as access to direct paths and more effective heating of the homes. On February 27, 1901, her own house caught fire; the house was saved, but Scott could not live in it for months afterward.

An even more serious disruption occurred in the spring of 1906, when she developed an acute case of rheumatoid arthritis. After that her ill health and her increasing deafness, which was apparent even in her Girton College days and was complete by the time anyone now living knew her, marred her life significantly. Her publications ceased for two decades, and the doctor recommended outside exercise. Gardening was compatible with her academic duties; and her garden was "brought, year after year, unbelievably, to greater beauty" (Maddison and Lehr 1932). She developed a new strain of chrysanthemum. Her correspondence reveals the zest with which she continued to live. "I am not a Vandal, as you know; but this tree is not good, it simply encourages visitors of objectionable kinds, beginning with scab and continuing accordingly, and any miserable little apples that it does produce are infected with maggots. My wish is to cut it down and dig it up, and then plant a less troublesome tree a few feet away, so as not to spoil the appearance of the slope" (Scott Papers).

Scott maintained her church membership in England for at least a decade after she came to the United States. American mathematicians joked about her leaving for Europe every spring as soon as exams were marked, but this was not literally true. Still, she crossed the Atlantic Ocean often, at a time when each voyage involved at least a week of discomfort and danger. She thus provided an invaluable link between the fledgling mathematical community of the United States and the established centers in Europe.

Scott officially retired in 1924 but remained an extra year at Bryn Mawr to help her seventh and last doctoral graduate complete her dissertation. Then she moved to a large house on the bus line halfway between Girton College and the center of Cambridge University. Her complete deafness made social interactions difficult, even with her next door neighbor, who also happened to be a retired mathematician. Her primary diversion was betting on horses, an activity to which she applied mathematicial statistics. Her doctor, who had introduced her to his own bookie, Mr. Cook, believed that she neither gained nor lost much money. However, he was amused how her Victorian outlook affected her view of Cook. One Christmas when he visited her home, she was extremely agitated. "Dr. Nourse, I am very worried. Do you see that umbrella in the corner? That has been sent to me by Mr. Cook. Of course I couldn't accept it!" The doctor explained that the bookie sent umbrellas to all his women clients and purses to the men and would feel hurt if the presents were returned. "Do you really think I can keep it?" Scott replied, obviously relieved that the umbrella was not an indication of moral turpitude.

On November 10, 1931, she died quietly in Cambridge. She was buried with Miss Nevins in St. Peter's part of the St. Giles's Churchyard in Cambridge near the northwest corner of the chapel. The inscription on a small stone gives only her date of death and age and no indication of her place in the history of mathematics.

Although she seems almost forgotten today, Scott received many honors in her lifetime. Rebière, writing in Paris in 1897, called Scott "one of the best living mathematicians" with no apparent need to justify its claim. She was the only woman starred in the first edition of *American Men of Science* (i.e. considered prominent in mathematics by her contemporaries) and the only mathematician included in *Notable American Women, 1607–1950*.

Her honors at Cambridge in 1880 were informal because she was female, but they had a lasting impact on women everywhere. Later honors by academic institutions were official. She was the chief examiner in mathematics of the College Entrance Examination Board in 1902 and 1903. In 1909 the alumnae of Bryn Mawr honored her with the college's first endowed chair. When she retired, the board of directors of Bryn Mawr College cited her contribution to the college in its first forty years as "second only to that of President Thomas."

On April 18, 1922, the American Mathematical Society met at Bryn Mawr, and about 200 people gathered in her honor. Alfred North Whitehead gave the featured talk on "Some principles of physical science." Although it was his first trip across the Atlantic Ocean, he refused invitations from Harvard and Columbia universities because he did not want competing attractions in Scott's "neighborhood." At the end of his talk Whitehead observed, "A friendship of peoples is the outcome of personal relations. A life's work such as that of Professor Charlotte Angas Scott is worth more to the world than many anxious efforts of diplomatists. She is a great example of the universal brotherhood of civilizations" (Putnam 1922).

Work

When the New York Mathematical Society opened its membership to people outside New York, Scott immediately responded, and she was one of the major organizers who developed the group into the American Mathematical Society (AMS) in 1891. She served on its council from 1891 to 1894 and again from 1899 to 1902, and was its vice-president in 1905–1906. When Thomas Fiske gave an anniversary talk in 1938 reviewing the first fifty years of the society, he cited the work of about thirty people, of whom Scott was the only woman.

She brought experience as a member of established European societies, including the London Mathematical Society, the Edinburgh Mathematical Society, the Deutsche Mathematiker-Vereinigung, the Circolo Matematico di

Palermo, and as an "honorary member" of the Amsterdam Mathematical Society. She was one of only seventeen Americans who attended the World Congress of Mathematicians in 1900, and she wrote an extensive report of it for the *Bulletin* of the AMS. Since Scott's field (algebraic geometry) was the same as that of both the father and the future dissertation advisor of Emmy Noether*, both of whom attended the congress and must have conferred with Scott there, perhaps it is not coincidence that Emmy Noether switched fields that summer from languages (more common for young women) to mathematics (still largely male-dominated).

In 1899 Scott became coeditor of the eminent *American Journal of Mathematics*, an influential position she held for twenty-seven years. Her own papers were published not only in American journals, but also in the more competitive European publications, where American mathematicians appeared extremely rarely.

Her book, *An Introductory Account of Certain Modern Ideas and Methods in Plane Analytical Geometry*, was published in 1894 and reprinted, essentially without change, thirty years later. Although its title includes the word "introductory," and it was indeed used by many beginners, it took its readers to the edges of research. It was used widely. Cole's review (1896) praised its inclusion of such recent concepts as groups, subgroups, invariants, and covariants. However, even more far-reaching than its subject matter was its obvious "distinction between a general principle and a particular example." Scott was one of the first textbook writers, especially those writing in English, to be "perfectly aware" of this distinction and to teach it to the next generations of college mathematics students. Her other book, a "school" book about plane geometry, was not well received, because she based her development on lines instead of points, an innovation that was not widely adopted.

F. S. Macaulay's obituary of her summarized, "Miss Scott was a geometer who whenever possible brought to analytical geometry the full resources of pure geometrical reasoning" (1932, 232). Her published research, like most mathematical writing of her time, consisted of discussions of various specific mathematical phenomena. Her specialty was the geometric interpretations of algebraic expressions in two variables of degree greater than two, that is, of plane curves neither linear nor quadratic. However, she had a keener sense of the difference between example and proof than most of her contemporaries, playing an important role in the transition to the twentieth-century custom of presenting mathematics via abstract proofs. Her most notable paper may be her 1899 "geometric" proof of a theorem of Max Noether, Emmy Noether's father. Unfortunately for Scott's fame, her particular field fell out of fashion in the twentieth century.

---

*Cross-reference to other women discussed in the volume is given by an asterisk following the first mention in a chapter of the individual's name.

She was hired by Bryn Mawr College to be department head, to teach ten or eleven hours a week of both graduate and undergraduate courses, and to supervise graduate research. Although her written offer in 1884 said that her hours of teaching would be diminished as her other duties grew, they were still at their original level thirty years later. Committees also absorbed much time. "She would··· sit through a long meeting··· and at just the right moment make a brief, incisive speech which—such was the respect with which her opinion was regarded—often turned the vote from the direction in which it was tending to the side which she supported."

Her impact on mathematics education in the United States was enormous. Although Harvard University had dropped its requirements that all freshmen take a course in addition, subtraction, and multiplication only fifty years earlier, her initial requirements for students entering Bryn Mawr College included passing examinations in arithmetic, plane geometry, and algebra through quadratic equations and geometric progressions. Students who did not pass admission examinations in solid geometry and trigonometry had to pass courses in these subjects before graduation. Mathematics majors were required to take one semester of algebra and the theory of equations, a year of differential and integral calculus, and a semester of differential equations and elements of finite differences. Early Bryn Mawr students took another sequence concurrently in "analytical geometry," one year in two dimensions and another in three.

During her early years at Bryn Mawr, she was distressed at the amount of time she spent writing and grading entrance examinations, so she worked for a nationwide testing service. The College Entrance Examination Board began in 1901, and she was its chief examiner in mathematics in 1902 and 1903, setting standards that have changed little in over eighty years, although, ironically, the name of their promulgator is rarely mentioned.

Scott was a special inspiration to women, who received three times the percentage of American Ph.D.'s in mathematics before 1940 than they did in the 1950s. She herself was the dissertation advisor of seven women, and Bryn Mawr conferred two other Ph.D.'s in mathematics while she was department head. During this time Bryn Mawr College was third only to the University of Chicago and Cornell University, both much larger institutions, in the number of doctorates in mathematics granted to women in mathematics. When she had delivered her talk to the AMS in 1905, nine of the forty-five listeners were women, only two of whom were from Bryn Mawr. It is difficult to measure influence by numbers, but her visibility, her conversations, and her preparation of many women to teach younger women clearly had a major impact on the academic and economic position of women in America.

Bibliography

Works by Charlotte Angas Scott

Mathematical Works

"The binomial equation $x^p - 1 = 0$." *American Journal of Mathematics* 8 (1886): 261–264.

"On the higher singularities of plane curves." *American Journal of Mathematics* 14 (1892): 301–325.

"The nature and effect of singularities of plane algebra curves." *American Journal of Mathematics* 15 (1893): 221–243.

"On plane cubics." *Philosophical Transactions of the Royal Society of London* 185(A) (1894): 247–277.

An Introductory Account of Certain Modern Ideas and Methods in Plane Analytical Geometry. London and New York: Macmillan, 1894. 2nd ed. New York: G. E. Stechert, 1924. 3rd ed. under the title *Projective Methods in Plane Analytical Geometry.* New York: Chelsea, 1961.
Review: F. N. Cole. *Bulletin of the American Mathematical Society* 2 (1896): 265–269.

"Arthur Cayley." *Bulletin of the American Mathematical Society* 1 (1895): 133–141.

"Note on adjoint curves." *Quarterly Journal of Pure and Applied Mathematics* 28 (1896): 377–381.

"Note on equianharmonic cubics." *Messenger of Mathematics* 25 (1896): 180–185.

"Sur la transformation des courbes planes." *Comptes rendus de l'Association Française, pour l'Avancement des Sciences* (*Congrès de St. Étienne*) (26) (1897): 50–59.

"On Cayley's theory of the absolute." *Bulletin of the American Mathematical Society* 3 (1897): 235–246.

"Studies in the transformation of plane algebraic curves." Parts I, II. *Quarterly Journal of Pure and Applied Mathematics* 29 (1898): 329–381; 32 (1901): 209–239.

"Note on linear systems of curves." *Nieuw Archief voor Wiskunde* (2)3 (1898): 243–252.

"On the intersections of plane curves." *Bulletin of the American Mathematical Society* 4 (1898): 260–273.

"A proof of Noether's fundamental theorem." *Mathematische Annalen* 52 (1899): 592–597.

"The status of imaginaries in pure geometry." *Bulletin of the American Mathematical Society* 6 (1900): 163–168.

"On von Staudt's Geometrie der Lage." *Mathematical Gazette* 1 (1900): 307–314, 323–331, 363–370.

"On a memoir by Riccardo de Paolis." *Bulletin of the American Mathematical Society* 7 (1900): 24–38.

"Report on the International Congress of Mathematicians in Paris." *Bulletin of the American Mathematical Society* 7 (1900): 57–79, Excerpts printed in *The Mathematical Intelligencer* 7 (4) (1985): 75–78.

"Note on the geometrical treatment of conics." *Annals of Mathematics* (2) 2 (1901): 64–72.

"On a recent method for dealing with the intersections of plane curves." *Transactions of the American Mathematical Society* 3 (1902): 216–263. Reprinted as a Bryn Mawr College Monograph, vol. 4, no. 2.

"On the circuits of plane curves." *Transactions of the American Mathematical Society* 3 (1902): 388–398. Reprinted as a Bryn Mawr College Monograph, vol. 4, no. 3.

"Note on the real inflexions of plane curves." *Transactions of the American Mathematicsl Society* 3 (1902): 399–400. Reprinted as a Bryn Mawr College Monograph, vol. 4, no. 4.

"Elementary treatment of conics by means of the regulus." *Bulletin of the American Mathematical Society* 12 (1905): 1–7. Reprinted as a Bryn Mawr Monograph, vol. 8, no. 3.

"Note on regular polygons." *Analls of Mathematics* 8 (1906): 127–134. Reprinted as a Bryn Mawr College Monograph, vol. 8, no. 8.

Cartesian Plane Geometry. Part I: Analytical Conics, London: J. M. Dent and Company, 1907.

"Higher singularities of plane algebraic curves." *Proceedings of the Cambridge Philosophical Society* 23 (1926): 206–232.

Scott's name appears in the list of contributors to problem-solving in the *Educational Times* until 1892.

Works about Charlotte Angas Scott

Bradbrook, M. C. *"That Infidel Place": A Short History of Girton College, 1869–1969*. London: Chatto & Windus, 1969.

Dobkin, Marjorie Housepain. *The Making of a Feminist; Early Journals and Letters of M. Carey Thomas*. Kent, Ohio: Kent State University Press, 1979.

Finch, Edith, *Carey Thoms of Bryn Mawr*. New York: Harper, 1947.

Jones, E. E. Constance. *Girton College*. London: Adam and Charles Black, 1913.

Katz, Kaila, and Patricia Kenschaft. "Sylvester and Scott." *The Mathematics Teacher* 75 (1982): 490–494.

Kenschaft, Patricia C. "Charlotte Angus [*sic*] Scott 1858–1931." *Association for Women in Mathematics Newsletter* 7(6) (November-December 1977): 9–10; 8(1) (April 1978): 11–12.

——. "The students of Charlotte Angas Scott." *Mathematics in College* (Fall 1982): 16–20.
Biographies of four of Scott's outstanding students.

——. "Women in mathematics around 1900." *Signs* 7 (4) (Summer 1982): 906–909.
Compares the participation of women in the United States research mathematical community in the era of Charlotte Scott to that of recent years and indicates that the modern feminist movement has just barely regained the position that women had at the turn of this century.

Lehr, Marguerite. "Charlotte Angas Scott." *Notable American Women, 1607–1950*, vol. 3, 249–250. Cambridge, Mass.: Belknap Press of Harvard University Press, 1971.

Macaulay, F. S. "Dr. Charlotte Angas Scott." *Journal of the London Mathematical Society* 7 (1932): 230–240.
Summary of Scott's research achievements written by a contemporary in her field.

Maddison, Isabel, and Marguerite Lehr. "Charlotte Angas Scott: An Appreciation." *Bryn Mawr Alumni Bulletin* 12 (1932): 9–12.
This article in two parts (one by each author) is probably the most personal published piece written by people who knew Scott. Maddison was one of her doctoral graduates who spent her career on the Bryn Mawr campus and lived with Scott for a while. Lehr was her last doctoral graduate, who also taught at Bryn Mawr for forty years.

Megson, Barbara, and Jean Olivia Lindsay. *Girton College, 1869–1959, An informal History*. London: W. Heffer, 1961.
"Ministers Deceased: Dr. Caleb Scott." *Manchester Guardian* (23 July 1919).

Putnam, Emily James. "Celebration in honor of Professor Scott." *Bryn Mawr Bulletin* 2 (1922): 12–14.

Rebière, A. *Les Femmes dans la Science*. 2nd ed. Paris: Nony, 1897.

Scott, Caleb, "An Address to the newly elected deacons." Delivered March 15, 1865. Unpublished records in the Lincoln Public Library, Lincoln, England.

Scott, Charlotte A. Papers. Bryn Mawr College Archives, Bryn Mawr, Pa.

Silver, John. Letter to author, March 29, 1981.

Thomas, M. Carey. Papers. Bryn Mawr College Archives, Bryn Mawr, Pa.

EDWARD BURR VAN VLECK

1863–1943

BY RUDOLPH E. LANGER AND MARK H. INGRAHAM

"THE ILLUSTRIOUS SON of a distinguished father and the distinguished father of an illustrious son" was the description given by Dean Holgate, of Northwestern University, of Professor Van Vleck at a dinner of the American Mathematical Society. This description was not only literally true but also symbolically true. Professor Van Vleck was a scholar who to a superlative degree inherited the intellectual and cultural riches of the ages and succeeded in his determination to transmit these enhanced to coming generations.

Edward Burr Van Vleck was born in Middletown, Connecticut, on June 7, 1863. His father, John Monroe Van Vleck, was Professor of Mathematics and Astronomy at Wesleyan University from which he had graduated in 1850 at the age of seventeen and where he taught from 1854 until his death in 1912. Moreover, he frequently acted as president of the University. The Van Vleck Observatory at Wesleyan was named after John Monroe Van Vleck. The Van Vlecks were an ancient family of Maastricht, Holland; and Tielman Van Vleck in 1658 came to America, where he became one of the founders of Jersey City after a period as a notary in New Amsterdam. The family, through the generations, like many other Dutch families, moved up the Hudson Valley—John Monroe Van Vleck being born at Stone Ridge, New York. There was also a large strain of French Huguenot blood in his ancestry. Professor Edward Burr Van Vleck's mother was born Ellen Maria Burr, of Middle-

Reprinted from *Biographical Memoirs*, Vol. 30, 1957, with permission of the National Academy Press, Washington, D.C.

Edward B. Van Vleck

(Reprinted from *Biographical Memoirs*, Vol. 30, 1957, with permission from the National Academy Press, Washington, DC.)

town, Connecticut, and was chiefly of English descent from stock that had come to New England as early as 1635.

Young Van Vleck's education was in the schools of Middletown and at Wilbraham Academy. He graduated from Wesleyan University in 1884. Endowed with a brilliant mind, blessed with good health, but being quite devoid of athletic skill, he early turned to highly intellectual interests. (His collector's instinct was also shown in youth in his enthusiastic acquisition of stamps.) He found it difficult to decide whether his major interest would be the classics, especially Greek literature, or mathematics. After graduating from Wesleyan, he studied mathematics and mathematical physics at Johns Hopkins University from 1885 to 1887 and taught at Wesleyan from 1887 to 1890. From 1890 to 1893, when he received his Ph.D. degree, he studied at Göttingen, where he formed lifelong friendships with his fellow students (some of them American) and with his major professor, Felix Klein, who had great influence upon Van Vleck as he had upon many others of his students. He always regretted that this period of study had not come somewhat earlier, as from it dates his career as a productive mathematician. The rest of his official career was spent at Wesleyan University, where he was Assistant Professor from 1895 to 1898 and Professor from 1898 to 1905, and at the University of Wisconsin, where he was Instructor from 1893 to 1895 and Professor from 1906 until his retirement as Professor Emeritus in 1929.

Upon his return from Germany in 1893 he married Hester L. Raymond, of Lyme, Connecticut. They had one son, John Hasbrouck Van Vleck, now Hollis Professor of Mathematics and Natural Philosophy and Dean of Applied Science, at Harvard University. Professor Van Vleck's home life was a well from which flowed the quality of his work, his cultural interests, and the influence he had upon his friends—an influence based on intellectual vigor tempered by a fundamental serenity of spirit. Mrs. Van Vleck had much to do not only with her husband's happiness but also with his effectiveness.

Note should be made of three other aspects of his life apart from his research: his interest in literature and the fine arts, his love of travel, and his quality as a teacher.

In connection with the first two of these it must be mentioned that his father late in life had been bequeathed by a brother a considerable estate, part of which Professor Van Vleck inherited. Hence he had means to live graciously, to collect books, etchings, and prints, and to travel extensively. With true Dutch characteristics he was able to combine the love of good living with meticulous care in money matters. He took joy both in giving generously and in investing wisely, but inexactitude, financial or otherwise, went against the grain.

Professor Van Vleck kept abreast of what was published in his field of mathematics, but in spite of this found time for much reading of literature. Often, however, he joked about doing his reading vicariously through Mrs. Van Vleck, who was a prodigious reader. In the graphic arts they shared consuming interest. The etchings of Rembrandt, Seymour Hayden, and Whistler adorned their walls, which however, were always the walls of a home—not those of a museum. Their collection of Japanese prints was notable, and Mrs. Van Vleck became expert in repairing these. Friends from all over America remember with pleasure the occasions when for an hour or so the Van Vlecks would show to small groups some selected prints from their collection.

Travel played a very large role in the life of the Van Vleck family. The guide book and the atlas were ever at hand. (A timetable was not needed in the presence of their son.) The galleries, the churches, and the mountains of Europe were equally familiar. It was perfectly natural for a conversation to turn from point sets to the comparative beauty of the north and south spires of Chartres cathedral. Professor Van Vleck's retirement at sixty-six was associated with both his love of art and of travel for, as he explained to his friends, he wished to retire while he could still enjoy a trip around the world and return to catalogue his Japanese prints. For each of these programs he set aside a year. He apologized for the fact that, because on his return he missed a connection in Chicago, the trip had taken a year and six hours instead of a year. However he had acquired so many prints during the journey that the cataloguing of this collection was prolonged well past the allotted time.

As a teacher Professor Van Vleck had both natural assets and liabilities. He had the gift of exact expression and of clear organization. However, it was difficult for him to understand a slow mind or to pace himself in accordance with the requirements of an average class. In quizzing a small group or an individual he was superb—discovering any lack of apprehension and clarifying difficult points. He was courteous, yet impatient—one of the few dichotomies of a remarkably integrated personality and related to the conflict between his great tolerance of spirit and his own almost puritanical standards of conduct. He was generous in the extreme with his time, but demanded that he see some results for his effort. He was a stimulating teacher and colleague of the gifted; others surpassed him in getting moderately satisfactory results from the average. As chairman of the Department of Mathematics of the University of Wisconsin, he constantly upheld the highest scholarly ideals.

The qualities of insight, exactitude, and consideration when there was a spark worth fanning made his work as editor of the *Transactions of the American Mathematical Society* in its formative years of great and beneficial influence. A mathematical result was not something to be transmitted haphazardly to the public. It should be a part of a great cultural structure and, as such, it should be expressed with precision and elegance. Many young authors gained much from his kindly but incisive suggestions. Moreover, such standards have been transmitted from scholar to scholar, to become traditional for the *Transactions*.

Not only did Professor Van Vleck believe strongly in the unity of mathematics, but he also believed in the unity of the scholars who dealt with that subject; and at the time when it seemed likely that they would divide themselves into regional groups, he was a potent force in keeping the American Mathematical Society a truly national organization.

Professor Van Vleck was interested both in the affairs of the University and in those of the community—an interest that was shown through generous giving and through active participation in committees, boards, etc.

There are many who, in their ideal of the scholar and what the life of the scholar should be, have acquired much from Professor Van Vleck and his family.

As a mathematician Van Vleck won his spurs with the completion of his doctoral dissertation in 1893. He had spent five semesters at Göttingen, where he had found his primary inspiration in Felix Klein. His thesis subject, "The Development of Hyperelliptic Integrals in Continued Fractions," was in the focal center of interest of the day. The hyperelliptic integrals are of the form

$$\int \frac{W(x)dx}{(x-a_1)^{1-\lambda_1}\cdots\cdots(x-a_n)^{1-\lambda_n}}$$

$W(x)$ being a polynomial of the degree $(n-2)$ and the $a_1, \ldots, a_n$, $\lambda_1, \ldots, \lambda_n$, being real or complex constants. Work in this field had been initiated by Gauss in connection with the hypergeometric functions, in particular in connection with the function

$\frac{1}{2}\log\frac{x-1}{x+1}$, which is represented by the integral

$$\int \frac{dx}{(x-1)(x+1)}$$

It had been carried forward by others in connection with studies of the polynomials of Lamé and Stieltjes. Such polynomials appear as solutions of linear differential equations of the form

$$\frac{d^2y}{dx^2}+\left(\frac{1-\lambda_1}{x-a_1}+\cdots\cdots+\frac{1-\lambda_n}{x-a_n}\right)\frac{dy}{dx}+\frac{W(x)}{(x-a_1)\cdots\cdots(x-a_n)}y=0$$

By an extensive and searching analysis Van Vleck greatly broadened and generalized the existing theory, and threw light upon it from several new angles. His approach was both analytic and geometric. From the analytic standpoint, the convergents of the continued fraction developments yield algebraic approximations to the integral. Van Vleck concerned himself with such approximations, both such as were valid in the neighborhood of a single branch point, and such as were simultaneously valid in the neighborhoods of several branch points. His geometric discussion, which was extensive, was based upon the theory of conformal mapping. The irregularities of the algebraic approximants and the distribution of the roots of the polynomial factors that figure in the integral representations of the remainder terms were investigated. The upshot was an extensive coordination and classification of the integrals, and revelations of some deeper lying connections of their theory with the theories of linear differential equations, of groups, of polynomials, etc.

With this important memoir Van Vleck had opened for himself a number of avenues along which investigations were to occupy him for the ensuing decade. The fruits of these researches were a succession of papers, on the roots of Bessel functions and Riemann P-functions, on the classification along group theoretic lines of differential equations that admit two solutions whose product is a polynomial, on criteria for the radii of convergence of power series, on the roots of hypergeometric series, and, most especially and extensively, on the theory of the convergence of continued fractions. Well-known theorems in this last field are his. His extended preoccupation with this field of analysis well qualified him for the role of "Colloquium lecturer" of the American Mathematical Society. Delivered in 1903, his lectures were on the subject of "Divergent Series and Continued Fractions."

In 1907 and 1908 Van Vleck published papers on point-set theory, his primary concern being the analysis of non-measurable sets. That his appreciation of this field of analysis was not transient is evidenced by the fact that he chose in 1915, as retiring president of the American Mathematical Society, to direct his address to the subject of "The Role of the Point-Set Theory in Geometry and Dynamics."

Between 1910 and 1916 Van Vleck's research was concerned with the functional equations of the sine and the theta functions, and with linear difference equations. Although he wrote only one paper on the latter subject, he also treated it in a lecture course at the University of Wisconsin, in a manner that was described by George D. Birkhoff, one of his auditors, as "suggestive and stimulating." Birkhoff and his students subsequently achieved notable advances in this field. It is therefore appropriate to observe Birkhoff's remarks, that "one must look upon Van Vleck as an essential factor in American contributions to linear homogeneous difference equations."

The properties and classifications of groups of linear substitutions in any number of variables were treated by Van Vleck in various papers at different times. Another subject of recurring interest to him was the location of the roots of polynomials. On that he wrote in 1899 and 1903, and again in 1925. He made it the subject of his "Symposium lectures" before the American Mathematical Society in 1929.

Van Vleck was a well-informed and discerning mathematician, and a clear and fluent writer. Some essays in which he reviewed various mathematical developments therefore deserve mention, since they were widely read and appreciated. Among these were his address on the role of point-set theory, which has already been mentioned above, his address on "The Influence of Fourier's Series upon the Development of Mathematics," delivered in 1913 on the occasion of his retirement from a vice presidency of the American Association for the Advancement of Science, and his address "Current Tendencies of Mathematical Research," delivered on the occasion of his investiture with the honorary Doctorate of Science by the University of Chicago in 1916.

His honors were numerous: the degrees of Doctor of Mathematics and Physics from Groningen, Doctor of Science from the University of Chicago, and Doctor of Laws from Clark University and Wesleyan University. He was made "Officier de l'instruction publique" by the French Republic; and, in addition to serving as editor of the *Transactions of the American Mathematical Society,* he was President of the Society, 1913-1915. He was elected to the National Academy of Sciences in 1911.

Dr. Van Vleck died in Madison, Wisconsin, on June 2, 1943, at the age of 80.

KEY TO ABBREVIATIONS

Am. J. Math. = American Journal of Mathematics
Ann. Math. = Annals of Mathematics
Bull. Am. Math. Soc. = Bulletin of the American Mathematical Society
Trans. Am. Math. Soc. = Transactions of the American Mathematical Society

BIBLIOGRAPHY

1894

Zur Kettenbruchentwickelung Hyperelliptischer und Ähnlicher Integrale. Inaugural dissertation, Göttingen, 1893. Am. J. Math., 16:1-92.

1897

On the Roots of Bessel- and P-Functions. Am. J. Math., 19:75-85.

1898

On the Polynomials of Stieltjes. Bull. Am. Math. Soc., ser. 2, 4:426-438.

1899

On Certain Differential Equations of the Second Order Allied to Hermite's Equation. Am. J. Math., 21:126-167.

On the Determination of a Series of Sturm's Functions by the Calculation of a Single Determinant. Ann. Math., ser. 2, 1:1-13.

1900

On Linear Criteria for the Determination of the Radius of Convergence of a Power Series. Trans. Am. Math. Soc., 1:293-309.

1901

On the Convergence of the Continued Fraction of Gauss and Other Continued Fractions. Ann. Math., ser. 2, 3:1-18.

On the Convergence of Continued Fractions with Complex Elements. Trans. Am. Math. Soc., 2:215-233.

On the Convergence and Character of a Certain Form of Continued Fraction

$$\frac{a_1z}{1} + \frac{a_2z}{1} + \frac{a_3z}{1} + \cdots.$$

Trans. Am. Math. Soc., 2:476-483.

1902

A Determination of the Number of Real and Imaginary Roots of the Hypergeometric Series. Trans. Am. Math. Soc., 3:110-131.

1903

A Sufficient Condition for the Maximum Number of Imaginary Roots of an Equation of the N-th Degree. Ann. Math., ser. 2, 4:191-192.

On an Extension of the 1894 Memoir of Stieltjes. Trans. Am. Math. Soc., 4:297-332.

Divergent Series and Continued Fractions. Am. Math. Soc. Boston Colloquium Lectures, 1903.

1904

On the Convergence of Algebraic Continued Fractions Whose Coefficients Have Limiting Values. Trans. Am. Math. Soc., 5:253-262.

1907

A Proof of Some Theorems on Pointwise Discontinuous Functions. Trans. Am. Math. Soc., 8:189-204.

1908

On Non-measurable Sets of Points with an Example. Trans. Am. Math. Soc., 9:237-244.

1910

A Functional Equation for the Sine. Ann. Math., ser. 2, 11:161-165.

1912

On the Extension of a Theorem of Poincaré for Difference Equations. Trans. Am. Math. Soc., 13:342-352.

One-Parameter Projective Groups and the Classification of Collineations. Trans. Am. Math. Soc., 13:353-386.

1914

The Influence of Fourier's Series upon the Development of Mathematics. Science, n.s., 39:113-124.

1915

The Role of the Point-Set Theory in Geometry and Dynamics. Bull. Am. Math. Soc., ser. 2, 21:321-341.

1916

With F. H'Doubler. A Study of Certain Functional Equations of the θ-Functions. Trans. Am. Math. Soc., 17:9-49.

Current Tendencies of Mathematical Research. Bull. Am. Math. Soc. ser. 2, 23:1-13.

1917

Haskin's Momental Theorem and Its Connection with Stieltjes's Problem of Moments. Trans. Am. Math. Soc., 18:326-330.

1919

On the Combination of Non-loxodromic Substitutions. Trans. Am. Math. Soc., 20:299-312.

1921

An Extension of Green's Lemma to the Case of a Rectifiable Boundary. Ann. Math., 22:226-237.

1922

Non-loxodromic Substitutions and Groups in N-Dimensions. Trans. Am. Math. Soc., 24:255-273.

1925

On Limits to the Absolute Values of the Roots of a Polynomial. Bulletin de la Société mathématique de France, 53:105-125.

1929

On the Location of Roots of Polynomials and Entire Functions. Bull. Am. Math. Soc., 35:643-683.

The Mathematical Work of R. L. Moore: Its Background, Nature and Influence

R. L. WILDER

Communicated by C. TRUESDELL

ROBERT LEE MOORE ("R. L.") was probably one of the most influential American mathematicians of the first half of the 20th century. Whether this was due more to his famous teaching method (the "MOORE Method") or to his creative work in mathematics is debatable; the current folklore seems to credit the former. On the other hand, careful scanning of his published work reveals that while, from a present point of view, it was narrowly oriented in scope, being confined to what he called "Point Set Theory," it contained the germs of a large portion of modern research in both general and algebraic topology.

MOORE did not himself venture into algebraic topology at all. Possessed by dogmatic prejudices, he eschewed algebraic methods, and while a preacher of the necessity of axiomatic foundations, he apparently based his personal ideas and beliefs about mathematics on some kind of absolute intuition whose decrees, once revealed, were not to be tampered with. To him, the Axiom of Choice was a matter of *truth*, not convenience, and to question it in his presence stirred him to anger.

But it is not my purpose to discuss here either his teaching methods or his general philosophy, except insofar as they influenced his mathematical work. My principal concern will be with the published materials outlined in the Bibliography which is appended hereto, and some of the circumstances surrounding and effecting it as revealed by his correspondence and other papers now available in the Archives of American Mathematics at the University of Texas. I shall be concerned both with the origin and evolution of MOORE's ideas as well as with their influence on events in the history of modern mathematics.

Reprinted from *Arch. History Exact Sci.* **26** (1982), 73–97, with permission of Springer-Verlag, Heidelberg.

PART I. THE BACKGROUND

When Moore, a native Texan, matriculated at the University of Texas in 1898, he encountered one of the most forceful personalities on the campus (as well as in Americal mathematics), namely GEORGE BRUCE HALSTED. Geometry was finally being put on a satisfactory basis, and HALSTED's greatest interest at the time was in geometry, particularly in HILBERT's recently published *Grundlagen der Geometrie* (1899). The outstanding characteristic of this work was its attempt to found the geometry of the plane and three-space on a rigorous axiomatic basis. Not only did HALSTED apparently acquaint R. L. MOORE with HILBERT's work, but MOORE was induced to check one of the axioms (Axiom II 4) for independence. MOORE's first piece of research embodied finding that this axiom was actually not independent. HALSTED communicated the result to E. H. MOORE, head of the mathematical group at Chicago—only to find, however, that E. H. had a few months before established the same result. Nevertheless, it turned out that R. L.'s proof was shorter and more elegant than E. H.'s, and the latter, in a note in the American Mathematical Monthly (vol. 9, 1902, pp. 152–153) termed it "delightfully simple." HALSTED wrote up R. L.'s proof in the form of a short note in the same journal.*

After earning his B.A. and M.A. degrees at Texas (both in 1901), and spending the subsequent two years first as teaching fellow at Texas and then as high school teacher in Marshall, Texas, R. L. spent the years 1903–1905 as a graduate student at the University of Chicago. Here he found an atmosphere of research that formed a natural continuation of that which HALSTED created. As already mentioned, E. H. MOORE was the head of the mathematics department at Chicago, and he had become thoroughly imbued with the ideas of the German school of mathematics, and particularly with the exploitation of the axiomatic method. He had spent a year of study in Göttingen and Berlin after receiving his doctorate at Yale in 1885, and according to two of his biographers (G. A. BLISS & L. E. DICKSON), "It seems that the work of Kronecker made the most lasting influence upon him, but in his habits of thought and his later work there are many indications of influences which might be traced to WEIERSTRASS and KLEIN."** Also at Chicago was MASCHKE, who is termed by the same authors (*loc. cit*) "one of the most delightful lecturers on geometry of all time." Inevitably these geometric interests helped to foster R. L.'s already formed interest in geometry.

At the time when R. L. arrived in Chicago, O. VEBLEN had just finished his doctoral work under E. H. MOORE and had been appointed an Associate.† In the latter capacity, he seems to have been enlisted by E. H. to assist in

*G. B. HALSTED, *The betweenness assumption*, Amer. Math. Mo., vol. **9** (1902), pp. 98–101.

G. A. BLISS & L. E. DICKSON, Nat'l Acad. of Sci. Memoirs, vol. **17, 1936, pp. 83–102.

†*Cf.* S. MACLANE, Nat'l Acad. of Sci. Memoirs, vol. **37**, 1964.

the supervision of R. L.'s thesis work. The association between VEBLEN and R. L. became quite close at this time, and R. L.'s thesis was, like VEBLEN'S, devoted to the axiomatic foundations of geometry. Much of R. L.'s early work was to be closely related to VEBLEN'S, *e.g.*, papers 1–3, 5.

As one scans his published papers, one is struck by MOORE'S predilection for the axiomatic method. All of his first 10 papers, with the exception of a discussion of DUHAMEL'S Theorem (paper 4) were concerned with some kind of axiomatic procedure. Aside from incidental use of the method (as in proving his classical theorem on upper semi-continuous collections of continua, paper 38), 15 of his 66 papers wer based on axioms; the same holds for his major work, his book entitled "Foundations of Point Set Theory." It was mainly in this respect that R. L., throughout his life, showed his Chicago background of the early 1900's. VEBLEN, who started his career at Princeton in 1905, continued his investigations in geometry throughout his life.

The first work that R. L. carried out, when he left Chicago and went to the University of Tennessee for a year, was concerned with axioms for the positive integers and their arithmetic. In this work, never published, he attempted to found the theory entirely on the undefined term *integer* and operations $\oplus$ and $\otimes$, thus skirting the notion of order entirely. Unfortunately one of the axioms was inordinately long and complicated (although quite easily grasped if one had at hand the tools of modern mathematical logic), and extant correspondence between MOORE and VEBLEN leads one to infer that the work was rejected by the *Annals of Mathematics*. In a letter dated April 9, 1906,* VEBLEN, who was now at Princeton and to whom MOORE and evidently sent his manuscript, wrote R. L. as follows:

"Your 'lists' of axioms came back from Huntington** the other day. I doubt if he understands A_6, [the axiom referred to above]. In consequence he was more impressed by the difficulty than by the value of your work.... You ought to write a preamble about your logical aims, condense the proofs as much as possible.... The avoidance of *ordinal counting* and order in any form, the replacing of 'class' by 'statement,' and some account of 'logic of propositions' which you presuppose ought to go into the preface. Have you tried to write postulates of logic?"

Another statement in the same letter from VEBLEN indicates that R. L. was also working at this time on some problems concerning curves: "Why not send me your curve business in its final form? If I can I will try to look up the literature better than you can in your town." What this refers to I have not been able to ascertain. Incidentally there follows a remark that will

*In the R. L. MOORE Collection at the Archives of American Mathematics at the University of Texas; quoted here by permission. For help in locating and obtaining materials from this collection, I am indebted to Professor LUCILLE WHYBURN and Dr. ALBERT C. LEWIS.

**E. V. HUNTINGTON was at that time one of the editors of the *Annals of Mathematics*. R. L. gave two alternative lists of axioms in his paper which he entitled "List 1" and "List 2."

strike a chord in everyone who has ever directed dissertations: "No doubt you are getting your geometry work into final form as quickly as possible? I am anxious to see that work come out as soon as possible." (The first two papers of MOORE, on geometry, were published within the next two years.)

R. L. also sent the "lists" of axioms for the positive integers to E. H. MOORE. A letter from the latter to R. L. indicates that E. H. pleaded the press of other duties and did not render any judgment on the work. However, like VEBLEN, he suggested to MOORE the possibility of delving into logic, specifically advising him to read the articles that had been appearing in the *Mathematische Annalen*. The articles referred to were chiefly concerned with the foundations of the theory of sets, especially with the Axiom of Choice and the Continuum Hypothesis.

Can it be that we have here and in VEBLEN's urgings, the origin of R. L.'s dislike for such investigations?* As any of his doctoral students can testify, he was a platonist in regard to the Axiom of Choice, regarding it as an absolute principle and not a matter for research regarding its consistency or admissibility. In his book, he did not indicate which theorems were dependent upon the Axiom, but stated it as a general principle in his Preface, to be used wherever needed in the text; this was quite contrary to his custom of giving clear indication in the book regarding which of his set theory axioms each theorem depended upon. This was also evidently one of the matters upon which he disagreed with VEBLEN. The latter, in his presidential address before the American Mathematical Society in 1924 stated, "The conclusion seems inescapable that formal logic has to be taken over by mathematicians. The fact is that there does not exist an adequate logic at the present time, and unless the mathematicians create one, no one else is likely to do so." (*Cf.* MACLANE, *loc. cit.*) It is interesting to note that exactly three years later, Alonzo Church, who may be considered the dean of mathematical logic in this country, received his Ph.D. under VEBLEN's direction with a dissertation concerning a set theory in which the Axiom of Choice is false.

Another colleague of R. L.'s younger days, who is not perhaps ordinarily thought of as forming one of the early influences on MOORE, was N. J. LENNES. He has probably been principally known as the coauthor, with VEBLEN, of *Introduction to Infinitesimal Analysis, Functions of One Variable*, published in 1907, a work popular among students of function theory and analysis in this country for many years.** Although LENNES was eight years older than R. L., and had received his M.S. degree at Chicago in 1903, he had taken time out to do high school teaching and did not receive his Ph.D. (Chicago) until 1907. He apparently kept in close touch with the Chicago

*See the comments below in Part IIIa concerning MOORE's feelings about such matters.

**See, for instance, R. C. ARCHIBALD, "A Semicentennial History of the American Mathematical Society, 1888–1938." N.Y., Amer. Math. Soc., 1938, p. 208. At the time when he wrote this book with VEBLEN, LENNES was a high school teacher in Chicago!

mathematical group during the time when R. L. was there, and he and R. L. corresponded with one another after R. L. left Chicago; LENNES was also one of those to whom R. L. set his "Lists" concerning the positive integers. But more important, LENNES was the creator of the topological definitions of *connected, arc* and *simple closed* (JORDAN) *curve* which played such an important part in MOORE's work. As will be noted later, LENNES' 1911 paper was drawn upon by MOORE in one of the latter's most important works.

Aside from these personal influences on R. L., there was of course the literary part of the mathematical environment, which at that time consisted almost predominantly of current papers of HILBERT, VEBLEN, LENNES, FRÉCHET and others, as well as the books of W. H. YOUNG and G. C. YOUNG (1906). A. Schoenflies (1908) and F. HAUSDORFF (1914) which inevitably helped to set the pattern that guided R. L. in his choice of work. Such influences can be more meaningfully brought out in our discussion of MOORE's own papers, to which we now turn.

PART II. THE MATHEMATICAL WORK

Reference will be made to MOORE's papers according to their numbers in the Bibliography; these numbers correspond to those which were assigned to them in my obituray of MOORE.* What follows directly below will essentially be an expansion of the discussion in the obituary, and I will therefore employ the same classification by subject that was used therein.

IIa. Geometry

Only six of MOORE's papers were devoted to what would today be called geomtry (*cf.* the remark in Part I contrasting VEBLEN's geometric work with R. L.'s). MOORE's interests seem to have undergone a gradual change during the period between his departure from Chicago in 1905 and the year 1915. This period turned out to be an almost sterile interval in R. L.'s life, probably partly induced by his apparent lack of success in finding, during this time, an environment that he could consider satisfactory and permanent.** Between 1908 and 1912, he published nothing, and likewise between 1912 and 1915 (he did publish papers on the specific dates mentioned, however); before

*R. L. WILDER, *Robert Lee Moore*, 1881–1974, Bull. Amer. Math. Soc., vol. **82** (1976), pp. 417–427.

**After his year at the University of Tennessee, R. L. held positions at the following institutions: Princeton University, 1906–1908; Northwestern University, 1908–1911; University of Pennsylvania, 1911–1920; University of Texas, 1920–. At Texas, he taught a full-time schedule until 1969, although nominally on half-time after age 70. For details, see the obituary, *loc. cit.* It can also be surmised that during this period, R. L. was coming to the conclusion that axiomatic foundations of geometry, to which he had devoted so much time, was not as fruitful a field of investigation as he would like.

1915, he published only four papers, one of which was his dissertation—not a very promising start for one whose later production belied these early portents. Both E. H. MOORE and O. VEBLEN urged him, in the letters cited from them in Part I, to get his dissertation into publishable form; he was obviously busy with the lists of axioms for the positive integers at that time, however.

By the year 1915, his thoughts had become focused on problems, outside classical geometry, that would lead him into the areas that were destined to form his life's work.

MOORE's work in geometry was chiefly devoted to its axiomatic foundations. His first two papers (1,2) were presented to the American Mathematical Society on April 22, 1905, in combined form under the same title, *Sets of metrical hypotheses for geometry.* DEHN had shown† that HILBERT's original axiom sets I, II, and IV, augmented by the assertion, S, that the sum of the angles of a triangle is two right angles, are not sufficient to yield III (parallels). In paper 1, R. L. showed that any space satisfying I, II, IV and S must nevertheless be a subspace (via the addition of ideal points) of a space in which III holds. It is interesting to note that in the proof MOORE made use of his former mentor HALSTED's book "Rational Geometry."

In his thesis, paper 2, R. L. gives axioms for Euclidean geometry using as primitive notions *point, order* and *congruence.* As already mentioned above, it is closely related to VEBLEN's dissertation*, whose axioms I and III–X it utilizes: alternative sets of axioms are considered, some of them being systems in which ordinary ruler and compass constructions are possible, as well as a set for Bolyai-Lobachevskian geometry. In showing that every circle is a JORDAN curve, he had to use a definition thereof given by VEBLEN in 1905** in terms of order and continuity conditions, the *Lennes* definition not being available at the time he wrote out his proofs. He also considered independence of his axioms. Incidentally, one of the most popular of R. L.'s courses, which he frequently gave during summer sessions at the University of Texas, utilized one of his systems of axioms for geometry (along with his now famous method of teaching).

The paper 5, published in 1915, contained a result surprising for the time. In the paper published by VEBLEN in 1905, which has just been cited above, a proof of the Jordan Curve Theorem was given which purported to hold in a non-metrizable space V satisfying the Axioms I–VIII, X of his thesis. In

†M. DEHN, *Die Legendre'schen Sätze über die Winkelsumme im Dreieck*, Math. Ann., vol. **53** (1900), pp. 404–439.

*O. VEBLEN, *A system of axioms for geometry*, Trans. Amer. Math. Soc., vol. **5** (1904), pp. 343–384.

O. VEBLEN, *Theory of plane curves in non-metrical analysis situs*, Trans. Amer. Math. Soc., vol. **6 (1905), pp. 83–98.

5, MOORE showed that the space V was actually metrizable, being topologically equivalent to the Euclidean coordinate plane. This paper gives some indication of the trend of MOORE's ideas toward the topological material and methods which were to occupy most of his time in after years.

In paper 17, MOORE applied his by then maturing familiarity with topological point set methods to give an axiomatic foundation of Euclidean and Bolyai-Lobachevskian plane geometry in terms of *point, region* and *motion* as primitives. HILBERT, in his fundamental paper *Grundzüge der Geometrie* of 1903,[†] had analyzed the transformation group, assuming the underlying space to be a number plane; MOORE's paper analysed the underlying space and the group simultaneously. The paper can be considered as a digression to his first mathematical love, but using the topological tools that had by now become part of his mathematical arsenal.

The review, 18, of the now classical VEBLEN-YOUNG work on projective geometry, seems to be the only review ever undertaken by MOORE. It is mainly devoted to a critical analysis of the foundations, especially as to whether one of the defined terms should really be treated as undefined.

IIb. Analysis

There is evidence that, while R. L. was coping with the problem of his major interests, he tried his hand as the field of Analysis. This was at a time when Analysis was passing from its "classical" stage to the modern form. Utilizing set-theoretic notions, such authors as BOREL and LEBESGUE had introduced newer and more general types of integration and more refined tools for attacking and analyzing problems in the theory of functions. During the period 1911–1912, MOORE presented papers to meetings of the American Mathematical Society under the titles "On the transformation of double integrals"[*] and "On sufficient conditions that an integral equation of the second kind shall have a continuous solution."[**] However, these were published only in abstract form. His published papers in Analysis can be considered as papers 4, 6, 7, 16, 25, 30, 34 and 42. The importance of the form of *Duhamel's* Theorem (of wider application than that due to OSGOOD), given in paper 4 was later emphasized by H. J. ETTLINGER, who also gave it generalizations and outlined its use in the study of summable functions.[‡] Papers 6 and 7 present sets of axioms in terms of *point* and *limit* for the linear continuum with emphasis in paper 6 on the question of complete independence of the axioms in the sense of E. H. MOORE. However, the statement that the axioms are categorical with respect to point and limit is retracted in paper 7,

[†]Math. Ann., vol. **56** (1902–1903), pp. 381–422.

[*]See Bull. Amer. Math. Soc., vol. **17** (1910–1911), p. 513, abstract No. 10.

[**]*Ibid.*, vol. **18** (1911–1912), pp. 217–218, abstract No. 5.

[‡]H. J. ETTLINGER, *R. L. Moore's principle and its converse*, Comptes Rendus des Séances de la Soc. des Sc. et de Lettres de Varsovie, XIX, 1927 Classe III, 455–460.

in which it is shown how to modify one of the axioms so that the statement concerning categoricalness becomes true.

In Paper 16, in which necessary and sufficient conditions are given that a certain type of FRÉCHET space be compact. MOORE demonstrates that by this time (1919) be has attained a maturity capable of dealing with the most abstract kind of mathematics. In paper 25, he corrects a proposition in the classic "Theory of Functions of a Real Variable" by E. W. HOBSON, and paper 30 is concerned with the relatively uniform convergence introduced by E. H. MOORE, with special reference to functions defined on a measurable set.

IIc. Point Set Theory

We use the term "Point Set Theory" to denote this section in difference to MOORE's own preference,‡ although current usage would dictate the term "Set-theoretic Topology."

Even though paper 10 is, according to our classification, the first paper of R. L.'s that we place in this category, the ideas and methods used are a natural evolution of both his own and other's previously published set-theoretic ideas. Besides G. CANTOR and HILBERT (particularly the "Grundzüge cited above), there were, in addition to MOORE himself, such mathematicians as O. VEBLEN, N. J. LENNES,* SCHOENFLIES, FRÉCHET and F. HAUSDORFF involved in this evolution. MOORE's own paper 5, discussed above, formed with the works of those just cited, a background of which MOORE's later work is a natural extension.

Because of its later influence, especially on MOORE's teaching, we consider paper 10 more in detail. It formed a basis for both his renowned style of teaching and for his later research methods. Its general format was by now classic: Primitive terms (a class S of elements called *points* and a class of subclasses of points called *regions*), axioms, development of the theory (of plane topology) therefrom, and finally independence examples for the axioms. The axiom system used for the proofs was denoted by Σ_1; in a final section he

‡R. L. had very strong feelings regarding terminology. If he felt a term was the "right" one, he adhered to its use regardless of majority opinion. This principle led in at least one instance to a terminology somewhat paradoxical after dimension theory had come in; he retained, throughout his life, the term "continuous curve" for spaces which could hardly be called "curves".

*See especially abstracts of LENNES' papers in Bull. Amer. Math. Soc., vol. **12** (1905–1906), as well as his paper of 1911, *Curves in Non-Metrical Analysis Situs with an Application in the Calculus of Variations*, Amer. Jour. Math., vol. **33** (1911), pp. 287–326. Such notions are *arc, simple closed curve, connected, accessibility*, as well as relations of a simple closed curve to its complement in the plane, all are introduced in this paper and play a prominent part therein.

discussed modification of Σ_1 denoted by Σ_2 and Σ_3. Of particular interest, both historically and for future developments, was axiom 1 (of both Σ_1 and Σ_2):

There exists an infinite sequence of regions, $K_1, K_2, K_3, \ldots$, such that (1) *if m is an integer and P is a point, there exists an integer n, greater than m, such that K_n contains P*; (2) *if P' and P are distinct points of a region R, then there exists an integer δ such that if $n > \delta$ and K_n contains P, then the closure of K_n is a subset of $R - P'$.*

In a footnote, MOORE remarks that there is a "certain amount of resemblance between Axiom 1 and Veblen's Postulate of Uniformity" stated in the latter's paper of 1905. *Definition in terms of order alone in the linear continuum and in well-ordered sets,* cited above. The Postulate of Uniformity was stated for the linear continuum, and asserted the existence for each point P and integer n, of a segment (= open interval) R_{nP} such that the set $\{R_{nP}\}$ satisfies the conditions:

1) For fixed P, and all n, $R_{nP} \supset R_{n+1P}$;

2) for fixed P, $P = \cap_n R_{nP}$;

3) for every segment R, there exists an integer n_R such that for no P does $R_{n_R P} \supset R$.

Although VEBLEN's postulate may have influenced MOORE's formulation of his Axiom 1, it should be noted how much stronger are the implications of the latter. In VEBLEN's postulate, there is required a sequence of segments (regions) for *every* point P—hence a non-denumerable class for the entire continuum, whereas MOORE's axiom postulates only a denumerable class of regions for the whole space. (Compare MOORE's Axiom 1′ of the axiom system Σ_3, however.) In particular, MOORE's axiom 1 implies the separability of the space, whereas VEBLEN's does not; VEBLEN postulated the separability in a separate axiom. Moreover, as E. W. CHITTENDEN pointed out some eleven years later, "The importance of the regular and perfectly separable, therefore metric, spaces in the analysis of continua is indicated by the fact that nine years before the publication of the discoveries of Urysohn, R. L. Moore assumed these properties in the first of a system of axioms for the foundations of plane analysis situs. This axiom is furthermore of particular interest historically since it yields when slightly modified a necessary and sufficient condition that a topological space be metric and separable."* The modification referred to here consisted only in adding an "Axiom 0" to the effect that "for every region R, there exists an integer n such that $R_n \subset R$." In the same connection, CHITTENDEN noted that MOORE had inferred from the

*E. W. CHITTENDEN, *On the metrization problem and related problems in the theory of abstract sets*, Bull. Amer. Math. Soc., vol. **33** (1927), pp. 13–34.

now well-known theorem of TYCHONOFF** that his Axiom was a sufficient condition for metrizability.†

Attention should also be called to Theorem 4, §3, of LENNES' paper of 1911, referred to above, concerning the existence of sequences of sets of regions closing down uniformly upon closed and bounded sets of points. There is no evidence that I have found, that this had a direct influence on MOORE's thinking in setting up his Axiom 1, although MOORE was thoroughly familiar with LENNES' paper. (For example, in the paper under discussion, MOORE not only uses the definitions presented in LENNES' paper, but makes use of LENNES' theorems 2–9 of §4, which are provable on the basis of MOORE's axioms,‡ and of LENNES' argument for the analysis of MOORE's Theorem 48 (the SCHOENFLIES converse of the JORDAN Curve Theorem in terms of accessibility). From this and other evidence,≠ it is clear that LENNES and MOORE were taking similar approaches to the topology of the plane.

Two year before the publication of paper 10, F. HAUSDORFF had given in his book§ his so-called second countability axiom. It is doubtful that R. L. had even seen this book before submission of paper 10 to the publisher, so that we cannot consider HAUSDORFF's axiom as having contributed to MOORE's thinking in the formulation of Axiom 1.

We shall call attention shortly to the later evolution of Axiom 1 as exemplified in MOORE's book of 1932 (item 51 in the Bibliography). To return to the paper (10) itself, it may be said to represent a culmination of the trends, so far as plane analysis situs is concerned, to be found in the previous works of VEBLEN, LENNES and MOORE himself. A central core of these works was the topological characterization of the basic Euclidean elements, *viz.*, the simple arc, the simple closed curve (S^1), the plane and 2-sphere. In paper 10, the topological characterization of the plane was achieved; in paper 14, MOORE showed that every space that satisfies either of the systems Σ_1 or Σ_2 is topologically equivalent to the number plane. It is interesting that MOORE did not use, in his proof, the result of paper 5, and show that the space satisfied VEBLEN's Axioms I–VIII, XI, but proceeded independently.

Before leaving paper 10, it should be pointed out that the independence examples given for axioms 6 and 7 are not valid—the discovery of which led

The TYCHONOFF theorem states that a necessary and sufficient condition that a perfectly separable HAUSDORFF space be metrizable is that it be regular. See Math. Ann., vol. **95 (1926), p. 139.

†Of importance from an evolutionary standpoint, we note that we shall see later that a replacement of Axiom 1, leading to the notion of "Moore Space," played a part in the discovery, by R. H. BING, of a new and important metrization theorem.

‡See p. 139 of LENNES' paper.

≠For instance, a letter of LENNES to MOORE in the Humanities Research Collection of the University of Texas, dated May 18, 1912.

§F. HAUSDORFF, "Grundzüge der Mengenlehre", Leipzig, Verlag von Vert. u. Comp., 1914, p. 263.

to the elimination of axiom 6, which can be proved from the other axioms.* Also, a notable reduction of MOORE's system Σ_1 was accomplished by ZIPPIN in connection with his characterization of the 2-sphere (see Part III). MOORE was to return later to the axiomatization of the plane (2-sphere) in both his book and paper 53, in the latter of which the undefined terms were *piece* (which may be interpreted as bounded, connected open set) and a relation which he called *imbedded in.*

IId. Continuous Curves

These configurations, defined analytically by C. JORDAN** in 1893, were quickly proved (PEANO, HILBERT, E. H. MOORE, PÓLYA) to encompass not only most of the "thin" geometric entities that were ordinarily considered to be curves, but to comprise a whole host of higher dimensional spaces, including all the Euclidean n-spheres ($n = 1, 2, 3, \ldots$). They very early became a subject for topological investigation, especially by SCHOENFLIES.† The basic problem of giving them a characterization in topological, rather than analytic, terms, was solved independently by H. HAHN and S. MAZURKIEWICA circa 1913‡ using concepts of general topology and in particular that of *local connectedness* ("connectedness im kleinen").

MOORE's first venture in this area—the "arc theorem" for continuous curves, represented by paper 11, turned out to be a "multiple" in that it was independently proved by S. MAZURKIEWICZ and H. TEITZE. However, it was quickly followed by a series of investigations which are reported on in the expository paper 27, *Report on Continuous Curves from the Veiwpoint of Analysis Situs*, published in 1923. Subsequent to this report, MOORE devoted less attention to continuous curves (papers 26, 44 and 45 are exceptions), leaving the field to his students, especially G. T. WHYBURN. To the latter is due the notion of *cyclic element* of a continuous curve, which proved to be a most useful device for analyzing the structure of these curves.≢

A weaker property than local connectedness, "semi-local-connectedness," was formulated by G. T. Whyburn and shown to be capable of replacing local connectedness in a number of situations; it is discussed by him in his book "Analytic Topology." Later, F. B. JONES introduced the notion of *aposyndetic continua.* Although the two notions (semi-locally-connected; aposyndetic) differ at a point, they are equivalent as applied at all points of a continuum.

*R. L. WILDER, *Concerning R. L. Moore's axioms for plane analysis situs*, Bull. Amer. Math. Soc. Vol. **34** (1928), pp. 752–760.

**C. JORDAN, *Cours d'Analyse*, 2 ed., Paris, Gauthier-Villars, 1893, vol. 1.

†See A. SCHOENFLIES, *Die Entwickelung der Lehre von den Punktmanningfaltigkeiten*, II, Leipzig, Teubner, 1908.

‡For references see paper 27, p. 202, footnote†.

≢For a report on the work of WHYBURN and others on cyclic element theory, *see* B. L. MCALLISTER, *Cyclic elements in topology, a history*, Amer. Math. Mo., **73** (1966), pp. 337–350.

A great deal of research has been done on such continua; two reports* of JONES may be consulted for descriptions and citations of results.

These notions can be generalized, using methods of algebraic topology, so that they appear as 0-dimensional cases of certain n-dimensional "avoidability" properties. For details, see my book "Topology of Manifolds," pp. 333f.**

IIe. The Structure of Continua

In the early twenties, work on the topology of general spaces and especially in the theory of continua began to take on a wider geographical spread, notably to Poland, where Sierpinski and others founded a school of set-theoretic topology; the new journal *Fundamenta Mathematicae* was founded there in 1920. As could be expected, duplication of effort was inevitable. In the case of MOORE's papers 28 and 31, which extend a theorem of SIERPINSKI, paper 31 turned out to be a "multiple" with MAZURKIEWICZ; *cf.* footnote 8 of paper 37.† Paper 39 duplicated results of W. GROSS and FRÉCHET. Paper 41 was a contribution to the theory of indecomposable continua, a type of topological configuration which began to receive much attention during the decade of the 1920's.‡

Papers 54–56 are of particular interest in that they display a system of axioms whose list of primitive terms, in addition to *point* and *region*, contains the term *contiguous to*, denoting a relation between points. Presumably a major reason for introducing this notion was for its application to structural properties of a continuum in terms of specialized subsets; for example, if the cyclic elements of a continuous curve C are regarded as "points" and two such points p and q are called "contiguous" if and only if one of the pair p, q is a point (in the ordinary sense) of the other, then C becomes an acyclic continuous curve in terms of its "points". One may wonder why this material has not led to more subsequent research than it has, since the notion of contiguous points could prove fruitful as both a mathematical and physical notion.≢

In papers 57, 59 and 64, MOORE continued his researches in the structure of continua, making special use of concepts such as continua of condensation

*F. B. JONES, *Concerning aposyndetic and non-aposyndetic continua*, Bull. Amer. Math. Soc., vol. **58** (1952), pp. 137–151; and *Aposyndetic continua*, Coll. Math. Soc. Janos Bolyai, 8, Topics in Topology, Keszthely, Hungary, 1972.

R. L. WILDER, "Topology of Manifolds," Providence, R.I., Amer. Math. Soc. Coll. Pub., vol. **32, 1949, 1963.

†Theorem 3 of paper 37 turned out to be false, and was corrected by paper 48.

‡Regarding paper 41, see B. KNASTER & C. KURATOWSKI, *Remark on a theorem of R. L. Moore*, Proc. Nat. Acad. Sci., vol. **13** (1927), pp. 647–649.

≢*Cf.* T. HAILPERIN, *On contiguous point spaces*, Bull. Amer. Math. Soc., vol. **45** (1939), pp. 172–174; E. C. KLIPPLE, *Two-dimensional spaces in which there exist contiguous points*, Trans. Amer. Math. Soc., vol. **41** (1938), pp. 250–276; and K. S. BUTCHER, *A homology theory for multiply connected contiguous point spaces*, Univ. of Michigan Dissertation, 1946.

and upper semicontinuous collections of continua. Upper semicontinuous collections had been introduced in paper 38, where it was shown that if such a collection, G, of disjoint bounded continua fills up a plane E^2 and none of its elements separates E^2, then it is itself a plane in terms of the elements of G considered as "points" and with "limit point" suitably defined. A similar statement holds for the 2-sphere, S^2, and in paper 50, MOORE showed that if the elements of G are allowed to separate S^2, then the resulting configuration, C, in terms of the elements of G as "points," is a *cactoid* (= a continuous curve whose maximal cyclic elements are 2-spheres). In view of the definition of limit for the elements of an upper semicontinuous collection, these elements may be considered as the counter-images of points of C under a monotonic continuous mapping of S^2 onto C. In such terms the theorem was later generalized not only to 2-manifolds and higher dimensional configurations,* but the notion of monotone mapping proved very fruitful in later set-theoretic investiagtions.**

The notion of triod was introduced in papers 46 and 49. One of the most striking results was the impossibility of imbedding an uncountable number of disjoint triods in the plane.

Prime part decompositions, which had been introduced by H. HAHN, were exploited in papers 29 and 35.† It was shown, for instance, that in terms of its prime parts, every bounded continuum is a continuous curve (possibly degenerate). The prime part notion was further extended and generalized by G. T. WHYBURN and the present writer.

Although MOORE did not venture far into the structure of point sets having no compactness properties (an exception is paper 43), some of his students, particularly M. E. (ESTILL) RUDIN, made notable discoveries in the area of connected point sets lacking compactness restrictions, as did also P. M. SWINGLE.‡ Such investigations, it turns out, can be expected to lead into questions of the foundations of set theory.

---

*Cf. J. H. ROBERTS & N. E. STEENROD, *Monotone transformations of two-dimensional manifolds*, Ann. of Math., vol. **39** (1938), pp. 851–862; R. L. WILDER, *Monotone mappings of manifolds*, Pacific Jour. Math., vol. **7** (1957), pp. 1519–1523; and *Monotone mappings of manifolds, II*, Mich. Math. Jour., vol. **9** (1958), pp. 19–23.

See, for instance, the report by L. F. MCAULEY, *Some fundamental theorems and problems related to monotone mappings*, Proc. Conf. On Monotone Mappings and Open Mappings, ed. L. F. MCAULEY, Binghamton, State Univ. of N.Y., 1970. pp. 1–36, as well as papers cited in the bibliography thereof. Also, R. C. LACHER, *Cell-like Mappings and their Generalization*, Bull. Amer. Math. Soc., vol. **83 91977), pp. 495–552.

†Paper 29 contains certain errors which were corrected in the footnote at the bottom of pp. 426–427 of paper 38.

‡See, for example, M. E. RUDIN, *A property of indecomposable connected sets*, Proc. Amer. Math. Soc., vol. **8** (1957), pp. 1152–1157; and M. E. RUDIN, *A primitive dispersion set of the plane*, Duke Math. Jour., vol. **19** (1952), pp. 323–328.

IIf. Positional Papers

Paper 12 was evidently principally inspired by (1) A. SCHOENFLIES' classic work *Die Entwickelung der Lehre von den Punktmannigfaltigkeiten* already referred to in IId, in which, among other results concerning positional properties of plane continuous curves, conditions were given under which the common boundary of two plane domains will be a simple closed curve, and (2) CARATHEODORY's work on prime ends. Like CARATHEODORY's condition, given for a similar purpose, MOORE's condition of "uniform connectedness im kleinen" applied to one domain alone; otherwise it is much simpler than the CARATHEODORY condition, and in the higher dimensional properties "ulc_n" and "ULC_n" has led to extensive generalizations.* Paper 21 gives examples in three-dimensional space for which neither MOORE's theorem nor the theorem of SCHOENFLIES holds.

In earlier work of ZORETTI, F. RIESZ, SCHOENFLIES and DENJOY, it developed that every closed, bounded totally disconnected plane point set is a subset of an arc. In paper 15, written jointly with J. R. KLINE,** necessary and sufficient conditions were given in order that a plane closed point set should be a subset of an arc. This was later extended to n-dimensional space by E. W. MILLER (*On subsets of a continuous curve which lie on an arc of the continuous curve*, Amer. Jour. Math., vol. **54** (1932), pp. 397–416).

The concept of equicontinuous systems of curves was introduced in paper 20, and in paper 24 was used to characterize both closed 2-cells and open surfaces in three-dimensional space. "Property S," a property weaker than uniform local connectedness yet stronger than local connectedness, was introduced in paper 22; a modification of a notion that SIERPINSKI had used to characterize continuous curves, it was used here to characterize those simply connected plane domains which have continuous curve boundaries.† And with reference to bounded plane domains that are complementary to continuous curves, MOORE proved in paper 23 that their outer boundaries are simple

---

*I have recently learned (see J. M. MCGREW, *The origins of connectedness im kleinen*, Dissertation, Mich. State Univ., 1976) that A. DENJOY proved the sufficiency part of MOORE's theorem of paper 12 in *Sur l'analysis situs du plan*, Comptes Rendus, Paris Acad., vol. **153** (1911), pp. 423–426. Also, L. E. J. BROUWER proved a theorem narrowly related to the necessity part of MOORE's theorem in *Über Jordansche Mannigfaltigkeiten*, Math. Ann., vol. **71** (1911), pp. 320–327.

**This seems to be the only jointly authored paper in which MOORE was involved.

†Property S was later generalized by the present author to a class of higher dimensional medial properties; for a report thereon, see R. L. WILDER, *A certain class of topological properties*, Bull. Amer. Math. Soc., vol. **66** (1960), pp. 205–239.

closed curves; from this he was able to show that if two points are separated by a continuous curve, C, then they are separated by a simple closed curve of C.

Spirals were introduced (in the plane) in paper 68 and certain results established concerning sets of points on which a spiral may close down; *e.g.*, if M is a compact, totally disconnected point set and p is a point not in M, then there exists an arc from p which spirals down on every point of M but on no point that is not in M. Several of MOORE's later doctoral students found further results concerning this notion.

Others of his positional papers, as their titles indicate, treat plane separation and accessibility. Paper 44 establishes an interesting theorem to the effect that any two points in the complement of a plane continuous curve M can be joined be an arc that does not separate M.

IIg. The Book, "Foundations of Point Set Theory"

From 1919 to 1932, inclusive, MOORE published 38 papers—more than half his total of 67—and his book, the first edition of which appeared in 1932 (see item 51 of the Bibliography). Any doubts about his creativity that may have been harbored by his mathematical colleagues during his earlier years, were certainly by now completely obliterated. His most productive period was from 1919–1926, during which 8 year period he published 32 papers. The book may be considered a kind of culmination of his work in topology, although he published some 17 papers thereafter. However, from 1930 on, he seems to have thrown most of his energies into teaching and the production of doctoral students (of whom 43 were awarded their Ph.D.'s in 1930 and after). Much of his teaching and hence much of the work done by his students after 1930 was based on the ideas of his book. In particular, the new "Axiom 1" introduced in the book is of special importance in future work, as well as of interest in connection with the Axiom 1 of paper 10 (from now on we denote this paper by FPAS). The new Axiom 1 is stated as follows:*

Axiom 1. *There exists a sequence* $G_1, G_2, G_3, \ldots$ *such that* (1) *for each* n, G_n *is a collection covering* S [*the set of all points*] *such that each element of* G_n *is a region,* (2) *for each* n, G_{n+1} *is a subcollection of* G_n, (3) *if* R *is a region,* X *is a point of* R *and* Y *is a point of* R, *whether identical with* X *or not, then there exists a natural number* m *such that if* g *is any region belonging to the collection* G_m *and containing* X *then the closure of* g *is a subset of* R *and, unless* Y *is* X, *the closure of* g *does not contain* Y, (4), *if* $M_1, M_2, M_3, \ldots$ *is a sequence of closed point sets such that for each* n, M_n *contains* M_{n+1} *and, for*

*We use the 1962, revised edition.

each n there exists a region g_n of the collection G_n such that M_n is a subset of the closure of g_n, then there is at least one point common to all the point sets of the sequence $M_1, M_2, M_3, \ldots$.

This axiom may seem to the uninitiated to be rather formidable, but MOORE's method of teaching was such that it was introduced to his students quite naturally and became "second nature" to them. It undoubtedly evolved, in MOORE's thinking, from the "Axiom 1" of FPAS, and was created as a means of accomplishing most of the purposes of the original, while not implying that the space S is metrizable or separable (both of which were derivable from the original axiom of FPAS). Evidently his earlier discovery that VEBLEN's proof of the JORDAN Curve Theorem, which was designed for non-metric spaces, was actually based on axioms which implied the underlying space to be the number plane and hence metrizable (see the discussion of paper 5 above) had some influence on MOORE's desire for such a new axiom, and he hoped that the latter might be sufficiently broad to serve as a basis for plane topology without implying the metrizability.* As with the system FPAS, the new system of axioms incorporating the new Axiom 1, served for years as a basis for his advanced course in point set theory, and, like FPAS, contributed to the evolution and success of his renowned method of teaching.

PART III. INFLUENCES OF MOORE'S RESEARCH ON FUTURE MATHEMATICS

Here we distinguish between the influence of MOORE's *teaching* and MOORE's *research*. The former has long been recognized through both the reputation of the "MOORE method" or, as it is sometimes called, the "Texas method," and the large number (50) of the doctorates awarded under his supervision. The influence of MOORE's research is a more subtle and difficult matter since in most cases a significant result in mathematics usually has a complex and intricate background, no particular element of which can be assigned as *the* influence that motivated the result. The best one can do is to indicate significant areas of mathematics in which the influence of MOORE's research is clearly indicated, even though not necessarily the sole reservoir from which the basic ideas involved were derived. The situation is further complicated by the fact that MOORE was, after all, only one of a growing number of mathematicians in this country, Poland, Russia, Germany, *etc.*, who took part in the building of the foundations of topology. And although one can expect that the early work of MOORE's own students was motivated by him, or by ideas that they got from him, as they developed they naturally

*For this information, I am indebted to Professor F. B. JONES. According to him, although MOORE discovered the axiom in 1926 (see Bull. Amer. Math. Soc., vol. **33** (1927), p. 141), and used it in his Boulder Colloquium Lectures in 1929, he did not succeed in proving the existence of non-metrizable spaces satisfying the axiom until, after the lectures, he was enroute home by rail.

adopted new ideas and methods from other sources. This becomes accentuated as one passes on to the work of later "generations". Even many of those mathematicians loosely designated as belonging to the "MOORE School" prove, on closer examination, to have adopted new philosophies and new interests some of which would not have met with MOORE's approval. Inevitably we shall omit much work, especially of recent years, that could be traced back to work of Moore.

Much of the direct influence of MOORE's individual results on further research has already been commented upon in Part II, so that in this part we shall usually omit further reference to such items.

*IIIa. Moore Spaces.***

The term "MOORE Space" for a topological space that satisfies Axiom 0 (which states that every region is a point set) and parts (1), (2) and (3) of Axiom 1 quoted above from MOORE's book, seems first to have been used by F. B. JONES.† The term "complete MOORE space" for a MOORE space which satisfies in addition part (4) of the axiom was justified by the proof by J. H. ROBERTS,* that every metric space which satisfies Axioms 0 and 1 is complete in some one of its compatible metrics. As mentioned above, there exist, however, complete MOORE spaces that are not metrizable (a fact first discovered by MOORE himself).

In the fifty years since their introduction, around 300 papers relating to MOORE spaces have been published.*** These have largely been stimulated by problems concerning metrizability, and particularly by the "JONES conjecture"‡ that every normal MOORE space must be metrizable, as well as by the question of the actual position of MOORE spaces among the geneal abstract spaces. JONES proved (*loc. cit.*) that if $2^{\aleph_0} < 2^{\aleph_1}$ (the so-called LUSIN hypothesis) then every separable normal MOORE space is metrizable. And in 1951, R. H. BING, in connection with his important work on the metrization problem, showed that if a MOORE space is collectionwise normal, then it is metrizable.†† It is a corollary that every paracompact MOORE space is metrizable.

**The writer is indebted to both Professor MARY ELLEN RUDIN and Professor F. BURTON JONES for advice and information in composing this section.

†See F. B. JONES, *Concerning normal and completely normal spaces*, Bull. Amer. Math. Soc., vol. **43** (1937), pp. 671–679.

*J. H. ROBERTS, *A property related to completeness*, Bull. Amer. Math. Soc., vol. **38** (1932), pp. 835–838.

***In 1968, a separate subject classification number was assigned to MOORE spaces by the abstracting journal *Mathematical Reviews*.

‡F. B. JONES, *loc. cit.*, p. 676, Here JONES raised the question, "Is every normal MOORE space metric?"

††R. H. BING, *Metrization of topological spaces*, Canadian Jour. Math., vol. **3** (1951), pp. 175–186. The theorem quoted is considered by many to be the most important theorem concerning MOORE spaces.

It may seem paradoxical, in view of MOORE's own feeling regarding logical foundations,[≠] that the "conjecture" has led into problems of mathematical logic. In particular, the independence of the normality of certain MOORE spaces from the usual axioms of set theory has been shown. Such involvement with set theory was, of course, already indicated by JONES' above cited result of 1937. Since then, research in MOORE spaces has proceeded along both topological and set-theoretic lines. Of the former character is a theorem proved by REED & ZENOR[§] to the effect that every locally connected, locally compact, normal MOORE space is metrizable.[§§] Regarding the latter see T. PRZYMUSINSKI & F. D. TALL, *The undecidability of the existence of a non-separable, normal Moore space satisfying the countable chain condition*, Fund. Math., vol. **85** (1974), pp. 291–297.

Recently it has been shown by W. G. FLEISSNER that, assuming the continuum hypothesis, one can construct a normal, non-metrizable MOORE space.[*]

Perhaps the most amazing sequence of results can be put together using theorems of FLEISSNER, P. NYIKOS, K. KUNEN and R. JENSEN: The existence of a strongly compact cardinal (a special kind of measurable cardinal) in a model for set theory implies the existence of an extension in which all normal MOORE spaces are metrizable. On the other hand, if there is a model in which all normal MOORE spaces are metrizable, there is an inner model in which there is a measurable cardinal.[**]

IIIb. Spheres and Manifolds

A subject that occupied much of MOORE's thinking was the characterization, in topological terms, of the plane and 2-sphere; his achievements in this

---

[≠]It has been well known among MOORE's students that he expressed a violent dislike of questions concerning logical foundations. The present writer recalls vividly one occasion on which he queried of one of MOORE's students whether he had used the Axiom of Choice in obtaining a certain result. MOORE, who was present, turned angrily and exclaimed, "I thought you'd ask that!" Curiously, he was urged in 1906 by both his colleagues E. H. MOORE and O. VEBLEN to study questions of logic. (*Cf.* Part I).

[§]G. M. REED & P. L. ZENOR, *A metrization theorem for normal Moore spaces*, Stud. Top. Proc. Conf. Charlotte, N.C., 1974, pp. 485–488.

[§§]It has been pointed out to me by F. B. JONES that P. S. ALEXANDROFF also discovered MOORE spaces, and that his uniform spaces are, in fact, metacompact MOORE spaces. See P. S. ALEXANDROFF, *Some results in the theory of topological spaces, obtained within the last twenty-five years*, Russian Math. Surveys, vol. **15** (1960), pp. 23–83. See also A. V. ARHANGELSKII, *On a class of spaces containing all metric and all locally bicompact spaces*, Soviet Math., vol. **4** (1963), pp. 1051–1055.

[*]W. G. FLEISSNER, *Normal non-metrizable Moore Space from Continuum Hypothesis or nonexistence of inner models with measurable cardinals*, to appear in Proc. Nat. Acad. Sci., USA.

[**]P. NYIKOS, *A provisional solution to the normal Moore space problem*, Proc. Amer. Math. Soc., vol. **78** (1980), pp. 424–435; W. G. FLEISSNER, *If all normal Moore spaces are metrizable, then there is an inner model with a measurable cardinal.*

regard have been discussed in Part II above. One of the notable and early characterizations of the 2-sphere was obtained by L. ZIPPIN, who showed that a continuous curve which contains at least one simple closed curve and is not disconnected by any arc of such a curve, but is separated by each of its simple closed curves must be a 2-sphere.† It may be noted that the JORDAN Curve Theorem constitutes one of the axioms (#4) in MOORE's book. Later the question was raised by J. R. KLINE as to whether a continuous curve which is separated by each of its simple closed curves, but not by any pair of distinct points in a 2-sphere. this proved a stubborn problem, but was eventually answered by R. H. BING in the affirmative.‡

Characterizations of 2-manifolds were obtained by I. GAWEHN (1927), J. H. ROBERTS (1932), E. R. VAN KAMPEN (1935) and G. S. YOUNG, Jr (1945). The VAN KAMPEN characterization utilized a localization of the ZIPPIN characterization of the 2-sphere, and the YOUNG characterization used the condition that each arc in a 2-manifold has two sides.≠

The difficulties encountered in achieving characterization of n-manifolds of dimension greater than 2 led to the introduction of "generalized manifolds," almost simultaneously and independently by E. ČECH, S. LEFSCHETZ and the present writer,* having the homology properties of the classical manifolds. For the classical dualities, this was shown by both ČECH and LEFSCHETZ; the present author went beyond this in his book** on manifolds, whose last five chapters are devoted to a verification that all the separation, accessibility, and other external and internal homology properties of the classical manifolds are possessed by the generalized manifolds. The generalized manifolds (now often called "homology manifolds") do not, of course, have

† L. ZIPPIN, *On continuous curves and the Jordan Curve Theorem*, Amer. Jour. Math., vol. **52** (1930), pp. 331–350.

‡ R. H. BING, *The Kline sphere characterization problem*, Bull. Amer. Math. Soc., vol. **52** (1946), pp. 644–653. For other characterization of the 2-sphere, see the references given by BING in the introduction to this paper.

≠ I. GAWEHN, *Über unberandete 2-dimensionale Mannigfaltigkeiten*, Math. Ann., vol. **98** (1927), pp. 321–354; J. H. ROBERTS, *A point set characterization of closed 2-dimensional manifolds*, Fund. Math., vol. **18** (1932), pp. 39–46; G. S. YOUNG, Jr., *Spaces in which every arc has two sides*, Ann. of Math., vol. **46** (1945), pp. 182–193.

* E. ČECH, *Théorie générale des variétés et de leurs théorèmes de dualité*, Ann. of Math., vol. **34** (1933), pp. 621–730; S. LEFSCHETZ, *On generalized manifolds*, Amer. Jour. Math., vol. **55** (1933), pp. 469–504; R. L. WILDER, *Generalized closed manifolds in n-space*, Ann. of Math., vol. **35** (1934), pp. 876–903. The general equivalence of all three types of generalized manifolds did not become evident until later; those of ČECH and LEFSCHETZ were constructed with the precise purpose of duplicating the homology properties of the classical manifolds; that of WILDER for the homological positional properties of the classical manifolds in n-space (see below).

** R. L. WILDER, "Topology of Manifolds," *loc. cit.*

the same homotopy properties as the classical manifolds. We forego later developments such as the applications in the theory of transformation groups.[†]

IIIc. Upper Semicontinuous Decompositions[‡]

One of MOORE's primary interests was upper semi-continuous decompositions. In Chapter V of his book (which devoted some 66 papers to their study), he modified Axiom 1 to a form which he called Axiom 1′ in order to derive the following theorem: *If G is an upper semi-continuous decomposition of the space, S, whose elements are compact (and closed) subsets of S, then G (given the quotient topology) also satisfies Axiom* 1′. WORRELL proved[≠] that MOORE could just as well have used Axiom 1 itself, since if S satisfies Axiom 1 and G is an upper semi-continuous decomposition of a compact (closed) subset of S, then G satisfies Axiom 1 also.

Upper semicontinuous decompositions were also discussed by G. T. WHYBURN in his book "Analytic Topology", and they continue to be of interest, especially in connection with problems relating to homeomorphisms of spaces and certain decompositions thereof.[§]

Notable work on upper semicontinuous collections was done by R. D. ANDERSON, who in his dissertation[*] under MOORE showed that for every continuous curve M (of any dimension whatsoever,) there exists a 1-dimensional continuous curve K in E^3 and an upper semi-continuous collection G of disjoint continua filling up K, which with respect to its elements as points is homeomorphic with M. Later, the same author established theorems concerning dimension-raising monotone mappings,[**] and also studied continuous collections of continuous curves. Another of MOORE's students, E. DYER, investigated dimension-lowering mappings, and J. H. ROBERTS studied two-to-one transformations.[***]

We recall that in paper 38, MOORE showed that an upper semi-continuous decomposition of the plane, E^2, into bounded continua that do not separate

[†] See A. BOREL *et al.*, "Seminar on Transformation Groups," Princeton, University Pr., 1960. (Annals of Math. Studies. No. 48).

[‡] For valuable help in writing sections IIIc and IIId, the writer is indebted to Professor C. E. BURGESS.

[≠] J. M. WORRELL, Jr., *Upper semi-continuous decompositions of developable spaces*, Proc. Amer. Math. Soc., vol. **16** (1965), pp. 485–490.

[§] See, for example, W. T. EATON, *Applications of a mismatch theorem to decomposition spaces*, Fund. Math., vol. **89** (1975), pp. 199–224, and the citations therein.

[*] R. D. ANDERSON, *Concerning upper semi-continuous collections of continua*, Trans. Amer. Math. Soc., vol. **67** (1949), pp. 451–460.

[**] See, for example, R. D. ANDERSON, *Monotone interior dimensionraising mappings*, Duke Math. Jour., vol. **19** (1952), pp. 359–366.

[***] See, for instance, E. DYER, *Certain transformations which lower dimension*, Annals of Math., vol. **63** (1956), pp. 15–19; and V. MARTIN & J. H. ROBERTS, *Two-to-one transformations on 2-manifolds*, Trans. Amer. Math. Soc., vol. **49** (1941), pp. 1–17.

E^2 yields the space E^2 again, and we have mentioned results of ROBERTS-STEENROD and WILDER constituting generalizations of MOORE's result to higher dimensions. In 1957, BING described a decomposition of E^3 into points and tame arcs such that the resulting decomposition space is not E^3 and the projection of the non-degenerate elements is zero-dimensional, and EATON, in 1973, described similar examples for all E^n, $n \geq 3$.[†] BING showed further that his example, which is not a manifold, is a Cartesian factor of E^4; that is, its Cartesian product with E^1 yields E^4.[‡] More recently, EATON & PIXLEY, and EDWARDS & MILLER[≢] have shown that any cell-like decomposition of E^3 is a Cartesian factor of E^4 provided the projection of the non-degenerate elements is closed and zero-dimensional.

In 1971, ARMENTROUT[§] extended the result of ROBERTS & STEENROD on cellular decompositions of 2-manifolds, to cellular decompositions of 3-manifolds that yield 3-manifolds. (Examples of BING and EATON, mentioned above, show that it is necessary to require, in dimensions higher than 2, that the decomposition space be a manifold.) SIEBENMANN then extended ARMENTROUT's result to n-manifolds, $n > 4$.[§§] These results on cellular decompositions of manifolds have involved the use of what has become known as "BING's shrinking criterion".[*] EATON developed a criterion called the "mismatch theorem" which has been very useful in determining whether certain types of decompositions of E^3 yield E^3, and CANNON & DAVERMAN have recently extended this concept to higher dimensions.[**]

Summaries of work on decompositions of manifolds, at various stages of its development, can be found in two papers by ARMENTROUT, and also in a paper by CANNON in which he develops properties of cell-like embedding

[†] R. H. BING, *A decomposition of E^3 into points and tame arcs such that the decomposition space is topologically different from E^3*, Ann. of Math., vol. **65** (1957), pp. 484–500; W. T. EATON, *A generalization of the dog bone space to E^n*, Proc. Amer. Math. Soc., vol. **39** (1973), pp. 379–387.

[‡] R. H. BING, *The cartesian product of a certain non-manifold and a line is E^4*. Ann. of Math., vol. **70** (1959), pp. 399–412.

[≢] W. T. EATON & CARL PIXLEY, *S^1 cross a* UV *decomposition of S^3 yields $S^1 \times S^3$*, Geometric Topology (Proc. Conf., Park City, Utah, 1974) Lecture Notes in Math., No. 438, Springer-Verlag, New York, 1975, pp. 166–194; R. D. EDWARDS & R. T. MILLER, *Cell-like closed* 0-*dimensional decompositions of R^3 are R^4 factors*, Trans. Amer. Math. Soc., vol. **215** (1976), pp. 191–203.

[§] S. ARMENTROUT, *Cellular decompositions of* 3-*manifolds that yield* 3-*manifolds*, Memoirs Amer. Math. Soc., No. 107 (1971).

[§§] L. C. SIEBENMANN, *Approximating cellular maps by homeomorphisms*, Topology, vol. **11** (1972), pp. 271–294.

[*] Originally formulated in R. H. BING, *A homoemorphism between the* 3-*sphere and the sum of two solid horned spheres*, Ann. of Math., vol. **56** (1952), pp. 354–362; and *A decomposition of E^3 into points and tame arcs such that the decomposition space is topologically different from E^3*, cited above.

[**] W. T. EATON, *Sums of solid spheres*, Mich. Math. Jour., vol. **19** (1972), pp. 193–207; J. W. CANNON & R. J. DAVERMAN, *Cell-like decompositions arising from mismatched sewings. Applications to* 4-*manifolds.* (Unpublished.)

relations.† In the latter paper, CANNON shows that problems on decompositions of manifolds are closely related to taming problems.

Decompositions of manifolds have played a significant role recently in the solution of the double suspension problem and the associated existence of noncombinatorial triangulations of manifolds. Bot CANNON and EDWARDS have shown that the double suspension of each homology n-sphere is a cellular decomposition of the $(n+2)$-sphere. CANNON‡ established a shrinking criterion for decompositions of manifolds general enough to establish that the double suspension of each homology n-sphere is an $(n+2)$-sphere. More recently, EDWARDS§ has given a more general condition characterizing the cellular decompositions of n-manifolds ($n \geq 5$) that have the same n-manifold for decomposition spaces.§§

IIId. Positional Properties

MOORE evinced great interest in positional properties of the simple closed curve in the plane and, more generally, of continuous curves in the plane. As pointed out in Part IIe above, this interest seems to have stemmed from a number of earlier works, particularly of SCHOENFLIES and N. J. LENNES. To such properties as accessibility, used by these earlier writers, MOORE added such positional properties as uniform local connectedness and Property S; and along with such investigators as SCHOENFLIES ("Die Entwickelung...") and F. HAUSDORFF ("Grundzüge der Mengenlehre," 1914, p. 335), MOORE expressed interest (Paper 21; also paper 27, pp. 301–302) in the extension to Euclidean 3-space of the positional properties which he and his predecessors had already studied in the plane. Such extensions were made by the present writer, not only to 3-dimensional Euclidean space but, using generalized manifolds, to dimensions greater than 3 (see my book, "Topology of Manifolds," *loc. cit.*).*

---

†S. ARMENTROUT, *Monotone decompositions of* E^3, Topology Seminar (Wisconsin, 1965), Ann. of Math. Studies, No. 60, Princeton, University Pr., 1966, pp. 1–25; and *A survey of results on decompositions*, Proc. Univ. of Oklahoma Topology Conf., Norman, Okla., Dept. of Math., Univ. of Oklahoma, 1972, pp. 1–12, J. W. CANNON, *Taming cell-like embedding relations*, Geometric Topology (Proc. Conf., Park City, Utah, 1974), Lecture Notes in Math., No. 438, Springer-Verlag, N.Y., 1975, pp. 77–118.

‡J. W. CANNON, *Shrinking cell-like decompositions of manifolds, codimension three*, Annals of Math., vol. **110** (1979), pp. 83–112.

§R. D. EDWARDS, *The topology of manifolds and cell-like maps*, Proc. Int'l Cong. of Mathematicians, Helsinki, 1978, pp. 111–127.

§§See also the summary by J. W. CANNON, *The recognition problem: What is a topological manifold*? Bull. Amer. Math. Soc., vol. **84** (1978), pp. 832–866.

*It is another amusing sidelight that this work ran into another of MOORE's dislikes. Analogous to his aversion to logical studies of the consistency and independence of principles such as the Axiom of Choice, he opposed strongly the introduction of algebraic methods into point set theory. There is nothing novel about this type of attitude, of course; one may recall the intense

These extensions were characterized by the usse of homological methods. Work of another type, more geometric in nature, may be said to have been inspired by the example of a wild 2-sphere in 3-dimensional Euclidean space E^3, discovered in 1924 by J. W. ALEXANDER.** The SCHOENFLIES extension theorem had established that there are no wild simple closed curves in E^2. The discovery of "wildness" in E^3 led to a considerable amount of work in the 1950's and 1960's, in which R. H. BING and his students had a leading role, on conditions under which a 2-sphere S is tamely embedded in E^3; that is, such that the embedding of S is equivalent with that of the round sphere. Much of this work has been summarized by BURGESS & CANNON and by BURGESS.† Some of the methods were initiated with MOISE's work, about 1950, on the triangulation of 3-manifolds.‡ Important key results, in the middle and late 1950's, were proofs of DEHN's lemma, the sphere theorem, and the loop theorem by PAPAKYRIAKOPOLOUS,≠ the development of approximation theorems for 2-spheres in E^3 by BING,§ and a solution of a SCHOENFLIES problem for $(n-1)$-spheres in E_n by BROWN.§§ Some of the fundamental properties of manifolds of dimension 2 and 3 are developed in a recent book by MOISE.***

Much of the work that was done for dimension 3 in the 1950's and 1960's has been extended to higher dimensions in the late 1960's and the 1970's. KIRBY & SIEBENMANN**** proved trinagulation theorem for n-manifolds, $n > 4$. DAVERMAN has recently presented a summary of work on embeddings on $(n-1)$-spheres in E^n, $n > 4$.*****

---

antagonism between advocates of "pure" geometric as opposed to analytic methods in geometry (resulting, for instance, in the famous geometer STEINER threatening to cease publishing in CRELLE's *Journal* if it continued to accept PLÜCKER's analytical papers).

J. W. ALEXANDER, *An example of a simply connected surface bounding a region which is not simply connected*, Proc. Nat. Acad. Sci., vol. **10 (1924), pp. 8–10.

†C. E. BURGESS & J. W. CANNON, *Embeddings of surfaces in E^3*, Rocky Mountain Jour. Math., vol. **1** (1971), pp. 259–344; C. E. BURGESS, *Embeddings of surfaces in Euclidean three-space*, Bull. Amer. Math. Soc., vol. **81** (1975), pp. 795–818.

‡E. E. MOISE, *Affine structures in 3-manifolds*, V. *The Triangulation theorem and Hauptvermutung*, Ann. of Math., vol. **56** (1952), pp. 96–114.

≠C. D. PAPAKYRIAKOPOULOS, *On Dehn's lemma and the asphericity of knots*, Ann. of Math., vol. **66** (1957), pp. 1–26.

§R. H. BING, *Approximating surfaces with polyhedral ones*, Ann. of Math., vol. **65** (1957), pp. 456–483; *Approximating surfaces from the side*, Ann. of Math., vol. **77** (1963), pp. 145–192.

§§M. BROWN, *A proof of the generalized Schoenflies theorem*, Bull. Amer. Math. Soc., vol. **66** (1960), pp. 74–76.

***E. E. MOISE, "Geometric Topology in Dimensions 2 and 3," Graduate texts in Mathematics, vol. **47** N. Y. Springer-Verlag, 1977.

****R. C. KIRBY & L. C. SIEBENMANN, *On the triangulation of manifolds and the Hauptvermutung*, Bull. Amer. Math. Soc., vol. **75** (1969), pp. 742–749.

*****R. J. DAVERMAN, *embeddings of $(n-1)$-spheres in n-space*, Bull. Amer. Math., vol. **84** (1978), pp. 377–405.

Some important key results are (1) proofs, developed independently by CERNAVSKII and DAVERMAN,† showing that an $(n-1)$-sphere is tame in E^n $(n > 4)$ if its complement is 1-ULC and (2) recent joint work by ANCEL & CANNON showing that any $(n-1)$-sphere in $E_n (n > 4)$ can be approximated with a tame sphere.‡

While many of the results in higher dimensions are similar to those in dimension 3, the methods in many cases are quite different. Proofs of many of the theorems on embeddings of 2-spheres in E^3 depended upon properties of E^2. On the other hand, much of the work for higher dimensions depended upon STALLINGS' engulfing theorem,≠ and generalizations of it.§ Thus, except for accessibility and separation properties and the SCHOENFLIES theorem, which are valid in all dimensions, work similar to what is mentioned above has not yet been done for 3-spheres in E^4.

CONCLUDING REMARKS

It is inevitable that the interpretations of historical events will reflect the prejudices of the historian. However, a good historian will try to avoid this, and will strive to be as impartial and factual as possible in the light of his own weaknesses. I am all to aware of the latter, and it is possible that my interpretation of MOORE's place in the history of topology is both inadequate and colored by my own acquaintance with the man. During my years of study and teaching at the University of Texas (1921–1924), I shared an office with Dr. MOORE for a whole year, and came to know his personality well. His was a forceful personality, but despite our areas of disagreement, we always retained a deep affection for one another as persons. When I left Texas in 1924, I felt I had a good idea of how MOORE would like to see the subjects, to which he had contributed, grow in the future.

In what I have written above, I have tried to emphasize the areas that MOORE liked best, although not neglecting at least to mention those in which methods that he disliked came into play. This will account for the greater detail that I have given to Parts IIIc and IIId, since I judge, possibly wrongly, that MOORE was rather intrigued by the outcome of these studies in the decomposition of continua and the positional properties in higher dimensions.

---

†A. V. CERNAVSKII, *Coincidence of local flatness and local simple-connectedness for embeddings of $(n-1)$-dimensional manifolds in n-dimensional manifolds when $n > 4$*, Mat. Sbornik, vol. **91** (133), (1973), 279–286 = Math. USSR Sbornik, vol. **20** (1973). 297–304; R. J. DAVERMAN, *Locally nice codimension one manifolds are locally flat*, Bull. Amer. Math. Soc., vol. **79** (1973), pp. 410–413.

‡F. D. ANCEL & J. W. CANNON, *The locally flat approximations of cell-like embedding relations*, Annals of Math., vol. **109** (1979), pp. 61–86.

≠J. R. STALLINGS, *The piece-wise linear structure of Euclidean space*, Proc. Cambridge Phil. Soc., vol. **58** (1962), pp. 481–488.

§R. H. BING, *Vertical general position*, Geometric Topology, (Proc. Conf., Park City, Utah, 1974), Lecture notes in Math., No. 438, N.Y., Springer-Verlag, 1975, pp. 16–41.

PUBLICATIONS OF R. L. MOORE

1. Geometry in which the sum of the angles of every triangle is two right angles, *Trans. Amer. Math. Soc.* **8** (1907), 369–378.

2. Sets of metrical hypotheses for geometry, *Trans. Amer. Math. Soc.* **9** (1908), 487–512.

3. A note concerning Veblen's axioms for geometry, *Trans. Amer. Math. Soc.* **13** (1912), 74–78.

4. On Duhamel's theorem, *Ann. of Math.* **13** (1912), 161–168.

5. On a set of postulates which suffice to define a number-plane, *Trans. Amer. Math. Soc.* **16** (1915), 27–32.

6. The linear continuum in terms of point and limit, *Ann. of Math.* **16** (1915), 123–133.

7. On the linear continuum, *Bull. Amer. Math. Soc.* **22** (1915), 117–122.

8. Concerning a non-metrical pseudo-Archimedean axiom, *Bull. Amer. Math. Soc.* **22** (1916), 225–236.

9. On the foundations of plane analysis situs, *Proc. Nat. Acad. Sci. U.S.A.* **2** (1916), 270–272.

10. On the foundations of plane analysis situs, *Trans. Amer. Math. Soc.* **17** (1916), 131–164.

11. A theorem concerning continuous curves, *Bull. Amer. Math. Soc.* **23** (1917), 233–236.

12. A characterization of Jordan regions by properties having no reference to their boundaries, *Proc. Nat. Acad. Sci. U.S.A.* **4** (1918), 364–370.

13. Continuous curves that have no continuous set of condensation, *Bull. Amer. Math. Soc.* **20** (1919), 174–176.

14. Concerning a set of postulates for plane analysis situs, *Trans. Amer. Math. Soc.* **20** (1919), 169–178.

15. (With J. R. KLINE) On the most general plane closed point set through which it is possible to pass a simple continuous arc, *Ann. of Math.* **20** (1919), 218–223.

16. On the most general class L of Fréchet in which the Heine-Borel-Lebesgue theorem holds true, *Proc. Nat. Acad. Sci. U.S.A.* **5** (1919), 206–210.

17. On the Lie-Riemann-Helmholtz-Hilbert problem of the foundations of geometry, *Amer. Jour. Math.* **41** (1919), 299–319.

18. The second volume of Veblen and Young's projective geometry, *Bull. Amer. Math. Soc.* **26** (1920), 412–425 (book review).

19. Concerning simple continuous curves, *Trans. Amer. Math. Soc.* **21** (1920), 333–347.

20. Concerning certain equicontinuous systems of curves, *Trans. Amer. Math. Soc.* **22** (1921), 41–45.

21. On the relation of a continuous curve to its complementary domains in space of three dimensions, *Proc. Nat. Acad. Sci. U.S.A.* **8** (1922), 33–38.

22. Concerning connectedness in kleinen and a related property, *Fund. Math.* **3** (1922), 232–237.

23. Concerning continuous curves in the plane, *Math. Zeit.* **15** (1922), 254–260.

24. On the generation of a simple surface by means of a set of equicontinuous curves, *Fund. Math.* **4** (1923), 106–117.

25. An uncountable, closed and non-dense point set each of whose complementary intervals abuts on another one at each of its ends, *Bull. Amer. Math. Soc.* **29** (1923), 49–50.

26. Concerning the cut-points of continuous curves and of other closed and connected point-sets, *Proc. Nat. Acad. Sci. U.S.A.* **9** (1923), 101–106.

27. Report on continuous curves from the viewpoint of analysis situs, *Bull. Amer. Math. Soc.* **29** (1923), 289–302.

28. An extension of the theorem that no countable point set is perfect, *Proc. Nat. Acad. Sci. U.S.A.* **10** (1924), 168–170.

29. Concerning the prime parts of certain continua which separate the plane, *Proc. Nat. Acad. Sci. U.S.A.* **10** (1924), 170–175.

30. Concerning relatively uniform convergence, *Bull. Amer. Math. Soc.* **30** (1924), 504–505.

31. Concerning the sum of a countable number of mutually exclusive continua in the plane, *Fund. Math.* **6** (1924), 189–202.

32. Concerning upper semi-continuous collections of continua which do not separate a given continuum, *Proc. Nat. Acad. Sci. U.S.A.* **10** (1924), 356–360.

33. Concerning the common boundary of two domains, *Fund. Math.* **6** (1924), 203–213.

34. Concerning sets of segments which cover a point set in the Vitali sense, *Proc. Nat. Acad. Sci. U.S.A.* **10** (1924), 464–467.

35. Concerning the prime parts of a continuum, *Math. Zeit.* **22** (1925), 307–315.

36. A characterization of a continuous curve, *Fund. Math.* **7** (1925), 302–307.

37. Concerning the separation of points sets by curves, *Proc. Nat. Acad. Sci. U.S.A.* **11** (1926), 469–476.

38. Concerning upper semi-continuous collections of continua, *Trans. Amer. Math. Soc.* **27** (1925), 416–428.

39. Concerning the relation between separability and the proposition that every uncoutable point set has a limit point, *Fund. Math.* **8** (1926), 189–192; also, An acknowledgement, *ibid.*, 374–375.

40. Conditions under which one of two given closed linear point sets may be thrown into the other one by a continuous transformation of a plane into itself, *Amer. Jour. Math.* **48** (1926), 67–72.

41. Concerning indecomposable continua and continua which contain no subsets that separate the plane, *Proc. Nat. Acad. Sci. U.S.A.* **12** (1926), 359–363.

42. Covering theorems, *Bull. Amer. Math. Soc.* **32** (1926), 275–282.

43. A connected and regular point set which contains no arc, *Bull. Amer. Math. Soc.* **32** (1926), 331–332.

44. Concerning paths that do not separate a given continuous curve, *Proc. Nat. Acad. Sci. U.S.A.* **12** (1926), 745–753.

45. Some separation theorems, *Proc. Nat. Acad. Sci. U.S.A.* **13** (1927), 711–716.

46. Concerning triods in the plane and the junction points of plane continua, *Proc. Nat. Acad. U.S.A.* **14** (1928), 85–88.

47. On the separation of the plane by a continuum, *Bull. Amer. Math. Soc.* **34** (1928), 303–306.

48. A separation theorem, *Fund. Math.* **12** (1928), 295–297.

49. Concerning triodic continua in the plane, *Fund. Math.* **13** (1929), 261–263.

50. Concerning upper semi-continuous collections, *Monatsh. Math. Phys.* **36** (1929), 81–88.

51. Foundations of point set theory, *Amer. Math. Soc. Coll. Pub.*, vol. **13**, Amer. Math. Soc., Providence, R.I., 1932; rev. ed. 1962; reprinted with corrections, 1970.

52. Concerning compact continua which contain no continuum that separates the plane, *Proc. Nat. Acad. Sci. U.S.A.* **20** (1934), 41–45.

53. A set of axioms for plane analysis situs, *Fund. Math.* **25** (1935), 13–28.

54. Foundations of a point set theory in which some points are contiguous to others, *Rice Institute Pamphlet* **23** (1936), 1–41.

55. Upper semi-continuous collections of the second type, *Rice Institute Pamphlet* **23** (1936), 42–57.

56. On the structure of continua, *Rice Institute Pamphlet* **23** (1936), 58–74.

57. Concerning essential continua of condensation, *Trans. Amer. Math. Soc.* **42** (1937), 41–52.

58. Concerning accessibility, *Proc. Nat. Acad. Sci. U.S.A.* **25** (1939), 648–653.

59. Concerning the open subsets of a plane continuum, *Proc. Nat. Acad. Sci. U.S.A.* **26** (1940), 24–25.

60. Concerning separability, *Proc. Nat. Acad. Sci. U.S.A.* **28** (1942), 56–58.

61. Concerning intersecting continua, *Proc. Nat. Acad. Sci. U.S.A.* **28** (1942), 544–550.

62. Concerning a continuum and its boundary, *Proc. Nat. Acad. U.S.A.* **28** (1942), 550–555.

63. Concerning domains whose boundaries are compact, *Proc. Nat. Acad. Sci. U.S.A.* **28** (1942), 555–561.

64. Concerning continua which have dendratomic subsets, *Proc. Nat. Acad. Sci. U.S.A.* **29** (1943), 384–389.

65. Concerning webs in the plane, *Proc. Nat. Acad. Sci. U.S.A.* **29** (1943), 389–393.

66. Concerning tangents to continua in the plane, *Proc. Nat. Acad. Sci. U.S.A.* **31** (1945), 67–70.

67. A characterization of a simpel plane web, *Proc. Nat. Acad. Sci. U.S.A.* **32** (1946), 311–316.

68. Spirals in the plane, *Proc. Nat. Acad. Sci. U.S.A.* **39** (1953), 207–213.

University of California
Santa Barbara

(*Received May 12, 1981*)

Anna Johnson Pell Wheeler (1883–1966)

LOUISE S. GRINSTEIN AND PAUL J. CAMPBELL

BIOGRAPHY

Anna Johnson Pell Wheeler was the daughter of Swedish immigrants, Andrew Gustav and Amelia (Friberg) Johnson, who came to the United States in 1872 from the same Swedish parish—Lyrestad in Skaraborglän, Wästergotland. Settling originally at Union Creek in Dakota Territory, they lived in a dugout hollowed from the side of a small hill, and the father tried to eke out a living as a farmer. In 1882 he moved his ever-growing family to the nearby town of Calliope (now Hawarden), Iowa, where Wheeler was born on May 5, 1883, the youngest of three surviving children. Her sister Esther, to whom she was very close, was four years older, and her brother Elmer was two years older. Around 1891 the Johnsons moved to Akron, Iowa, where her father became a furniture dealer and undertaker.[1]

The earliest extant records indicate that Wheeler was sent to the Akron public school. Though there appears to have been no tradition of academic achievement in the family, in the fall of 1899 Wheeler enrolled at the University of South Dakota, where her sister had already been studying for a year. After one year as a "sub-freshman" making up entrance requirements, she fulfilled the degree requirements in three years. Her main interest—mathematics—was evinced early in her college career. One of her mathematics professors at South Dakota, Alexander Pell, recognized her talent for mathematics and actively coached her into a mathematical career.

Obtaining an A.B. degree from South Dakota in 1903, Wheeler won a scholarship to the University of Iowa. She completed a master's degree the

[1]Louise S. Grinstein and Paul J. Campbell, "Anna Johnson Pell Wheeler (1883–1966)," in *WOMEN OF MATHEMATICS A Biographic Sourcebook*, Louise S. Grinstein and Paul J. Campbell, eds. (Greenwood Press, Inc., Westport, CT, 1987), pp. 241–246. Copyright ©1987 by Louise S. Grinstein and Paul J. Campbell. Reprinted with permission.

following year, taking five mathematics courses and a philosophy course. Simultaneously, she taught a freshman mathematics course and wrote her master's thesis, "The extension of the Galois theory to linear differential equations." The quality of her work was high, and she was elected to the Iowa chapter of the scientific society Sigma Xi. Winning a scholarship to Radcliffe, she earned a second master's degree in 1905. She stayed at Radcliffe an additional year on scholarship, enrolling in courses with such noted mathematicians as Maxime Bôcher, Charles Bouton, and William Osgood.

In 1906 she applied for and won the Alice Freeman Palmer Fellowship offered by Wellesley College to a woman graduate of an American college. A stipulation of the fellowship was that she agree to remain unmarried throughout the fellowship year. Wheeler used the funds to finance a year's study at Göttingen University, then the worldwide center of intense mathematical activity. While at Göttingen, Wheeler attended lectures given by the mathematicians David Hilbert, Felix Klein, Hermann Minkowski, and Gustav Herglotz, and the astronomer Karl Schwarzschild. Of these professors, she was most influenced by Hilbert and his work.

Throughout Wheeler's years of graduate study at Iowa, Radcliffe, and Göttingen, her former teacher, Alexander Pell, kept in touch with her. He was very proud of her progress and achievements. His first wife having died in the interim, he and Wheeler finally decided to marry, despite her family's objections to the twenty-five-year age differential. In July 1907, when her fellowship expired, they were married in Göttingen. They then returned to South Dakota, where Pell had been promoted to the position of first dean of the College of Engineering. During the fall term of 1907–1908, the young wife taught two courses at South Dakota—theory of functions and differential equations. Still, she wanted the Ph.D.; and in the spring of 1908, she decided to return to Göttingen alone to complete her doctoral work.

By the late fall of 1908, Wheeler had almost completed the requirements. The final examination for the Ph.D. was imminent. Evidently, some conflict of unknown origin arose between her and Hilbert, and she returned to America in December 1908 with a thesis (written independently of Hilbert) but no degree. She rejoined her husband in Chicago, where he had moved after academic policy disagreements forced his resignation from the University of South Dakota. His new position involved teaching at the Armour Institute of Technology.

Undeterred by the turn of events in Göttingen, Wheeler enrolled immediately at the University of Chicago. After a year's residency, during which she studied under the mathematician E. H. Moore, the astronomer Forest Moulton, and the astronomer/mathematician William Macmillan, she received a Ph.D. magna cum laude. The thesis accepted by her advisor, Professor Moore, was the one she had written initially for the Göttingen degree.

After receiving the Ph.D., she sought a full-time teaching position. Unfortunately, the large midwestern universities were reluctant to hire women. In the fall of 1910, she taught part-time at the University of Chicago. When Pell suffered a paralytic stroke in the spring of 1911, she substituted for him at the Armour Institute of Technology, another institution that did not want to hire women on a full-time basis.

In the fall of 1911, a vacancy opened at Mount Holyoke College. She applied for it and was accepted. Hired initially as an instructor, she was promoted to associate professor in 1914. However, Wheeler's years at Mount Holyoke (1911–1918) were not easy ones. Teaching loads were heavy. She felt compelled at all costs to continue her research work, and she had to take care of her husband, who never fully recovered from his stroke.

In 1918 Wheeler decided to resign from her position at Mount Holyoke College and accept an associate professorship at Bryn Mawr College. She felt that Bryn Mawr offered great potential for her career advancement. The possibility of teaching advanced mathematics to graduate students intrigued her, and there was the prospect of being promoted to chairperson when Charlotte Angas Scott* retired. Professionally, her career at Bryn Mawr was successful. She became chairperson in 1924 and full professor in 1925. Except for brief periods, Wheeler remained at Bryn Mawr as chairperson and teacher until her own retirement in 1948.

Wheeler's personal life during the Bryn Mawr years was not a consistently happy one. She lost her father in 1920 and her husband several months later. There was a brief but happy second marriage, followed by the death of her second husband in 1932. In 1935 her mother died. Later that same year, Emmy Noether*, her colleague and new-found friend, also died suddenly. All of these events took their toll on Wheeler.

During Wheeler's second marriage, to Arthur Leslie Wheeler, a classics scholar, the coupled lived in Princeton. Wheeler gave up her administrative duties at Bryn Mawr but continued lecturing on a part-time basis. She had more time to devote to her own research and could participate in the stimulating mathematical environment at Princeton University. Summers the Wheelers spent in the Adirondacks at a place they built and called "Q.E.D.," a name appropriate in the light of both of their careers. Following her husband's death, Wheeler returned to live and work full-time at Bryn Mawr.

Retirement for Wheeler in 1948 did not mean withdrawal from all mathematical activity. Despite recurring severe bouts of arthritis, she kept abreast of new developments and attended mathematical meetings. She remained in contact with many of her students, taking great pride in their achievements.

---

*Cross-reference to other women discussed in the volume is given by an asterisk following the first mention in a chapter of the individual's name.

She traveled, spending most of her summers in the Adirondacks, where she enjoyed various outdoor activities.

Wheeler suffered a stroke early in 1966. Never recovering, she died a few months later, on March 26, at the age of eight-two. According to her wishes she was buried beside Alexander Pell, in the Lower Merion Baptist Church Cemetery at Bryn Mawr.

Wheeler was highly respected professionally during her lifetime. Of the 211 mathematicians ever starred in *American Men of Science*, only three were women. One of them was Wheeler. Such starring was an honor reserved for those considered prominent in their field of activity by their contemporaries. In 1926 she was elected to Phi Beta Kappa. She received honorary doctorates from the New Jersey College for Women (now Douglass College of Rutgers University) (1932) and Mount Holyoke College (1937). In 1940 she was singled out as one of the one hundred American women to be acclaimed by the Women's Centennial Congress as having succeeded in careers not open to women a century before.

WORK

When Wheeler was studying at Göttingen, the most influential mathematician there was David Hilbert. In the early 1900s, Hilbert's work and interest evolved around integral equations, and he attached a great deal of importance to the subject. As a result, many mathematicians at Göttingen and throughout the world, among them Wheeler, were inspired to pursue further investigations in this area. Numerous papers were published. As the years passed, interest declined, and many of the results obtained passed into relative obscurity. An outgrowth of the work on integral equations was the development of a field in mathematics known as functional analysis, dealing with transformations, or operators, acting on functions.

Wheeler's research work spanned this period when the study of integral equations per se was at its peak of popularity and functional analysis was in its infancy. She regarded her work as being centered on "linear algebra of infinitely many variables." Her interest derived from possible applications of linear algebra to both differential and integral equations. Particularly noteworthy were her results on biorthogonal systems of functions. Some of the results she published were extended and generalized in the work of her own doctoral students at Bryn Mawr.

In 1927 Wheeler herself attempted to summarize her work and its overall importance in a series of invited lectures on the theory of quadratic forms in infinitely many variables. Unfortunately, these so-called Colloquium Lectures, presented during an American Mathematical Society meeting, were never published; but a detailed outline of the topics covered is found in an abstract written by T. H. Hildebrandt. In all the years that the Colloquium

Lectures have been given at American Mathematical Society meetings, only three lecturers have been women: Wheeler in 1927, Julia Robinson* in 1980, and Karen K. Uhlenbeck in 1985.

Wheeler drew accolades for her teaching throughout her career. Despite personal pressures and research commitments, she found time and even money to give to her students. Frequently she would invite graduate students to visit her summer home, where she provided them with encouragement and research time. Students felt free to talk to her about both personal and academic problems. Often she would take students to professional meetings at neighboring colleges and universities and urge them to participate actively.

As an administrator, Wheeler strove to enhance the national and worldwide reputation of the Bryn Mawr mathematics department. She tried to create an atmosphere in which students and faculty had ample opportunity for professional growth and development. When the Depression cut into available funds at the college, she nonetheless reduced teaching loads whenever possible so that faculty could find time for research.

Wheeler was instrumental in offering professional and political asylum at Bryn Mawr to the eminent German-Jewish algebraist Emmy Noether. A group of qualified Bryn Mawr students was assembled to take part in advanced algebraic seminars with Noether. Wheeler laid plans to involve Noether in an exchange of graduate mathematics courses with the Univeristy of Pennsylvania. Unfortunately, these plans never materialized because of Noether's unexpected death following surgery in 1935, less than two years after her arrival in America.

Wheeler did not confine her professional activities to her own research or to Bryn Mawr College. She was an active participant in such national professional organizations as the American Mathematical Society and the Mathematical Association of America. From 1927 to 1945 she served as an editor of the *Annals of Mathematics*. She worked on a College Entrance Examination Board committee which formulated basic guidelines for testing the mathematical potential of college-bound students (1933–1935). She was among those who petitioned for the establishment of the *Mathematical Reviews* in 1939, when the German abstract and review journal *Zentralblatt für Mathematik und ihre Grenzgebiete* became a victim of Nazi policy.

Bibliography

Works by Anna Johnson Pell Wheeler

Mathematical Works

"The extension of the Galois theory to linear differential equations." Master's thesis, University of Iowa, 1904.

*See the previous footnote.

"On an integral equation with an adjoined condition." *Bulletin of the American Mathematical Society* 16 (1909/1910): 412–415.

"On an integral equation with an adjoined condition." *Bulletin of the American Mathematical Society* 16 (1909/1910): 412–415.

"Existence theorems for certain unsymmetric kernels." *Bulletin of the American Mathematical Society* 16 (1909/1910): 513–515.

"Biorthogonal systems of functions." *Transactions of the American Mathematical Society* 12 (1911): 135–164.
Part I of Wheeler's doctoral thesis.

"Applications of biorthogonal systems of functions to the theory of integral equations." *Transactions of the American Mathematical Society* 12 (1911): 165–180.
Part II of Wheeler's doctoral thesis.

"Non-homogeneous linear equations in infinitely many unknowns." *Annals of Mathematics* (2) 16 (1914/1915): 32–37.

(with R. L. Gordon) "The modified remainders obtained in finding the highest common factor of two polynomials." *Annals of Mathematics* (2) 18 (1916/1917): 188–193.

"Linear equations with unsymmetric systems of coefficients." *Transactions of the American Mathematical Society* 20 (1919): 23–39.

"A general system of linear equations." *Transactions of the American Mathematical Society* 20 (1919): 343–355.

"Linear equations wth two parameters." *Transactions of the American Mathematical Society* 23 (1922): 198–211.

"Linear ordinary self-adoint differential equations of the second order." *American Journal of Mathematics* 49 (1927): 309–320.

"Spectral theory for a certain class of nonsymmetric completely continuous matrices." *American Journal of Mathematics* 57 (1935): 847–853.

Works about Anna Johnson Pell Wheeler

Case, Bettye, ed. "Anna Johnson Pell Wheeler (1883–1966), Colloquium Lecturer, 1927. Proceedings of the Symposium held on August 20, 1980, at Ann Arbor, MI." *Association for Women in Mathematics Newsletter* 12 (4) (July-August 1982): 4–13.
Summary of a symposium at which Wheeler's life and achievement were described and glowing tributes from former students and colleagues were presented. Several previously unpublished photographs are included.

Grinstein, Louise S., and Paul J. Campbell, "Anna Johnson Pell Wheeler: Her life and work." *Historia Mathematica* 9 (1982): 37–53.
Detailed account of Wheeler's life and achievements. An earlier version was published in the Association for *Women in Mathematics Newsletter* 8 (3) (September 1978): 14–16, 8 (4) (November 1978): 8–12.

Hildebrandt, T. H. "Abstract of 'The theory of quadratic forms in infinitely many variables and applications,'" *Bulletin of the American Mathematical Society* 33 (1927); 664–665.
Summary of Wheeler's Colloquium Lectures.

Pesi Rustom Masani is University Professor in Mathematics at the University of Pittsburgh. He formerly held professorships in Bombay and at Indiana University. He received his B.S. from the University of Bombay, and his Ph.D. from Harvard University in 1946 *as a student of Garrett Birkhoff. His main interest is in probabilistic functional analysis and cybernetics, specifically in prediction theory, vector measures, and positive-definiteness. He collaborated with Norbert Wiener and is editor of Wiener's collected works in four volumes (MIT Press). He is the author of a biography:* Norbert Wiener 1894–1964, *(Vita Mathematica 5), Birkhäuser Basel, 1989.*

Norbert Wiener:
A Survey of a Fragment of His Life and Work

P. R. MASANI*

Contents

*I am grateful to Professor J. Benedetto of the University of Maryland for his helpful suggestions in reducing the size of a longer version of this paper.

Norbert Wiener
1894–1964
Photograph taken in the late 1920s.
(Photograph courtesy of the MIT Museum.)

1. Wiener the Man

Wiener used to lunch at the Faculty Club in the Sloan Building of MIT. Around noon he would walk from his office to the club and back again. During one such walk he encountered an old friend whom he had not seen for a long time. It was a balmy day. They chatted amiably, admired the trees, the Charles and its sailboats. At last, they said goodbye. But as the friend departed, Wiener, looking bewildered, stood still. "By the way," he asked, "which way was I headed when we met?" "Why, Norbert, you were headed towards your office," the friend replied. "Thanks," said Wiener, "that means I have finished lunch."

With the substitution "Walker Memorial" for "Sloan Building," and minus my embellishments (balmy weather, etc.), this story is true. The encounter occurred in 1929 with Ivan A. Getting, then a physics freshman and an organist, whom Wiener had met previously at a demonstration of a new electric organ.[1] Dr. Getting also tells me of a tennis practice in which, after failing to connect with any of nearly 100 serves from him, Wiener suggested that they might exchange racquets.

At a garden party at the Statistical Institute in Calcutta, we were standing near a table when someone (whom I had not met) approached to pick up some refreshments. Wiener introduced himself, and got the response, "I am Abraham Matthai." "Matthai," said Wiener, "that's the name Matthew in Malayalam." Dr. Matthai was a statistician. Believing that he had just met Wiener, I approached him later to tell him of Wiener's lectures. Matthai laughed: it had been his third encounter with Wiener, and the third time he had learned about "Matthew".[2]

Such absent-mindedness, quirkishness and idiosyncrasy, amusing and even endearing, were punctuated unfortunately by recurrent manifestations of petulance, emotional instability, and irrational insecurity and anxiety. This was the source of his uneven relationships with some colleagues. Three rather distinct descriptions of Wiener have been penned by colleagues. A good "first look" is portrayed by Hans Freudenthal:

> In appearance and behavior, Norbert Wiener was a baroque figure, short, rotund, and myopic, combining these and many qualities in extreme degree. His conversation was a curious mixture of pomposity and wantonness. He was a poor listener. His self-praise was

[1]During World War II, Dr. Getting was appointed Director of the Radar and Fire Control Division at the MIT Radiation Laboratory, and he enlisted the services of mathematicians such as Ralph Phillips, Witold Hurewicz and others. Several allied victories in the air war are attributed to his work on radar. He retired a few years ago as Vice-president of the Aerospace Corporation in Los Angeles.

[2]For more reminiscences of Wiener, both amusing and serious, see Dr. Brockway McMillan's recent article in this series {**M10**}. Numbers in braces refer to the list of (non-Wiener) references at the end of the paper.

> playful, convincing, and never offensive. He spoke many languages but was not easy to understand in any of them.[3] {**F2**, p. 344}

N. Levinson has described his experiences as a student in one of Wiener's postgraduate courses:

> As soon as I displayed a slight comprehension of what he was doing, he handed me the manuscript of Paley–Wiener for revision. I found a gap in a proof and proved a lemma to set it right. Wiener thereupon sat down at his typewriter, typed my lemma, affixed my name and sent it off to a journal.... He convinced me to change my course from electrical engineering to mathematics. He then went to visit my parents, unschooled immigrant working people living in a rundown ghetto community, to assure them about my future in mathematics. He came to see them a number of times during the next five years to reassure them until he finally found a permanent position for me. {**L5**, pp. 24–25}

This little story is more telling of Wiener, the man, than the earlier description. But Levinson hastens to add:

> If this picture of extreme kindness and generosity seems at odds with Wiener's behavior on other occasions, it is because Wiener was capable of childlike egocentric immaturity on the one hand and extreme idealism and generosity on the other. {**L5**, p. 25}

When from personality and character we turn to Wiener's mind, D. J. Struik's observations are germane:

> ... the first impression was that of an enormous scientific vitality, which the years did not seem to affect. The second was to a certain extent complementary, and that was of extreme sensitivity. Complementary indeed, since a man with heart and mind so close to nature and the technique of his time must have had very fine antennae; he sees, or believes he sees, he feels, or believes he feels, where others remain unresponsive. {**S4**, p. 35}

The historian of science, G. de Santillana, said of Wiener, "In his reactions he was a child, in his judgements a philosopher." Indeed, the transformation was striking. One recalls Leonard Bernstein's talks on television about that disorderly, unhappy and irritable individual called Beethoven, and of the mental metamorphosis that occurred when he picked up his musical pen. With Wiener too, all traces of immaturity and eccentricity vanished when he picked up his scholarly pen. This writer has had the good fortune to

[3] In this otherwise apt description, the ambiguous term "wantonness" is totally inappropriate, and is perhaps indicative of inadequate acquaintance with the English language.

study all the 250-odd publications of Wiener. He can raise his right hand and say that in this corpus he has found only three, viz. [**47b**, **48d**, **49h**][4] in which Wiener-noise damps out the Wiener message. Wiener's Manuscript Collection (MC)[5] in the MIT Archives comprises 900 folders, among which is Wiener's correspondence with about a thousand individuals, ranging from an Attica prisoner to leaders of industry and labor, and some of the world's great minds.[6] Among the few letters this writer has scanned, he came across only one (to Dr. Frank Jewett, in September 1941, in which Wiener tenders his resignation from the National Academy of Sciences) that was intellectually confusing.

Wiener's life work, its enormous range notwithstanding, exhibits a coherence of thought from start to finish reminiscent of a great work of art. Unfortunately, because of space limitations, we shall be able to convey only a fragment of this piece of art. Nonetheless, writing this paper has been a pleasure.

2. The Nurturing Intellectual Environment

The climate at home was extraordinarily conducive. Wiener's father, Leo (1862–1939), a Tolstoian romantic and humanist who left Russia in his youth, was a genius, a scholar, a great linguist who spoke forty languages, and a Harvard professor. Leo had very definite ideas as to how children had to be trained so as to bring out their fullest potentialities. Norbert, being a precocious child, was subjected to a most vigorous and intense training primarily at home under Leo's direct tutelage, but he was also encouraged to read on his own, to have the run of libraries and museums and explore the countryside.

Unfortunately, certain prejudices and tempermental weaknesses of the parents affected this training with rather devastating effects on Wiener's emotional life. Apart from unnecessary harshness in training, he was led to believe that he was a gentile. The sudden revelation at age fifteen (1911) that this was a lie was shattering:

> The wounds inflicted by the truth are likely to be clean cuts which heal easily, but the bludgeoned woulds of a lie draw and fester. [**53h**, p. 147]

The "black year of my life" was his description of 1911.

---

[4]The numbers in square brackets refer to the Wiener papers cited at the end of the paper. Other references are in braces. For the papers [**47b**, **48d**], and [**49h**] a book review he wrote for *The New York Times*, see Wiener's *Coll. Works*, IV, pp. 748–750, 764–766 and 996–1000, cf. {**M5**}.

[5]Numerals prefixed by MC are to this Collection.

[6]It is good to report that this correspondence, now on microfilm, is being studied by Dr. Albert C. Lewis of The Bertrand Russell Editorial Project, McMaster University, Ontario, Canada, and may see the light of day within a few years.

Even so, the overall benefits were enormous. By the time he had finished school at age eleven and college at age fifteen, he had done a colossal amount of reading, had exposed himself to some of the world's greatest minds, and had formed a more or less coherent attitude towards the external world. He had also come to love American democracy, especially as it is practiced in the New England small towns, and was deeply patriotic, as his futile attempts to enlist in World War I testify. Furthermore, Wiener acquired a strong sense of duty that he retained throughout his life, cf. e.g. [**60e**]. The year 1911 notwithstanding, he received a Ph.D. from Harvard in philosophy in 1913 at the age of eighteen, and was awarded a John Thorton Kirkland Traveling Fellowship by Harvard. Without doubt Leo Wiener was Norbert Wiener's first great mentor.

The second very favorable factor in Wiener's environment was the extremely healthy intellectual climate that prevailed in world science during his postdoctoral and later years, roughly between 1914 and 1933. This had much to do with the publication of the *Principia Mathematica* (PM) by Bertrand Russell and Alfred North Whitehead during the years 1910 and 1913 {**W3**}. Let us see how this affected Wiener.

The German logician G. Frege's attempts in 1893 to reduce arithmetic to logic had failed: his system allowed the antinomy concerning the Russell class $R = \{X : X \notin X\}$. In 1910 Whitehead and Russell succeeded in attaining Frege's objective: they kept the antinomies at bay by adhering to the canons of type that Russell had introduced in 1903 {**R2**}. In fact the PM salvaged the entire Cantorian theory of sets and the Dedekind theory of numbers. Moreover, to use Gödel's words, the subject "was enriched by a new instrument, the abstract theory of relations," on which is based the theory of measurement {**G2**, p. 448}. Russell's use of recursive definitions brought to the forefront the idea of *recursion*, which when set is a proper metamathematical footing by K. Gödel, A. Church, A. M. Turing and others had revolutionary ramifications on mathematical philosophy and, via the work of C. Shannon, J. von Neumann, Wiener and others, on automata theory and industrial technology. It ushered in the age of automatization.

Wiener's Harvard thesis was on "A comparison between the treatment of the algebra of relatives by Schroder and that by Whitehead and Russell." Wiener decided to spend his Harvard overseas traveling fellowship to study mathematical philosophy with Russell at Cambridge. In his thesis Wiener had missed the philosophical import of the theory of types. Relearning it from Russell's lectures was an eye-opener:

> For the first time I became fully conscious of the logical theory of types and of the deep philosophical considerations it represented. [**53h**, p. 191]

In Russell, Wiener had found his second great mentor.

To Wiener, the Russellian hierarchy was more than a convenient tool to keep off the antinomies. Indeed Zermelo (1908) and later von Neumann (1925) and Bernays (1937) kept off the paradoxes more effectively by cutting down the hierarchy to just two: "element and non-element," "set and class," "sets formally expressible, and sets not so expressible." For Wiener, however, it was the hierarchical classificatory attitude behind the Russellian doctrine that was stimulating. For instance, in later years he assigned types 1, 2, 3, to automatons A, B, C, in case B could evaluate the performance of A, and C that of B. Likewise, he assigned types 1, 2, 3,... to the military categories: tactics, strategy, general considerations to use in framing strategy,....

Wiener was wont to see the Russell antinomy behind many a situation where most of us might see none. Roughly he saw it whenever trouble ensued from two variables becoming equal or near equal. He took just as readily to the more Cantorian "paradoxes of the superlative," which yield self-contradictory concepts such as the set of all sets. A favorite self-contradictory concept was "the totally efficient slave."[7] Wiener used such paradoxical concepts tellingly to illustrate phenomena such as the Roman household in which the Greek philosopher-slave becomes the real master.

This love notwithstanding, Wiener was not able to marshall the full potency of the paradoxes in the disciplined and creative way in which K. Gödel was able. His only contribution to axiomatic set-theory was his type-theoretical definition of the ordered pair as a set **[14a]** (age nineteen). This simplification completed a line of thought of C. S. Peirce: it showed that three primitive constants, $\downarrow$, $\forall$, $\in$ suffice for logic and mathematics. It also simplied the theory of types, cf. Quine {**Q1**, p. 163}.

During the years 1914–1920 Wiener did a lot of work on mathematical and general philosophy. The best of this extends the theory of measurement in the PM, Vol. III, to quantities, the range of whose values is bounded, e.g. the intensity of "redness" of a red patch, and to relations such as "seem louder than" [**14b**, **15a**, **21a**]. There was also a 101-page paper on Kant's theory of space [**22a**].

A remarkable sequel to the PM appeared in 1919 in the *Tractatus logico-philosophicus* by Ludwig Wittgenstein {**W6**}. The novel ideas in this work led the philosophers of the Vienna Circle to formulate the main theses of *logical empiricism*: (i) the analyticity, or devoidness of factual content, of all logical and mathematical statements; (ii) the hypothetical character of all empirical ones; (iii) the paramount importance of mathematical concepts in the formulation of general hypotheses of the sciences, and of logical and mathematical theorems in the transition from such hypotheses to verifiable

[7] Self-contradictory, because to be fully efficient one has to be free.

experimental and observational statements. The resulting faith in the intimacy of logic-mathematics on the one hand and physics in the wide sense on the other, despite a clear-cut separation between the two, is perhaps best illustrated by the words of Einstein:

> As far as the laws of mathematics refer to reality they are not certain; and as far as they are certain they do not refer to reality. {**E2**, p. 28}[8]

This was the faith which dominated the intellectual climate of the period in which Wiener began research. It affected him in concrete ways. Russell urged him to adopt the broadest standpoint, to concentrate not just on the foundations but also to look at the frontiers of mathematics as well as of theoretical physics. This advice brought Wiener into contact with G. H. Hardy, then a young don. Hardy was without question Wiener's third and perhaps last great mentor. It also exposed Wiener to Bohr's atomic theory, the work of J. W. Gibbs on statistical mechanics and the Einstein–Smoluchowski papers on the Brownian motion.

In this free and clean atmosphere a good physicist could extol the virtues of mathematics without a feeling of having let down his regiment. Thus what the French physicist and Nobel Prize-winner, Jean Perrin, wrote in 1913 was music to Wiener's ears:

> Those who hear of curves without tangents or of functions without derivations often think at first that Nature presents no such complications nor even suggests them. The contrary, however, is true and the logic of the mathematicians has kept them nearer to reality than the practical representations employed by physicists. This assertion may be illustrated by considering certain experimental data without preconception. {**M2**, pp. 5, 6}[9]

These then were the kind of messages that entered into the nonlinear transducer we call Wiener, messages which are hard to come by today. Before considering later inputs, and there were many, let us see the messages that emerged from this transducer.

3. The Leap from Postulate Theory to the Brownian Motion and Potential Theory

After Wiener had left the U. S. Army in early 1919, and joined the Mathematics Department of MIT in the fall of 1919, his intellectual interests began

[8]It should be clear from this quotation that we are interpreting logical empiricism in the broadest way. This is necessary in the light of the criticism levelled by Professor Quine {**Q2**} and others against narrower interpretations of the thesis, cf. Carnap {**C2**}.

[9]For the rest of this quotation, see Mandelbrot {**M2**}.

to move away from the philosophical foundations of mathematics towards its superstructure. He finalized his long paper on his 1915 Docent Lectures at Harvard on geometry and experience [**22a**]. But concurrently he began to focus on the postulates of specific systems then engaging the curiosity of mathematicians.

Space permits us to comment only very briefly on the papers on postulation [**17a**, **20a–e**, **21b**, **22b**,**c**, **23g**]. They bear the impress of E. V. Huntington and M. Fréchet. Wiener's objective was to do what Sheffer (and in fact C. S. Peirce) had done for the truth-functional sentential calculus, viz. have just one primitive operation, and then study "all the sets of postulates in terms of which the system may be determined" [**21b**, p. 1]. Thus in [**20b**] Wiener introduced a single binary connective $*$ subject to 7 postulates, and showed its equipollence to the usual postulate system for a field $\mathbb{F}$. In the paper [**22c**] on topology, his object was to place postulates on the primitives X, $\sum$, that would make X a topological space, and $\sum$ the group of its homeomorphisms. Thus, a "limit-point" a of a set E is defined by

$$f \in \sum \ \& \ \forall x \in E \backslash \{a\}, \ f(x) = x \ \Rightarrow \ f(a) = a.$$

In [**22b**] Wiener characterized the linear continuum in this manner, departing thereby from Huntington, Veblen and R. L. Moore. Wiener's postulational interests also included metric affine and vector spaces [**20e**, **22c**]. In [**23g**] he assumed for the first time that the metric is complete, thus defining a Banach space, but he focused on the analysis of vector-valued functions, puny stuff in relation to the deep work that Banach had begun a little earlier. There is no need to start speaking of "espaces du type BW".

Among the propellants that steered Wiener away from such work into deeper waters, was his reading of the treatises of Osgood, Volterra, Fréchet and Lebesgue during the summer of 1919, his meetings with Fréchet in Europe, and above all his conversations with the young mathematician I. A. Barnett. From the latter he learned of the potential importance of probabilistic questions in which the events are curves, such as the paths traced by a swarm of flying bees, and of the possibility of using infinite-dimensional vector spaces in their analysis.

Wiener began in earnest. Spotting the papers of Daniell, he tried to adapt them to his needs. He started with a sequence $(\pi_n, w_n)_{n=1}^{\infty}$, where π_n is a finite partition of a fixed set X, and w_n is a function on π_n to $\mathbb{R}_+$. He then took the class

$$\mathscr{L} := \{F \colon F \in \mathbb{R}^X \ \& \ \exists n \geq 1 \ni F \text{ is } \pi_n\text{-simple}\},$$

and for any F in $\mathscr{L}$, defined its mean-value $M(F)$ by

$$M(F) := \left\{ \sum_{t\in\Delta\in\pi_n} f(t)w(\Delta) \right\} \Big/ \left\{ \sum_{\Delta\in\pi_n} w(\Delta) \right\}.$$

With the assumption that π_{n+1} is a refinement of π_n, and w is finitely additive, $M(F)$ becomes independent of n, and is unique. Wiener subjected the π_n, w_n to further conditions (Kolmogorov's marginals in disguise) and showed that $(\mathscr{L}, M)$ then fulfills Daniell's conditions. Consequently, there is an extension $(\overline{\mathscr{L}}, \overline{M})$, $\mathscr{L} \subseteq \overline{\mathscr{L}}$, $M \subseteq \overline{M}$, $\overline{\mathscr{L}}$ being the class of "summable" functions and $\overline{M}$ the "Daniell" integral. Wiener showed that F is in $\overline{\mathscr{L}}$ if F is "uniformly continuous," i.e., $\inf_{n\geq 1} \sup_{\Delta\in\pi_n} \operatorname{Osc}(F, \Delta) = 0$. This work **[20f]** was to provide a firm mathematical footing for a venture into physics.

Wiener's first object of attack, to wit turbulence, was suggested by the then fresh paper of Sir Goeffrey Taylor {**T1**}. When these attempts failed, Wiener tried his hand on something else he knew, vaguely akin to turbulence, viz. the Brownian movement as conceived by Einstein in his fundamental 1905 paper {**E1**}.

Recall what Einstein had done. He had assumed that there is a positive number τ such that a time-interval of length τ is, in his words,

> ... very small compared to the observed interval of time, but nevertheless, ... such ... that the movement executed by a (colloidal) particle in consecutive intervals of time τ are ... mutually independent phenomena. {**E1**, p. 13}

From this premise Einstein derived the result that the displacements in disjoint intervals are normally distributed, that "the mean (square) displacement is ... proportional to the square root of the time" {**E1**, p. 17}, being given by the equation

$$\bar{d}_t^2 = \frac{RT}{3\pi a\mu N} \cdot t, \tag{1}$$

where T is the temperature, R is the gas constant ($pv = RT$), μ is the viscosity of the liquid, N is Avogadro's number, and a is the radius of the colloidal particle.

Wiener's concern, unlike Einstein's, was with the nature of the curve followed by a single particle. He therefore made the idealization that Einstein's conditions prevail for all positive lengths τ. To this idealized Brownian motion, "an excellent surrogate for the cruder properties of the true Brownian motion" **[56g**, p. 39], Wiener was able to apply the theorem proved in **[20f]**. The nexus is clear from §§3, 4 of **[24d]**. Here Wiener defined the sequence $(\pi_n, w_n)_{n=1}^\infty$ so that it not only fulfills the premises of the **[20f]** theorem, but the extensions $(\overline{\mathscr{L}}, \overline{M})$ of the resulting $(\mathscr{L}, M)$ also have the following additional properties. Write

$$\text{(2)} \qquad \begin{cases} \mathfrak{X} := \{x \colon x \text{ is continuous on } [0,1] \text{ to } \mathbb{R} \;\&\; x(0) = 0\} \\ \mathscr{B}(\mathfrak{X}) := \{B \colon B \text{ is a Borel subset of } \mathfrak{X}\}. \end{cases}$$

Then $\overline{\mathscr{L}}$ is the class of $\mathscr{B}(\mathfrak{X})$ measurable functions on $\mathfrak{X}$ to $\mathbb{R}$, and writing $\mu(B) := \overline{\mathscr{L}}(1_B)$ for $B \in \mathscr{B}(\mathfrak{X})$, μ has all the properties of what today we call *Wiener measure.* (We cannot, unfortunately, state Wiener's ingenious definitions of π_n and w_n since they involve a lot of notation.)

Excellent commentaries on the papers [**21c**, **21d**, **23d**, **24d**] on the Brownian motion by K. Ito (*Coll. Works*, I), M. Kac {**K1**} and J. Doob {**D3**} are available. However, so overwhelming have been the effects of this work on the development of analysis and probability theory, and later on communication theory, that a little more must be said.

It was only in the 1930s that Wiener was able to unearth what is buried in these papers, and there is a lot, as Kolmogorov's important 1933 work {**K3**} suggested. By mapping the space $\mathfrak{X}$ (cf. (2)), with Wiener measure onto the interval $[0,1]$ with Lebesgue measure, Wiener characterized the Brownian motion as the *stochastic process* $\{x(t,\alpha) \colon t \in [0,1],\ \alpha \in [0,1]\}$, governed by the conditions: (i) the increments $x(b,\cdot) - x(a,\cdot)$ are normally distributed random variables with mean 0 and variances $\sigma^2(b-a)$, i.e.,

$$\text{(3)} \qquad \int_0^1 |x(b,\alpha) - x(a,\alpha)|^2 \, d\alpha = \sigma^2(b-a), \qquad \sigma = \text{const.},$$

the abstract formulation of Einstein's equation (1); (ii) for nonoverlapping intervals $[a,b]$, $[c,d]$ the increments $x(a,\cdot) - x(b,\cdot)$ and $x(c,\cdot) - x(d,\cdot)$ are stochastically independent, cf. {**D2**}. Equivalently we may characterize it as the process for which $x(t,\cdot)$ is normally distributed with zero mean and such that

$$\text{(4)} \qquad \int_0^1 x(s,\alpha)x(t,\alpha)\, d\alpha = s \wedge t, \qquad s,t \in [0,1].$$

Wiener showed that for almost all α in $[0,1]$, the trajectories $x(\cdot,\alpha)$ are continuous everywhere but differentiable nowhere. Although the functions $x(\cdot,\alpha)$ are of "extremely sinusoidity," and definitely not of bounded variation, Wiener was able to define for any f in $L_2[a,b]$, a "Stieltjes" type integral

$$g(\cdot) = \int_a^b f(t)\, dx(t,\cdot) \quad \text{on } [0,1],$$

and to enunciate the beautiful properties of the random variables $g(\cdot)$ so obtained. Thus Wiener opened up the whole area of probability theory we nowadays call *stochastic integration.* These developments occur in the later works [**33a**, **34a**, **34d**] done in collaboration with Paley and Zygmund.

The Brownian motion also penetrates deeply into the nonstochastic parts of mathematical analysis. An interesting example is afforded by the initial value problem of the one-dimensional heat equation with potential term. Its solution can be expressed as an integral over $C[0,\infty)$ with respect to Wiener measure. This was discovered in 1948 by Mark Kac under the stimulus of R. P. Feyman's cognate result in nonrelativistic quantum theory, and has led to widespread use of functional integration in both mathematical analysis and field physics.

Since 1950 the ideas of Perrin and Wiener have been receiving considerable enlargement in the researches of Dr. Benoit Mandelbrot and others into phenomena marked by intrinsic irregularities that persist even as we improve the accuracy of the scale of observation. Such are the jagged lines of cracks in rock filaments, for instance. To deal with such irregularities, Dr. Mandelbrot has singled out sets whose Hausdorff dimension exceeds the topological dimension, calling them *fractals*, {**M2**, **M3**}. But it seems clear that this definition is too restrictive. Roughly speaking, fractals emerge after an infinite number of iterations of a step which involves breaking up a set as well as changing the scale. It would seem best to treat the term "fractal" as an undefined, governed by certain postulates. Some significant hints as to this appear in Professor Cannon's 1982 lectures {**C1**}, unfortunately still unpublished. Cannon has found that cognate ideas are useful in the theory of topological manifolds, especially of the hyperbolic type. Thus Wiener's idealized Brownian motion has turned out to be the progenitor of a growing variety of fractals encountered in physics as well as in pure mathematics.

During the early 1920s Wiener often consulted O. D. Kellogg, the Harvard authority on potential theory. Within a very short time these conversations brought him to the frontiers of the subject. Wiener then wrote six papers within a space of three years, which revolutionized the field. In the words of the French authority M. Brelot, he "initiated a new period for the Dirichlet problem and potential theory" {**B8**, p. 41}.

The *Dirichlet problem* is to determine the steady-state temperature distribution $u(\cdot)$ in region R, given its distribution $\phi(\cdot)$ on the boundary S, i.e., to find the function $u(\cdot)$ satisfying the Laplace equation

$$\Delta u = \frac{\partial^2 u}{\partial x^2} + \frac{\partial^2 u}{\partial y^2} + \frac{\partial^2 u}{\partial z^2} = 0$$

on R, and such that for any s in S, $u(r) \to \phi(s)$ as $r \to s$. The solution $u(\cdot)$, which depends both on the shape of S and the boundary-distribution ϕ, was known for smooth surfaces S and continuous ϕ. But, as Wiener learned from Kellogg, the issues are extremely complicated when the surface S has sharp dents and corners. This mathematical complexity is reflected in the physical instabilities which occur when chambers have such crooked surfaces, as Wiener noted. An illustration for the electric potential is the glowing of nails and pointed objects observed by sailors during thunderstorms.

Wiener's great contribution was to show that no matter how rough the surface S, the Dirichlet problem has a "solution" in a genuine but nonclassical sense, and to introduce several ideas of lasting value to accomplish this. As this work is very technical and has been commented on by Brelot {**B8**} and in greater detail in the *Collected Works*, I, it will suffice to say just the following. In his papers Wiener introduced the now central and crucial concepts of *capacity* for arbitrary sets and a *generalized solution* for the Dirichlet problem, as well as the criterion of regularity of the solution. To solve the Dirichlet problem, Wiener and H. B. Phillips considerably advanced the finite-difference technique of solving partial differential equations that has become standard forty years later with the advent of computers, cf. §7B.

Mrs. L. Lumer, commenting on [**24a**], writes, "the notion of capacity is perhaps Wiener's most important and long lasting contribution to potential theory" {*Coll. Works*, I, p. 393}. Indeed, this notion has been repeatedly generalized by Frostman, Choquet and others. It may therefore be worthwhile to recall what Wiener showed, viz. if B is a compact set in $\mathbb{R}^q$, $q \geq 2$, then there exists a $\mathbb{R}_{0+}$-valued countably additive measure μ on the Borel subsets of B with the property that $\int_B \mu(dy)/|x-y|^{q-2}$, $x \in \mathbb{R}^q$, tends to 1 on B and tends to 0 as $x \to \infty$. Wiener defined the capacity of B by cap $B := \mu(B)$.

An important link between Wiener's work on Brownian motion and potential theory was discovered by S. Kakutani in 1944 {**K2**}. In $\mathbb{R}^2$, for instance, it expresses the solution u of the Dirichlet problem as an integral $\int_0^1 \dots d\alpha$, where the integrand involves, apart from the boundary function ϕ, the time $\tau(x, y, \alpha)$ at which the Brownian path initially at $(x, y) \in R$ crosses S. It has initiated a new approach to potential theory.

4. From Communications Engineering to Generalized Harmonic Analysis and Tauberian Theory

Wiener was extremely fortunate in finding at MIT a forward-looking electrical engineering department, led by Professor Dougald C. Jackson in the early 1920s and Dr. Vannevar Bush later on. Wiener had an early flair for things electrical, and got along splendidly with the engineers who often sought his advice on mathematical methodology.

The first theoretical task that Wiener undertook at the behest of the engineers was the rigorization of the 1893 operational calculus of Oliver Heaviside. The thought underlying his rigorization is that "when applied to the function e^{nit}, the operator $f(d/dt)$ is equivalent to multiplication by $f(ni)$" [**26c**, p. 550]. Given an arbitrary function f, he dissected it into a number of frequency ranges, and applied to each range that expansion of $f(d/dt)$ which converged on this range.

By making these moves, Wiener in effect produced an embryonic form of the *theory of distributions* that was to come twenty-five years later. In §8 of his paper, which deals with the operational solution of second-order linear partial differential equations in two variables, Wiener wrote:

> ... there are cases where u must be regarded as a solution of our differential equation in a general sense without possessing all the orders of derivatives indicated in the equation, and indeed without being differentiable at all. It is a matter of some interest, therefore, to render precise the manner in which a non-differentiable function may satisfy in a generalized sense a differential equation. [**26c**, p. 582]

In this he anticipated Laurent Schwartz. Moreover, as Professor Schwartz tells us, by 1926 Wiener had seen farther than what all others had seen before 1946:

> Il est amusant de remarquer que c'est exactement cette idée qui m'a poussé moi-même à introduire les distributions![10] Elle a tourmenté de nombreux mathématiciens, comme le montrent ces quelques pages. Or Wiener donne une très bonne définition d'une solution généralisée; j'en avais, dans mon livre sur les Distributions, attribué les premières définitions à Leray (1934), Sobolev (1936), Friedrichs (1939), Bochner (1946), la définition la plus génerale étant celle de Bochner; or la définition de Wiener est la même que celle de Bochner, et date donc de ce mémoire, c'est-à-dire de 1926, elle est antérieure à toutes les autres. {*Coll. Works*, II, p. 427}

In [**26c**] and its sequel [**29c**], Wiener also took the important step of introducing the concept of *retrospective* or *causal operator* thus initiating the theory of *causality and analyticity*: the study of how one-sided dependence in the time domain leads to holomorphism in the spectral domain. In essence he defined an operator T on a space of signals f on $\mathbb{R}$ to be *causal*, if and only if for each t,

$$f_1 = f_2 \text{ on } (-\infty, t) \Rightarrow T(f_1) = T(f_2) \text{ on } (-\infty, t).$$

It follows easily that the transfer operator of a time-invariant linear filter with convolution weighing W, i.e., the filter which yields for input f the output g:

$$g(t) := (W * f)(t) := \int_{-\infty}^{\infty} W(t-s) f(x)\, ds, \tag{1}$$

[10] Voir l'introduction de mon livre sur les distributions, p. 4.

will be causal, if and only if $W = 0$ on $(-\infty, 0]$. Thus in the causal case, (1) gets amended to

$$g(t) := (W * f)(t) = \int_{-\infty}^{t} W(t-s) f(s)\, ds, \quad t \text{ real.} \tag{2}$$

Equivalently, the Fourier transform $\hat{W}$ is holomorphic on the lower half-plane ($\hat{W}(\lambda) := (1/2\pi) \int_{-\infty}^{\infty} e^{-it\lambda} W(t)\, dt$).

Causal operators are important in engineering, since the transfer operator of a physically realizable filter must obviously be causal. In recent years Wiener's basic ideal of causality has been extended to cover more general "time domains," cones and the like, cf. Foures and Segal {**F1**} , Saeks {**S1**}, and it has also shown up in the so-called *dispersion relations* of quantum mechanics.

The mid 1920s also saw Wiener's embarkation on generalized harmonic analysis. This came from his discernment of a certain parallel between the developments of electrical engineering and of mechanics. In the latter the consideration of uniform motion gave way to that of simple harmonic motion and then periodic motion, notably planetary motion, which, with the rise of statistical mechanics, in turn gave way to the study of highly random movements such as the Brownian motion. In electrical engineering there was first the direct current ("uniform level"), and then came the alternating current of one or several frequencies ("periodic level"). This corresponded to the stage of *power engineering*, the study of generators, motors and transformers, in which the central concept is *energy*. For this study fairly classical mathematics sufficed. But with the advent of the telephone and radio came *communications engineering* in which the central entity is the irregularly fluctuating current and voltage, which carries the *message* ("everything from a groan to a squeak"), and which is neither periodic nor pulse-like (i.e., in L_2). Thus the voltage curve of a busy telephone line has the same kind of local irregularity and overall persistence that Wiener had encountered in the Brownian motion, and he began to associate the communication phase of electrical engineering with the statistical phase of mechanics. For these phases, new and more difficult mathematics was required. He set about to find it, spurred on by his engineering friends.

A Fourierist at heart, Wiener assigned to the notions of orthogonal expansion, linearity and pure tone, a central place. We have harmonic analysis and synthesis when the pure tones are identified with the sinusoidal functions $\cos \lambda t$, $\sin \lambda t$, of different frequencies λ, i.e., in complex notation with the continuous characters e_λ:

$$e_\lambda(t) := e^{i\lambda t} = \cos \lambda t + i \sin \lambda t, \quad t \in \mathbb{R},$$

of the additive group $\mathbb{R}$ of real numbers, their acoustical realizations being the sounds of tuning forks. Wiener attributed great significance to such analysis for the following reasons.

Our faith is that the laws of Nature are invariant under time and space translations. Such invariant laws together with Huygens' Principle give PDEs with time-independent coefficients, and the propagators $U(t, s)$ they yield are invariant under time translations; i.e., $U(t, s) = T(t - s)$, T being a function on $\mathbb{R}$, and the $T(t)$ commute with the translation operators in the space variables. For small oscillations both the PDEs and the $T(t)$ are linear. The characters of the group $\mathbb{R}^3$ are eigenfunctions of linear operators having this commutation property; i.e., with $e_\lambda(s) := e^{i(\lambda' s)}$, $\lambda, s \in \mathbb{R}^3$, we have

$$T(t)(e_\lambda) = \alpha(t, \lambda)e_\lambda, \ \lambda \text{ in } \mathbb{R}^3; \quad \alpha(t, \lambda) := \{T(t)(e_\lambda)\}(0),$$

where $\alpha(t, \lambda)$ is a real or complex number. If the initial ($t = 0$) disturbance f of the medium can be represented as a linear combination of the characters, $f = \sum c_\lambda e_\lambda$, then from the linearity of $T(t)$, it follows at once that the disturbance at instant t is given by the elegant formula:

$$T(t)(f) = \sum c_\lambda \alpha(t, \lambda) e_\lambda.$$

Thus the problems of expressing functions as combinations of characters, and finding the "Fourier coefficients" c_λ for a given f—in short, *harmonic analysis and synthesis*—are exceedingly important.

Wiener believed that signals of wide varieties are harmonically analyzable, and that for this, the wider class of irregular and persisting curves must be properly demarcated and the averaging operations drastically altered. In this research, the earlier work on the rigorous demarcation of functions having convergent Fourier series being of no avail, Wiener sought his ideas from nonestablishment "radicals" such as Lord Kelvin, Lord Raleigh, Sir Oliver Heaviside, Sir Arthur Schuster and Sir Geoffrey Taylor, who were interested in the harmonic analysis of allied random phenomena in acoustics, optics and fluid mechanics.

To leave history aside, Wiener considered the class S of complex-valued measurable functions f on the real axis $\mathbb{R}$ for which the *auto-covariance function* ϕ:

$$\phi(t) = \lim_{T \to \infty} \frac{1}{2T} \int_{-T}^{T} f(s + t)\overline{f(s)}\, ds \tag{3}$$

exists and is continuous on $\mathbb{R}$. This very large class S includes the almost periodic functions of H. Bohr and A. S. Besicovitch and of course the periodic functions originally analyzed by Fourier. Wiener, spurred by the needs of communication engineers, set out to develop a harmonic analysis for this class [**30a**], guided by Schuster's work.

This generalized analysis has two parts. In the first, one seeks the Fourier associate of ϕ, and in the second, the Fourier associate of f itself. Today, after Bochner's theorem on positive-definite functions, the first part is rather elementary. We have

$$\phi(t) = \int_{-\infty}^{\infty} e^{it\lambda}\, dF(\lambda), \tag{4}$$

where F is a nonnegative distribution function on $\mathbb{R}$, called the *spectral distribution* of f. In the second, Wiener defined for each f in S a *generalized Fourier transform* s by

$$\begin{aligned}s(\lambda) := \lim_{A\to\infty} \frac{1}{\sqrt{2\pi}} \left(\int_{-A}^{-1} + \int_{1}^{A}\right) \frac{f(t)e^{-it\lambda}}{-it}\, dt \\ + \frac{1}{\sqrt{2\pi}} \int_{-1}^{1} f(t) \frac{e^{-it\lambda} - 1}{-it}\, dt.\end{aligned} \tag{5}$$

Wiener, however, defined F, not by (4) but by the analogue of (5) with ϕ replacing f, and only with much difficulty recovered (4), cf. [**30a**, (5.40)]. Both F and s are clearly defined in [**30a**], after much groping extending back to [**25c**].

Very ingeniously, Wiener used the stochastic integral

$$f(t,\alpha) = \int_{-\infty}^{\infty} W(t-\tau)\, d_\tau x(\tau,\alpha), \quad \alpha \in [0,1], \quad W \in L_2(R),$$

of the Brownian motion $x(\cdot,\cdot)$ to show that for almost all α, $f(\cdot,\alpha) \in S$ and for this f the spectral distribution F is absolutely continuous with $F'(\lambda) = \sqrt{2\pi}|\widetilde{W}(\cdot)|^2$, a.e. where $\widetilde{W}$ is the (indirect) Fourier–Plancherel transform of W [**30a**, §13].

The concepts of covariance ϕ of a signal f, the spectral distribution F and their interconnection have an interesting history. The fact that Einstein was a participant was revealed only in 1985, when a remarkable two-page heuristic note of his, dealing with f, ϕ, F', and carrying a version of (4), came to light, cf. {**M6**}. Einstein was unaware of the work of Schuster. The genesis of the different ideas went as follows:

Spectral density F' ("periodogram"): Schuster, 1889; Einstein, 1914.

Covariance ϕ: Einstein, 1914; Taylor, 1920.

Spectral distribution F: Wiener [**28a**].

Interconnections: inverse of (4) with F', Einstein, 1914

(4) itself with F, Wiener [**30a**].

All the work was done independently.[11]

[11]For an interesting, stochastic process interpretation of Einstein's note, and its links to Khinchine's work of 1934, see A. M. Yaglom {**Y1**}.

The s function was exclusively Wiener's, and he felt certain that the following Bessel-type identity between f and s should prevail:

$$\lim_{T\to\infty} \frac{1}{2T} \int_{-T}^{T} |f(t)|^2\, dt = \lim_{h\to 0} \frac{1}{4\pi h} \int_{-\infty}^{\infty} |s(\lambda+h) - s(\lambda-h)|^2\, d\lambda. \tag{6}$$

Wiener showed that its correctness hinged on that of the simpler identity for nonnegative g:

$$\lim_{T\to\infty} \frac{1}{T} \int_{0}^{T} g(t)\, dt = \lim_{h\to 0} \frac{2}{\pi h} \int_{0}^{\infty} g(t) \frac{\sin^2 ht}{t^2}\, dt, \tag{7}$$

an identity of which he seems to have become aware as early as 1925. But Wiener had a hard time proving it. The stalemate broke dramatically during 1926 when Wiener was at Göttingen and Copenhagen on a Guggenheim Fellowship. At Göttingen he met an acquaintance, the British number-theorist I. A. Ingham, and learned from him for the first time that the identity (7) was, as they say, "Tauberian" in nature, and that his mentors Hardy and Littlewood were authorities on such matters. This was news to Wiener. Another important contact was with the Tauberian theorist Dr. Robert Schmidt of Kiel, whom Wiener met at a German mathematical meeting at Düsseldorf after the summer of 1926.

Following Ingham's advice, Wiener studied the work of Hardy and Littlewood, and noticed that they too were concerned with the equality of two means. Then, he suddenly saw that a simple logarithmic change of variables would reduce the integrals occurring in Schmidt's theorems to convolution integrals with which he was familiar from his electrical studies, and that the real problem was to find a theorem for convolution integrals. He then proceeded to find such a theorem, by a novel and hard attack in which no classical Tauberian theorem was used **[28b, 32a]**. Wiener's final result has a beautifully simple enunciation. Write

$$(W * f)(t) := \int_{-\infty}^{\infty} W(t-s) f(s)\, ds, \quad t \text{ real}. \tag{8}$$

Here $W \in L_1$ and $f \in L_\infty$. Regard $g = W * f$ as the response of an ideal convolution filter W^* with weighing W subject to the input signal f. When will the output g tend to a limit as $t \to \infty$? Wiener's Tauberian theorem gives the answer: *feed the signal f into another filter W_0^* with a nowhere vanishing frequency response, i.e., where the weighing W_0 in L_1 is such that its Fourier transform $\hat{W}_0$ vanishes nowhere. If the resulting response g_0 has a limit as $t \to \infty$, so will the original response g.* Wiener also proved a "Stieltjes" form of this theorem. It may be looked upon as its analogue when the input is a real or complex-valued measure μ over $\mathbb{R}$ and the filter performs the convolution

$$(W * \mu)(t) := \int_{-\infty}^{\infty} W(t-s) \mu(ds). \tag{9}$$

All the classical Tauberian theorems, even the deepest, can be recovered from Wiener's two theorems: one has only to pick W, W_0, f and μ intelligently and change variables. This applies also to (7) and therefore to the Bessel-type identity (6) on which rests the appropriateness of Wiener's generalized Fourier transformation. Thus generalized harmonic analysis was put on a sound footing. But many new theorems were also uncovered. Among these, perhaps the most interesting was the one on the Lambert series, which bears on the analytic theory of prime numbers. It led Wiener and his Japanese former student, Professor S. Ikehara, to a simpler proof of the celebrated Prime Number theorem, to wit

$$\lim_{n\to\infty} \frac{\pi(n)\cdot\log n}{n} = 1,$$

where $\pi(n)$ is the number of primes not exceeding n. For this proof, all one has to know about the Riemann zeta function is that it has no zeros on the line $x = 1$ in the complex plane, and the traditional appeal to contour integration is avoided.

As the enormous ramifications of the memoir **[32a]** (e.g. on Banach algebras) are dealt with extensively in the *Collected Works*, II, we shall say no more about them. The memoir **[30a]** on G.H.A., however, has only recently been assimilated in the framework of abstract analysis. The Wiener class S is a *conditionally* linear subspace (cf. *Coll. Works*, II, pp. 333–379) of the Marcinkiewicz Banach space:

$$\mathfrak{M}_2(\mathbb{R}) := \left\{ f\colon f \in L_1^{\text{loc}}(\mathbb{R}) \ \&\ \|f\|^2 := \overline{\lim_{T\to\infty}}\, \frac{1}{2T}\int_{-T}^{T} |f(t)|^2\, dt < \infty \right\}.$$

K. S. Lau {**L1**} has shown that the f in S with $\|f\| = 1$ are extreme points of the unit ball of $\mathfrak{M}_2(\mathbb{R})$.[12] Wiener's generalized Fourier transform s of a function f in S gets a nice interpretation in terms of helices in the Hilbert space $L_2(\mathbb{R})$. Since the middle 1960s, J. P. Bertrandias in France has subjected $\mathfrak{M}_2(\mathbb{R})$ to extensive analysis from the standpoint of S and its subspaces, cf. {**B3**} and also {**B1**}. J. Benedetto, Benke and Evans have just announced a full-fledged generalization of the Tauberian identity (6) of G.H.A. to functions on $\mathbb{R}^n$ {**B2**}.

Wiener went on to show that G.H.A. has a propaedeutic role in optics **[28d, 30a** (§9), **53a]**. In Maxwell's theory, the flux of electromagnetic energy at a fixed point P in a medium traversed by light, through a small surface at P perpendicular to the direction of propagation, is proportional to $|E(t)|^2$, where $E(t)$ is the electric vector at P at the instant t in question. This led

[12]Recently, Lau has obtained an interesting extension of BMO spaces by considering the class in which the Marcinkiewicz $\overline{\lim}$ is replaced by sup, cf. {**L2**}.

Wiener to regard $|f(t)|^2$ as the energy at instant t of the signal f in the class S and to regard

$$\phi(0) = \lim_{T\to\infty} \frac{1}{2T} \int_{-T}^{T} |f(t)|^2 \, d\tau, \tag{10}$$

cf. (3),as the *total-mean-power* or "brightness" of the signal f.

Central to Wiener's clarification of optical ideas was his tacit interpretation of the *photometer* as an instrument which, when impinged with a light signal f in S gives the reading $\phi(0)$. The justification of this came in 1932 when von Neumann {**V1**} noted that instruments have a response time T, and measure not the input $f(t)$ but time-averages $(1/T)\int_{t-T}^{t} f(\tau)\,d\tau$. In the case of the photometer, the enormity of T from the standpoint of atomic chaos allows us to let $T \to \infty$, and equation (10) follows from the ergodic theorem. Next Wiener considered the *Michelson interferometer*, equipped at the output end with a photometer (camera, eye, etc.). A simple calculation shows that when the input $f = E$ (electric field), the observed intensity at the output is

$$\phi(0) + \phi(\Delta l/c),$$

where Δl is the difference between the lengths of the two arms and c is the speed of light. By turning the screws, i.e., changing Δl, we can find $\phi(x)$, for any given (not too large) x. Thus, as Wiener noted, the Michelson interferometer is an analogue computer for the covariance function of light signals.

By extending G.H.A. to vector-valued signals, Wiener defined the *coherency matrix* of a set of light signals $(f_1, \dots, f_q)$, as the spectral distribution matrix F of its matricial covariance functions

$$\Phi = [\phi_{ij}], \quad \phi_{ij}(t) = \lim_{T\to\infty} \frac{1}{2t} \int_{-T}^{T} f_i(t+\tau)\overline{f_j(\tau)}\, d\tau.$$

This also allowed him to deal with polarization (cf. [**30a**, §9] for details).

Elsewhere, we have traced how these ideas bear on the polarization of light and on the question as to why two candles are twice (and not four times) as bright as one {**M4**, §4}. With the reasonable hypothesis that (macroscopically observed) light signals are trajectories of a stationary stochastic process, Wiener was able to justify what physicists such as Schuster and Raleigh knew intuitively but could not formulate rigorously. His ideas have found a place in the standard repertory in optics, e.g. in the treatise by Born and Wolf {**B7**}. But the full significance of some of this work, begun in 1928, emerged only with the advent of lasers, masers and holograms. Sir Dennis Gabor spoke of the coherency matrix as a "philosophically important" idea, adding that "it was entirely ignored in optics until it was reinvented ... by Dennis Gabor in England in 1955 and Hideya Gamo in Japan in 1956." He pointed out that the matrix theory of light propagation had been initiated by Max von Laue in 1907, and covers the transmission of information, and that "the entropy in optical information is a particularly fine illustration of the role of entropy in the Shannon–Wiener theory of communication" (*Coll. Works*, III, pp. 490–491).

5. Max Born and Wiener's Thoughts on Quantum Mechanics

An inequality in classical harmonic analysis, sometimes referred to as "the time-frequency uncertainty principle," reads: if f is in $L_2(\mathbb{R})$ and its L_2-norm $|f|_2$ is 1, and if the integrals $\int_{-\infty}^{\infty} |tf(t)|^2\,dt$ and $\int_{-\infty}^{\infty} |\lambda \tilde{f}(\lambda)|^2\,d\lambda$ are finite, $\tilde{f}$ being the (indirect) Fourier–Plancherel transform of f, then

$$(1) \qquad \int_{-\infty}^{\infty} |tf(t)|^2\,dt \cdot \int_{-\infty}^{\infty} |\lambda \tilde{f}(\lambda)|^2\,d\lambda \geq \frac{1}{4},$$

cf. e.g. Weyl {**W1**, p. 393}. It follows that if a sound oscillation f of intensity (or loudness) l, and centered around $t = 0$, is of short duration, then the first factor on the left being small, the second factor will have to be large, i.e., the oscillation f will comprise a whole range of frequencies, and will not be a pure tone. Conversely, if the oscillation is approximately pure, i.e., its frequencies are all clustered around a single frequency λ_0, then the second factor will be small; the first factor will now have to be large, i.e., the oscillations spread over a long interval of time.

In practical terms, a pure tone of only momentary duration cannot be created. Likewise in optics, it is impossible to produce a light ray passing through a definite point A in a definite direction. (For to ensure that the light passes through A, we will have to interpose in its path a screen with pin hole at A; but the latter will cause the emergent light to diffract, i.e., to spread out in a conical beam, and not along a definite straight line.) Wiener found other such instances in which precision in the determination of a quantity inexorably results in an uncertainty in the value of another.

These ideas formed the contents of Wiener's seminar talk on harmonic analysis at Göttingen in the summer of 1924 [**56g**, p. 106]. At that very time Max Born and Werner Heisenberg were grappling with the failure of the classical laws in atomic radiation, and were becoming gradually aware of the limitations afflicting the simultaneous determination of complementary quantities such as the position and momentum of atomic particles. This interest in uncertainty at Göttingen led Wiener to learn the quantum theory, and to collaborate with Max Born in the fall of 1925 when the latter came to MIT as Foreign Lecturer.

Heisenberg had just enunciated his matrix mechanics to cover the motion of a closed system with discrete energy levels or frequencies, and Born raised the question as to its substitute for nonquantizable systems with continuous spectra, such as rectilinear motion, in which no periods are possible. They wrote a joint paper on this [**26d**]. In it appears for the first time the idea that *physical quantities correspond to linear operators on function space.* They introduced the operator Q defined by

$$Q(f)(\cdot) = \lim_{T\to\infty} \frac{1}{2T} \int_{-T}^{T} q(s,\cdot) f(s)\, ds$$

as a replacement for the Heisenberg matrix for configuration, and obtained commutation laws governing differential operators; they then proposed more general operators on function space. Unfortunately, the function space is left unspecified. It appears to be a subclass of S comprising at least the Besicovitch almost periodic functions on $(-\infty, \infty)$.

In two more papers [**28d**, **29e**] on the subject as well as in G.H.A. [**30a**], Wiener pointed out that the linear-quadratic relationship prevailing in quantum theory also occurs in the branches of classical physics, e.g. white light optics or communication engineering, that demand generalized (rather than ordinary) harmonic analysis. In the theory of white light, for instance, the fundamental Maxwell equations are linear, but pertain to the intrinsically nonobservable quantities $E(t)$ and $H(t)$, whereas the observable intensities are defined in terms of the squares of their amplitudes. Since all observation at this sub-Hertzian level is necessarily photometric, to take an observation amounts to reading $\phi(0)$, where ϕ is the covariance function of the light signal f, cf. §4(10). But an apparatus which reads covariances destroys phase relations. For instance, if $f(t) = \sum_1^n a_k e^{i\lambda_k t}$, then $\phi(t) = \sum_1^n |a_k|^2 e^{i\lambda_k t}$. The phases of the complex numbers a_k are gone. Feeding this "observed light" into another optical instrument will not produce the same response as feeding in the unobserved light f. Thus observation affects the signal and thwarts prediction, much as in quantum mechanics. If we think of the light beam as a vector-signal, then in Wiener's words:

> ... if two optical instruments are arranged in series, the taking of a reading from the first will involve the interposition of a ground-glass screen or photographic plate between the two, and such a plate will destroy the phase relations of the coherency matrix of the emitted light, replacing it by the diagonal matrix with the same diagonal terms. Thus the observation of the output of the first instrument alters the output of the second. [**30a**, p. 194]

Thus in vague analogy with the quantum situation, in Wiener's white light optics, observation has the effect of diagonalizing an operator, and of enlightening the mind only by killing off a lot of information.

The role of the Born–Wiener paper [**26d**] in the history of quantum mechanics is alluded to in Whittaker's history {**W4**, vol. II, p. 267}, and discussed much more fully in J. Mehra and H. Rechenberg's recent comprehensive history of the subject {**M10**, Ch. 5}. The idea of a Hilbert space, still embryonic, was just not in Wiener's consciousness at that time. But the paper [**26d**], its limitations notwithstanding, had "an immediate impact on Heisenberg" as Mehra and Rechenberg point out, and they conclude:

> At a time when just a few physicists struggled to develop a consistent theory of quantum mechanics, the Born–Wiener collaboration not only indicated the way for handling the problem of aperiodic motion but also contributed to the physical interpretation of the theory. {**M11**, p. 246}

In the late 1940s an interesting use of Wiener's Brownian motion in quantum theory was revealed by M. Kac's analysis of the "path integral" defined in R. P. Feynman's important thesis. This stems from the deep resemblance between the "path integration" employed in quantum mechanics and integration over the space $C[0, \infty)$ with respect to the "Wiener measure" induced over this space by the Brownian motion stochastic process. With imaginary time it allows us to use the Brownian motion to prove theorems on Hilbert spaces germane to quantum field theory. We refer the reader to E. Nelson's commentary in the *Collected Works*, III, pp. 565–579.

In the early 1950s Wiener himself realized that the intriguing appearance of probabilities as squares of amplitudes in quantum mechanics was explainable in Gibbsian terms by use of his Brownian motion. This work, done in collaboration with A. Siegel and J. Della Riccia, is complicated and remains unfinished [**55c**, **56c**, **63a**, **66a**]. Since it is hardly understood, a short justification of its validity for pure quantum states may be in order.

Recall that the *states* of a quantum mechanical system are countably additive probability measures μ on the lattice $\mathscr{L}$ of projections P on a separable complex Hilbert space $\mathscr{H}$ (cf. G. Mackey {**M1**}). It follows from Gleason's fundamental theorem {**G1**} that to any pure state μ (i.e., extreme point of the state space) corresponds a unit vector ψ in $\mathscr{H}$ such that

$$\mu(P) = |P\psi|^2_{\mathscr{H}}, \quad P \in \mathscr{L}. \tag{1}$$

Given any quantum mechanical system for which the Hilbert space $\mathscr{H}$ is $L_2(\mathbb{R}^q)$, Wiener and Siegel exhibit a probability space $(\Omega, \mathfrak{A}, \rho)$ and for any pure state μ, a function M_μ on the lattice $\mathscr{L}$ to the σ-algebra $\mathfrak{A}$ such that

$$\mu(P) = \rho\{M_\mu(P)\}, \quad P \in \mathscr{L}. \tag{I}$$

Unfortunately, we do not have the space to define this crucial mapping $M_\mu(\cdot)$.[13] With it Wiener and Siegel fulfill their goal of showing that the probability appearing in quantum mechanics, as the square of the absolute value of the complex-valued function $P\psi$ in $L_2(\mathbb{R}^q)$, is the probability of a

---

[13]It will appear in a new book on Wiener by this writer.

well-defined subset of a well-understood probability space $(\Omega, \mathfrak{A}, \rho)$, viz. the space of q-parametrized Brownian movement—a generalized Wiener space.

Unfortunately, it is not clear from the Wiener–Siegel work if an equality of the type (I) is available for impure states μ. Furthermore, Wiener and Siegel do not discuss the physical relevance of the Brownian motion as yielding "hidden parameters" in quantum mechanics, although this notion is central to their approach. Nor do they discuss how their use of the Brownian motion fits in with that of Bohm, de Broglie, Vigier and others.

Efforts to pass beyond the present viewpoint of quantum physics are most desirable, for as de Broglie has pointed out, the history of science teaches us

> that the actual state of our knowledge is always provisional and that there must be, beyond what is actually known, immense new regions to discover. {**B4**, p. x}.

Professor Nelson points out that it is "a deep drive within science" that impels such efforts, for without them science would die. See §C for Nelson's commentary, *Collected Works*, III, pp. 575–576, for more.

6. Ergodic Theory, Homogeneous Chaos, Statistical Mechanics, Information, and Maxwell's Demon

> It is one of the greatest triumphs of recent mathematics in America, or elsewhere, that the correct formulation of the ergodic hypothesis and the proof of the theorem on which it depends have both been found by the elder Birkhoff of Harvard. [**38e**, p. 63]

It was Eberhard Hopf, of Potsdam Observatory, who during his visit to MIT in 1931, aroused Wiener's interest in Birkhoff's theorem. Wiener's fascination with the theorem becomes understandable, for recasted in Wienerian terms, it reads:

Ergodic Theorem. *Let $\{f(t,\cdot) : -\infty < t < \infty\}$ be a complex-valued strictly stationary stochastic process such that $f(0,\cdot) \in L_2(\Omega, \mathscr{B}, P)$. Then for P almost all ω in Ω, the signal $f(\cdot,\omega)$ belongs to the Wiener class S, and its covariance function $\phi(\cdot,\omega)$ satisfies the equality*

$$\phi(\tau,\omega) = E_{\mathfrak{J}}(f(\tau,\cdot)\overline{f(0,\cdot)})(\omega), \quad \tau \in \mathbb{R}, \tag{1}$$

where $E_{\mathfrak{J}}\{\cdot\}$ is the conditional expectation with respect to the σ-algebra $\mathfrak{J}$ of invariant sets in $\mathscr{B}$. In case the process is ergodic, i.e. $\mathfrak{J}$ is trivial,

$$\phi(\tau,\omega) = E\{f(\tau,\cdot)\overline{f(0,\cdot)}\}(\omega), \quad \tau \in \mathbb{R}, \tag{2}$$

where $E\{\cdot\}$ is the (unconditional) expectation.

Apart from the equalities (1) and (2), Birkhoff's Theorem validated Wiener's long cherished belief that

$$\text{(3)} \qquad \lim_{T\to\infty} \frac{1}{T} \int_{-T}^{0} f(t+\tau)\overline{f(t)}\,dt = \lim_{T\to\infty} \frac{1}{2T} \int_{-T}^{T} f(t+\tau)\overline{f(t)}\,dt.$$

Thus past observations of the signal f suffice for the estimation of its covariance function ϕ, and this function gives information intrinsic to the stochastic process from which the signal hails. The fact that, for this, ergodicity has to be postulated, did not deter Wiener, for the assumption of ergodicity was just one of many idealizations that his scientific philosophy permitted. Moreover, von Neumann's celebrated 1932 theorem on the distintegration of regular measure-preserving flows over complete metric spaces into ergodic sections {**V2**}, gave Wiener a good excuse to deal almost exclusively with the ergodic case, see for instance [**61c**, pp. 55–56].

Wiener's own research in this area began when Paley and Wiener demonstrated the existence of a flow T_t, $t \in \mathbb{R}$, which preserves Lebesgue measure over $[0,1]$ and is ergodic, and such that for almost all α in $[0,1]$,

$$x(b+t,\alpha) - x(a+t,\alpha) = x(b, T_t\alpha) - x(a, T_t\alpha), \quad t \in \mathbb{R},$$

where $x(\cdot,\cdot)$ is the Brownian motion stochastic process, cf. the book [**34d**, §40]. The combined use of this result with Birkhoff's theorem appreciably simplifies certain proofs in the memoir [**30a**] on G.H.A., but it is also very significant in several other contexts.

In the paper [**39a**] Wiener, apart from deducing Birkhoff's theorem from von Neumann's mean ergodic theorem, extended the ergodic theorems to measure-preserving flows with several parameters, i.e., flows T_λ, where $\lambda \in \mathbb{R}^n$, $n > 1$, thereby making them available in the study of spatial or spatio-temporal *homogeneous random fields*. For the latter fields, $\lambda = (x, y, z, t)$ represents the space-time coordinates of an evolving random process, in a certain quantity $f(\lambda, \omega)$ of which we are interested. In collaboration with A. Wintner, Wiener also proved that almost all signals f, which emanate from an ergodic stationary stochastic process in L_2, are cross-correlated with the characters e_λ, $e_\lambda(t) = e^{it\lambda}$, i.e., possess generalized Fourier coefficients [**41a**, **b**]; more fully, for P almost all ω in Ω and *every* λ in $\mathbb{R}$,

$$\lim_{T\to\infty} \frac{1}{2T} \int_{-T}^{T} f(t,\omega) e^{-i\lambda t}\,dt$$

exists. They also gave conditions under which such a signal $f(\cdot,\omega)$ will be a Besicovitch almost periodic function. These results are significant in view of the presence of lots of functions in the unrestricted Wiener class S that are not cross-correlated with the characters e_λ.

In an earlier paper [**38a**] Wiener extended the Birkhoff theorem to homogeneous chaoses. A *homogeneous chaos* (nowadays termed a *stationary random* *measure*) is a finitely additive measure μ on a ring $\mathscr{R}$ of subsets of a group X, such that all its values $\mu(A)$, (A in $\mathscr{R}$) are real- or complex-valued random variables over a probability space $(\Omega, \mathscr{B}, P)$, and furthermore the random variables $\mu(A)$ and $\mu(A+x)$, have the same probability distribution over $\mathbb{R}$ or $\mathbb{C}$, for different x in X. A simple instance is afforded by the measure μ defined by

$$\mu(a,b](\alpha) = x(b,\alpha) - x(a,\alpha), \quad \alpha \in [0,1],\ a \leq b. \tag{4}$$

where $x(\cdot,\cdot)$ is Wiener's Brownian motion. For any real-valued homogeneous chaos μ on a ring $\mathscr{R}$ over $X = \mathbb{R}^n$, and any measurable function f on $\mathbb{R}$, Wiener proved that if

$$\int_\Omega f\{\mu(A)(\omega)\} \cdot \log^+ |f\{\mu(A)(\omega)\}| P(d\omega) < \infty,$$

then for all A in $\mathscr{R}$ and P almost all ω in Ω,

$$\lim_{r\to\infty} \frac{1}{v(r)} \int_{V(r)} f\{\mu(A+t)(\omega)\}\, dt$$

exists, where $v(r)$ is the volume of the ball $V(r)$, center 0, radius r, in $\mathbb{R}^n$, $t = (t_1, \ldots, t_n)$ and $dt = dt_1 \ldots dt_n$. Furthermore, this limit is equal to the expectation $E[f\{\mu(A)\}]$ in case the chaos μ is "ergodic," which Wiener defined to mean

$$\lim_{|t|\to\infty} P\{\omega\colon \mu(A)(\omega) \in G\ \&\ \mu(A+t)(\omega) \in H\} = P\{\omega\colon \mu(A)(\omega) \in H\},$$

for any Borel subsets G and H of $\mathbb{R}$. This was a far-reaching extension of the Ergodic Theorem.

Wiener now felt that he had the right viewpoint and equipment to tackle the problems of statistical mechanics, in particular the problem of turbulence that had evaded him in 1920. He had written the paper [**38a**], we spoke of, in this hope. In it he represented arbitrary random functions by sums of multiple stochastic integrals of Brownian motion. A joint paper with A. Wintner on the discrete chaos [**43a**] followed. While the importance of the pure mathematical side of these papers is beyond question, their import in statistical mechanics is still in doubt. These issues are discussed in *Collected Works*, I, especially in the comprehensive survey by Drs. McMillan and Deem.[14] On the other hand, a present school of thought, let by Professor J. Bass in Paris, holds that it is more pertinent to regard a turbulent velocity simply as a function in the Wiener class S rather than as a trajectory of some hypothetical stochastic process. This has revived an interest in [**30a**] and has brought to light some hitherto unnoticed connections of Wiener's G.H.A. to H. Weyl's earlier work on *equidistribution*, and to the so-called Monte-Carlo method. We would refer the interested reader to the *Collected Works*, II, pp. 359–372, and the references therein, and to Bass {**B1**} and Bertrandias, et al. {**B3**}.

[14]For more on these questions, see Dr. McMillan's recent article in this series {**M10**}.

At about this time, the concept of entropy began to permeate Wiener's work. From the Second Law of Thermodynamics (Inaccessibility Principle, in Carathéodory's elegant treatment), it follows that for a thermodynamic system α with a state space $\mathcal{S}$ and empirical temperature $\theta_\alpha(s)$ in state s, the Pfaffian equation $dQ = 0$ ($Q =$ heat) has an integrating divisor of the form $T\{\theta_a(\cdot)\}$; i.e., there exists a function $T(\cdot)$ on $\mathbb{R}$ such that

$$dQ/T\{\theta_\alpha(\cdot)\} = dS_\alpha(\cdot) \quad \text{on } \mathcal{S},$$

where $S_\alpha(\cdot)$ is a function on $\mathcal{S}$. The function $T\{\theta_\alpha(\cdot)\}$ on $\mathcal{S}$ is called the *absolute temperature* and $S_\alpha(\cdot)$ the corresponding *entropy* of the system. With the standardization, $T(t) = c \cdot t$, $c =$ const., obtained by taking the empirical temperature $\theta_\alpha(s)$ as that measured by the perfect gas thermometer, it can be shown *that the entropy $S_\alpha(\cdot)$ of the system α cannot decrease in any adiabatic transformation, and must increase in all non-quasi-static adiabatic ones.*

The law of increasing entropy imposes on events an ordering, past $\to$ future, determined by increase in entropy. It thus provides the foundation for the objectivity of *anisotropic time*, or "time with an arrow" as Eddington used to say. Many phenomena that interested Wiener, such as controlled experiment, communication, memory and learning, hinge on the anisotropy of time.

Recall that the molecular kinetic theory asserts first that disorder at the atomic level engenders at the microscopic level phenomena governed by probabilistic laws such as the Brownian motion, and second that what the causal phenomena at the macroscopic level exemplify are statistical stabilities emerging from the cooperation of an enormous number of irregular impulses. Guided by this and by the intrinsically stochastic aspect of subatomic phenomena (quantum theory), Wiener held that there is a random element in the very texture of Nature, and that *the orderliness of the world is incomplete.* We can no longer regard the universe as a strictly deterministic system, the state of which at any instant t is determined exclusively by its states at all previous instances $t' < t$. It was Wiener's position that we still have a cosmos: the Principle of the Uniformity of Nature still reigns, but at a stochastic level. It is the probability measures, engendered by the statistical aspect of Nature, that remain invariant under time-translations. Ergodic considerations become paramount. (See [**50j**, *Introd.*] and [**55a**, pp. 251–252].)

Wiener also realized that the time-concept that emerged from the contingent nature of the cosmos had more to it than mere anisotropy. The indeterminism of the new physics opened up the possibilities of noise and orderliness, freedom, innovation, growth and decay, error and learning. Stochastic prediction and filtering rest on the possibilities of contingency, innovation and noise, as does modern control theory, military science, meteorology and a host of other fields. The writings of the French philosopher H. Bergson on time and evolution, though somewhat diffuse, where poignant in stressing these nonextensive (nonspatial) novelty-creating aspects of time. Wiener therefore spoke of *Bergsonian time* in contrast to *Newtonian time*, when he wanted to emphasize the last aspect of time, cf. [**61c**, Ch. I].

The transmission of messages via a medium (or channel) is a statistical phenomenon in Bergsonian time, for we have to deal with a collection of messages (such as those that cross a telephone exchange) not prescribed by definite laws but only by a few statistical rules. Generally speaking, the more informative a message, the longer it will be, and the more the energy needed to transmit it. Clearly, a proper numerical measure of the *informative-value of a transmitted message* is needed. Wiener and Shannon provided such measures suitable for telephony and telegraphy, respectively.

For a recipient of a message, who knows the probability distribution p of the different outcomes of a repeatable experiment being performed far away, the *informative value* of the message "the outcome x has occurred" is deemed to be

$$I(x) := -\log p\{x\} = \log[1/p\{x\}].$$

(This definition of $I(x)$ meets the reasonable requirement that the more a message removes uncertainty, the greater its informative value.) Let X be the set of all atomic outcomes of the experiment. Then what concerns the transmission engineer are not the individual values $I(x)$, but rather their average:

$$\text{(1)} \qquad \operatorname{Inf}(p) := E[-\log p\{\cdot\}] = -\sum_{x \in X} p\{x\} \cdot \log p\{x\}.$$

We may call this *the average informative value of the probability distribution p*. It gives the average energy (and average cost) of transmission. This definition, in which X is finite or at most countable, is due to Dr. C. E. Shannon in 1947 or 1948 {**S3**}, who was at the Bell Telephone Laboratories and was concerned with the energy-efficient coding of telegraphic messages over noisy channels. It presaged his deep work on channel capacities, encoding and decoding.

In the summer of 1947 Wiener was led to the same problem for an absolutely continuous probability distribution p over the real line $\mathbb{R}$, by the needs of filter theory. Since an infinite sequence of binary digits is required to transmit a real number, a limiting approach, starting with the Shannon concept, will assign an infinite average information to such a p. Wiener's starting point was the observation that we may forget all digits after a fixed number because of noise. By an argument, very obscurely presented, he arrived at the following definition for the *average informative value of p*:

$$\text{(2)} \qquad \operatorname{Inf}(p) = E\{-\log p'(\cdot)\} = -\int_{\mathbb{R}} p'(x) \cdot \log p'(x)\, dx,$$

where $p'(\cdot)$ is the probability density, cf. **[61c**, p. 62, (3.05)].[15] He too took logarithms to the base 2. This definition is used in the theory of information processing for continuous time. Wiener's Inf, unlike Shannon's, can become negative. Nevertheless both Inf's have essential common features.

The nexus between communications engineering and statistical mechanics, which Wiener had dimly discerned in the mid 1920s (cf. §4) is deep indeed, for the Shannon–Wiener concept of information has turned out to be a disguised version of the *statistical entropy* to which Boltzman was driven seventy years earlier. The demonstration is hard, and the final result linking the Boltzman entropy of the gas in the complexion λ to the Shannon average information of an associated probability measure p_λ, is impossible to state here since we would have to define Boltzman's concepts of the complexion of a gas and of its statistical entropy. Be it noted, however, that to prove his H-theorem, $d\,\mathrm{Ent}(s_t)/dt \geq 0$, Boltzman had to replace his summation by an integral. This integral is precisely the Wiener average information of a multidimensional probability distribution, cf. **[61c**, p. 63] and Born {**B6**, pp. 57 (6.25), 165}.

Whereas the Boltzman entropy of a gas is best interpreted as a measure of "internal disorder," Shannon's average information is most naturally interpreted as a measure of "uncertainty removed". Their equality has suggested the term *negentropy*, or *measure of internal order*, as a substitute for the term *information* in certain contexts.

The cogency of this viewpoint has become clear from the work of Szilard, Brillouin and also Wiener **[50g, 52a]** on the Maxwell demon.[16] To perform its miracles, the demon must receive information about impending molecular movements, and for this, electromagnetic radiation must be available in the gas. But each time the demon draws information from a photon of light, it degrades its energy and (by Planck's law) also its frequency, and creates an equal amount of entropy. The entropy of the matter-radiation mixture is not reduced. For the mixture itself, nothing miraculous occurs.

But Wiener's imaginative mind was not satisfied with this rather easy disposal of the demon. While the demon fails in its overall mission, it still scores a local success: it enhances the negentropy in its immediate neighborhood by degrading the photons of light. Does not such enhancement occur when a piece of green leaf uses sunlight to produce portions of a molecule of a

[15]Wiener omitted the minus sign, but this is of little consequence since the integral can take any real value for different p.

[16]Dr. J. R. Pierce {**P1**, pp. 198–200, 290} defines this demon as "a hypothetical and impossible creature, who without expenditure of energy, can see a molecule coming in a gas which is all at one temperature, and act on the basis of this information". Seated in a cup of water with an insulating partition having a tiny door, it can by intelligently opening and shutting the door warm the water on one side of the partition and cool it on the other.

carbohydrate from the carbon dioxide and water in its midst, and to release a molecule of oxygen, following the chemical formula:

$$\text{Light} + nCO_2 + nH_2O \xrightarrow[\text{phyll}]{\text{chloro-}} (CH_2O)_n + nO_2?$$

If so, a piece of green leaf in an environment of carbon dioxide and water, irradiated by the sun, is a thermodynamic machine studded with "Maxwell demons" (particles of chlorophyll), all of whom enhance negentropy *locally* by degrading the sunlight impinging on them.

Briefly, no demons, no life. But Wiener also noticed the temporariness of the demon's local successes. It could perform locally only as long as it had usable light, i.e., light from a source at a temperature higher than that of the gas. In a gas-radiation mixture in equilibrium, it will be as helpless as in a gas devoid of light. Ultimately, it too will fall into equilibrium and its intelligent activities cease. In short, it will die. Wiener felt that these reflections on the demon impinged on the biological issues of life, decay and death. See [**50g**], and Brillouin {**B9**}.

7. The Limitations of this Survey

What we have surveyed so far is roughly 70 percent of Wiener's mathematical work and 25 percent of his work in the empirical realms. What has been left out comprises work that has had a marked impact on contemporary life and thought. Because of space limitations, only its synoptic description is possible.

A. Work on the **Hopf–Wiener integral equation**, in collaboration with E. Hopf [**31a**], and the further exploration of the underlying idea of *causality and analyticity* with R.E.A.C. Paley [**33a**, **33e**, **34d**]. The factorization and other techniques introduced in [**31a**] and [**34d**] have had enormous ramifications, cf. the commentary of J. Pincus is in *Collected Works*, III.

B. Work on **electrical networks and analogue and digital computers** from 1926–1940. This was preceded by an early flair for things electrical, a fascination with Leibniz's *Ars Characteristica*, and by practical experience as computer at the U.S. Army Proving Grounds in Aberdeen, Maryland, in 1918–1919. In 1926 he conceived the *optical integraph*, an analogue computer for convolutions $y = \int_0^a f(t)g(x-t)\,dt$. This was put in-the-metal with better and better designs by Bush's junior colleagues, starting with K. E. Gould {**G3**} in 1929 and ending with Hazen and Brown {**H2**} in 1940. In the early 1930s came the *Lee–Wiener network* {**L3**}.

More remarkable was Wiener's 1940 letter and memorandum to Bush on *mechanical solution of PDEs*. Filed away, it began to surface only in the late 1970s. It is printed in *Collected Works*, IV [**85a**, **b**] and in the *Annals of the History of Computing* {**M7**} with comments by B. Randell and by S. K.

Ferry and R. E. Saeks. Wiener's proposed machine is digital, and embodies a discrete quantized numerical algorithm, Turing machine architecture, binary arithmetic and data storage, an electronic arithmetic logic unit, and a multitrack magnetic tape. Wiener was about fifteen years ahead of his time in both his recommendation of magnetic taping, and his emphasis on attaining "several thousand times the present speed" with only a slight increase in cost. Such speeds were attained only by the first generation of transistorized computers (IBM 7090, etc.) in the late 1950s.

C. The work on **anti-aircraft fire control** with Mr. Julian Bigelow (1940–1943). The shift from deterministic to stochastic prediction culminated in Wiener's book on *time-series* [**49g**]. The understanding of its relationship to Kolmogorov's monumental work on stationary sequences {**K4**}, and subsequent work by Wiener and others on the multivariate and nonlinear cases brought into being the **theory of prediction**. This subject has wide ramifications in functional analysis, as the commentaries of P. Muhly and H. Salehi in the *Collected Works*, III, show.

The Wiener–Bigelow work has another large component, on *filtering, control and regulation*. This, as is well known, has revolutionized the field of communications engineering, cf. the extensive commentaries of T. Kailath in *Collected Works*, III, and Y. W. Lee {**L4**}.

D. In an important military document in 1942 appear the words:

> ... we realized that the "randomness" or irregularity of an airplane's path is introduced by the pilot; that in attempting to force his dynamic craft to execute a useful maneuver, such as a straight-line flight, or a 180-degree turn, the *pilot behaves like a servomechanism*, attempting to overcome the intrinsic lag due to the dynamics of his plane as a physical system, in response to a stimulus which increases in intensity with the degree to which he has failed to accomplish his task.[17] (emphasis added)

These observations of Bigelow and Wiener, when integrated with the thought of the neurophysiologist Dr. Arturo Rosenblueth, led to the joint paper on *teleology* [**43b**] which opened the field of **cybernetics** [**48f**, cf. **61c**]; cf. also Ashby {**A1**}. This led to further work often in conjunction with Drs. W. McCulloch and W. Pitts in the following areas:

(a) Self-learning and reproducing servomechanisms; organization and homeostasis [**58i**, **62b**];

(b) Neural nets, and the proximity of the brain and the electronic computer {**S2**, **M8**}, and [**61c**, **53d**];

[17] N. Wiener: *A. A. Directors, Summary Report*, June 1942, Department of Defense, see p. 6, para. 1, cf. *Coll. Works*, IV, p. 170.

(c) Pattern recognition ("Gestalt") [**61c**, pp. 133–139];

(d) Brain-rhythms and electro-encephalography;

(e) Sensory and muscular-skeletal prosthesis.

These contributions are commented on extensively by Drs. John Barlow, R. W. Mann, W. Ross Ashby, H. von Foerster and this writer in *Collected Works*, IV, and all the relevant Wiener papers are cited therein.

E. **Physiological work** on muscle clonus, heart flutter and fibrillation, spike potential of axons and synaptic excitation, done in collaboration with Dr. A. Rosenbleuth and others. It will suffice to refer to the survey of Dr. Garcia Ramos in *Collected Works*, IV.

F. In his book, *The Nerves of Government*, the political scientist, K. Deutsch, quotes the following words of Wiener:

> Communication is the center that makes organizations. Communication alone enables a group to think together, to see together, and to act together. All sociology requires the understanding of communication. {**D1**, p. 819}

Wiener's work on the **cybernetical aspects (communication and control) of social organization** can be classified as follows:

(a) *The difference between long-time and short-time institutions.* Low-probability events (benevolent "acts of Grace," or malevolent "acts of God" in insurance parlance) become important in long-time prediction and planning. A greater faith in the benevolence of God, and a different system of investment is required in the planning of long-time institutions (cities, universities, cathedrals) than in the management of short-time ones. This led Wiener to the concept of the *long-time State* [**62c**]. The *control of the means of communication* being "the most effective and most important" of all "homeostatic factors in society" (cf. [**61c**, p. 160]), Wiener felt that their control should be entrusted to the long-time institutions: the churches, universities, academies, etc. Their entrustment to short-sighted profit makers is pernicious [**61c**, pp. 161, 162], [**50j**, pp. 131–135].

(b) *The capitalist market as an n-person game.* Far from being a homeostatic process, the capitalist market is a highly volatile one, with recurrent down-sides and a propensity to misuse the channels of communication, and to accept mass gullibility and an indifferent system of education [**61c**, pp. 158–160], [**50j**, pp. 132–134].

(c) *Military theory.* The von-Neumann-Morgenstern game theory {**V3**} is of service in military contests only at the lowest level. Operations at higher levels proceed by strategic evaluations based on the analysis of enemy time-series [**60d**]. Continual reconnaisance, essential to good military planning, is not possible with units that operate on different "time scales" (cf. [**60d**,

p. 721]). The atomic bomb is a bad weapon from this standpoint [**UP1**, p. 6]. This evaluation of Wiener's ideas on military theory differs somewhat from that of S. Heims {**H3**}.

(d) *Observer-observed coupling in the social fields.* Wiener's sharp perception of the difference between such coupling in the natural and social sciences (cf. Bohr {**B5**}) led to his skepticism concerning much of the economics and sociology that is dressed up in classical mathematical garb. He felt that non-classical branches of mathematics such as game theory and fractals were more appropriate to these fields [**61c**, p. 163], [**64e**, pp. 90–91].

Wiener was more than a great theoretician. Ever since 1943, when he first surmised the occurrence of growing automatization in the American economy ("the Second Industrial Revolution"), he made efforts to alert American labor to its social consequences and emphasize the need for ever-expanding education and retraining. But his efforts fell on deaf ears. His eventual correspondence with Walter Reuther of the United Auto Workers led to the formation of a Council on Science and Labor in 1952. It never got off-ground. Today, American society is paying the penalty for its disregard of Wiener's far-sighted wisdom, and for its pathetic condescension to the debasement of its schools, cf. Lynd {**L6**}.

8. Wiener's Place in the Philosophia Perennis

Wiener's cybernetically inspired conception of history enabled him to lay bare the illusory aspects of certain basic beliefs now in vogue, and to unveil more balanced and realistic attitudes on issues of human survival.

Thus Wiener saw the speciousness of the belief in "unlimited human progress" that came from the French Enlightenment and Marxism. Wiener's religious thought rests on the analogy he drew [**50j**, p. 11] between entropy and St. Augustine's negative evil {**A2**, vol. 1}. "The paradox of homeostasis is that it always collapses in the end" [**UP2**, p. 103]. The life of man is further afflicted by his corrupt inclinations. In crying over spilt milk, beating about the bush and venting greed, man, his cerebral cortex notwithstanding, is less intelligent than the puppy and the elephant. Man's murderousness has grown with his knowledge and understanding, and so-called "rational self-interest," far from redounding to the common good, becomes a gateway to avarice and to the spoilation of man and earth [**61c**, p. 158]. In the words of T. S. Eliot, "Sin grows with doing good."[18] Wiener's writings wisely emphasize the fact that man is not "animal rationale" but "animal symbolicum," and a corrupt one at that, and wisely observe that the well-balanced tragic attitude depicted in Greek mythology is more conducive to the human welfare than

[18]For an interesting early exchange of views between T.S. Eliot and Wiener, whose backgrounds have interesting parallels, see *Coll. Works*, IV, pp. 68–75.

the anxiety-ridden attitude of many a modern man hankering after success and progress, [**50j**, pp. 40, 41, 183, 184].

The fact that the human species is severely handicapped led Wiener to view the chief function of science in human life as prosthetic: "That of maintaining a rapport with the environment, which will enable us to face our environment and its changes as we come to them" [**UP2**, p. 102]. The insights of Pythagoras, Plato, Aristotle, Aquinas, Newton, Kant, and (after non-Euclidean geometry and mathematical logic) of Whitehead and Einstein have enlarged our understanding of the scientific methodology, and of the organic enterprise we call science. But its evolution notwithstanding, this enterprise has an enduring integrity stemming from its all-time prosthetic value. From this standpoint, the savage, who formulates his observations of Nature in animistic terms, is trying to understand Nature in order to overcome his handicaps, and is thus pursuing science. This wholesome concept of science is entirely antithetical to the view that it is a "game against Nature"—a sad confusion of the disparate activities of inquiry and contest.

The arts too, Wiener felt, subserve a homeostasis in human life, and he did not attribute much significance to their differences with the sciences. Indeed, for mathematics the difference vanishes:

> Mathematics is every bit as much an imaginative art as it is a logical science. [**23a**, p. 269]

cf. also [**29h**]. Unlike Halmos {**H1**}, however, Wiener was speaking of all mathematics, the pure corresponding to the *presentative* aspect of art, and the applied to its *representative* aspect. He cited Einstein's general theory (which Halmos disqualifies as "matho-physics") as a magnificent example of both forms of art:

> This double aspect of Einstein's work, and indeed of all physics, may serve as a final link between mathematics and the arts. As is well known, most of the arts possess both a presentative and a representative aspect. A painting has beauty not merely as a study in abstract design but as a representation of the outer world. ... Thus mathematics, too besides the beauty of inner structure, has a further beauty as a representation of reality. This is most clear in mathematical physics but even in the purest of pure mathematics, mathematical physics often serves as a valid if unconscious guide. Many a pure mathematical study is an impression of some chord of the physical world. [**29h**, p. 162]

To Wiener the creative activities in both fields appeared as manifestations of a spirit seeking objectification, cf. [**29h**, pp. 130–131]. This attitude towards the aesthetic impulse brought Wiener very close to the religious view of art of the scholastics Dante, Meister Eckehart, and St. Bonaventura, cf. Coomaraswamy {**C3**}.

In religion also Wiener saw a homeostatic and prosthetic factor. The survival of physics depends on its ceaseless quest for ideal concepts: particles without volume, perfect liquids, the electromagnetic field, the momentum-energy tensor, and so on. The belief that daily living too is enhanced by inclusion of ideal, nonphysically representable elements is the religious view of life. The average family placed in a social system, almost invariably exploitative, needs "acts of grace" for its successful survival, no less than a long-time institution such as a city, cf. §7F(a). Uncorrupted religiosity promotes individual acts of grace and thus serves a very high homeostatic and prosthetic purpose. But Wiener was wary of rigid creeds, and of course saw in corrupted religious establishments an anti-homeostatic factor.

Thus, unlike many contemporaries, Wiener did not let the revolutions of thought that make (and partition) the history of science obscure the vision of its fundamental continuity. There was nothing "anti-Euclidean" in non-Euclidean geometry or Mengenlehre or fractals, and nothing "anti-Newtonian" in relativistic mechanics.[19] Indeed, the notion of fractal has roots extending to Aristotle, as Mandelbrot has indicated {**M3**, p. 406}, and Wiener's own cybernetical ideals go back at least to Leibniz, if not to Plato's *Georgias*, as S. Watanabe has suggested (cf. *Coll. Works*, IV, p. 215). A similar perception of continuity marked Wiener's vision of history as a whole. An admirer of both the sixteenth-century Renaissance and the eighteenth-century French Enlightenment, he was blinded by neither. The removal of medieval teleology from post-Renaissance science was a boon, but Wiener contributed to its useful restoration in a modern scientific setting [**43b**]. The same remark applies to his view of the long-time State [**62c**], cf. {**A2**, vol. 2, **C4**}. The facile division of history into three ages, viz. of superstition, religion and science, is shallow. The symbiotic relation of religion and science has been well expressed by Einstein: "Science without religion is lame, religion without science is blind {**E4**, p. 26}, cf. also H. Weyl {**W2**, pp. 89, 214}.

A unique and significant aspect of Wiener's writings is the underlying thought that the incomplete and contingent cosmos revealed by science merits the same feeling of awe that Einstein expressed in the words, "Intelligence is manifested throughout all Nature" and in his references to Spinoza's God {**E3**}. The stochastic aspect is not an impairment. This Pythagorean faith

---

[19]Wiener saw no parallels between developments in twentieth-century science and those in twentieth-century art. His disdain of the latter is indicated by his favorite title "The Emperor's New Clothes" for many a piece of modern painting. On the other hand, Wiener did sense a resemblance between nineteenth-century mathematics and the German Romantic movement, cf. [**29h**].

sustained Wiener's ability to fuse the transcendent and the abstract with the practical and concrete. In his words:

> ... so often in my work, the motivation that has led me to the study of a practical problem has also induced me to go into one of the most abstract branches of pure mathematics. [**56g**, p. 192]

Unlike Wigner {**W5**}, Wiener found nothing perplexing or "unreasonable" about the efficacy of mathematics in the sciences. In fact, the investigations of the remarkable medical men surrounding him (all disciples of Russell, cf. {**R3**}), showed that mathematical relation-structure is all that is preserved in the course of sense-observation and the subsequent neurological transitions that constitute cognition {**M8**, **M9**, **R1**}. The paramountcy that Pythagoras, Plato, Roger Bacon, and Galileo accorded to mathematics was not misplaced.

Wiener's towering stature in the history of science rests not only on his unusual ability to discern so much unity amid such wide apparent diversity—he lived, in Struik's words, "the life of the unity of science"—but in his appreciation of its continuity. He was a revolutionary-traditionalist in the best sense of the word. He incorporated in the edifice of human wisdom the new stochastic storey without impairing the total architecture.

Norbert Wiener References

Note: The numbering adopted tallies with that in the *Bibliography of Norbert Wiener* appearing in his *Collected Works*, (Ed. P. Masani), The MIT Press, Cambridge, Mass., 1976, 1979, 1982, 1985. A reference such as [**14a**] indicates that the paper was written in 1914. The Roman numeral at the end indicates the volume of the *Collected Works* in which the paper is reprinted.

[**14a**] "A simplification of the logic of relations," *Proc. Cambridge Philos. Soc.* **17** (1914), pp. 387–390. I

[**14b**] "A contribution to the theory of relative position," *Proc. Cambridge Philos. Soc.* **17** (1914), pp. 441–449. I

[**15a**] "Studies in synthetic logic," *Proc. Cambridge Philos. Soc.* **18** (1915), pp. 14–28. I

[**17a**] "Certain formal invariances in Boolean algebras," *Trans. Amer. Math. Soc.* **18** (1917), pp. 65–72. I

[**20a**] "Bilinear operations generating all operations rational in a domain Ω," *Ann. of Math.* **21** (1920), pp. 157–165. I

[**20b**] "A set of postulates for fields," *Trans. Amer. Math. Soc.* **21** (1920), pp. 237–246. I

[**20c**] "Certain iterative characteristics of bilinear operations," *Bull. Amer. Math. Soc.* **27** (1920), pp. 6–10. I

[20d] "Certain iterative properties of bilinear operations," *G. R. Strasbourg Math. Congress*, 1920, pp. 176–178. I

[20e] "On the theory of sets of points in terms of continuous transformations," *G. R. Strasbourg Math. Congress*, 1920, pp. 312–315. I

[20f] "The mean of a functional of arbitrary elements," *Ann. of Math.* (2) **22** (1920), pp. 66–72. I

[21a] "A new theory of measurement: A study in the logic of mathematics," *Proc. London Math. Soc.* **19** (1921), pp. 181–205. I

[21b] "The isomorphisms of complex algebra," *Bull. Amer. Math. Soc.* **27** (1921), pp. 443–445. I

[21c] "The average of an analytic functional," *Proc. Nat. Acad. Sci. U.S.A.* **7** (1921), pp. 253–260. I

[21d] "The average of an analytic functional and the Brownian movement," *Proc. Nat. Acad. Sci. U.S.A.* **7** (1921), pp. 294–298. I

[22a] "The relation of space and geometry to experience," *Monist* **32** (1922), pp. 12–60, 200–247, 364–394. I

[22b] "The group of the linear continuum," *Proc. London Math. Soc.* **20** (1922), pp. 329–346. I

[22c] "Limit in terms of continuous transformations," *Bull. Soc. Math. France* **50** (1922), pp. 119–134. I

[23a] "On the nature of mathematical thinking," *Austral. J. Psych. and Phil.* **1** (1923), pp. 268–272. I

[23b] "Nets and the Dirichlet problem," (with H. B. Phillips), *J. Math. and Phys.* **2** (1923), pp. 105–124. I

[23d] "Differential-space," *J. Math. and Phys.* **2** (1923), pp. 131–174. I

[23g] "Note on a paper of M. Banach," *Fund. Math.* **4** (1923), pp. 136–143. II

[24a] "Certain notions in potential theory," *J. Math. and Phys.* **3** (1924), pp. 24–51. I

[24d] "The average value of a functional," *Proc. London Math. Soc.* **22** (1924), pp. 454–467. I

[25c] "On the representation of functions by trigonometrical integrals," *Math. Z.* **24** (1925), pp. 575–616. II

[26c] "The operational calculus," *Math. Ann.* **95** (1926), pp. 557–584. II

[26d] "A new formulation of the laws of quantization of periodic and aperiodic phenomena," (with M. Born), *J. Math. and Phys.* **5** (1926), pp. 84–98. III

[28a] "The spectrum of an arbitrary function," *Proc. London Math. Soc.* (2) **27** (1928), pp 483–496. II

[28b] "A new method of Tauberian theorems," *J. Math. and Phys.* **7** (1928), pp. 161–184. II

[28d] "Coherency matrices and quantum theory," *J. Math. and Phys.* **7** (1928), pp. 109–125. III

[29c] "Fourier analysis and asymptotic series," Appendix to V. Bush, *Operational Circuit Analysis*, John Wiley, New York, 1929, pp. 366–379. II

[29e] "Harmonic analysis and the quantum theory," *J. Franklin Inst.* **207** (1929), pp. 525–534. III

[29h] "Mathematics and art (Fundamental identities in the emotional aspects of each)," *Tech. Rev.* **32** (1929), pp. 129–132, 160, 162. IV

[30a] "Generalized harmonic analysis," *Acta. Math.* **55** (1930), pp. 117–258. II

[31a] "Uber eine Klasse singularer Integralgleichungen" (with E. Hopf), Sitzber. Preuss. Akad. Wiss. Berlin, Kl. *Math. Phys. Tech.*, 1931, pp. 696–706. III

[32a] "Tauberian theorems," *Ann. of Math.* **33** (1932), pp. 1–100. II

[33a] "Notes on random functions," (with R.E.A.C. Paley and A. Zygmund), *Math. Z.* **37** (1933), pp. 647–668. I

[33e] "Notes on the theory and application of Fourier transforms," (with R.E.A.C. Paley) I, II, *Trans. Amer. Math. Soc.* **35** (1933), pp. 348–355; III, IV, V, VI, VII, *Trans. Amer. Math. Soc.* **35** (1933), pp 761–791. II

[34a] "Random functions," *J. Math. and Phys.* **14** (1934), pp. 17–23. I

[34d] *Fourier transforms in the complex domain* (with R.E.A.C. Paley), Amer. Math. Soc. Colloq. Publ. **19**, Amer. Math. Soc., Providence, R.I., 1934

[38a] "The homogeneous chaos," *Amer. J. Math.* **60** (1938), pp. 897–936. I

[38e] "The historical background of harmonic analysis," Amer. Math. Soc. Semicentennial Publications, vol. II, Semicentennial Addresses, Amer. Math. Soc., Providence, R.I., 1938, pp. 513–522. II

[39a] "The ergodic theorem," *Duke Math. J.* **5** (1939), pp. 1–18. I

[43a] "The discrete chaos," (with A. Wintner), *Amer. J. Math.* **65** (1943), pp. 279–298. I

[43b] "Behavior, purpose, and teleology," (with A. Rosenblueth and J. Bigelow), *Philos. Sci.* **10** (1943), pp. 18–24. IV

[47b] "A scientist rebels," *Atlantic Monthly* **179** (1946), p. 46; *Bull. Atomic Scientists* **3** (1947), p. 31. IV

[48d] "A rebellious scientist after two years," *Bull. Atomic Scientists* **4** (1948), pp. 338–339. IV

[48f] *Cybernetics, or control and communication in the animal and the machine*, The MIT Press, Cambridge, Mass., 1948.

[49g] *Extrapolation, interpolation, and smoothing of stationary time series with engineering applications*, The MIT Press, Cambridge, Mass.; Wiley, New York; Chapman & Hall, London, 1949.

[49h] Review of Philipp Frank, *Modern Science and its Philosophy, New York Times Book Review*, August 14, 1949, sec. 7, p. 3. IV

[50g] "Entropy and information," *Proc. Sympos. Appl. Math.*, vol. **2**, Amer. Math. Soc., Providence, R.I., 1950, p. 89. IV

[50j] *The human use of human beings*, Houghton Mifflin, Boston, 1950.

[52a] "Cybernetics (Light and Maxwell's demon)," *Scientia* (Italy) **87** (1952), pp. 233–235. IV

[53a] "Optics and the theory of stochastic processes," *J. Opt. Soc. Amer.* **43** (1953), pp. 225–228. III

[53d] "Les machines a calculer et la forme (Gestalt), Les machines a calculer et la pensée humaine," Colloques Internationaux du Centre National de la Recherche Scientifique, Paris, 1953, pp. 461–463. IV

[53h] *Ex-prodigy: my childhood and youth*, Simon and Schuster, New York, 1953.

[55a] "Nonlinear prediction and dynamics" (Proc. 3rd Berkeley Symp. on Mathematical Statistics and Probability), University of California Press, Berkeley, Calif., 1954–1955, pp. 247–252. III

[55c] "The differential-space theory of quantum systems," (with A. Siegel), *Nuovo Cimento* (10) **2** (1955), pp. 982–1003, No. 4, Suppl. III

[56c] "'Theory of measurement' in differential-space quantum theory," (with A. Siegel), *Phys. Rev.* **101** (1956), pp. 429–432. III

[56g] *I am a mathematician: the later life of a prodigy*, Doubleday, Garden City, New York, 1956.

[58i] *Nonlinear problems in random theory*, The MIT Press, Cambridge, Mass., and Wiley, New York, 1958.

[60d] "Some moral and technical consequences of automation," *Science* **131** (1960), pp. 1355–1358. IV

[60e] "The duty of the intellectual," *Tech. Rev.* **62** (1960), pp. 26–27. IV

[61c] *Cybernetics*, Second edition of **[48f]** (revisions and two additional chapters), The MIT Press, Cambridge, Mass., and Wiley, New York, 1961.

[62b] "The mathematics of self-organizing systems," in *Recent developments in information and decision processes*, Macmillan, New York, 1962, pp. 1–21. IV

[62c] "Short-time and long-time planning" (originally presented at 1954 ASPO National Planning Conference), Jersey Plans, An ASPO Anthology (1962), pp. 29–36. IV

[63a] "Random theory in classical phase space and quantum mechanics," (with Giacomo Della Riccia), (Proc. Int'l. Conf. on Functional Analysis, Massachusetts Institute of Technology, Cambridge, Mass., June 9–13, 1963.) III

[64e] *God and golem, inc.—A comment on certain points where cybernetics impinges on religion*, The MIT Press, Cambridge, Mass., 1964.

[66a] "Wave mechanics in classical phase space, Brownian motion, and quantum theory," (with G. Della Riccia), *J. Math. Phys.* **7** (1966), pp. 1372–1383). III

[85a] Letter covering the memorandum on the scope, etc., of a suggested coupling machine (September 21, 1940), pp. 122–124 of {**M5**} vol. IV. IV

[85b] Memorandum on mechanical solution of partial differential equations, pp. 125–134 of {**M5**} vol. IV. IV

Quoted Unpublished Papers of Wiener

UP1 "Automatic control techniques in industry" (Industrial College of the Armed Forces, Washington D.C., 1952–1953.)

UP2 "Prelegomena to theology," 1961, (MC. 877–881).

Other References

{**A1**} W. R. Ashby, *An introduction to cybernetics*, Wiley, New York, 1963.

{A2} St. Augustine, "Concerning the nature of the good" (vol. 1, 1948), "The city of God" (vol. 2, 1948), in *Basic writings of St. Augustine*, Edited by W. J. Oates, Random House, New York.

{**B1**} J. Bass, *Fonctions de correlation, fonctions pseudo aléatoires et applications*, Masson, Paris, 1984.

{**B2**} J. Benedetto, G. Benke, and W. Evans, "An n-dimensional Wiener–Plancherel formula," *Advances in Applied Math.* (to appear).

{**B3**} J. P. Bertrandias, et al. *Espaces de Marcinkiewicz correlations—measures, systèmes dynamiques*, Masson, Paris, 1987.

{**B4**} D. Bohm, *Causality and chance in modern physics*, Harper & Row, New York, 1957.

{**B5**} N. Bohr, *Atomic physics and human knowledge*, Wiley, New York, 1958.

{**B6**} M. Born, *The natural philosophy of cause and chance*, Clarendon Press, Oxford, 1949.

{**B7**} ——, and J. Wolf, *Principles of optics*, 5th ed., Pergamon Press, Oxford, 1975.

{**B8**} M. Brelot, "Norbert Wiener and potential theory," *Bull. Amer. Math. Soc.* **72** (1966) (No. 1, Part II), pp. 39–41.

{**B9**} L. Brillouin, "Life, thermodynamics and cybernetics," *Amer. Sci.* **37** (1949), pp. 554–568.

{**C1**} J. W. Cannon, "Topological, combinational and geometric fractals," Hedrick Lectures of Math. Assoc. of Amer., 1982 (unpublished).

{**C2**} R. Carnap, *Philosophical foundations of physics*, Basic Books Inc., New York, 1966.

{**C3**} A. K. Coomaraswamy, *The transformation of nature in art*, Harvard University Press, Cambridge, Mass., 1935.

{**C4**} ——, *Spiritual authority and temporal power in the Indian theory of government*, American Oriental Society, New Haven, 1942.

{**D1**} K. Deutsch, *The nerves of government*, Free Press, New York, 1966.

{**D2**} J. L. Doob, *Stochastic processes*, Wiley, New York; Chapman & Hall, London, 1953.

{**D3**} ——, "Wiener's work in probability theory," *Bull. Amer. Math. Soc.* **72** (1966) (No. 1, Part II), pp. 69–72.

{**E1**} A. Einstein, *Investigations on the theory of the Brownian movement*, (1905), Methuen, London, 1926.

{**E2**} ——, *Sidelights on relativity*, Methuen, London, 1922.

{**E3**} ——, *The New York Times*, April 25, 1929, p. 60, column 4.

{**E4**} ——, "Science, philosophy and religion," (Proc. of a Conf. on Science, Philosopohy and Religion in New York, 1941). Reprinted in *Out of my later years*, Citadel Press, Secaucus, New Jersey, 1950, pp. 24–30.

{**F1**} Y. Foures, and I. E. Segal, "Causality and analyticity," *Trans. Amer. Math. Soc.* **98** (1955), pp. 384–405.

{**F2**} H. Freudenthal, Norbert Wiener, *Dictionary of scientific biography*. vol. XIV, Chas. Scribner's Sons, New York, 1976, pp. 344–347.

{**G1**} A. M. Gleason, "Measures on the closed subspaces of a Hilbert space," *J. Rat. Mech. Analysis* **6** (1957), pp. 885–894.

{**G2**} K. Gödel, "Russell's mathematical logic," pp. 447–469, in *Philosophy of mathematics*, Eds. P. Benacerrof & H. Putnam, Cambridge Univ. Press, 1986.

{**G3**} K. E. Gould, "A new machine for integrating a functional product," *J. Math. & Physics* **17** (1929), pp. 305–316.

{**H1**} P. R. Halmos, "Mathematics as a creative art," *American Scientist,* **56** (1968), pp. 375–389.

{**H2**} H. L. Hazen and G. S. Brown, "The cinema integraph. A machine for integrating a parametric product integral," *J. Franklin Institute,* **230** (1940), pp. 19–44, 183–205.

{**H3**} S. J. Heims, *John von Neumann and Norbert Wiener: from mathematics to the technologies of life and death,* MIT Press, Cambridge, Mass., 1980.

{**K1**} M. Kac, "Wiener and integration in function space," *Bull. Amer. Math. Soc.* **72** (1966) (No. 1, Part II), pp. 52–68.

{**K2**} S. Kakutani, "Two-dimension Brownian motion and harmonic functions," *Proc. Imp. Acad. Tokyo* **20** (1944), pp. 706–714.

{**K3**} A. N. Kolmogorov, *Foundations of the theory of probability,* Chelsea Publishing Company, New York, 1933.

{**K4**} ——, "Stationary sequences in Hilbert space" (Russian), *Bull. Math. Univ.,* Miscou, **2**, No. 6 (1941), 40 pp. (English translation by Natasha Artin).

{**L1**} K. S. Lau, "On the Banach space of functions with bounded upper means," *Pac. J. Math.* **91** (1980), pp. 153–172.

{**L2**} ——, "On some classes of Hardy spaces," *J. Functional Analysis* (to appear).

{**L3**} Y. W. Lee, "Synthesis of electric networks by means of Fourier transforms of Laguerre's functions," *J. Math. Phys.* **11** (1932), pp. 83–113.

{**L4**} ——, "Contributions of Norbert Wiener to linear theory and nonlinear theory in engineering," in *Selected Papers of Norbert Wiener,* MIT Press, Cambridge, Mass., 1964.

{**L5**} N. Levinson, "Wiener's life," *Bull. Amer. Math. Soc.* **72** (1966) (No. 1, Part I), pp. 1–32.

{**L6**} Albert Lynd, *Quackery in the public schools,* Grosset & Dunlap, New York, 1953.

{**M1**} G. W. Mackey, *Unitary group representations in physics, probability and number theory,* Benjamin, Reading, Mass., 1978.

{**M2**} B. Mandelbrot, *Fractals.* Freeman, New York, 1977.

{**M3**} ——, *The fractal geometry of nature,* Freeman, New York, 1983.

{**M4**} P. Masani, "Wiener's contributions to generalized harmonic analysis, prediction theory and filter theory," *Bull. Amer. Math. Soc.,* **72** (1966) (No. 1, Part II), pp. 73–125 and 135–145.

{**M5**} P. Masani, *Norbert Wiener: Collected works,* vol. 1, 1976; vol. II, 1979; vol. III, 1981; vol. IV, 1985; MIT Press, Cambridge, Mass.

{**M6**} ——, "Einstein's contribution to generalized harmonic analysis and his intellectual kinship with Norbert Wiener," *Jahrbuch Uberblicke Mathematatik 1986,* vol. **19** (1986), pp. 191–209.

{**M7**} ——, B. Randell, D. K. Ferry, and R. Saeks, "The Wiener memorandum on the mechanical solution of partial differential equations," *Ann. History of Computing* **9** (1987), pp. 183–197.

{**M8**} W. McCulloch and W. Pitts, "A logical calculus of the ideal immanent in nervous activity," *Bull. Math. Biophys.* **5** (1943), pp. 115–133.

{**M9**} ——, "How we know universals: the perception of auditory and visual forms," *Bull. Math. Biophys.* **9** (1947), pp. 127–147.

{**M10**} Brockway McMillan, "Norbert Wiener and chaos," in *History of mathematics, vol. 2, A century of mathematics in America, Part* II, Edited by Peter Duren, American Mathematical Society, Providence, R.I., pp. 479–492.

{**M11**} J. Mehra and H. Rechenberg, *The historical development of quantum theory.* vol. III, Springer-Verlag, New York, 1982.

{**P1**} J. R. Pierce, *Signals and noise: the nature and process of communication*, Harper & Row, New York, 1961.

{**Q1**} W. V. Quine, *Mathematical logic*, Harvard Univ. Press, Cambridge, Mass., 1947.

{**Q2**} ——, *From a logical point of view*, Harvard Univ. Press, Cambridge, Mass., 1953.

{**R1**} A. Rosenblueth, *Mind and brain: a philosophy of science.* MIT Press, Cambridge, Mass., 1970.

{**R2**} B. Russell, *Principles of mathematics*, (1903), W. Norton & Company, New York, 1937.

{**R3**} ——, *Human knowledge, its scope and limits*, Simon & Schuster, New York, 1948.

{**S1**} R. Saeks, "Causality in Hilbert space," *SIAM Rev.* **12** (1970), pp. 357–383.

{**S2**} C. E. Shannon, "A symbolic analysis of relay and switching circuits," *Trans. Am. Inst. Electr. Eng.* **57** (1938), pp. 713–723.

{**S3**} ——, "The mathematical theory of communication," *Bell Syst. Techn. Journ.* **27** (1948), pp. 379–423 and 623–656.

{**S4**} D. J. Struik, "Wiener: colleague and friend," *American Dialogue*, March–April 1966.

{**T1**} G. I. Taylor, "Diffusion by continuous movements," *Proc. Lond. Math. Soc.* **20** (1920), pp. 196–212.

{**V1**} J. von Neumann, "Physical applications of the ergodic hypothesis," Proc. Nat. Acad. Sci. **18** (1932), pp. 263–266.

{**V2**} ——, "Zur Operatorenmethoden der Klassischen Mechanik," *Ann. of Math.* (2), **33** (1932), pp. 587–642.

{**V3**} ——, and O. Morgenstern, *Theory of games and economic behavior*, Princeton University Press, Princeton, N.J., 1944.

{**W1**} H. Weyl, *The theory of groups and quantum mechanics*, Dover Publications, New York, 1931.

{**W2**} H. Weyl, *Philosophy of mathematics and natural science*, Princeton Univ. Press, Princeton, N.J., 1949.

{**W3**} A. N. Whitehead and B. Russell, *Principia mathematica*, vol. I, II, III, (1910–1913), Cambridge Univ. Press, 1925–1927.

{**W4**} Sir Edmund Whittaker, *A history of the theories of aether and electricity*, vol. II, Harper, New York, 1953.

{**W5**} E. P. Wigner, "The unreasonable effectiveness of mathematics in the natural sciences," *Commun. Pure Appl. Math.* **13** (1960), pp. 1–14.

{**W6**} L. Wittgenstein, *Tractatus logico philosophicus*, Routledge & Kegan Paul, New York, 1961.

{**Y1**} A. M. Yaglom, "Einstein's 1914 paper on the theory of irregularly fluctuating series of observations," *IEEE ASSP Magazine*, vol. **4**, pp. 7–11, October 1987.

The School of Antoni Zygmund

RONALD R. COIFMAN AND ROBERT S. STRICHARTZ
WITH THE HELP OF GINA GRAZIOSI AND JULIA HALLQUIST

To most mathematicians, the words "harmonic analysis" bring to mind a narrow subfield of analysis dedicated to very technical and classical subjects involving Fourier series and integrals. In fact, it is a very broad field that draws from, inspires, and unifies many disciplines: real analysis, complex analysis, functional analysis, differential equations, differential geometry, topological groups, probability theory, the theory of special functions, number theory, ... Several mathematicians have contributed to the breadth and influence of harmonic analysis. We mention only a few names of those who were active in this century before the second World War: Bernstein, Besicovitch, Bochner, Bohr, Denjoy, Fejér, Hardy, Kaczmarz, Kolmogorov, Lebesgue, Littlewood, Lusin, Menschov, Paley, Plancherel, Plessner, Privalov, Rademacher, F. and M. Riesz, Steinhaus, Szegö, Titchmarsh, Weyl, Wiener, G. C. and W. H. Young. Perhaps it is even appropriate to mention that Cantor's theory of transfinite numbers has its origin in a problem involving trigonometric series. The present status and prominence of harmonic analysis, however, is due in large part to Antoni Zygmund and the school that he created in the United States.

We shall first say a few words about Antoni Zygmund and try to explain why he was able to establish such a large and influential school. By doing this, we shall also describe, briefly, the field of harmonic analysis and the vision Zygmund had for this discipline. We then present a two-generation "mathematical genealogy" of Zygmund's students and their students. We do this for two reasons. First, we believe that this is the most concrete evidence we can provide for gauging the influence Zygmund had. Second, such a compilation may be a most useful document for a historian of mathematics.

Antoni Zygmund was born in Warsaw, Poland, on December 26, 1900. After completing high school, he enrolled in the University of Warsaw in 1919. A few months later, he enlisted in the Polish army where he served during the creation of the state of Poland. He returned to Warsaw when

Antoni Zygmund
1987
(Photograph courtesy of University of Chicago News and Information.)

the fighting ceased and graduated from the University in 1923. He studied with Aleksander Rajchman and devoted himself to the study of trigonometric series. He and Rajchman wrote some joint papers on summability theory. Another of his teachers was Waclaw Sierpiński with whom he published a paper in 1923. While still a student, he met Saks, who was three years older. Saks had a significant influence on Zygmund. They wrote some joint papers and later produced an excellent text on the theory of functions.

He began his teaching career at the Warsaw Polytechnical School. From 1926 to 1930, he held the position of "Privat Dozent" at the University of Warsaw. During these years in his native city, Zygmund's mathematical activity (mostly in the field of trigonometric series) was intense. He spent the academic year 1929–1930 in England as a Rockefeller Fellow at the Universities of Oxford and Cambridge. There he met both Hardy and Littlewood as well as others who shared his scientific interests. In particular, it was there that the seeds of an important collaboration with R. E. A. C. Paley were sown. He also met Norbert Wiener with whom he and Paley later wrote a seminal paper that showed the important relationship probability has with the theory of Fourier series. During the ten months in England, he wrote ten papers.

In the summer of 1930, Zygmund was appointed Associate Professor of mathematics at the University of Wilno. He stayed there until March 1940, when, together with his wife and son, he managed to escape from occupied Poland. The ten-year period in Wilno was a remarkably productive one. His unique ability to integrate ideas from many fields and his sense of direction on various subjects are evident from his publications during this decade. His collaboration with Paley pointed the way to the many connections between the theory of functions and the study of Fourier series. With Paley and Wiener, he showed the important ties between this last topic and probability theory. In Wilno, he discovered a brilliant youth, Josef Marcinkiewicz. It is one of the many tragedies of the second World War that this very talented man died in the spring of 1940 when he was serving as an officer in the Polish army. Together with Marcinkiewicz, Zygmund explored and pioneered in other fields of analysis. This effort included an important paper on the differentiability of multiple integrals (another young mathematician, Jessen, was involved in this research as well). Much of the subsequent study of functions of several real variables depends on the ideas in this work. Perhaps the most important achievement of this period was the publication of the first edition of his famous book *Trigonometrical Series.* In this book, one can find practically all the important results that were known on this subject, as well as its connections with other disciplines. In addition to the topics we have already mentioned, the book includes subjects and points of view that were new at that time. In particular, one should keep in mind that it was during this period that much of modern functional analysis was developed in Poland by Banach and others. In Zygmund's book, one can find the treatment of function spaces and operators on them that is much in the spirit of this new topic. It was in this work that the importance of the M. Riesz Convexity Theorem, as a tool for studying operators, was made evident.

Thanks to the efforts of J. D. Tamarkin, Norbert Wiener, and Jerzy Neyman, in 1940 he received an offer of a visiting professorship at M.I.T. as well as a visa to the United States. The American academic world, at that time, was facing many problems. Zygmund had to start his American career from the beginning. From 1940 to 1945, he was an assistant professor at Mount Holyoke College. During this period, he was also granted a leave of absence to spend the academic year 1942–1943 at the University of Michigan. This, too, was a prolific period for Zygmund. He produced eleven papers. His collaboration with Raphael Salem began at this time. A little-known fact is that one of these papers, with Tamarkin, contains the elegant proof of the M. Riesz Convexity Theorem that is known as the "Thorin proof." This proof gave birth to the "complex method" in the theory of interpolation of operators. Thorin did obtain his proof earlier (in 1942), but he did not publish it until 1947. Zygmund acknowledged Thorin's priority and always referred to the result involved as the "Riesz-Thorin Theorem." All this was done despite the very heavy teaching schedule (by modern standards) of nine hours per week. We should add that often, during his career in Poland, Zygmund had comparably heavy teaching duties.

In 1945, Zygmund accepted an associate professorship at the University of Pennsylvania where he stayed until 1947. In that year, he was invited to join the faculty at the University of Chicago where he spent the rest of his career. This was the beginning of an exceptional period for Zygmund and, more generally, for mathematics. Under the leadership of its chancellor, Robert M. Hutchins, the University of Chicago became a world leader in many academic fields. In particular, Hutchins hired Marshall H. Stone who built an exceptional department of mathematics in the ensuing years. In addition to Zygmund, he brought many distinguished mathematicians to this department. S. Mac Lane, S. S. Chern, and A. Weil were some of the senior men that joined well-known professors already in the department: A. Adrian Albert, E. P. Lane, and L. M. Graves. The more junior newcomers who came developed into well-known leaders in their fields. I. Kaplansky, P. Halmos, and I. E. Segal were some of these. Distinguished visitors from all over the world spent various periods of time at the University of Chicago. J. E. Littlewood, M. Riesz, L. Hormander, S. Smale, and R. Salem represent only a very small and arbitrarily chosen sample of this group. In addition to all this, a large number of extraordinary graduate students came to Chicago to study with this illustrious group.

Zygmund flourished in this atmosphere. Many of the talented young people who came to study in Chicago became his students. In addition, he went to Argentina in 1949 on a Fulbright fellowship where he discovered two outstanding students, Alberto Calderón and Mischa Cotlar. Both went to Chicago and soon earned their Ph.D.'s with him. Calderón soon became Zygmund's collaborator, and their joint work is of such importance that many refer to the school we are discussing as the "Zygmund–Calderón school." Though this name appropriately classifies an important portion of harmonic analysis, it does not cover all that should be referred to as the "Zygmund school."

It is important to realize the following unique features of this school. When Zygmund came to Chicago, the "trend" in mathematics was very much influenced by the Bourbaki school and other forces that championed a rather abstract and algebraic approach for all of mathematics. Zygmund's approach toward his mathematics was very concrete. He felt that it was most important to extend the more classical results in Fourier analysis to other settings, to show the connections of this field to others (as we have already indicated in this article) and to discover methods for carrying this out. He realized that fundamental questions of calculus and analysis were still not well understood. In a sense, he was "bucking the modern trends." In retrospect, his approach proved to be very successful. This is seen not only by what we state here (his achievements and the two-generation genealogy that includes more than 170 names), but by the fact that the very concrete problems posed by Zygmund, with well-defined scope, attracted many of the very gifted students in Chicago to work with him.

Zygmund continued making important contributions. Perhaps the most significant is the second edition of his book *Trigonometrical Series*. This two-volume work, published in 1959, includes all that was in the earlier edition in addition to most of the development in the field that occurred in the twenty-five years after the first edition was written. This was a tremendous effort for Zygmund. He complained to J. E. Littlewood that writing this book cost him at least thirty research papers. Littlewood replied that the book was worth more than twice that many good papers. His work with Calderón, of course, was of paramount importance. Even before he met Calderón, he often said that "the future of harmonic analysis lies in several dimensions." The Calderón–Zygmund theory is a giant step in this direction. They developed a theory of "singular integral operators" that has led to many advances in the theory of partial differential equations and many other fields.

By 1956, Zygmund had trained the three students, Calderón, Elias M. Stein, and Guido Weiss, who were to form the backbone of the Zygmund school, not only because of their research contribution, but because of the large number of students they have trained, a total of seventy-three to date (a number that will probably increase to seventy-seven by the time this article is printed). He continued having students until 1971. Even after that date, however, he was active mathematically. Soon after coming to Chicago, he organized a weekly seminar that consisted of a one-hour presentation of a current topic followed by an informal hour of discussion. This discussion was open to anyone who wanted to present an idea or formulate a problem. This "Zygmund Seminar" continued under his leadership through the seventies and early eighties.

We have described, briefly, some of Zygmund's work, vision, and influence in the study of Fourier series and integrals. We indicated that he was a pioneer in showing how this field was connected with the theory of functions, probability theory, functional analysis, analysis in higher-dimensional Euclidean spaces, and partial differential equations. A more thorough biography would indicate an even broader vision. He showed the importance of certain function spaces: $L \log L$, the weak type spaces, the space of smooth functions (he was most proud of this creation). He paved the way to other topics in higher dimensions by being the first to establish important results in the theory of Hardy spaces involving analytic functions of several variables. By writing a beautiful paper on the Marcinkiewicz Interpolation Theorem (after Marcinkiewicz's death), he led the way to "the real method" in the theory of interpolation of operators. His collected works have been compiled and include more than 150 publications. We give a precise reference to this volume at the end of this article, where we cite some other works containing relevant historical material.

Zygmund's personality contributed greatly to the influence he had on his students and colleagues. He was gentle, generous, and friendly. His interests always extended way beyond mathematics. Literature and current events occupied a considerable amount of his attention. The beginning of each day was devoted to a thorough reading of the New York Times, and he ended the day engrossed in a book; but mathematics was his passion. His outlook on life and his considerable sense of humor almost always were connected with mathematics. Once when walking past a lounge in the University of Chicago that was filled with a loud crowd watching TV, he asked one of his students what was going on. The student told him that the crowd was watching the World Series and explained to him some of the features of this baseball phenomenon. Zygmund thought about it all for a few minutes and commented, "I think it should be called the World Sequence." On another occasion, after passing through several rooms in a museum filled with the paintings of a rather well-known modern painter, he mused, "Mathematics and art are quite different. We could not publish so many papers that used, repeatedly, the same idea and still command the respect of our colleagues." His judgements of others, however, was usually kind. Once, when discussing the philosophy of writing letters of recommendation, he said to one of his students, "Concentrate only on the achievements, and ignore the mistakes. When judging a mathematician you should only integrate f+ (the positive part of his function) and ignore the negative part. Perhaps this should apply more generally to all evaluations of your fellow men." Despite his considerable achievements, he always considered others as his equal and made his students feel at ease with him. He was always easy to approach and encouraged students to come and talk with him. His office was often filled with students and colleagues.

The Genealogy

The following is a list of all of Zygmund's Ph.D. students in the U.S. in chronological order. Under each student, indented, is a list of all his or her students (through 1987), also in chronological order. Each entry lists the current affiliation if known, the date the Ph.D. was granted, the university granting the Ph.D., and the thesis title. Zygmund also had four Ph.D. students in Poland: L. Jasmanowicz, L. Lepecki, J. Marcinkiewicz, and K. Sokol-Sokolowski; the last three are deceased.

Before presenting this list, let us make a few observations about such a genealogy. Such a list has to be terminated somewhere. We have chosen to limit ourselves to the second generation since the influence of Zygmund as a teacher would be quite diluted by the third generation. We are aware that there are quite a few mathematicians who either totally or partially retrained under Zygmund and his students, but do not show up on our list. One of us (Coifman), for example, was a student of Karamata, but studied intensively under Guido Weiss and, later, Calderón and Zygmund. We are also aware that a Ph.D. student may have more than one advisor. For example, when Calderón and Zygmund were at the University of Chicago together, they had common students. A consequence is that those officially listed as Zygmund students have their students on our list, while those listed as Calderón students do not. A similar situation occurred at Washington University between Coifman and Weiss (the Coifman students do not appear on our list). To the best of our knowledge, our list reflects the advisor-student relation that was given to us by the departments of mathematics involved. We know that there are many who have made significant contributions to the Zygmund school but who are not mentioned here. We offer our apologies to them for this and ask for their understanding.

The students of Zygmund are listed in boldface. The second generation's names are indented and are listed below the name of their advisor.

Acknowledgments

We are grateful to Deena Berton and Thomas Mahr, who participated in the early stages of this work. We are grateful to the many students of Zygmund who provided us with information for the genealogy and personal remembrances and who suggested improvements in the text. In particular, we want to thank Mischa Cotlar, Eugene Fabes, Benjamin Muckenhoupt, Cora Sadosky, Eli Stein, Daniel Waterman, Guido Weiss, and Richard Wheeden.

Zygmund's Ph.D. Students in the U.S

Nathan J. Fine
Retired, Pennsylvania State University
Ph.D. 1946, University of Pennsylvania
"On the Walsh Functions"

Justin J. Price
Purdue University
Ph.D. 1956, University of Pennsylvania
"Some Questions about Walsh Functions"

Anthony W. Hager
Wesleyan University
Ph.D. 1965, Pennsylvania State University
"On the Tensor Product of Function Rings"

William A. Webb
Washington State University
Ph.D. 1969, Pennsylvania State University
"Automorphisms of Formal Puiseux Series"

Ching-Tsu Loo
Ph.D. 1948, University of Chicago
"Note on the Properties of Fourier Coefficients"

Alberto Calderón
Buenos Aires, Argentina
Ph.D. 1950, University of Chicago
I. "On the Ergodic Theorem"
II. "On the Behavior of Harmonic Functions at the Boundary"
III. "On the Theorem of Marcinkiewicz and Zygmund"

Robert T. Seeley
University of Massachusetts, Boston
Ph.D. 1959, M. I. T.
"Singular Integrals on Compact Manifolds"

Irwin S. Bernstein
City College, CUNY
Ph.D. 1959, M.I.T.
"On the Unique Continuation Problem of Elliptic Partial Differential Equations"

Israel Norman Katz
Washington University, Dept. of Systems, Science and Math.,
St. Louis, Missouri
Ph.D. 1959, M.I.T
"On the Existence of Weak Solutions to Linear Partial Differential Equations"

Jerome H. Neuwirth
University of Connecticut
Ph.D. 1959, M.I.T.
"Singular Integrals and the Totally Hyperbolic Equation"

Earl Berkson
University of Illinois
Ph.D. 1961, University of Chicago
I. "Generalized Diagonable Operators"
II. "Some Metrics on the Subspaces of a Banach Space"

Evelio Tomas Oklander
Deceased
Ph.D. 1964, University of Chicago
"On Interpolation of Banach Spaces"

Cora S. Sadosky
Howard University
Ph.D. 1965, University of Chicago
"On Class Preservation and Pointwise Convergence for Parabolic Singular Operators"

Stephen Vági
DePaul University
Ph.D. 1965, University of Chicago
"On Multipliers and Singular Integrals in L_p Spaces of Vector Valued Functions"

Nestor Rivire
Deceased
Ph.D. 1966, University of Chicago
"Interpolation Theory in S-Banach Spaces"

John C. Polking
Rice University
Ph.D. 1966, University of Chicago
"Boundary Value Problems for Parabolic Systems of Differential Equations"

Umberto Neri
University of Maryland
Ph.D. 1966, University of Chicago
"Singular Integral Operators on Manifolds"

Miguel De Guzmán
Universidad Complutense de Madrid
Ph.D. 1967, University of Chicago
"Singular Integral Operators with Generalized Homogeneity"

Carlos Segovia
Universidad de Buenos Aires
Ph.D. 1967, University of Chicago
"On the Area Function of Lusin"

Keith William Powers
Ph.D. 1972, University of Chicago
"A Boundary Behavior Problem in Pseudo-differential Operators"

Alberto Torchinsky
Indiana University
Ph.D. 1972, University of Chicago
"Singular Integrals in Lipschitz Spaces of Functions and Distributions"

Robert R. Reitano
Senior Financial Officer for John Hancock
Ph.D. 1976, M.I.T.
"Boundary Values and Restrictions of Generalized Functions with Applications"

Josefina Dolores Alvarez Alonso
Florida Atlantic University
Ph.D. 1976, Universidad de Buenos Aires
"Pseudo Differential Operators with Distribution Symbols"

Telma Caputti
Universidad de Buenos Aires
Ph.D. 1976, Universidad de Buenos Aires
"Lipschitz Spaces"

Carlos Kenig
University of Chicago
Ph.D. 1978, University of Chicago
"H_p Spaces on Lipschitz Domains"

Angel Eduardo Gatto
DePaul University
Ph.D. 1979, Universidad de Buenos Aires
"An Atomic Decomposition of Distributions in Parabolic H_p Spaces"

Cristian E. Gutierrez
Temple University
Ph.D. 1979, Universidad de Buenos Aires
"Continuity Properties of Singular Integral Operators"

Kent Merryfield
California State Univ., Long Beach
Ph.D. 1980, University of Chicago
"H_p Spaces in Poly-Half Spaces"

F. Michael Christ
UCLA
Ph.D. 1982, University of Chicago
"Restriction of the Fourier Transform to Submanifolds of Low Codimension"

Gerald Cohen
Ph.D. 1982, University of Chicago
"Hardy Spaces: Atomic Decompostion, Area Functions, and Some New Spaces of Distributions"

Maria Amelia Muschietti
National University of La Plata, Argentina
Ph.D. 1984, National University of la Plata
"On Complex Powers of Elliptic Operators"

Marta Urciuolo
National University of Cordoba, Argentina
Ph.D. 1985, University of Buenos Aires
"Singular Integrals on Rectifiable Surfaces"

Bethumne Vanderburg
Ph.D. 1951, University of Chicago
"Linear Combinations of Hausdorff Summability Methods"

Henry William Oliver
Professor Emeritus Williams College (Retired 1981)
Ph.D. 1951, University of Chicago
"Differential Properties of Real Functions"

George Klein
Ph.D. 1951, University of Chicago
"On the Approximation of Functions by Polynomials"

Richard P. Gosselin
University of Connecticut
Ph.D. 1951, University of Chicago
"The Theory of Localization for Double Trigonometric Series"

Richard Montgomery
University of Connecticut, Groton
Ph.D. 1973, University of Connecticut
"Closed Sub-algebra of Group Algebra"

Leonard D. Berkovitz
Purdue University
Ph.D. 1951, University of Chicago
I. "Circular Summation and Localization of Double Trigonometric Series"
II. "On Double Trigonometric Integrals"
III. "On Double Sturm–Liouville Expansions"

Harvey Thomas Banks
Brown University
Ph.D. 1967, Purdue University
"Optimal Control Problems with Delays"

Lian David Sabbagh
Sabbagh Associates, Inc.
Ph.D. 1967, Purdue University
"Variational Problems with Lags"

Thomas Hack
Ph.D. 1970, Purdue University
"Sufficient Conditions in Optimal Control Theory and Differential Games"

Jerry Searcy
Ph.D. 1970, Purdue University
"Nonclassical Variational Problems Related to an Optimal Filter Problem"

Ralph Weatherwax
Ph.D. 1972, Purdue University
"Lagrange Multipliers for Abstract Optimal Control Programming Problems"

William Browning
Applied Math. Inc.
Ph.D. 1974, Purdue University
"A Class of Variational Problems"

Gary R. Bates
Murphy Oil
Ph.D. 1977, Purdue University
"Hereditary Optimal Control Problems"

Negash G. Medhim
Atlanta University
Ph.D. 1980, Purdue University
"Necessary conditions for Optimal Control Problems with Bounded State by a Penalty Method"

Jiongmin Yong
University of Texas, Austin
Ph.D 1986, Purdue University
"On Differential Games of Evasion and Pursuit"

Victor L. Shapiro
University of California at Riverside
Ph.D. 1952, University of Chicago
"Square Summation and Localization of Double Trigonometric Series"
"Summability of Double Trigonometric Integrals"
"Circular Summability C of Double Trigonometric Series"

Aaron Siegel
Deceased
Ph.D. 1958, Rutgers University
"Summability C of Series of Surface Spherical Harmonics"

Robert Fesq
Kenyon College
Ph.D. 1962, University of Oregon
"Green's Formula, Linear Continuity, and Hausdorff Measure"

Richard Crittenden
Portland State University
Ph.D. 1963, University of Oregon
"A Theorem on the Uniqueness of (C_{11}) Summability of Walsh Series"

Lawrence Harper
University of California at Riverside
Ph.D. 1965, University of Oregon
"Capacity of Sets and Harmonic Analysis on the Group 2^{ω}"

Lawrence Kroll
Ph.D. 1967, University of California at Riverside
"The Uniqueness of Hermite Series Under Poisson–Abel Summability"

Robert Hughes
Boise State University
Ph.D. 1968, University of California at Riverside
"Boundary Behavior of Random Valued Heat Polynomial Expansions"

William R. Wade
University of Tennessee
Ph.D. 1968, University of California at Riverside
"Uniqueness Theory of the Haar and Walsh Series"

Stanton P. Phillip
University of California at Santa Cruz
Ph.D. 1969, University of California at Riverside
"Hankel Transforms and Generalized Axially Symmetric Potentials"

James Diederich
University of California at Davis
Ph.D. 1970, University of California at Riverside
"Removable Sets for Pointwise Solutions of Elliptic Partial Differential Equations"

Gary Lippman
California State University, Hayward
Ph.D. 1970, University of California at Riverside
"Spherical summability of Conjugate Multiple Fourier Series and Integrals at the Critical Index"

Richard Escobedo
Ph.D. 1971, University of California at Riverside
"Singular Spherical Harmonic Kernels and Spherical Summability of Multiple Trigonometric Integrals and Series"

Joseph A. Reuter
Ph.D. 1973, University of California at Riverside
"Uniqueness of Laguerre Series Under Poisson-Abel Summability"

John Basinger
Lockheed, Ontario, California
Ph.D. 1974, University of California at Riverside
"Trigonometric Approximation, Fréchet Variation, and the Double Hilbert Transform"

Charles Burch
Ph.D 1976, University of California at Riverside
"The Dini Condition and a Certain Nonlinear Elliptic System of Partial Differential Equations"

Lawrence D. DiFiore
Ph.D. 1977, University of California at Riverside
"Isolated Singularities and Regularity of Certain Nonlinear Equations"

David Holmes
TRW, San Bernardino, California
Ph.D. 1981, University of California at Riverside
"An Extension to n-dimensions of Certain Nonlinear Equations"

John C. Fay
California State University, San Bernardino
Ph.D. 1986, University of California at Riverside
"Second and Higher Order Quasilinear Ellipticity on the N-torus"

Mischa Cotlar
Universidad Central de Venezuela
Ph.D. 1953, University of Chicago
"On the Theory of Hilbert Transforms"

Rafael Panzone
Universidad Nacional del Sur, Bahia Blanca, Argentina
Ph.D. 1958, University of Buenos Aires
"On a Generalization of Potential Operators of the Riemann–Liouville Type"

Cora Ratto de Sadosky
Deceased (1980)
Ph.D. 1959, University of Buenos Aires
"Conditions of Continuity of Generalized Potential Operators with Hyperbolic Metric"

Eduardo Ortiz
Imperial College, London
Ph.D. 1961, University of Buenos Aires
"Continuity of Potential Operators in Spaces with Weighted Measures"

Rodrigo Arocena
Mathematics Institute, Montevideo, Uruguay
Ph.D. 1979, Universidad Central de Venezuela

George W. Morgenthaler
University of Colorado
Ph.D. 1953, University of Chicago
I. "The Central Limit Theorem for Orthonormal Systems"
II. "The Walsh Functions"

Daniel Waterman
Syracuse University
Ph.D. 1954, University of Chicago
I. "Integrals Associated with Functions of L_P"
II. "A Convergence Theorem"
III. "On Some High Indicies Theorems"

Syed A. Husain
Ph.D. 1959, Purdue University
"Convergence Factors and Summability of Orthonormal Expansions"

Dan J. Eustice
Ohio State University
Ph.D. 1960, Purdue University
"Summability of Orthogonal Series"

Donald W. Solomon
University of Wisconsin, Milwaukee
Ph.D. 1966, Wayne State University
"Denjoy Integration in Abstract Spaces"

Jogindar S. Ratti
Ph.D. 1966, Wayne State University
"Generalized Riesz Summability"

George Gasper, Jr.
Northwestern University
Ph.D. 1967, Wayne State University
"On the Littlewood–Paley and Lusin Functions in Higher Dimensions"

James R. McLaughlin
Ph.D.1968, Wayne State University
"On the Haar and Other Classical Orthonormal Systems"

Cornelis W. Onneweer
University of New Mexico, Albuquerque, NM
Ph.D. 1969, Wayne State University
"On the Convergence of Fourier Series Over Certain Zero-Dimensional Groups"

Sanford J. Perlman
Ph.D. 1972, Wayne State University
"On the Theorem of Fatou and Stepanoff"

Elaine Cohen
University of Utah
Ph.D. 1974, Syracuse University
"On the Degree of Approximation of a Function by Partial Sums of its Fourier Series"

David Engles
Ph.D. 1974, Syracuse University
"Bounded Variation and its Generalizations"

Arthur D. Shindhelm
Ph.D. 1974, Syracuse University
"Generalizations of the Banach–Saks Property"

Michael J. Schramm
LeMoyne College, Syracuse, NY
Ph.D. 1982, Syracuse University
"Topics in Generalized Bounded Variation"

Pedro Isaza
Ph.D. 1986, Syracuse University
"Functions of Generalized Bounded Variation and Fourier Series"

Lawrence D'Antonio, Jr.
SUNY at New Paltz
Ph.D. 1986, Syracuse University
"Functions of Generalized Bounded Variation.
Summability of Fourier Series"

Izaak Wirszup
University of Chicago
Ph.D. 1955, University of Chicago
"On an Extension of the Cesàro Method of Summability to the Logarithmic Scale"

Elias M. Stein
Princeton University
Ph.D. 1955, University of Chicago
"Linear Operators on L_p Spaces"

Stephen Wainger
University of Wisconsin, Madison
Ph.D. 1962, University of Chicago
"Special Trigonometrical Series in K-Dimensions"

Mitchell Herbert Taibleson
Washington University in St. Louis
Ph.D. 1963, University of Chicago
"Smoothness and Differentiability Conditions for Functions and Distributions on E_n"

Robert S. Strichartz
Cornell University
Ph.D. 1966, Princeton University
"Multipliers on Generalized Sobolev Spaces"

Norman J. Weiss
Queens College, CUNY
Ph.D. 1966, Princeton University
"Almost Everywhere Convergence of Poisson Integrals on Tube Domains Over Cones"

Daniel A. Levine
Ph.D. 1968, Princeton University
"Singular Integral Operators on Spheres"

Charles Louis Fefferman
Princeton University
Ph.D. 1969, Princeton University
"Inequalities for Strongly Singular Convolution Operators"

Stephen Samuel Gelbart
Weizmann Institute of Science, Israel
Ph.D. 1970, Princeton University
"Fourier Analysis on Matrix Space"

Lawrence Dickson
Ph.D. 1971, Princeton University
"Some Limit Properties of Poisson Integrals and Holomorphic Functions on Tube Domains"

Steven G. Krantz
Washington University in St. Louis
Ph.D. 1974, Princeton University
"Optimal Lipschitz and L_p Estimates for the Equation $\overline{\partial} u = F$ on Strongly Pseudo-Convex Domains"

William Beckner
University of Texas, Austin
Ph.D. 1975, Princeton University
"Inequalities in Fourier Analysis"

Robert A. Fefferman
University of Chicago
Ph.D. 1975, Princeton University
"A Theory of Entropy in Fourier Analysis"

Israel Zibman
Ph.D. 1976, Princeton University
"Some Characteristics of the n-Dimensional Peano Derivative"

Gregg Jay Zuckerman
Yale University
Ph.D. 1975, Princeton University
"Some Character Identities for Semisimple Lie Groups"

Daryl Neil Geller
SUNY at Stony Brook
Ph.D. 1977, Princeton University
"Fourier Analysis on the Heisenberg Group"

Duong Hong Phong
Columbia University
Ph.D. 1977, Princeton University
"On Hölder and L_p Estimates for the $\overline{\partial}$ Equation on Strongly Pseudo-Convex Domains"

David Marc Goldberg
Sun Microsystems, Palo Alto, CA
Ph.D. 1978, Princeton University
"A Local Version of Real Hardy Spaces"

Juan Carlos Peral
Facultad de Ciencias, Bilbao, Spain
Ph.D. 1978, Princeton University
"L_p Estimates for the Wave Equation"

Meir Shinnar
Ph.D. 1978, Princeton University
"Analytic Continuation of Group Representations"

Robert Michael Beals
Rutgers University
Ph.D. 1980, Princeton University
"L_p Boundedness of Certain Fourier Integral Operators"

David Saul Jerison
M.I.T.
Ph.D. 1980, Princeton University
"The Dirichlet Problem for the Kohn Laplacian on the Heisenberg Group"

Charles Robin Graham
University of Washington
Ph.D. 1981, Princeton University
"The Dirichlet Problem for the Bergman Laplacian"

Allan T. Greenleaf
University of Rochester
Ph.D. 1982, Priniceton University
"Prinicipal Curvature and Harmonic Analysis"

Andrew Granville Bennett
Kansas State University
Ph.D. 1985, Princeton University
"Probabilistic Square Functions, Martingale Transforms and A Priori Estimates"

Christopher Sogge
University of Chicago
Ph.D. 1985, Princeton University
"Oscillatory Integrals and Spherical Harmonics"

Robert Grossman
University of California, Berkeley
Ph.D. 1985, Princeton University
"Small Time Local Controllability"

Katherine P. Diaz
Texas A & M University
Ph.D. 1986, Princeton University
"The Szegö K Kernel as a Singular Integral Kernel on a Weakly Pseudo-Convex Domain"

Peter N. Heller
Ph.D. 1986, Princeton University
"Analyticity and Regularity for Nonhomogeneous Operators on the Heisenberg Group"

C. Andrew Neff
IBM, Watson Research Center, Yorktown Heights, NY
Ph.D. 1986, Princeton University
"Maximal Function Estimates for Meromorphic Nevanlinna Functions"

Der-Chen Chang
University of Maryland
Ph.D. 1987, Princeton University
"On L_p and Hölder Estimates for the $\overline{\partial}$-Neumann Problem on Strongly Pseudoconvex Domains"

Sundaram Thangavelu
Tata Institute, Bangalore, India
Ph.D. 1987, Princeton University
"Riesz Means and Multipliers for Hermite Expansions"

Hart F. Smith
Massachusetts Institute of Technology
Ph.D. 1988, Princeton University
"The Subelliptic Oblique Derivative Problem"

William J. Riordan
Ph.D. 1955, University of Chicago
"On the Interpolation of Operations"

Vivienne E. Morley
Ph.D. 1956, University of Chicago
"Singular Integrals"

Guido Leopold Weiss
Washington University in St. Louis
Ph.D. 1956, University of Chicago
"On Certain Classes of Function Spaces and on the Interpolation of Sublinear Operators"

Jimmie Ray Hattemer
Southern Illinois University, Edwardsville
Ph.D. 1964, Washington University
"On Boundary Behavior of Temperatures in Several Variables"

Richard Hunt
Purdue University
Ph.D. 1965, Washington University
"Operators Acting on Lorentz Spaces"

Robert Ogden
Southwest Texas State University
Ph.D. 1970, Washington University
"Harmonic Analysis on the Cone Associated with Noncompact Orthogonal Groups"

Robert William Latzer
Ph.D. 1971, Washington University
"Non-Directed Light Signals and the Structure of Time"

Richard Rubin
Florida International University
Ph.D. 1974, Washington University
"Harmonic Analysis on the Group of Rigid Motions of the Euclidean Plane"

Roberto Macias
PEMA, Sante Fe, Argentina
Ph.D. 1974, Washington University
"Interpolation Theorems on Generalized Hardy Spaces"

Roberto Gandulfo
Universidade de Brasilia, Brasil
Ph.D. 1975, Washington University
"Multiplier Operators for Expansions in Spherical Harmonics and Ultraspherical Polynomials"

Minna Chao
Ph.D. 1976, Washington University
"Harmonic Analysis of a Second Order Singular Differential Operator Associated with Non-Compact Semi-Simple Rank-One Lie Groups"

Michael Hemler
The Fuqua School of Business, Duke University
Ph.D. 1980, Washington University
"The Molecular Theory of $H^{p,q,s}(H^n)$"

José Dorronsoro
Universidad Autonoma de Madrid
Ph.D. 1981, Washington University
"Weighted Hardy Spaces on Hermitian Hyperbolic Spaces"

Eugenio Hernandez
Universidad Autonoma de Madrid
Ph.D. 1981, Washington University
"Topics in Complex Interpolation"

Leonardo Colzani
Universita degli Studi di Milano
Ph.D. 1982, Washington University
"Hardy and Lipschitz Spaces on Unit Spheres"

Fernando Soria
Universidad Autonoma de Madrid
Ph.D. 1983, Washington University
"Classes of Functions Generated by Blocks and Associated Hardy Spaces"

Han Yong Shen
Peking University; presently on leave at Washington University,
Ph.D. 1984, Washington University
"Certain Hardy-Type Spaces that can be Characterized by Maximal Functions and Variations of the Square Functions"

Anita Tabacco Vignati
Politecnico di Torino, Torino, Italy
Ph.D. 1986, Washington University
"Interpolation of Quasi-Banach Spaces"

Marco Vignati
Politecnico di Torino, Torino, Italy
Ph.D. 1986, Washington University
"Interpolation: Geometry and Spectra"

Ales Zaloznik
University of Ljubljana, Yugoslavia
Ph.D. 1987, Washington University
"Function Spaces Generated by Blocks Associated with Spheres, Lie Groups and Spaces of Homogeneous Type"

Mary Bishop Weiss
Deceased
Ph.D. 1957, University of Chicago
"The Law of the Iterated Logarithm for Lacunary Series and Applications to Hardy-Littlewood Series"

Paul Joseph Cohen
Stanford University
Ph.D. 1958, University of Chicago
"Topics in the Theory of Uniqueness of Trigonometric Series"

Peter Sarnak
Stanford University
Ph.D. 1980, Stanford University
"Prime Geodesic Theorems"

Benjamin Muckenhoupt
Rutgers University
Ph.D. 1958, University of Chicago
"On Certain Singular Integrals"

Eileen L. Poiani
Saint Peter's College, Jersey City, NJ
Ph.D. 1971, Rutgers University
"Mean Cesàro Summability of Laguerre and Hermite Series and Asymptotic Estimates of Laguerre and Hermite Polynomials"

Hsiao-Wei Kuo
Ph.D. 1975, Rutgers University
"Mean Convergence of Jacobi Series"

Ernst Adams
Ph.D. 1981, Rutgers University
"On Weighted Norm Inequalities for the Riesz Transforms of Functions with Vanishing Moments"

Efrem Herbert Ostrow
California State University, Northridge
Ph.D. 1960, University of Chicago
"A Theory of Generalized Hilbert Transforms"

Richard O'Neil
SUNY at Albany
Ph.D. 1960, University of Chicago
"Fractional Integration and Orlicz Spaces"

Jack Bryant
Texas A & M University
Ph.D. Rice University

Geraldo S. de Souza
Auburn University
Ph.D. 1980, SUNY at Albany
"Spaces Formed by Special Atoms"

Marvin Barsky
Beaver College, Glenside, PA
Ph.D. 1964, University of Chicago
"On Repeated Convergence of Series"

Chao Ping Chang
Retired - University of Auckland, New Zealand
Ph.D. 1964, University of Chicago
"On Certain Exponential Sums Arising in Conjugate Multiple Fourier Series"

Eugene Barry Fabes
University of Minnesota
Ph.D. 1965, University of Chicago
"Parabolic Partial Differential Equations and Singular Integrals"

Max Jodeit
University of Minnesota
Ph.D. 1967, Rice University
"Symbols of Parabolic Singular Integrals and Some L_p Boundary Value Problems"

Julio Bouillet
Instituto Argentino de Matematica, Buenos Aires, Argentina
Ph.D 1972, University of Minnesota
"Dirichlet Problem for Parabolic Equations with Continuous Coefficients"

Stephen Sroka
Department of Defense, Fort Meade, MD
Ph.D. 1975, University of Minnesota
"The Initial-Dirichlet Problem for Parabolic Partial Differential Equations with Uniformly Continuous Coefficients and Data in L_p."

Angel Gutierrez
Universidad Autonoma de Madrid, Madrid, Spain
Ph.D. 1979, University of Minnesota
"A Priori L_p-Estimates for the Solution of the Navier Equations of Elasticity, Given the Forles on the Boundary"

Gregory Verchota
University of Illinois at Chicago
Ph.D. 1982, University of Minnesota
"Layer Potentials and Boundary Value Problems for Laplace's Equation on Lipschitz Domains"

Patricia Bauman
Purdue University
Ph.D. 1982, University of Minnesota
"Properties of Non-Negative Solutions of Second Order Elliptic Equations and Their Adjoints"

Russell Brown
University of Chicago
Ph.D. 1987, University of Minnesota
"Layer Potentials and Boundary Value Problems for the Heat Equation in Lipschitz Domains"

Richard Lee Wheeden
Rutgers University
Ph.D. 1965, University of Chicago
"On Trigonometirc Series Associated with Hypersingular Integrals"

Edward P. Lotkowski
Ph.D. 1975, Rutgers University
"Lipschitz Spaces with Weights"

Russell T. John
Ph.D. 1975, Rutgers University
"Weighted Norm Inequalities for Singular and Hypersingular Integrals"

Douglas S. Kurtz
New Mexico State University
Ph.D. 1978, Rutgers University
"Littlewood–Paley and Mulitplier Theorems on Weighted L_p Spaces"

J. Marshall Ash
DePaul University
Ph.D. 1966, University of Chicago
"Generalizations of the Riemann Derivative"

P. J. O'Connor
Ph.D. 1969, Wesleyan University
"Generalized Differentiation of Functions of a Real Variable"

I. Louis Gordon
Retired, University of Illinois, Chicago
Ph.D. 1967, University of Chicago
"Perron's Integral for Derivatives in L_r"

Yorham Sagher
University of Illinois at Chicago
Ph.D. 1967, University of Chicago
"On Hypersingular Integrals with Compex Homogeneity"

Michael Cwikel
Israel Institute of Technology

Sim Lasher
University of Illinois at Chicago
Ph.D. 1967, University of Chicago
"On Differentiation and Derivatives in L^r"

Leo Frank Ziomek
Deceased
Ph.D. 1967, University of Chicago
"On the Boundary Behavior in the Metric L_p of Subharmonic Functions"

William C. Connett
University of Missouri at St. Louis
Ph.D. 1969, University of Chicago
"Formal Multiplication of Trigonometric Series and the Notion of Generalized Conjugacy"

Thomas Walsh
University of Florida
Ph.D. 1969, University of Chicago
"Singular Integrals of L^1 functions"

Marvin J. Kohn
Brooklyn College, CUNY
Ph.D. 1970, University of Chicago
"Riemann Summability of Multiple Trigonometric Series"

Styllanus C. Pichorides
University of Crete
Ph.D. 1971, University of Chicago
"On the Best Values of the Constants in the Theories of M. Riesz, Zygmund, and Kolmogorov"

References

1. C. Fefferman, J. P. Kahane, and E. M. Stein, "O dor naukowym Antoniego Zygmunda" (Polish), *Wiadomosci Matematyczne* (Series 2) **19** (1976), 91–126. Includes a list of Zygmund's publications and Ph.D. students.

2. W. Beckner et al., editors, *Conference on Harmonic Analysis*, in honor of A. Zygmund, Wadsworth, 1983. Contains a brief biography of Zygmund by Calderón (pp. xii–xv), and "The development of square functions in the work of A. Zygmund" by E. M. Stein.

3. L. S. Grinstein and P. J. Campbell, editors, *Women of Mathematics*, Greenwood Press, 1987. Contains an article on Mary Weiss by Guido Weiss (pp. 236–240).

4. D. T. Haimo, editor, *Orthogonal Expansions and their Continuous Analogues*, Southern Illinois Univ. Press, 1968. Contains an article on Mary Weiss by A. Zygmund (pp. xi–xviii).

5. J. Marcinkiewicz, *Collected Papers*, Panstwowe Wydawnictwo Naukowe, Warsaw, 1964. Contains an article on Marcinkiewicz by A. Zygmund (pp. 1–30).

Richard Askey started to work with special functions with I. I. Hirschman while an undergraduate at Washington University. He received his Ph.D. in 1963 from Princeton University as a student of S. Bochner. After two years at the University of Chicago, learning more analysis from A. Zygmund and his colleagues, he went to the University of Wisconsin. There he progressed backwards from proving norm inequalities, to proving positivity, and is now trying to evaluate and transform series and integrals, and so discover new results for future handbooks.

Handbooks of Special Functions

RICHARD ASKEY

1. Background

Special functions are functions that satisfy certain differential equations, or difference equations, or are given by certain series or integrals. To be a special function, the function must arise often enough so that someone gives it a name, and then others use this name.

Some special functions were discovered so long ago it is probably impossible to determine who discovered them. For example, there are early books dealing with spherical trigonometry, but I do not know who introduced trigonometric functions and derived their fundamental properties. These functions and logarithms were once widely represented in books of tables, but these have now been replaced by pocket calculators.

In the eighteenth century, a number of other special functions were discovered. Euler found the gamma function and used it to evaluate a beta integral. He also considered the differential equation that is now called the hypergeometric equation and found both integral representations for solutions and series expansions for one solution. Bessel functions were studied by a number of people. See Watson [**45**] for some references.

Elliptic integrals were also studied by a number of people, from Fagnano and Euler to the systematic treatment by Legendre.

Starting with Gauss, systematic treatments of a number of functions were given. Gauss considered the hypergeometric function

$$y = {}_2F_1\left(\begin{matrix} a, b \\ c \end{matrix}; x\right) = \sum_{n=0}^{\infty} \frac{(a)_n (b)_n}{(c)_n n!} x^n \tag{1.1}$$

where the shifted factorial $(a)_n$ is defined by

$$(a)_n = \Gamma(n+a)/\Gamma(a). \tag{1.2}$$

The function defined in (1.1) is also written as ${}_2F_1(a, b; c; x)$.

Euler had studied this function and discovered some instances of what Gauss called contiguous relations. Two series given by (1.1) are said to be contiguous if they have the same power series variable, if two of their parameters agree and if the third differs by one. Euler discovered some contiguous relations when he found a continued fraction representation for

$${}_2F_1\left(\begin{matrix} a, b+1 \\ c+1 \end{matrix}; x\right) \Big/ {}_2F_1\left(\begin{matrix} a, b \\ c \end{matrix}; x\right). \tag{1.3}$$

He stated this for integrals rather than series, but eventually he discovered the identity of these different representations of this function. However, he did not systematically explore these contiguous relations. Gauss did. He showed that ${}_2F_1(a, b; c; x)$ and any two functions contiguous to it are linearly related. There are $15 = (6 \cdot 5)/2$ such relations, and he stated all of them. Kummer read Gauss's paper very carefully, even uncovering the existence of quadratic transformations from a list of expansions Gauss gave. He systematically studied the equation (1.4),

$$x(1-x)y'' + [c - (a+b+1)x]y' - aby = 0. \tag{1.4}$$

First he treated (1.4) with all three parameters free, then with one restriction when quadratic transformations exist, then with one free parameter when these transformations can be iterated, and finally in the confluent case of

$${}_1F_1\left(\begin{matrix} a \\ c \end{matrix}; x\right) = \lim_{b\to\infty} {}_2F_1\left(\begin{matrix} a, b \\ c \end{matrix}; \frac{x}{b}\right). \tag{1.5}$$

For historical accounts of hypergeometric series see **[5, 7, 12]**.

Gauss treated theta functions and some special elliptic functions just as systematically, but he did not publish his work. Thus it was Abel and Jacobi who first published on elliptic functions in 1827 and 1828. This work was so striking, and so obviously important, that it is not surprising that many people studied these functions and wrote their own accounts of them. Eventually, so many results were found that it became useful to include a compilation of results as well as the formal development. For example, volume 4 of Tannery and Molk **[41]** contains more than seventy pages of formulas in addition to the more than sixty pages of formulas in volume 2 of **[41]**.

Shortly after the turn of the century, E. T. Whittaker wrote *A Course in Modern Analysis* [**46**]. The second edition was coauthored with G. N. Watson [**47**], and with minor revisions this is still in print. This book has two parts. The first is a text in complex variables. The second is a treatment of the special functions that seemed important at the turn of the century. In the next section, I will contrast the treatment of special functions by Whittaker and Watson and that contained in *Higher Transcendental Functions* [**13, 14, 15**].

The first of the real handbooks was *Funktionentafeln mit Formeln und Kurven* by E. Jahnke and F. Emde [**20**]. This book has gone through many editions and is also still in print. In §3 this will be compared with [**1**], which was thought by the authors to be a modern version of Jahnke and Emde [**21**].

2. Whittaker and Watson and the Bateman Project

Harry Bateman was an English applied mathematician who spent most of his professional life in the United States, first at Bryn Mawr College, then five years at Johns Hopkins University, and finally at California Institute of Technology. Truesdell has written a very interesting account of Bateman's life and work [**44**]. The only aspect of his work that concerns us here is his work on special functions. In much of his work, Bateman regularly used special functions. He was a collector of facts about special functions and recorded useful facts about them on cards which he stored in shoe boxes. He had planned to write a many-volume work on special functions, treating their properties in many different ways. This project was so large that it did not really get started.

After Bateman's death, someone at California Institute of Technology approached E. T. Whittaker to ask for the recommendation of a person who could look at Bateman's cards and notebooks to see if they could be reworked for publication. Whittaker recommended a younger colleague, Arthur Erdélyi, who went to Pasadena for the year 1947–1948 to study the material. His conclusion was that it would be possible to write a series of useful books, but not on the scale proposed by Bateman. After a year in which he returned to Edinburgh, he returned to Cal. Tech. to head a large project which led to the publication of five books. Two of these were tables of integral transforms, and while this type of table is useful, none of the many that have been done have been very influential. The other three books, under the title *Higher Transcendental Functions*, have been very influential, so it is worthwhile considering their contents in some detail. To aid in this, the material will be compared with the material in the second half of Whittaker and Watson. To quote Erdélyi's first sentence in the first volume: "The work of which this book is the first volume might be described as an up-to-date

Arthur Erdélyi
ca. 1935
(Photograph courtesy of The Archives, California Institute of Technology.)

version of *Part II. The Transcendental Functions* of Whittaker and Watson's celebrated 'Modern Analysis'."

The chapters in Part II of Whittaker and Watson and the number of pages are:

	Chapter	Pages
XII	The Gamma Function	30
XIII	The Zeta Function of Riemann	16
XIV	The Hypergeometric Function	21
XV	Legendre Functions	35
XVI	The Confluent Hypergeometric Function	18
XVII	Bessel Functions	31
XVIII	The Equations of Mathematical Physics	18
XIX	Mathieu Functions	25
XX	Elliptic Functions. General Theorems and the Weierstrassian Functions	33
XXI	The Theta Functions	29
XXII	The Jacobian Elliptic Functions	45
XXIII	Ellipsoidal Harmonics and Lamé's Equation	43

Each of these chapters deals with a specific class of functions except for Chapter XVIII. This one primarily deals with Laplace's equation, and one of the main results is a proof of the addition formula for Legendre polynomials.

The general form of these chapters is the following. The specific functions being treated are introduced, and a systematic and careful treatment of some of their main properties is given. The chapter closes with a few references and a large number of problems. As many people have observed (but not in print as far as I know), most references to Whittaker and Watson are to a problem in one of these chapters. Some of these problems were Tripos problems, but most were taken from papers. The facts given in these problems are often very important. It is really this aspect of Whittaker and Watson, the listing of important facts, that Erdélyi and his coauthors use as a model for most of the chapters in *Higher Transcendental Functions* **[13, 14, 15]**. Also, most chapters have an outline of the development of the functions being treated, and many more references are given.

Here is a listing of the chapters and their lengths in these three books. The lengths of the corresponding chapters in these two works are not a good indicator of the amount of material contained in each. However the length of the treatment in each of these works is a good indication of the relative importance as seen in the 1910s and around 1950.

Higher Transcendental Functions

	Chapter	Pages
I	The Gamma Function	55
II	The Hypergeometric Function	64
III	Legendre Functions	62
IV	The Generalized Hypergeometric Series	20
V	Further Generalizations of the Hypergeometric Function	46
VI	Confluent Hypergeometric Functions	48
VII	Bessel Functions	48
VIII	Functions of the Parabolic Cylinder and of the Paraboloid of Revolution	18
IX	The Incomplete Gamma Functions and Related Functions	20
X	Orthogonal Polynomials	79
XI	Spherical and Hyperspherical Harmonic Polynomials	32
XII	Orthogonal Polynomials in Several Variables	30
XIII	Elliptic Functions and Integrals	90
XIV	Automorphic Functions	43
XV	Lamé Functions	47
XVI	Mathieu Functions, Spheroidal and Ellipsoidal Wave Functions	76
XVII	An Introduction to the Functions of Number Theory	39
XVIII	Miscellaneous Functions	20
XIX	Generating Functions	55

The last chapter was "based on an extensive list of generating functions compiled by the late Professor Harry Bateman" [**15**, p. 228], and so provides an indication of one type of book that Bateman had planned. I have owned this volume since its publication in 1955, and have never found this chapter particularly helpful. Not every topic can be appropriately treated in a handbook, and this chapter is a good illustration of one that does not work, at least the way it was organized here.

The most striking change from Whittaker and Watson to *Higher Transcendental Functions* is the greatly expanded treatment of hypergeometric functions. A hypergeometric series is a series

$$\sum_{n=0}^{\infty} c_n \tag{2.1}$$

with term ratio a rational function of n. Explicitly, this is usually taken as

$$\frac{c_{n+1}}{c_n} = \frac{(n+a_1)\cdots(n+a_p)x}{(n+b_1)\cdots(n+b_q)(n+1)} \tag{2.2}$$

and the series (2.1) is usually written as

$${}_pF_q\left(\begin{matrix}a_1,\dots,a_p\\ b_1,\dots,b_q\end{matrix};x\right) = \sum_{n=0}^{\infty}\frac{(a_1)_n\cdots(a_p)_n}{(b_1)_n\cdots(b_q)_n}\frac{x^n}{n!}. \tag{2.3}$$

A hypergeometric function is the analytic continuation of (2.3). Chapter II treats the case $p = 2$, $q = 1$. Legendre functions are the special case of this case when one of the parameters has been restricted so that a quadratic transformation exists. Confluent hypergeometric functions are the case $p = 1$, $q = 1$ or $p = 2$, $q = 0$. The case $p = 2$, $q = 0$ comes from a series that diverges, but there are integral representations that satisfy the appropriate differential equation, and are limits of the case $p = 2$, $q = 1$. Bessel functions come from $p = 0$, $q = 1$. Parabolic cylinder functions are sums of two confluent hypergeometric functions, and incomplete gamma functions are special cases of confluent hypergeometric functions, as is the error function. The chapters on orthogonal polynomials, spherical harmonics and orthogonal polynomials in several variables are also about hypergeometric functions. There are a few pages in the chapter on generalized hypergeometric series that deal with basic hypergeometric series, but the rest of this chapter and the chapter on further generalizations of the hypergeometric function deal with hypergeometric functions in one or several variables. Thus much more than half of these two books deals with hypergeometric functions. When one reads accounts of the development of special functions in books on the history of mathematics, one does not see the important role played by hypergeometric functions, and most mathematicians are unaware as well. Their importance was starting to be appreciated by the end of the last century. For example, in his 1893 lectures at Evanston, F. Klein [**27**] wrote:

> Next to the elementary transcendental functions the elliptic functions are usually regarded as the most important. There is, however, another class for which at least equal importance must be claimed on account of their numerous applications in astronomy and mathematical physics, these are the hypergeometric functions, so called owing to their connection with Gauss' hypergeometric series.

Klein was just referring to the case $p = 2$, $q = 1$. There are now many more applications of these functions in mathematics, and some other hypergeometric functions are also very useful. I will illustrate this by considering orthogonal polynomials. In Whittaker and Watson, the only orthogonal polynomials that are treated in detail are Legendre polynomials, with short sections on ultraspherical polynomials $C_n^\nu(x)$, Hermite polynomials which are not called by name and are denoted by $D_n(x)$, and one problem giving the interior asymptotics of Jacobi polynomials without mentioning their name. Jacobi polynomials are now given by

$$(2.4) \qquad P_n^{(\alpha,\beta)}(x) = \frac{(\alpha+1)_n}{n!}\, {}_2F_1\left(\begin{matrix} -n, n+\alpha+\beta+1 \\ \alpha+1 \end{matrix}; \frac{1-x}{2}\right)$$

and satisfy the orthogonality relation

$$(2.5) \quad \int_{-1}^{1} P_n^{(\alpha,\beta)}(x) P_m^{(\alpha,\beta)}(x)(1-x)^\alpha(1+x)^\beta\, dx = 0, \quad m \neq n,\ \alpha, \beta > -1.$$

Legendre polynomials are the special case $\alpha = \beta = 0$, and ultraspherical polynomials are the polynomials when $\alpha = \beta = \nu - \frac{1}{2}$ after they have been renormalized. Hermite polynomials are the limiting case when $\alpha = \beta \to \infty$ after the change of variables $x \to x\alpha^{-1/2}$. They are orthogonal with respect to $\exp(-x^2)$ on $(-\infty, \infty)$.

All of these polynomials and Laguerre polynomials, which are orthogonal on $(0, \infty)$ with respect to $x^\alpha e^{-x}$, are the main polynomials treated in the chapter on orthogonal polynomials in *Higher Transcendental Functions*. The authors had Gabor Szegö's great book *Orthogonal Polynomials* **[37]** to draw on, and so they had the work of a real expert on this subject to use as Whittaker and Watson did not. However, the chapter on orthogonal polynomials contains information about some other sets of polynomials which the authors thought would be useful. They were right. To explain these polynomials, and why they thought they might be useful, here is a very brief account of the classical polynomials of Jacobi, Laguerre, and Hermite, and their discrete analogues.

Jacobi polynomials and their limiting cases of Laguerre and Hermite polynomials have a number of common properties. They satisfy second-order Sturm-Liouville differential equations of the form

$$(2.6) \qquad a(x)y'' + b(x)y' + \lambda_n y = 0, \qquad y = p_n(x)$$

where $a(x)$ and $b(x)$ are independent of n and λ_n is independent of x. The derivatives $q_n(x) = p'_{n+1}(x)$ are also a set of orthogonal polynomials. Finally, they satisfy a Rodrigues' type formula

$$(2.7) \qquad w(x)p_n(x) = K_n \frac{d^n}{dx^n}\{w(x)[A(x)]^n\}$$

where K_n is independent of x and $A(x)$ is a polynomial which is independent of n.

Each of these three properties along with orthogonality with respect to a positive measure can be shown to lead to the same polynomials, Jacobi, Laguerre, and Hermite, after a linear change of variable and renormalization. These facts are often taken to mean that these are the only orthogonal polynomials with enough structure to be really useful. However, discrete versions of these polynomials had been found, starting with a discrete extension of Legendre polynomials found by Tchebycheff [**42**]. The polynomials found by Tchebycheff and others can be represented by hypergeometric series with the polynomial variable now appearing in a parameter spot rather than as the power series variable. Here are the discrete polynomials known before 1940, and an orthogonality.

Hahn polynomials (discovered by Tchebycheff [**43**])

(2.8)
$$Q_n(x) = Q_n(x;\alpha,\beta,N) = {}_3F_2\left(\begin{matrix}-n, n+\alpha+\beta+1, -x\\ \alpha+1, -N\end{matrix};1\right),\qquad x,n = 0,1,\dots,N,$$

$$\sum_{x=0}^{N} Q_n(x)Q_m(x)\binom{x+\alpha}{x}\binom{N-x+\beta}{N-x} = 0,\qquad m\neq n\leq N,\ \alpha,\beta>-1.$$

Meixner polynomials

(2.9)
$$M_n(x) = M_n(x;\beta,c) = {}_2F_1\left(\begin{matrix}-n,-x\\ \beta\end{matrix};1-c^{-1}\right),\quad \beta>0,\ 0<c<1,$$
$$\sum_{x=0}^{\infty} M_n(x)M_k(x)\frac{(\beta)_x}{x!}c^x = 0,\quad k\neq n.$$

Krawtchouk polynomials

(2.10)
$$K_n(x) = K_n(x;p,N) = {}_2F_1\left(\begin{matrix}-n,-x\\ -N\end{matrix};\frac{1}{p}\right),\qquad n,x = 0,1,\dots,N,\ 0<p<1,$$
$$\sum_{x=0}^{N} K_m(x)K_n(x)\binom{N}{x}p^x(1-p)^{N-x} = 0,\qquad m\neq n\leq N.$$

Charlier polynomials

(2.11)
$$C_n(x) = C_n(x;a) = {}_2F_0\left(\begin{matrix}-n,-x\\ -\end{matrix};-\frac{1}{a}\right),\quad a>0,$$
$$\sum_{x=0}^{\infty} C_n(x)C_m(x)\frac{a^x}{x!} = 0,\quad m\neq n.$$

The analogue of the derivative is

$$\Delta f(x) = f(x+1) - f(x). \tag{2.12}$$

Then

$$\Delta Q_n(x;\alpha,\beta,N) = \frac{-n(n+\alpha+\beta+1)}{N(\alpha+1)}Q_n(x;\alpha+1,\beta+1,N-1) \tag{2.13}$$

is an analogue of

$$\frac{d}{dx}P_n^{(\alpha,\beta)}(x) = \frac{n+\alpha+\beta+1}{2}P_{n-1}^{(\alpha+1,\beta+1)}(x). \tag{2.14}$$

Chapter 10 in [**14**] contains enough on most of these polynomials so that people who looked for information about one of these polynomials would probably become aware that others had studied them before. In particular, the three term recurrence relation which all orthogonal polynomials satisfy was included for each of these polynomials except for the general Hahn polynomials, where it was given only in the case $\alpha = \beta = 0$. For the general Hahn polynomials, the discrete Rodrigues' formula was given, as was an explicit formula and a reference to a paper that contained the recurrence relation. The orthogonality was given in a slightly too general form, since the claimed measure only has finitely many moments.

As an illustration that enough was included so that people became aware of these polynomials, one only needs to look at some papers of Karlin and McGregor. See [**23**] and [**24**], and also the long and impressive paper of Karlin and Szegö [**25**]. Of course Szegö had been aware of most of these polynomials, although he seems not to have known that Tchebycheff found the general Hahn polynomials as well as the special case when $\alpha = \beta = 0$ which Szegö mentioned in [**37**]. See [**2**] and my comments in [**40**, pp. 866–869] for more information and references about these polynomials.

Markoff found another discrete analogue of Legendre polynomials, and there were other polynomials of a related nature found by Stieltjes, Wigert and Geronimus. In 1949, W. Hahn [**18**] found a wider class of orthogonal polynomials where theorems like those mentioned above hold. He used the q-difference operator Δ_q defined by

$$\Delta_q f(x) = \frac{f(x) - f(qx)}{(1-q)x}. \tag{2.15}$$

He found all sets of orthogonal polynomials $p_n(x)$ for which

$$r_n(x) = \Delta_q p_{n+1}(x) \tag{2.16}$$

is a set of orthogonal polynomials. These polynomials are basic hypergeometric analogues of all the polynomials mentioned above.

A basic hypergeometric series is a series $\sum c_n$ with c_{n+1}/c_n a rational function of q^n for a fixed number q. There are two choices of q that occur most frequently, $|q| < 1$ and q an integer power of a prime.

The most general set of polynomials in the class of Hahn is

$$Q_n(x; a, b, c) = \sum_{k=0}^{n} \frac{(q^{-n}; q)_k (q^{n+1}ab; q)_k (x; q)_k}{(aq; q)_k (cq; q)_k (q; q)_k} q^k \tag{2.17}$$

where

$$(a; q)_k = \prod_{j=0}^{k-1} (1 - aq^j). \tag{2.18}$$

Hahn found the second order q-difference equation all the polynomials satisfy, but only worked out the orthogonality for the polynomials

$$(2.19)\qquad p_n(x;a,b:q)=\sum_{k=0}^{n}\frac{(q^{-n};q)_k(q^{n+1}ab;q)_k}{(qa;q)_k(q;q)_k}(qx)^k.$$

There are uses of these polynomials that are quite important. Here are two types of uses that are not mentioned in *Higher Transcendental Functions* with the exception of the classical case of symmetric Jacobi polynomials arising in the study of spherical harmonics, which occurs in Chapter XI.

Let S^{n-1} be the surface of the unit ball in R^n. Consider the functions

$$u(x_1,\dots,x_n)=r^kU\left(\frac{x_1}{r},\dots,\frac{x_n}{r}\right)$$

where u is harmonic in R^n and $r=(x_1^2+\cdots+x_n^2)^{1/2}$. Thus U is a function defined on the unit sphere. If u is a polynomial of degree k, then U is called a spherical harmonic. When $n=2$ and $k\geq 1$, there are two linearly independent choices for U, say $\cos k\theta$ and $\sin k\theta$. When $n=3$, there are $(2k+1)$ linearly independent functions U, one being $P_k^{(0,0)}(\cos\theta)=P_k(\cos\theta)$ and the others being functions of two variables which are a product of a function that is the jth derivative of $P_k(x)$ times an algebraic function of x, $j=1,2,\dots,k$ and then times two linearly independent functions of one variable which arise in the same way on the unit circle. This pattern continues for spheres of all dimensions, and a nice outline is given in Chapter XI. The zonal spherical harmonic, or spherical function, is the symmetric Jacobi polynomial $P_k^{((n-3)/2,\,(n-3)/2)}(\cos\theta)$.

The essential property that makes this work is that the sphere is a two-point homogeneous space. On the sphere there is a metric which is invariant under the rotation group, and given two points on the sphere with distance d between them, and a second pair also separated by distance d, there is an element of the rotation group that takes the first pair to the second. There are other compact connected two-point homogeneous spaces, and on each it is possible to construct similar zonal harmonics. These spaces are real, complex, and quaternionic projective spaces, and a two-dimensional projective space over the Cayley numbers. These spaces go back to work in the last century, primarily by E. Cartan. The theorem that there are no others was proved by Wang. In 1929, E. Cartan [**8**] found the zonal spherical functions on complex projective spaces. His spherical functions are the Jacobi polynomials $P_k^{(n,0)}(x)$, $k=0,1,\dots$, when the space is $(n+2)$-dimensional complex projective space. This was not widely known since most people who knew something about Jacobi polynomials knew little about complex projective spaces, and those who knew Cartan's work knew very little about hypergeometric functions and Jacobi polynomials. That situation has changed, and there are now a number of people who are comfortable with both of these.

One thing this has led to is an addition formula for Jacobi polynomials. I used to say that it was about ninety years between discovery of the addition formula for Legendre polynomials and Gegenbauer's discovery of the addition formula for ultraspherical polynomials, and about another ninety years until the discovery of the addition formula for general Jacobi polynomials, so one could guess how long it would be before another addition formula was found. The obvious guess was far too large, for it was less than ten years from the Jacobi case, which was found by Šapiro **[35]** in the case $(\alpha, \beta) = (\alpha, 0)$, and by Koornwinder **[28]** independently in this case and then extended to the general case, and when Dunkl **[11]** found the addition formula for symmetric Krawtchouk polynomials. He was able to do this because there are other important compact two-point homogeneous spaces, and they have spherical functions that not only can be found, but they are orthogonal polynomials and can be found among the polynomials outlined above. The appropriate space for the symmetric Krawtchouk polynomials is the unit cube in R^N, i.e., the space of sequences $(\varepsilon_1, \varepsilon_2, \ldots, \varepsilon_N)$ with $\varepsilon_i = 0$ or 1, and the distance to $(0, 0, \ldots, 0)$ is defined to be the number of 1's. This is called the Hamming distance, and this space is the natural setting to start the study of coding theory. It took many years before people developing coding theory realized that the polynomials they had constructed were not new, but once this realization was made then it was only a short while before the analogy with Legendre and Laplace's work on spherical harmonics was recognized and used as a model of what to look for. Next, this was done for other discrete two-point homogeneous spaces, including a finite field version of some of the spaces, so that many Krawtchouk, Hahn, q-Krawtchouk, and q-Hahn polynomials arise in this way. See Stanton **[36]** for a nice survey and many references.

There is a second way in which Legendre polynomials arise in a group setting. They are elements of matrix representations of $SU(2)$. Jacobi polynomials and Krawtchouk polynomials also arise in this fashion. When the tensor product of two representations of $SU(2)$ are decomposed, the coefficients can be written as a ${}_3F_2$ times a more elementary function, and this ${}_3F_2$ can be transformed to be either a Hahn polynomial or a dual Hahn polynomial (the polynomials that arise when n and x are interchanged). In physics, these coefficients are called $3 - j$ symbols, Clebsch-Gordan coefficients, or Wigner coefficients. The $6 - j$ symbols, or Racah coefficients, can also be transformed to be orthogonal polynomials times a more elementary function, and these polynomials fit into a slightly wider class of classical type orthogonal polynomials. Instead of using the finite difference operators Δ or Δ_q, divided difference operators need to be used. The resulting orthogonal polynomials are either ${}_4F_3$'s or ${}_4\phi_3$'s or special or limiting cases of them. In particular, all the polynomials mentioned above and some ${}_2F_1$'s and ${}_3F_2$'s not mentioned are also in this wider class of classical polynomials.

This is not the end, for recently a Hopf algebra extension of $SU(2)$ has been discovered. It is called a quantum group and is denoted by $SU_q(2)$. The matrix representations have elements that are Hahn's q-extension of Legendre and Jacobi polynomials, and the Clebsch-Gordan and Racah coefficients are basic hypergeometric extensions of the classical results. Koornwinder has used this setting to obtain an addition formula for the q-Legendre polynomials of Hahn. See **[29]** and references in this paper. Drinfield **[10]** and Jimbo **[22]** discovered quantum groups, and Woronowicz **[48, 49]** was the first to realize that one could compute explicitly in $SU_q(2)$.

I have outlined this material to point out what has to be done to write a really useful handbook. One needs to know the applications of the past, so that the formulas that were useful in the past are contained. They will probably be useful in the future. However, to make the book more useful, one has to include some results that have not been used yet, or have not been used very much yet. Erdélyi and his coauthors, W. Magnus, F. Oberhettinger, and F. Tricomi, did a better job of predicting the future than Whittaker and Watson did, but then their aims were different. Whittaker and Watson were writing a text, and while not everyone appreciates it **[19]**, there are many others who learned analysis from it and have gone on to make many important discoveries. However, one needs to realize that the second half of Whittaker and Watson is the part that was closest to both Whittaker and Watson's main interests, and so it is the part where their detailed knowledge shows to best advantage.

Szegö **[39]** wrote a review of **[13]** and **[14]**. After starting with the sentence: "These two volumes compiled by the Bateman Manuscript Project represent a stupendous accomplishment," he went on to write:

> The difficulties of such a compilation as was planned by Bateman and is carried out in the present work are enormous. They are due not so much to the vastness of the pertinent material but rather to the intrinsic difficulty of formulating and following clear and consistent principles in organizing it.

Szegö also mentioned that it seemed likely that Erdélyi played a central role in these books, even though Erdélyi's Foreword is modestly silent about Erdélyi's share. I wrote a comment about this work as an addendum to an article on Erdélyi written by A. G. Mackie **[30]**. I sent a copy to W. Magnus for comments. His reply included the following:

> You are absolutely right in assuming that Erdélyi was not only the editor but simply the soul of the Bateman Project. Nevertheless he never interfered directly with the work of his coauthors. He simply put in the final touches where necessary. May I add that it was Erdélyi's idea to include functions of number theory

> and automorphic functions into the Project. I wrote these chapters upon his request — but it would not have occurred to me to include them. What you call Erdélyi's foresight was, in part, his strong sense of responsibility. He knew that the Project was a rare opportunity to serve the mathematical community, and he took his task very seriously.

In a number of articles in these three volumes, *A Century of Mathematics in America*, there have been comments about the supremacy of mathematics in the United States. One should remember that there were areas where knowledge in the United States was not very strong, even with the many immigrants. Special functions was one. Erdélyi was brought over from Scotland. In 1943, Magnus and Oberhettinger had written a smaller handbook of special functions **[33]** which was a transitional book from Whittaker and Watson to *Higher Transcendental Functions*. They were brought from Germany. Tricomi came over from Italy. He was the only one to return to Europe at the end of the Bateman Project, although Erdélyi eventually returned to Edinburgh. Mathematics is such a broad field, and certain traditions are probably absent in every country, so it is not surprising that there are gaps in the detailed knowledge of some parts of mathematics in every country. In particular, consider multiple hypergeometric functions. In the United States, the most important work was started by some physicists. In particular, the work of Biedenharn and Louck and various coworkers is important. See [**6**] for references. Eventually some mathematicians started to work in this area. See Milne [**34**] and Gustafson [**17**]. These papers deal with well poised multiple hypergeometric series and other series with a lot of structure. They arise from various groups. In England, Ian Macdonald has used affine Lie algebras to discover some important multiple theta functions [**31**], and some multidimensional beta integrals [**32**]. In Japan, Aomoto has introduced some important multiple hypergeometric integrals [**3**]. There is a lot of other work on multiple hypergeometric functions in Japan, most of it connected with partial differential equations. Surprisingly, as far as I know, the only course on multiple hypergeometric functions given in the United States in at least twenty years was given by a Japanese mathematician who was a visiting professor at the University of Minnesota. See [**26**] for the notes on this course.

In the Soviet Union, Gelfand and some coworkers have introduced a more general class of multiple hypergeometric integrals via integrals over Grassmannians. See [**16**] and many later papers.

There is a connection between Macdonald's work and some of the work in the United States, but otherwise, except for one important paper of Aomoto [**4**], the groups in Japan, the United States, and in USSR are working independently, and no group really understands what the others are doing. There are some young people in the Netherlands starting to work in a different way. Each group knows the reasons why they care about their work and is usually ignorant about the reasons behind the other work. All of this work is very important, but I am the first to say that I understand little of it. The next generation of handbooks will be much richer because of it, and also because of all the work on basic hypergeometric functions. Much of the work on basic series goes back to work done by Ramanujan and earlier by Rogers. A year and a half ago, Gelfand told me he expects work on q-problems (including q-series) to be at the center of mathematics in fifteen years. He is working hard to try to learn it. His statement is probably too strong, but this area is clearly much more important than Erdélyi and his coworkers thought that it would be. However, they included an outline of Hahn's work, and that was what led me to be interested in this topic.

My copies of [**13**] and [**14**] have had to be rebound, which is one of the strongest statements I can make about how useful I have found them. They are not perfect, but they were a significant improvement on what was done before. This was partly because of added knowledge, and it was partly the decision of the authors to include material they thought would be useful in the future. Their guesses about what would be useful were quite good, given the few hints they had to work with.

3. Jahnke and Emde and the Bureau of Standards Handbook of Mathematical Functions

Jahnke and Emde wrote the first handbook of special functions. It was written for a different group of users than *Higher Transcendental Functions.* This is best seen by looking at the book and seeing the numerical tables and graphs. The lack of graphs is a weakness in [**13, 14, 15**], but there were other books to use for numerical data. That was much less true in 1909, so it was worthwhile using half the space for numerical tables and another fifteen percent for graphs. This left only slightly more than one third of the 174 pages for text and formulas. However, the graphs have led to one very nice theorem which mathematicians discovered and proved years before it would be rediscovered by a couple of physicists. In both the first edition [**20**] and in the one which was widely available in the United States after 1945 [**21**], there is the following picture of Legendre polynomials. See Figure 26 on page 82 in [**20**] and Figure 64 on page 118 in [**21**]

This graph has been recomputed by Paul Nevai using Mathematica. Nevai's graph is given in Figure 1. Almost forty years after the first publication, J. Todd looked at this graph carefully and suggested that the kth maximum of $|P_n(x)|$ is a decreasing function of n. Here the maximums are counted from the right. This was proved by G. Szegö [**38**], and many years

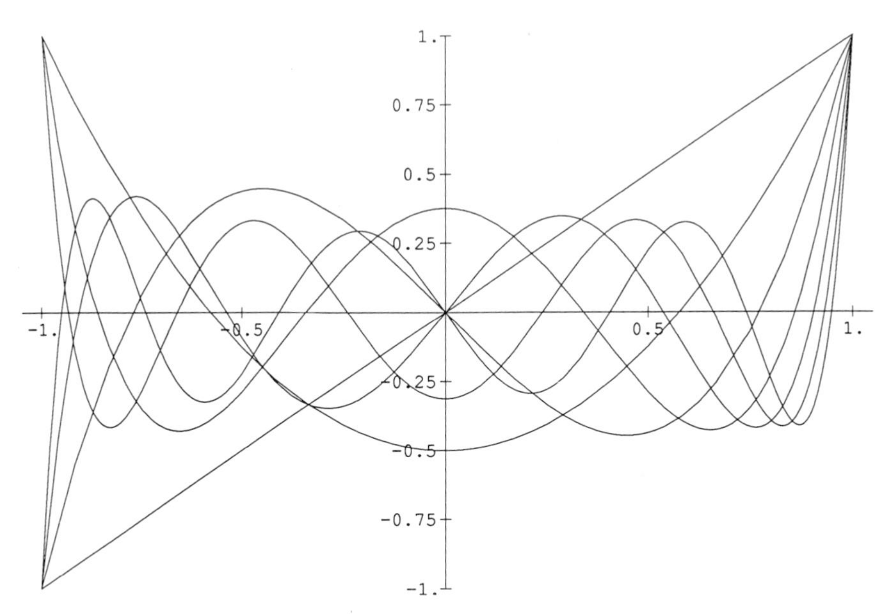

FIGURE 1. Graph of $P_n(x)$, $n = 1, 2, \ldots, 7$.

later Corneille and Martin [9] rediscovered this set of inequalities. It is reassuring that eventually someone looked seriously at this graph, but dismaying that it took almost forty years. Once Szegö proved Todd's conjecture, Todd and Szász obtained similar results for related functions. About twenty years ago, I tried to extend these results to the Jacobi polynomials

$$\frac{P_n^{(\alpha,\beta)}(x)}{P_n^{(\alpha,\beta)}(1)}.$$

Since these inequalities were known for $\alpha = \beta \geq -\frac{1}{2}$ and for $\alpha \geq \beta = -\frac{1}{2}$, it seemed clear they should be true for $\alpha > \beta > -\frac{1}{2}$. That is still unproven. The case $\beta = -1$, $\alpha = 0$ turns out to be very interesting. First

$$\frac{P_n^{(0,-1)}(x)}{P_n^{(0,-1)}(1)} = \frac{P_n^{(0,0)}(x) + P_{n-1}^{(0,0)}(x)}{2},$$

so it is just the average of two adjacent Legendre polynomials. Second, a graph of these functions up to $n = 7$ suggests that the monotonicity in Figure 1 is reversed. This partly explains why the cases when $-\frac{1}{2} < \beta < \alpha$ are so hard. The corresponding graph, again computed by Nevai, is in Figure 2.

With this as background, consider the graphs contained in the chapter on orthogonal polynomials in [1]. Figures 22.4 and 22.8 seem similar. In fact, they are identical. In Figures 22.2 and 22.3, it is very hard to see the monotonicity of the zeros as a function of the varying parameter α or β. This can be seen in Figure 22.5. I find it impossible to determine even by looking at

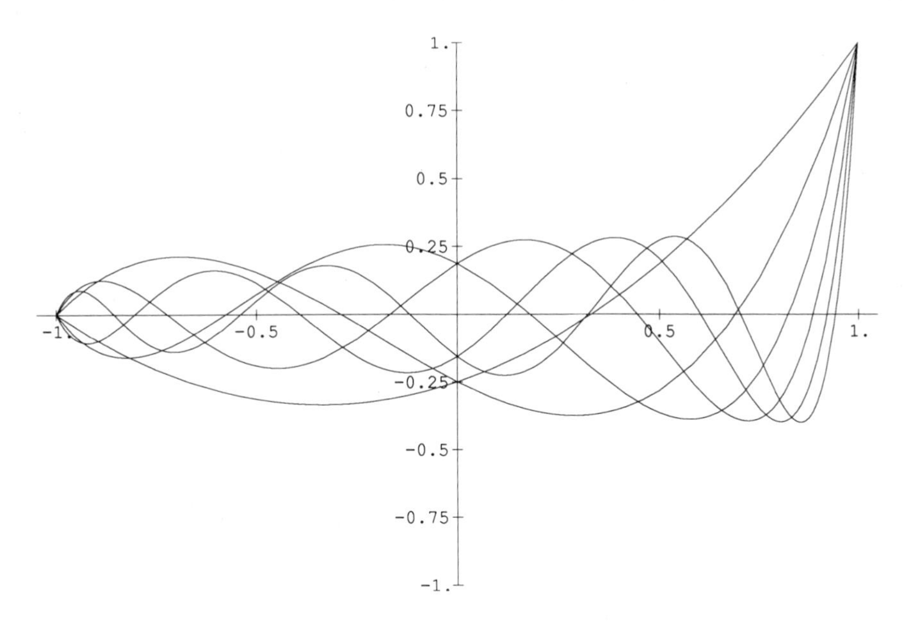

FIGURE 2. $\frac{P_n(x)+P_{n-1}(x)}{2}$, $n = 2, 3, \ldots, 7$.

the graphs that the left minimum values in Figures 22.2 and 22.3 are negative, as they are. The rest of the graphs are reasonable, although I suspect that Figure 22.5 would have been more informative qualitatively if the polynomials had been normalized to be 1 at $x = 1$ and only the right-hand side had been printed, which is all that is necessary by symmetry. Graphs are very useful to give qualitative information, but much less useful for quantitative information.

The chapter on orthogonal polynomials in [**1**] was written by someone who was not an expert on them, and it shows. The chapter starts with the definition of orthogonal polynomials but restricted to an absolutely continuous measure. Then there is the sentence:

> These polynomials satisfy a number of relationships of the same general form.

Four are listed, a differential equation, a three-term recurrence relation, a Rodrigues' formula, and their derivatives forming an orthogonal set. Unfortunately, only one of these holds for general orthogonal polynomials, the three-term recurrence relation. The others hold only for Jacobi, Laguerre, and Hermite polynomials, as was stated in §2.

There are a number of problems with the list of formulas. Some are incorrect; others are stated in a way that is inappropriate; and others are not interesting enough to justify space in a book where important results were omitted because of a lack of space. Here is one of each type. Formula 22.13.5 is

$$\int_{-1}^{1}(1-x^2)^{-1/2}P_n(x)dx = \frac{2^{3/2}}{2n+1}.$$

This is clearly wrong, since $P_n(x)$ satisfies

$$P_n(-x) = (-1)^n P_n(x),$$

and thus the integral vanishes when n is odd. When n is even, the integral is

$$\int_{-1}^{1}(1-x^2)^{-1/2}P_{2n}(x)dx = \left[\frac{\Gamma(n+\frac{1}{2})}{\Gamma(n+1)}\right]^2.$$

The correct form of 22.13.5 is

$$\int_{-1}^{1}(1-x)^{-1/2}P_n(x)dx = \frac{2^{3/2}}{2n+1}.$$

Formula 22.13.1 is given as

$$2n\int_0^x (1-y)^\alpha(1+y)^\beta P_n^{(\alpha,\beta)}(y)dy$$
$$= P_{n-1}^{(\alpha+1,\beta+1)}(0) - (1-x)^{\alpha+1}(1+x)^{\beta+1}P_{n-1}^{(\alpha+1,\beta+1)}(x).$$

It should be given as

$$2n\int_x^1 (1-y)^\alpha(1+y)^\beta P_n^{(\alpha,\beta)}(y)dy = (1-x)^{\alpha+1}(1+x)^{\beta+1}P_{n-1}^{(\alpha+1,\beta+1)}(x)$$

since this is more compact, easily implies the stated formula, and $P_n^{(\alpha,\beta)}(0)$ can be summed only when $\alpha=\beta$ or $\alpha+\beta=0$. Thus one does not want to use $P_n^{(\alpha,\beta)}(0)$ unless one has to. Unfortunately, this formula first appeared in **[14]**.

Finally, 22.14.10 is

$$P_n^2(x) - P_{n-1}(x)P_{n+1}(x) < \frac{2n+1}{3n(n+1)}, \qquad -1 \le x \le 1.$$

The real interest in inequalities about $P_n^2(x) - P_{n-1}(x)P_{n+1}(x)$ is Turán's inequality

$$P_n^2(x) - P_{n-1}(x)P_{n+1}(x) > 0, \qquad -1 < x < 1$$

and extensions of it. From asymptotics it is easy to see that

$$P_n^2(x) - P_{n-1}(x)P_{n+1}(x) = O(n^{-1}),$$

and a more exact upper bound is unlikely to be useful unless it is a sharp bound, and even then I doubt it will be useful. The asymptotic behavior is of interest, but that is easy to determine from asymptotics of the separate terms.

There is a simple test one can use to check the quality of a chapter on the classical orthogonal polynomials. The ultraspherical polynomials can be defined by

$$(1-2xr+r^2)^{-\lambda} = \sum_{n=0}^{\infty} C_n^{\lambda}(x) r^n.$$

This reduces to $1 = 1$ when $\lambda = 0$, so

$$C_n^0(x) = 0, \qquad n = 1, 2, \dots .$$

Here is how Szegö **[37]** handled the case $\lambda = 0$. He wrote:

$$\lim_{\lambda \to 0} \frac{C_n^{\lambda}(x)}{\lambda} = T_n(x), \quad n = 1, 2, \dots, \tag{4.7.8}$$

where

$$T_n(\cos\theta) = \cos n\theta.$$

He does not introduce $C_n^0(x)$.

In *Higher Transcendental Functions* the solution is as follows, skipping irrelevant material. The authors start with:

$$C_n^{\lambda}(1) = \frac{(2\lambda)_n}{n!}. \tag{3}$$

The standardization (3) fails when 2λ is zero or a negative integer. The only exception in the range $\lambda > -\frac{1}{2}$ is $\lambda = 0$, and for this we standardize according to

$$C_0^0(1) = 1, \quad C_n^0(1) = \frac{2}{n} \tag{5}$$

and we have

$$C_n^0(x) = \lim_{\lambda \to 0} \lambda^{-1} C_n^{\lambda}(x). \tag{6}$$

In **[1]**, the solution is

$$C_n^0(\cos\theta) = \frac{2}{n}\cos n\theta. \tag{22.3.14}$$

I think Szegö's solution is better than either of the others, since there is no reason to use $C_n^0(x)$, and it is probably confusing to introduce it with a different normalization. Actually, I like

$$\lim_{\lambda \to 0} \frac{C_n^{\lambda}(x)}{C_n^{\lambda}(1)} = T_n(x), \qquad n = 0, 1, \dots$$

or

$$\lim\nolimits_{\lambda \to 0} \tfrac{n+\lambda}{\lambda} C_n^{\lambda}(x) = 1, \quad n = 0,$$

$$= 2T_n(x), \quad n = 1, 2, \dots ,$$

better than Szegö's limit (4.7.8), but this is a matter of preference. The worst solution was formula 22.3.14 in **[1]**.

Some of the chapters in [**1**] are first rate, while others are poor. The problem is partly the choice of some of the authors, but more the absence of someone to direct the whole project who had the wide and detailed knowledge of Arthur Erdélyi. This book has been a best seller, both for the U. S. Federal Government, which is surprising considering their poor distribution system (the very low price led to the high sales), and also in the paperback edition published by Dover. It is a shame that the quality was not uniformly high as, for example, it was in Olver's chapter on Bessel functions.

4. Summary

In his article on Bateman [**44**, p. 429], Truesdell comes close to asserting that Erdélyi and his coworkers made an error in writing a handbook rather than a treatise. I agree completely with Truesdell when he laments the loss of Bateman's cards from the famous shoeboxes, but disagree with him about the relative importance of handbooks and treatises. Both handbooks and treatises are needed, and treatises are usually more restricted in topic, and so are easier to write. However, they are not easy to write, or we would have more good ones, but the same goes for good handbooks. I use special functions in most of my mathematics, and a fairly large percentage of my work is directly on special functions themselves. Of necessity, I have a fairly good knowledge of what exists in both the systematic treatises of special functions and the best handbooks. My copies of Szegö's *Orthogonal Polynomials* and Watson's *Bessel Functions* have had to be rebound, just as have the first two volumes of *Higher Transcendental Functions*, as was mentioned earlier. These two types of books serve different purposes even for a heavy user and not just for an occasional user. Once the amount of useful knowledge becomes so large that one cannot remember it all, or even remember where it is located, then it is necessary to have help in trying to find the useful facts one needs. Handbooks are one solution. There is talk about trying to make all this material available in a large computer system. This would be very useful, but it should not be the only source. Paper in books often becomes brittle and information is lost as the book disintegrates. The rate of disintegration of computer systems will almost surely be much faster than that of paper since systems change so rapidly. So for the foreseeable future, handbooks of special functions will be useful, and because of scientific and mathematical developments, they need to be redone every so often. Thirty to fifty years is probably the right time interval, so it is time to consider what we can do to help make this useful information more accessible to the mathematical and scientific communities.

George Andrews made another comment about computers versus books. Books permit easier browsing and so more easily lead to unexpected discoveries, interactions, and comparisons. I agree completely.

REFERENCES

1. Milton Abramowitz and Irene A. Stegun, eds., *Handbook of Mathematical Functions with Formulas, Graphs and Mathematical Tables*, National Bureau of Standard Applied Math. Ser. 55 (1964), Washington, D.C., paperback edition, Dover, New York, 1965.

2. G. Andrews and R. Askey, "Classical Orthogonal Polynomials," in *Polynômes Orthogonaux et Applications* (C. Brezenski et al., eds.), Lecture Notes in Math., no. 1171, Springer-Verlag, Berlin, 1985, pp. 36–62.

3. K. Aomoto, "Configurations and Invariant Theory of Gauss-Manin Systems," in *Group Representations and Systems of Differential Equations* (K. Okamoto, ed.), Kinokuniya, Tokyo and North-Holland, Amsterdam, 1984, pp. 165–179.

4. K. Aomoto, "Jacobi Polynomials Associated with Selberg's Integral," *SIAM J. Math. Anal.* **18** (1987), 545–549.

5. R. Askey, "Ramanujan and Hypergeometric and Basic Hypergeometric Series," in *Ramanujan International Symposium on Analysis* (N. K. Thakare, ed.), Proc. Ramanujan Birth Centenary Year International Symposium on Analysis held at Pune, December 26–28, 1987 (to appear).

6. L. C. Biedenharn, R. A. Gustafson, M. A. Lohe, J. D. Louck, and S. C. Milne, "Special Functions and Group Theory in Theoretical Physics," in *Special Functions: Group Theoretical Aspects and Applications* (R. Askey, T. H. Koornwinder, and W. Schempp, eds.), Reidel, New York, 1984, pp. 129–162.

7. W. K. Bühler, "The Hypergeometric Function—a Biographical Sketch," Math. Intelligencer **7** (1985), 35–40.

8. E. Cartan, "Sur la Detérmination d'un Système Orthogonal Complet dans un Espace de Riemann Symmétrique Clos," Rend. Circ. Mat. Palermo (2) **53** (1929), 217–252.

9. H. Corneille and A. Martin, "Constraints on the Phase of Scattering Amplitudes Due to Positivity," Nuclear Phys. **B 49** (1972), 413–440.

10. V. G. Drinfield, "Quantum Groups," in *Proc. International Congress of Mathematicians*, Berkeley, Vol. 1, 1986, Amer. Math. Soc., Providence, RI, 1988, pp. 798–820.

11.C. Dunkl, "A Krawtchouk Polynomial Addition Theorem and Wreath Products of Symmetric Groups," Indiana Univ. Math. J. **25** (1976), 335–358.

12. J. Dutka, "The Early History of the Hypergeometric Function," Arch. Hist. Exact Sci. **31** (1984), 15–34.

13. A. Erdélyi, W. Magnus, F. Oberhettinger, and F. Tricomi, *Higher Transcendental Functions, Vol. I*, McGraw Hill, New York, 1953, reprinted Krieger, Melbourne, Florida, 1981.

14. A. Erdélyi et al., *Higher Transcendental Functions, Vol. II*, McGraw Hill, New York, 1953, reprinted Krieger, Melbourne, Florida, 1981.

15. A. Erdélyi et al., *Higher Transcendental Functions, Vol. III*, McGraw Hill, New York, 1955, reprinted Krieger, Melbourne, Florida, 1981.

16. I. M. Gelfand, "General Theory of Hypergeometric Functions," Dokl. Akad. Nauk SSSR **288** (1986) (Russian), translation in Soviet Math. Dokl. **33** (1986), 573–577.

17. R. Gustafson, "The Macdonald Identities for Affine Root Systems of Classical Type and Hypergeometric Series Very-Well-Poised on Semisimple Lie Algebras" (to appear).

18. W. Hahn, "Über Orthogonalpolynome, die q-Differenzengleichungen Genügen," Math. Nachr. **2** (1949), 4–34.

19. P. R. Halmos, "Some Books of Auld Lang Syne," in *A Century of Mathematics in America, Part I* (P. Duren et al., eds.), pp. 131–174 (esp. pp. 166–168).

20. E. Jahnke and F. Emde, *Funktionentafeln mit Formeln und Kurven*, Teubner, Leipzig, 1909.

21. E. Jahnke and F. Emde, *Tables of Functions with Formulas and Curves*, fourth edition, Dover, New York, 1945.

22. M. Jimbo, "A q-Difference Analogue of $U(g)$ and the Yang-Baxter Equations," Lett. Math. Phys. **11** (1986), 247–252.

23. S. Karlin and J. McGregor, "Many Server Queueing Processes with Poisson Input and Exponential Service Times," Pacific J. Math. **8** (1958), 87–118.

24. S. Karlin and J. McGregor, "On Some Stochastic Models in Genetics, in *Stochastic Models in Medicine and Biology* (J. Gurland, ed.), Univ. of Wisconsin Press, 1964, pp. 245–271.

25. S. Karlin and G. Szegö, "On Certain Determinants Whose Elements are Orthogonal Polynomials," J. Analayse Math. **8** (1960–1961), 1–157, reprinted in [**40**, pp. 605–761].

26. T. Kimura, *Hypergeometric Functions of Two Variables*, Lecture Notes, Univ. of Minnesota, winter quarter, 1971–1972.

27. F. Klein, *The Evanston Colloquium, Lectures August 28 to September 9, 1893*, Macmillan, New York, 1911.

28. T. H. Koornwinder, "The Addition Formula for Jacobi Polynomials. I, Summary of Results," Indag. Math. **34** (1972), 188–191.

29. T. H. Koornwinder, "The Addition Formula for Little q-Legendre Polynomials and the $SU(2)$ Quantum Group" (to appear).

30. A. G. Mackie, "Arthur Erdélyi," Appl. Anal. **8** (1978), 1–10. Pages 5–10 are an addendum by R. Askey.

31. I. G. Macdonald, "Affine Root Systems and Dedekind's η-Function," Invent. Math. **15** (1972), 91–143.

32. I. G. Macdonald, "Some Conjectures for Root Systems," SIAM J. Math. Anal. **13** (1982), 988–1007.

33. W. Magnus and F. Oberhettinger, *Formeln und Sätze für die Speziellen Funktionen der Mathematischen Physik*, Springer, Berlin, 1943. English translation, Chelsea, New York, 1949.

34. S. Milne, "A q-Analog of a Whipple's Transformation for Hypergeometric Series in $U(n)$" (to appear).

35. R. L. Šapiro, "Special Functions Related to Representations of the Group $SU(n)$, of Class I with Respect to $SU(n-1)$ ($n \geq 3$)," Izv. Vyssh. Uchebn. Zaved. Mat. (71) no. 4 (1968), 97–107.

36. D. Stanton, "Orthogonal Polynomials and Chevalley Groups," in *Special Functions: Group Theoretic Aspects and Applications* (R. Askey, T. H. Koornwinder, and W. Schempp, eds.), Reidel, New York, 1984, pp. 87–128.

37. G. Szegö, *Orthogonal Polynomials*, Colloq. Publ., Vol. 23, Amer. Math. Soc., Providence, R.I., first edition 1939, second edition 1958, third edition 1967, fourth edition 1975.

38. G. Szegö, "On the Relative Extrema of Legendre Polynomials," Boll. Un. Mat. Ital. (3) **5** (1950), 120–121, reprinted in [**40**, pp. 219–220].

39. G. Szegö, "Review of Higher Transcendental Functions, Vols. 1 and 2," Bull. Amer. Math. Soc. **60** (1954), 405–408, reprinted in [**40**, pp. 435–438].

40. G. Szegö, *Collected Papers*, Vol. 3 (R. Askey, ed.), Birkhäuser, Boston, 1982.

41. J. Tannery and J. Molk, *Éléments de la Théorie des Fonctions Elliptiques*, 4 volumes, Paris, 1893, 1896, 1898, 1902, reprinted Chelsea, New York, 1972.

42. P. L. Tchebycheff, "Sur une Nouvelle Série," Bull. Phy. Math. Acad. Impériale St. Pétersbourg **17** (1858), 257–261, reprinted in *Oeuvres, 1*, Chelsea, New York, 1961, pp. 381–384.

43. P. L. Tchebycheff, "Sur l'Interpolation des Valeurs Équidistantes," Zapiski Imperatorskoi Akad. Nauk (Russia), Vol. 25, suppl. 5, 1875 (Russian), French translation in *Oeuvres, 2* Chelsea, New York, 1961, pp. 219–242.

44. C. Truesdell, "Genius and the Establishment at a Polite Standstill in the Modern University: Bateman," in C. Truesdell, *An Idiot's Fugitive Essays on Science*, Springer, New York, 1984, pp. 403–438.

45. G. N. Watson, *A Treatise on the Theory of Bessel Functions*, second edition, Cambridge Univ. Press, Cambridge, 1944.

46. E. T. Whittaker, *A Course in Modern Analysis*, Cambridge Univ. Press, Cambridge, 1902.

47. E. T. Whittaker and G. N. Watson, *A Course in Modern Analysis*, Cambridge Univ. Press, Cambridge, second edition 1915, third edition 1920, fourth edition, 1927.

48. S. L. Woronowicz, "Compact Matrix Pseudogroups," Comm. Math. Phys. **111** (1987), 613–665.

49. S. L. Woronowicz, "Twisted $SU(2)$ Groups. An Example of a Non-Commutative Differential Calculus," Publ. Res. Inst. Math. Sci. **23** (1987), 117–181.

Patrick Suppes received his Ph.D. in Philosophy in 1950 at Columbia University, where he worked with Ernest Nagel. He has been at Stanford since 1950 and is now Professor of Philosophy and Statistics. Suppes' main research interests include the philosophy of science, theory of measurement, decision theory and probability, and computer-assisted education. He is a member of the National Academy of Sciences. Jon Barwise received his Ph.D. in Mathematics in 1967 at Stanford University, where he worked with Solomon Feferman. After teaching at U.C.L.A., Yale, and Wisconsin, he returned to Stanford as Professor of Philosophy in 1983. Barwise's main research interests include mathematical logic, especially model theory, set theory, and generalized recursion theory; and applications of logic to the semantics of natural language. At Stanford, Barwise has been Director of the Center for Study of Language and Information, and is currently chairman of the Program in Symbolic Systems. Solomon Feferman received his Ph.D. in Mathematics in 1957 at the University of California, Berkeley, where he studied with Alfred Tarski. He has been at Stanford since 1956, where he is Professor of Mathematics and Philosophy. His research interests are in mathematical logic and the foundations of mathematics, especially proof theory and constructive and semiconstructive mathematics, as well as in the history of modern logic. Feferman is Editor-in-Chief of the Collected Works of Kurt Gödel; *he is currently Chairman of the Department of Mathematics.*

Commemorative Meeting for Alfred Tarski
Stanford University—November 7, 1983

PATRICK SUPPES, JON BARWISE,
AND SOLOMON FEFERMAN, SPEAKERS

INTRODUCTION

The meeting was held in memory of Professor Alfred Tarski, who died at the age of 82 on October 28, 1983. Tarski was one of the most important logicians of the twentieth century and his influence stretched over a period of more than fifty years. He established his reputation in the latter part of the

Alfred Tarski
(Photograph by Steve Givant, 1972.)

1920s through the 1930s at the University of Warsaw, came to the United States in 1939, and obtained a position in the Department of Mathematics at U.C. Berkeley in 1942, where he became a professor of mathematics in 1946. At Berkeley, Tarski established a leading center for the research and teaching of mathematical logic. Commemorative meetings were held not long after his death at both Stanford and Berkeley. The meeting at Stanford was sponsored by the Departments of Mathematics and Philosophy, where it was chaired by Professor Solomon Feferman, a former student of Tarski and a member of both departments. After some brief opening remarks, he introduced Patrick Suppes, Professor of Philosophy and Statistics, and Director of the Institute for Mathematical Studies in the Social Sciences, who told of his long experience with Tarski as a friend and colleague; Professor Suppes' remarks, which emphasized Tarski's personal characteristics and scientific style, are summarized below. He was followed by Professor Jon Barwise of the Department of Philosophy and (then) Director of the Center for the Study of Language and Information, who spoke about Tarski's work on the theory of models and model-theoretic semantics, which has been important in logic, philosophy, and linguistics. Feferman concluded the meeting with reminiscences about his experiences in the 1950s as a student of Tarski, and he enlarged on the topics of Tarski's work and interests. The talks of Barwise and Feferman are reproduced essentially as presented, though edited for this publication.

Patrick Suppes [Summary]

Sitting in on Tarski's seminars in the 1950s was a vicarious learning experience. Tarski had a passion for clarity, and he would halt seminar reports by students if at any point they failed to meet his standards. He would not let them proceed until they could present the material in a completely satisfactory, clear, and exact form; this could be very painful, though it was never a personal matter, and most students benefited by the experience.

Also to be emphasized is the elegance of his thought and talk, the strongly aesthetic feelings that came forth in his presentations and writings. He was really a dazzling lecturer and could explain technical subjects to wide audiences in a very clear and accessible way, starting with very simple ideas, gradually building up a full picture. His papers seem much simpler than they really are, because of his passion for organization and clarity. English was Tarski's fifth language (actually seventh, if one counts Greek and Latin); in Russian-occupied Poland he had studied in Russian at the classic *Gymnasium*, and also gained a command of French and German at the same time. His prewar papers were in Polish, French, and German, while he published entirely in English after his emigration to the United States. Tarski had the unusual ability to both lecture and write superbly well. He was also a pleasure to talk to in person because of the breadth of his interests, not only in mathematics, but also in philosophy, art, literature, and politics; his ideas were always well developed and articulated.

The final point to be emphasized—and which accounts for so much of Tarski's influence—was his unflagging energy and enthusiasm for work and ideas. He communicated this to his students, along with a positive attitude in stimulating a large body of work. Many of his students have gone on to become well-known in their own right, working in a variety of directions, but all carrying the distinctive stamp of Tarski's concerns for clarity and organization.

Jon Barwise

I first met Alfred Tarski in the summer of 1963. I was just on my way to graduate school here at Stanford, and was lucky enough to be able to attend the Berkeley International Model Theory Symposium. That symposium was quite an eye-opener for me. As an undergraduate I had been brought up on one view of logic; at the Berkeley symposium I found something dramatically different—namely, Tarski's kind of logic.

I met Tarski at the symposium and heard him lecture there. In fact, that was the only lecture I ever heard him give. I talked to him perhaps a half dozen times over the years, for about five minutes each. I'm far from an expert on Alfred Tarski, either personally or in terms of any kind of scholarly knowledge of his work, so I felt somewhat overwhelmed when I was asked to speak here. But upon reflection I realized that Alfred had an enormous impact on me and on my whole generation of logicians, and that this impact should not go unacknowledged here. Tarski's view of logic has changed the way all of us think about the subject. Indeed, together with Kurt Gödel and Steve Kleene, Tarski was one of the founders of modern mathematical logic. And Tarski is the person who turned model theory into a theory.

The term "mathematical logic" is rather confusing. Sometimes it is used to mean the logic of mathematics, that is, the logic of mathematical activity. Another name for this subject is "meta-mathematics." But "mathematical logic" can also be used to refer to a branch of applied mathematics—the use of mathematical tools to study logic per se, say as it arises in computer science or wherever. Then, of course, you can combine the two meanings and look at using the tools of mathematics to study the logic of mathematical activity itself. That, in fact, is the way the phrase is most commonly used: most mathematical logic consists in the use of mathematical tools to study the logic of mathematical activity. And it's that subject that I think Gödel, Kleene, and Tarski created. Gödel, working largely in isolation, made his contribution through a relatively small number of unquestionably seminal papers, papers which laid the foundations for the whole enterprise. Logicians just didn't think about things the same way after these papers. Kleene and Tarski made their contributions in other ways.

In thinking about the contributions of Kleene and Tarski, it seems to me that there is a remarkable parallel between the two men. But I would like to mention three or four aspects of this parallel.

The first is their concern for, and contributions to, conceptual analysis, that is, to the mathematical analysis of some given intuitive concept. Tarski's main tool was set theory. As an example of his contributions to conceptual analysis, probably the most famous is his work on the notion of truth. In this work, Tarski is a crucial link between logic before the thirties, and modern logic. The older logic focused on "logical systems," axioms and rules of proof. It was clear in much of the work that axioms and rules of proof were about something—that is, they were about mathematical objects. The axioms were supposed to be "true" and the rules of proof were supposed to preserve "truth." But it was Tarski who singled out this notion of truth and gave it a mathematical formulation in its own right.

It is impossible to mention Tarski's work on truth without mentioning Gödel's. For Gödel's work on the Completeness Theorem was also part of the bridge to the past. But, Gödel used the notion of truth implicitly. Tarski pulled the notion up out of the background and made it a core notion in mathematical logic.

You see, in mathematics there is the idea that a given mathematical discourse is about some particular mathematical domain, not about everything there is. Typically it is about something like the natural numbers, or the real numbers, or the elements of some field. Bringing out these domains, isolating them as objects in their own right, and developing the notion of truth in a domain, is what Tarski accomplished in his analysis.

This is the basis of model theory. Only after this piece of work can one ask the kinds of questions that Tarski asked, questions about cardinalities of models of some theory (for example, the Löwenheim-Skolem-Tarski Theorem), or about preservation of truth between different models of some theory (for example, the Łos-Tarski Theorem). It's only after you make the notion of domain and truth precise that there is any hope of proving such results. Thanks to Tarski's work on conceptual clarification, notions and results which were once very confusing to think about have achieved the ultimate compliment: they are either proved or assigned as homework exercises in every course on model theory.

This element of conceptual analysis in Tarski's work is one of the first contributions toward making any theory a branch of mathematics. The second, of course, is asking the right questions and getting answers, that is, proving results. I've mentioned two of Tarski's important results already, results which are basic tools in the tool box of any logician. But these are just two of a host of theorems due to Alfred Tarski, theorems in all parts of mathematical logic. For example, parallel to the definition of truth is his theorem on the Undefinability of Truth. On the one hand, he shows how to define truth for a mathematical structure by stepping outside that structure. But he also shows that if you try to have a single universe of mathematics, where things are either true or not, then the theory will not be adequate for defining its own notion of truth. These are two sides of truth, both of which Tarski helped explicate, through conceptual analysis, on the one hand, and through an important theorem, on the other.

Besides their work on conceptual analysis, and on proving theorems, there is a third and equally important aspect to the work of Tarski and Kleene, something that sets them apart from their contemporaries in logic. Each of them built up a school, in two senses of the word: a school of mathematical thought, and a center of research in logic. In Tarski's case, the former is model theory, the latter is the school of mathematical logic at the University of California at Berkeley. (For Kleene, it was recursion theory and logic at the University of Wisconsin.) Tarski's influence on both of those is enormous. If you look at a list of Tarski's students (which number more than 20), their students, and their students' students, and so on, you'll find an enormous number of the currently practicing logicians. And if you look across the bay at Berkeley itself, you'll find the logic group he founded still flourishing; indeed, its leadership in model theory and set theory has never been in serious jeopardy.

What made Tarski so special? What led him to become one of the three founders of modern logic? I think three things: his work on conceptual analysis, his asking (and often answering) the right mathematical questions once notions were precise, and his unselfish dedication to building up a school of logic. Lurking behind all of these we find the same thing, and another trait shared with Kleene: a boundless passion for mathematical clarity and rigor, both in his own work and of those around him. If there is anything he would want to pass on to those who follow, I think it would be that passion.

Solomon Feferman

I'll conclude this meeting with some remarks about my experiences in the fifties as a student of Tarski and subsequently as a colleague, and about my growing appreciation of his fundamental role in the development of the field of mathematical logic.

I began work as a graduate student in Berkeley in 1948 and before that, I had had just one (rather odd) course in logic. But I felt that it was a subject that I could be interested in—except that I had no idea of what one really did in studying logic, what there was to be done, or who the people were that one did it with. One reason I went to Berkeley was that I got a teaching assistantship there; I had applied to several places. I had heard dimly of Tarski. I really had no idea that I'd be coming to a place where one of the leaders of the field was working away. But there he was, and he was offering his course on metamathematics that year so I enrolled in it. I think this was an experience which regrettably happens only rather rarely, that you find right then and there what it is that you've been looking for all along. There was no question in my mind that this was the subject that I wanted to be working on, that this was the way of viewing it that I was looking for, and that this was what I would want to be following in the years to come.

Tarski was an extremely effective and powerful lecturer. Pat Suppes gave some sense of the personal passion and energy he always conveyed. If you've never seen him, I'd like to give a bit of a physical description: he was a small man, compact, with very intense eyes, balding, and a very prominent forehead with marked veins. If you have a picture of Picasso, that gives a kind of approximation to what he was like, a small person having enormous intensity and vitality, and prodigiously productive. He wrote in a very big, bold hand and spoke with a Polish accent. He was a bit old-fashioned, one had the sense; he stood out in that respect in Berkeley, as a kind of master of the old school.

Before long, I started attending seminars that Tarski gave and I worked very hard in those. I made one contribution that I spent a lot of time working up, and looking back it was probably some minor exercise on Boolean algebras, but he complimented me on my presentation, which was quite encouraging. Pretty soon it became clear to me and my fellow students that Tarski would be the one that I would want to work with for my Ph.D., though everybody said that Tarski was a very difficult person to work for. He was indeed a very demanding professor and had very high standards for his students. With all the students that did succeed under him, there are also a number who unfortunately were left by the wayside. So one had to have a certain amount of courage to enter into this course of studies, and it did take me a while to approach him. But when I did, he was very nice and made it easy for me, and said "Of course"; it seemed like quite a natural and normal thing that we would proceed in this way, and he proposed things that I would read and study. And so in the following years, besides regularly attending his seminars, I took more of his courses, including set theory and algebra and eventually became course assistant to him in some of these same courses and so followed his method of organization and development. I never ceased to be impressed with what an extraordinary lecturer he was and how he managed to start off so simply in certain ways and gradually build up, putting each brick in place to end up with a solid edifice. He seemed to have an endless fund of knowledge about all parts of logic and other fields of mathematics, particularly algebra. There was no subject that came up on which he wouldn't have some information and views. Especially in informal gatherings he loved to talk about literature and art, and politics as well, a favorite subject of his on which he had very strong feelings.

Besides Tarski's "superhuman" aspects, there was also a humanity about him, and I want to tell some little anecdotes that give a flavor of that side. He lectured frequently in Berkeley's Dwinelle Hall, many of whose rooms had small podiums; he always seemed to have an uncertain relationship with material objects and among them were these podiums which he'd constantly be backing into or almost backing off of, and one was always afraid of what would happen. But even though he'd teeter there, he never did fall off. And often, because of the forcefulness with which he wrote on the board, the chalk would explode in his hand. And then there was the business about the cigarettes. Since he was an inveterate chain smoker, he smoked while lecturing and there was always the cigarette and the chalk—and it looked like he was going to smoke the chalk and write with the cigarette! But somehow he always managed to put each one in the right place....

In the period that I was a student I saw a field being transformed in front of my eyes; it was quite amazing. Tarski's interests then were primarily in model theory, but I soon learned that he didn't just sit with one subject. While I'd be thinking, "Well, we're doing such and such these days," suddenly he'd come in and start talking about something entirely different. What I discovered was that he had a series of maybe a dozen or fifteen topics over the years that he kept circulating through, in set theory and model theory and algebra, particularly, but in other fields as well, even geometry. And he'd just go from one to another. He would work on a group of problems—and push them—by himself, with his colleagues, with students. He would see them tied up to a certain point, and then when he was satisfied with that, he'd just move on to the next thing that he had sitting around. And he'd just keep pulling things out of his desk drawers, all sorts of notes on topics that he had developed to some extent or another in the past.

But in the fifties especially, model theory was a very strong interest of his and that was a period in which the sort of things that Jon Barwise was talking about—the fundamentals of model theory—were being built up by him and his students and colleagues. Among these was Leon Henkin who came to Berkeley, and among others elsewhere, one should mention Abraham Robinson who was at the same time very influential in helping to give model theory the importance it has today. One of the main results that Tarski obtained back in the thirties but didn't publish until 1948 (with the help of J.C.C. McKinsey), was on a decision procedure for real algebra and geometry. This was really a paradigm solution to a problem of applying logic to questions of algebraic interest, and it turned out to be extremely important in various ways in model theory and applications of model theory to algebra and in computational uses of algebra today.

I want to say something in general terms about Tarski's scientific style and interests; some of this will overlap, of course, with what the previous speakers had to tell us. Besides model theory, he was interested in all kinds of algebras of logic, in Boolean algebra, relation algebras (which go back to Peirce and Schröder), and his own cylindric algebras; then there were algebras of topology, such as closure algebras, which turned out to be very useful for novel interpretations of intuitionistic logic. He wrote two books—one on cardinal algebras and one on ordinal algebras, which are less familiar and certainly not in fashion now, but which I think hold a lot of very useful and interesting material. In general, he liked the approach taken to the subject of universal algebra along the lines developed by people like Garrett Birkhoff.

In the thirties he had been particularly interested in developing metamathematics as a body of mathematical work. That is typified in his volume of selected papers, *Logic, semantics, metamathematics* (now in a paperback second edition), in which you will find besides the famous paper on the concept of truth and papers on definability, a number of papers on the calculus of systems, which I think were quite important. Also in the thirties he applied a lot of effort to set theory, particularly the role of the axiom of choice in cardinal arithmetic, and the set-theoretical structure of ideals in Boolean algebras. Finally, he kept returning over the years to the application of the method of quantifier elimination in order to obtain decision procedures for a variety of algebraic and mathematical theories. By way of complementary work, in 1953 he published the very influential *Undecidable theories*, with A. Mostowski and R.M. Robinson.

Overall in terms of describing his scientific style and his approach, the thing that I would mainly emphasize is that unlike people like Russell and Hilbert and Brouwer, he had no philosophical prejudices about the foundations of mathematics; he wanted to use mathematics fully in the development of mathematical logic, and he did this to complete advantage, by working within set theory without restriction. Now there's a curious side to this, and that concerns the question as to what his own philosophy of mathematics was. In conversation with me and others, he seemed to say that he really did not believe in set theory—that he really did not believe it was about something—and he treated it rather formally. If you read the article about Tarski in the *Encyclopedia of philosophy* by his first student Andrzej Mostowski, he brings out the same point. Mostowski says he's puzzled about this and doesn't know what to make of it but there it is and maybe eventually we'll find out. Well, I don't know if we ever will but I think what's quite amazing is that you could not tell he had that viewpoint from his own work, since here was someone who used set theory to its fullest and for all one knew, really believed everything he did with it. Maybe it was more a pragmatic attitude, but he certainly did very well with that.

On the other hand, this set theoretical emphasis limited Tarski's understanding in certain respects; he really had no feeling for proof theory and none for constructivity that I ever observed. What he did have was a very strong motivation to make logic mathematical, and at the same time to make it of interest to mathematicians. He struggled with that in many ways; sometimes he tried to force things into a certain mold that he thought would be the only way in which mathematicians would accept the material. Though it wasn't always necessarily the right way to go, one way or another he did certainly help attract the interest of mathematicians. He had a very strong feeling—I would almost call it ideological—for axiomatics and for the algebraic approach to logic. He would axiomatize and algebraicize whenever he could. And it's amazing how much of that he did. He had an extraordinary sense for rigor, exactitude, and organization, and he kept working and reworking his papers. By helping him with a few of these I could see the process he went through to bring them to final form. Yet at the same time, he was extraordinarily prolific, and his papers number in the hundreds. To get a sense of the extent of his contributions you should look at his *Collected works* published by Birkhäuser in four thick volumes; the final volume has a complete bibliography.

Also worth looking at is the volume published by the AMS of the *Proceedings of the Tarski Symposium* held in 1971 for his seventieth birthday. The variety of presentations to be found there gives a sense of how influential he was on so many people in so many directions. And also to be emphasized in terms of his influence were his energy, his drive, the fact that he kept pushing people, and particularly, that he had an enormous fund of problems to suggest and that his choice of problems was extremely good. He didn't just say, "Well, try this or that." He really thought about what the problems were that one ought to work on at a given time and how they ought to be pursued. I think almost all of them have been attacked or solved in one way or another. There are only a few that come to mind to which I think the answer is still unknown—one of them has to do with the decision problem for free groups with at least two generators, and another with the decision problem for the real field with exponentiation. Tarski was certainly instrumental in building up logic in Berkeley as one of the top centers in the world for the study of mathematical logic and in assembling there a faculty which was quite exceptional and continues to be exceptional, having leaders in many areas of mathematical logic.

Finally, Tarski was a prime mover behind a series of very important conferences. He was a leading participant in one extremely important conference in 1957 held at Cornell, which brought together people from all parts of logic. Then in 1960 we had here at Stanford the first Congress for Logic, Methodology, and Philosophy of Science that he, Pat Suppes and Ernst Nagel instituted, and that has met regularly at international points ever since. Just this last year [1983] we had the seventh such congress in Salzburg, and it is an increasingly important ongoing affair.[1] In 1963 there was a theory of models conference in Berkeley, and in 1967 an enormous set theory conference in UCLA. The sixties were a time of great development in the field of set theory and infinitary logic, to which Tarski and his students and colleagues contributed a great deal.

To conclude, I want to say that I feel Tarski was a leader in the best sense of the word. It is true that he maintained his dominance in his own school and in the group of people around him. But he did not suppress anybody; rather, he encouraged them and helped them develop the best they could offer to the field. He valued their contributions and gave everybody their proper share of the territory that they had helped to explore together.

[1] The eighth Congress of Logic, Methodology, and Philosophy of Science was held in Moscow in 1987.

Julia Bowman Robinson (1919–1985)

CONSTANCE REID WITH RAPHAEL M. ROBINSON

BIOGRAPHY

Julia Bowman Robinson was the first woman mathematician to be elected to the National Academy of Sciences and the first woman to be president of the American Mathematical Society (AMS). Her mathematical work was most often centered on the border between logic and number theory.

"I think that I have always had a basic liking for the natural numbers," she once said, recalling that her earliest memory was of arranging pebbles in the shadow of a giant saguaro on the Arizona desert, where she lived as a small child. "We can conceive of a chemistry which is different from ours, or a biology, but we cannot conceive of a different mathematics of numbers. What is proved about numbers will be a fact in any universe."

She was born Julia Bowman on December 8, 1919, in St. Louis, Missouri, the second daughter of Ralph Bowers Bowman and Helen Hall Bowman. Shortly after her second birthday, her mother died. Her father found that he had lost interest in his machine tool and equipment business, and a year later, when he remarried, he decided to retire. The family lived first in Arizona and then in San Diego.[1]

When Julia was nine years old, she contracted scarlet fever, which was followed by rheumatic fever. After several relapses she was forced to spend a year in bed at the home of a practical nurse. She had been in the fifth grade when she fell ill, and by the time she recovered she had missed two additional years of school. After a year of tutoring, she returned as a ninth grader.

[1]Constance Reid, with Raphael M. Robinson, "Julia Bowman Robinson (1919–1985), in *WOMEN OF MATHEMATICS A Biographic Sourcebook*, Louise S. Grinstein and Paul J. Campbell, eds. (Greenwood Press, Inc., Westport, CT, 1987), pp. 182–189. Copyright ©1987 by Louise S. Grinstein and Paul J. Campbell. Reprinted with permission.

She now knew that mathematics was the school subject which she liked above all others, and she persisted with it at San Diego High School in spite of the fact that by her junior year all the other girls had dropped the subject. When she graduated in 1936, she was awarded the honors in mathematics and the other sciences which she had elected to take, as well as the Bausch–Lomb medeal for all-around excellence in science.

At the age of sixteen, she entered San Diego State College, now San Diego State University. It had recently been a teachers' college and, before that, a normal school. Emphasis was still largely on preparing teachers. By this time the savings that her father had counted on to support his family in his retirement had been almost completely wiped out in the Depression of the 1930s. At the beginning of Julia's sophomore year, he took his own life. In spite of the family's straitened circumstances, she was able to continue her education, tuition at that time being only $12 a semester. When her older sister was hired as a teacher in the San Diego school system, money became available for Julia to transfer to the University of California at Berkeley for her senior year.

"I was very happy, really blissfully happy, at Berkeley," she later recalled.

> In San Diego there had been no one at all like me. If, as Bruno Bettelheim has said, everyone has his or her own fairy story, mine is the story of the ugly duckling. Suddenly, at Berkeley, I found that I was really a swan. There were lots of people, students as well as faculty members, just as excited as I was about mathematics. I was elected to the honorary mathematics fraternity, and there was quite a bit of departmental social activity in which I was included. Then there was Raphael.

"Raphael" was assistant professor R. M. Robinson, who taught the number theory course which she took during her first year at Berkeley. In the second semester there were only four students in the class—she was again the only woman—and he began to invite her to go on walks with him. In the course of these he told her about various interesting things in modern mathematics, including Kurt Gödel's results: "I was very impressed and excited by the fact that things about numbers could be proved by symbolic logic. Without question what had the greatest mathematical impact on me at Berkeley was the one-to-one teaching that I received from Raphael."

At the end of the first semester of her second graduate year at Berkeley, a few weeks after Pearl Harbor, she and Raphael Robinson were married. There was a rule at Berkeley that members of the same family could not teach in the same department. Since Julia already had a mathematics department teaching assistantship—she was teaching statistics for Jerzy Neyman—this rule did not immediately apply. Later, the prohibition did not concern her, since, now that she was married, she expected and very much wanted to have a family. In the meantime, while the United States was engaged in World War II, she and other mathematics faculty wives worked for Neyman in the Berkeley Statistical Laboratory on secret projects for the military.

When Julia finally learned that she was pregnant, she was delighted—and very disappointed when, after a few months, she lost the baby. She was then advised that because of the buildup of scar tissue in her heart (a result of the rheumatic fever), she should under no circumstances become pregnant again.

For a long time she was very depressed because she could not have children, but during the year 1946–1947, when she and Raphael were in Princeton, she took up mathematics again at his suggestion. The following year, back in Berkeley, she began to work toward a Ph.D. with Alfred Tarski, the noted Polish-born logician, who had joined the Berkeley faculty during the war. Her thesis, "Definability and decision problems in arithmetic," was accepted in June 1948.

The same year that she received her Ph.D., she began to work on the Tenth Problem on David Hilbert's famous list: to find an effective method for determining if a given Diophantine equation is solvable in integers. The problem was to occupy the largest portion of her professional career. As in the case of her thesis problem, the initial impetus came indirectly from Tarski, who had discussed casually with Raphael the problem whether, possibly using induction, one could show that the powers of 2 cannot be put in the form of a solution of a Diophantine equation. Not realizing, initially, the connection with the Tenth Problem, which she said later would have frightened her off, she began to work on solving Tarski's problem. When she found that she could not do so, she turned to related problems of existential definability.

During 1949–1950, when Raphael had a sabbatical, she worked at the RAND Corporation in Santa Monica. It was there that she solved the widely discussed "fictitious play" problem (see below). See did not, however, stop working on problems of existential definability relevant to Hilbert's Tenth Problem, and in 1950 she presented her results in a ten-minute talk at the International Congress of Mathematicians in Cambridge, Mass.

Following a frustrating and unsuccessful experience with a problem in hydrodynamics for the Office of Naval Research, she threw herself into Adlai Stevenson's presidential campaigns (1952 and 1956) and Democratic party politics for the next half dozen years.

In the summer of 1959, Martin Davis and Hilary Putnam proved a theorem which turned out to be an important lemma in the ultimate solution of the Tenth Problem. They sent a copy of their work to Julia, some of whose methods they had utilized.

"Her first move, almost by return mail, was to show how to avoid the messy analysis," Davis recalls. "A few weeks later she showed how to replace the unproved hypothesis about primes in arithmetic progression by the prime number theorem for arithmetic progressions... [She] then greatly simplified the proof, which had become quite intricate. In the published version, the proof was elementary and elegant."

By the time that the Davis-Putnam-Robinson paper appeared in 1961, she was forced by the deterioration of her heart to undergo surgery for the removal of the buildup of scar tissue in the mitral valve. After the operation her health improved dramatically. During the years that followed, she was able to enjoy many outdoor activities, particularly bicycling, which she had had to forego since childhood. She still found, however, that teaching one graduate course a quarter at Berkeley, as she did on occasion, was about all she could manage.

With Yuri Matijasevič's unexpected solution of Hilbert's Tenth Problem at the beginning of 1970 and the recognition of the crucial importance of Julia's work in the solution, many honors began to come to her. In 1975 she became the first woman mathematician to be elected to the National Academy of Sciences and, somewhat tardily, a full professor at Berkeley (with the duty of teaching just one-fourth time). In 1978 she became the first woman officer of the AMS and in 1982 its first woman president. She was also elected president of the Association of Presidents of Scientific Societies, a position she later had to decline because of ill health. In 1979 she was awarded an honorary degree by Smith College, and the following year she was asked to deliver the Colloquium Lectures of the AMS. It was only the second time a woman had been so honored (Anna Pell Wheeler* was the first, in 1927). In 1983 she was awarded a MacArthur Fellowship of $60,000 a year for five years in recognition of her contributions to mathematics. In 1984 she was elected to the American Academy of Arts and Sciences.

Even after Matijasevič's solution, Hilbert's Tenth Problem continued to pose interesting questions. She collaborated on two papers with Matijasevič, whom she had come to know personally on a 1971 trip to Leningrad. For the Symposium on Hilbert's Problems at De Kalb, Illinois, in May 1974, she also collaborated with Davis and Matijasevič on a paper concerning the positive aspects of the negative solution to the problem. It was her last published paper, the business of the AMS occupying most of her time and energy during the next decade. She was also frequently active during this period with problems of human rights.

At the 1984 summer meeting of the AMS in Eugene, Oregon, over which she was presiding, she learned that she was suffering from leukemia. After a

*Cross-reference to other women discussed in the volume is given by an asterisk following the first mention in a chapter of the individual's name.

remission of several months in the spring of the following year, she died on July 30, 1985.

WORK

Julia Robinson's dessertation was written under the direction of Alfred Tarski. He characteristically suggested many problems in class and in conversation, and she pursued those that particularly interested her. Her dissertation contained several results, the most interesting of which will be discussed here.

It follows from the work of Gödel that there can be no algorithm for deciding which sentences of the arithmetic of natural numbers are true. The sentences referred to in this context are those using the concepts of elementary logic, variables, and the operations of addition and multiplication. Since the theorem of Lagrange that every natural number is the sum of four squares can be used as a definition of natural numbers in the ring of all integers, it follows that the arithmetic of integers is also undecidable. On the other hand, Tarski had previously shown that the arithmetic of real numbers is decidable. In all three of these cases the same sentences are used; only the range of the variables is different.

The question raised by Tarski was whether the arithmetic of the rational numbers is decidable or undecidable. If an arithmetical definition of the integers in the field of rational numbers could be given, the undecidability would be proved. Such a definition was given in Julia Robinson's thesis (1949).

The first breakthrough was the observation that if M is a rational number, expressed as a fraction in lowest terms, then the denominator of M is odd if and only if $7M^2 + 2$ can be expressed as a sum of three squares of rational numbers. This follows easily from the classical result hat a natural number is the sum of three squares of integers if and only if it does not have the form $4^a(8b + 7)$.

This result led her to study the theory of quadratic forms. If one quadratic form could be used to eliminate the prime 2 from the denominator, perhaps other forms could be used to eliminate other prime factors. (If all prime factors could be eliminated from the denominator, the rational number would be an integer.) Other ternary quadratic forms were located which served this purpose. In the end the prime 2 was handled in a different way, in combination with other primes, so that the original observation does not appear in the dissertation.

There remained the problem of combining all the required conditions in one formula. It was impossible, in the language used, to describe the various quadratic forms which were needed. She resolved this difficulty by using a larger class of forms which could be described but which would not eliminate any integers.

In this way she used the theory of ternary quadratic forms in a successful attack on a problem of logic. In a later paper, she extended the result to fields of finite degree over the rationals ("The undecidability..." 1959).

Her dissertation exemplifies the fact that her main field of interest lay on the borderline between logic and number theory; however, she wrote two papers completely outside of this field. One was a small paper on statistics (1948), written before her dissertation when she was working in the Berkeley Statistical Laboratory. The other was an important paper on game theory (1951), written when she was working at the RAND Corporation. This latter paper solved one of a list of problems for which RAND had offered monetary prizes (although as an employee she was not eligible for the prize).

George W. Brown had proposed a method of finding the value of a finite two-person zero-sum game, sometimes called the method of fictitious play. Two players are imagined as playing an infinite sequence of games, using in each game the pure strategy which would yield the optimal payoff against the accumulated mixed strategy of the opponent. Brown noted that the value of the game lay between these optimal payoffs for the two players and conjectured that they would converge to the value of the game as the number of plays increased. Julia's paper, "An iterative method of solving a game," verified Brown's conjecture. It is still considered a basic result in game theory.

Several of her papers played an essential role in the negative solution of Hilbert's Tenth Problem, which asked for an algorithm to decide whether a Diophantine equation has a solution. The first of these was "Existential definability in arithmetic" (1952). The problem studied was whether various sets are existentially definable in the arithmetic of natural numbers. The set of composite numbers is existentially definable, but at the time it was not known whether the set of primes is, as was later established. In this paper she proved that binomial coefficients, factorials, and the set of primes are existentially definable in terms of exponentiation, and that exponentiation in turn is existentially definable in terms of any function of roughly exponential growth.

At the time these results seemed somewhat fragmentary, but they took on added importance after the publication of a joint paper with Davis and Putnam (1961). In this paper it is proved that every recursively enumerable set is existentially definable in terms of exponentiation. It follows that there is no algorithm for deciding whether an exponential Diophantine equation (that is, a Diophantine equation in which exponentiation as well as addition and multiplication is allowed) has a solution in natural numbers. In view of her earlier proof that exponentiation is existentially definable in terms of any function of roughly exponential growth, the negative solution of Hilbert's Tenth Problem was reduced to finding an existential definition of such a function. That was finally done by Matijasevič at the beginning of 1970.

Later she collaborated with Matijasevič (1975) in proving that there is no algorithm for deciding whether a Diophantine equation in thirteen variables has a solution in natural numbers. (Matijasevič has since reduced the number of variables to nine.)

Among her other works are two papers dealing with general recursive functions (1950, 1968), as well as one on primitive recursive functions (1955) and one on recursively enumerable sets (1968). The 1950 paper on general recursive functions was her first paper after the dissertation. In it she starts from the characterization of general recursive functions as those obtained by adjoining the μ-rule to the rules used to obtain primitive recursive functions, and then asks what restrictions can be placed on the defining schemes. One result is the proof that all general recursive functions of one variable can be obtained from two special primitive recursive functions (one of which is rather complicated) by composition and inversion. In the later paper, she showed that this same class of functions can be obtained from the zero and successor functions by composition and a new scheme which she calls general recursion.

Other papers include one giving an expository treatment of the class of hyperarithmetical functions (1967) and one giving a finite set of axioms for number-theoretic functions from which the Peano axioms can be derived (1973).

Her Colloquium Lectures, delivered in 1980, have not been published. The first, which was introductory, discussed Gödel's work and the concept of computability. The second dealt with work related to Hilbert's Tenth Problem and included a new proof, due to Matijasevič, of the undecidability of exponential Diophantine equations. The third treated the decision problem for various rings and fields; and the fourth, nonstandard models of arithmetic.

Bibliography

Works by Julia Bowman Robinson

Mathematical Works

"A note on exact sequential analysis." *University of California Publications in Mathematics* (N.S.) 1 (1948): 241–246.

"Definability and decision problems in arithmetic." *Journal of Symbolic Logic* 14 (1949): 98–114.
Doctoral Thesis.

"General recursive functions." *Proceedings of the American Mathematical Society* I (1950): 703–718.

"An iterative method of solving a game." *Annals of Mathematics* 54 (1951): 296–301.

"Existential definability in arithmetic." *Transactions of the American Mathematical Society* 72 (1952): 437–449.

"A note on primitive recursive functions." *Proceedings of the American Mathematical Society* 6 (1955): 667–670.

"The undecidability of algebraic rings and fields." *Proceedings of the American Mathematical Society* 10 (1959): 950–957.

"Problems of number theory arising in metamathematics." *Report of the Institute in the Theory of Numbers*, 303–306. Boulder, Colo.: 1959.

(with Martin Davis and Hilary Putnam) "The decision problem for exponential Diophantine equations." *Annals of Mathematics* 74 (1961): 425–436.

"On the decision problem for algebraic rings." *In Studies in Mathematical Analysis and Related Topics: Essays in Honor of George Polya*, edited by Gabor Szegö et al., 297–304. Stanford, Calif.: Stanford University Press, 1962.

"The undecidability of exponential Diophantine equations." *In Logic, Methodology, and Philosophy of Science: Preceedings of the* 1960 *International Congress*, edited by E. Nagel. P. Suppes, and A. Tarski, 12–13. New York: North-Holland, 1963.

"Definability and decision problems in rings and fields." In *The Theory of Models*, edited by J. W. Addison et al., 299–311. New York: North-Holland, 1965.

"An introduction to hyperarithmetical functions." *Journal of Symbolic Logic* 32 (1967): 325–342.

"Recursive functions of one variable." *Proceedings of the American Mathematical Society* 19 (1968): 815–820.

"Finite generation of recursively enumerable sets." *Proceedings of the American Mathematical Society* 19 (1968): 1480–1486.

"Diophantine decision problems." In *Studies in Number Theory*, MAA Studies in Mathematics, vol. 6 (1969), pp. 76–116.

"Finitely generated classes of sets of natural numbers." *Proceedings of the American Mathematical Society* 21 (1969): 608–614.

"Unsolvable Diophantine problems." *Proceedings of the American Mathematical Society* 22 (1969): 534–538.

"Hilbert's Tenth Problem." In *Proceedings of the* 1969 *Summer Institute on Number Theory...*, edited by Donald J. Lewis, 191–194. Proceedings of Symposium in Pure Mathematics, vol. 20, Providence, R.I.: American Mathematical Society, 1971.

"Solving Diophantine equations." *Proceedings of the Fourth International Congress for Logic, Methodology and Philosophy of Science*, edited by Patrick Suppes et al., 63–67. New York: North-Holland, 1973.

"Axioms for number theoretic functions." In *Selected Questions of Algebra and Logic*, edited by A. I. Shirshov et at., 253–263. Novosibirsk: Izdat. "Nauka" Sibirsk. Otdel., 1973.

(with Yuri Matijasevič) "Two universal three-quantifier representations of enumerable sets" (in Russian). *Theory of Algorithms, and Mathematical Logic* (in Russian), edited by B. A. Kushner and N. M. Nagornyi, 112–123, 216. Moscow: Vychisl. Centr Akad. Nauk SSSR, 1974.

(with Yuri Matijasevič) "Reduction of an arbitrary Diophantine equation to one in 13 unknowns." *Acta Arithmetica* 27 (1975): 521–553.

(with Martin Davis and Yuri Matijasevič) "Hilbert's 10th Problem. Diophantine equations: Positive aspects of a negative solution." In *Mathematical Developments Arising from Hilbert Problems*, edited by Felix F. Browder, 323–378 + loose erratum. Proceedings of Symposia in Pure Mathematics, vol. 28. Providence, R.I.: American Mathematical Society, 1976.

Works about Julia Bowman Robinson

Gaal, Lisl. "Julia Robinson's thesis." *Association for Women in Mathematics Newsletter* 16 (3) (May–June 1986): 6–8.

"Julia Bowman Robinson: 1919–1985." *Notices of the American Mathematical Society* 32 (1985): 739–742.

Obituary. *New York Times* (2 August 1985): D–15.

Reid, Constance. "The autobiography of Julia Robinson." *College Mathematics Journal* 17 (1986): 2–21.

Smoryński, C. "Julia Robinson, *In Memoriam*." *The Mathematical Intelligencer* 8 (2) (1986): 77–79.

John Wermer received his Ph.D. at Harvard in 1951, and his thesis advisor was George Mackey. His mathematical interests were strongly influenced by Arne Beurling, with whom he studied at Harvard and in Sweden. He taught at Yale from 1951 to 1954, and since then he has been at Brown. He has worked on operator theory, Banach algebras and complex function theory.

Function Algebras in the Fifties and Sixties[1]

JOHN WERMER[2]

1. Introduction

This essay is a very personal survey of a chapter of mathematical history in which I participated, the study of Function Algebras in the U.S. in the period 1950–1970. For obvious reasons the survey is very incomplete, as is the bibliography. For a balanced view of the subject the interested reader can consult three excellent works: *Introduction to Function Algebras* by A. Browder, W. A. Benjamin, Inc. (1969), *Uniform Algebras* by T. W. Gamelin, Prentice Hall, Inc. (1969), and *The Theory of Uniform Algebras* by E. L. Stout, Bogden and Quigley, Inc. (1971).

Starting in the early 1950s a band of American mathematicians went to work on some questions in complex analysis which came from two sources: the theory of polynomial approximation on compact sets in the complex plane, and the theory of commutative Banach algebras. The American mathematicians included Richard Arens at UCLA, Charles Rickart at Yale, Ken Hoffman and Iz Singer at MIT, Andy Gleason at Harvard, Hal Royden at Stanford, Errett Bishop at Berkeley, Irv Glicksberg at the University of Washington, Walter Rudin at Rochester and the University of Wisconsin, and the author at Brown. They and their students began to develop a theory of Function Algebras which formed a new link between classical Function Theory and Functional Analysis. Their inspiration came largely from the Soviet Union.

[1] A good discussion of many of the topics of this article, as well as a very extensive bibliography, is given in the article by G. M. Henkin and E. M. Čirka, *Boundary Properties of Holomorphic Functions of Several Complex Variables*, Plenum Publishing Corporation (1976).

[2] I am grateful to Andy Browder and Peter Duren for helpful comments for this article.

In the 1940s I. M. Gelfand and his coworkers had built a theory of commutative Banach algebras in which they had shown that such an algebra, if it has a unit and its radical is zero, is isomorphic to an algebra $\mathfrak{A}$ of continuous complex-valued functions on a compact Hausdorff space $\mathfrak{M}$. The points of $\mathfrak{M}$ are identified with the maximal ideals of $\mathfrak{A}$. G. Šilov had shown that among all closed subsets of $\mathfrak{M}$ there exists a smallest set $\check{S}$ with the property that if m is in $\mathfrak{M}$, then for each f in $\mathfrak{A}$

$$|f(m)| \leq \max |f(x)| \text{ taken over } \check{S}.$$

$\check{S}$ is called the *Šilov boundary* of the algebra.

A simple model for this is given by the *disk algebra* $A(D)$ consisting of all functions which are analytic in the open unit disk: $|z| < 1$ and continuous in the closed disk D: $|z| \leq 1$. Here the maximal ideal space $\mathfrak{M}$ can be identified with D and the Šilov boundary with the unit circle: $|z| = 1$. The natural norm on $A(D)$ is given by $\|f\| = \max |f(z)|$, taken over D.

The question arises: let $\mathfrak{A}$ be an arbitrary semi-simple commutative Banach algebra with unit, such that $\check{S}$ is nontrivial, i.e., $\check{S}$ is strictly smaller than $\mathfrak{M}$. Does there exist an *abstract function theory* for $\mathfrak{A}$, i.e., do the functions in $\mathfrak{A}$ behave on $\mathfrak{M}\backslash\check{S}$ like analytic functions (as in the example of the disk algebra)? Furthermore, does $\mathfrak{M}\backslash\check{S}$ possess *analytic structure*, i.e., can we find subsets of $\mathfrak{M}\backslash\check{S}$ which can be made into complex manifolds on which the functions in $\mathfrak{A}$ are analytic? If enough such analytic structure could be shown to exist, this would explain the Šilov boundary in terms of the maximum principle of analytic function theory.

In 1952 a brilliant achievement by the Soviet Armenian mathematician S. N. Mergelyan provided a second source of inspiration. Mergelyan showed in **[48]** that if X is a compact set in the z-plane $\mathbb{C}$ such that $\mathbb{C}\backslash X$ is connected, then every function which is continuous on X and analytic on the interior of X can be uniformly approximated on X by polynomials in z. This result can be read as a statement about a certain Banach algebra. We let $P(X)$ denote the uniform closure on X of the polynomials in z and we put on $P(X)$ the supremum norm over X. Then $P(X)$ is a Banach algebra, the maximal ideal space $\mathfrak{M}$ coincides with X, and the Šilov boundary $\check{S}$ coincides with the topological boundary of X. Mergelyan's theorem yields that a function φ defined and continuous on $\mathfrak{M}$ belongs to $P(X)$ if and only if φ is analytic on $\mathfrak{M}\backslash\check{S} = \text{int}(X)$ in the natural analytic structure which $\text{int}(X)$ inherits from $\mathbb{C}$.

2. Uniform Algebras

For the problems mentioned above, of constructing an abstract function theory for $\mathfrak{A}$ and of exhibiting analytic structure on $\mathfrak{M}\backslash\check{S}$, it seemed natural to take the norm on the algebra $\mathfrak{A}$ to be a uniform norm. The "Function

Algebras" to be studied where then as follows: we fix a compact Hausdorff space X and an algebra $\mathfrak{A}$ of continuous functions on X such that $\mathfrak{A}$ is closed in the algebra $C(X)$ of all continuous functions on X, contains the constants, and separates the points of X. If we put on $\mathfrak{A}$ the uniform norm over X, $\mathfrak{A}$ is then a Banach algebra. $\mathfrak{M}$ is a compact space in which X lies embedded, as proper subset in general, and $\check{S}$ is a closed subset of X.

Such algebras were baptised *uniform algebras* by Errett Bishop in 1964. He thought the name sounded good, and it has stuck. One says that $\mathfrak{A}$ is a uniform algebra *on* X. Uniform algebras are plentiful in nature. Here are some examples:

(i) Let Y be a compact set in $\mathbb{C}^n$, the space of n complex variables. Let $P(Y)$ denote the uniform closure on Y of polynomials in the complex coordinates $z_1, \ldots, z_n$. Then $P(Y)$ is a uniform algebra on Y.

The disk algebra is a special case. For $n = 1$ and so $Y \subset \mathbb{C}$, Mergelyan's theorem tells us which functions belong to $P(Y)$.

(ii) Let Σ be a finite Riemann surface with boundary $\partial\Sigma$ and denote by $A(\Sigma)$ the algebra of functions continuous on Σ and analytic on $\Sigma \backslash \partial\Sigma$. $A(\Sigma)$ is a uniform algebra on Σ.

(iii) Let K be a compact set in $\mathbb{C}$ and let $R_0(K)$ denote the space of rational functions whose poles lie in $\mathbb{C}\backslash K$. Let $R(K)$ denote the uniform closure of $R_0(K)$ on K. Then $R(K)$ is a uniform algebra on K.

(iv) Let H^∞ denote the algebra of all bounded analytic functions on the open unit disk. By Fatou's theorem, H^∞ is embedded in L^∞ of the unit circle, and L^∞, in turn, is isomorphic to $C(X)$ for a (complicated) space X. H^∞ is a uniform algebra on X.

(v) The Stone–Weierstrass theorem yields that the only uniform algebra on a compact space X which is closed under complex conjugation is the full algebra $C(X)$.

A first indication that it might be possible to do abstract function theory on a uniform algebra A was the proof that *representing measures* always exist. By a representing measure for a point m in $\mathfrak{M}$ is meant a probability measure μ on the Šilov boundary $\check{S}$ such that for all f in A

$$f(m) = \int f\, d\mu.$$

Arens and Singer in **[5]** and John Holladay in his Yale thesis (1953) proved that such a μ exists.

In the case of the disk algebra $A(D)$, μ is unique for a given m and is the Poisson measure on the circle, corresponding to m. In general, μ is far from unique.

A representing measure μ is multiplicative on A, i.e.,

$$\int fg\,d\mu = \left(\int f\,d\mu\right)\cdot\left(\int g\,d\mu\right) \quad \text{for all } f, g \text{ in } A,$$

and conversely, each multiplicative probability measure is the representing measure for some point m in $\mathfrak{M}$.

In 1953 in **[64]** Šilov made another fundamental contribution to Banach algebra theory by introducing the use of analytic functions of several complex variables into the theory. Let $\mathfrak{A}, \mathfrak{M}$ be as above. Suppose that $\mathfrak{M}$ is disconnected, i.e., $\mathfrak{M} = \mathfrak{M}_1 \cup \mathfrak{M}_2$ where $\mathfrak{M}_1, \mathfrak{M}_2$ are disjoint closed sets. Šilov showed that $\exists$ e in $\mathfrak{A}$ with $e^2 = e$ such that $e = 1$ on $\mathfrak{M}_1$ and $e = 0$ on $\mathfrak{M}_2$.

Not long after, Arens and Calderon in **[4]** and L. Waelbroeck in **[68]** developed a functional calculus for analytic functions of n variables acting on n-tuples of elements of a commutative Banach algebra.

Another application of several complex variables to Banach algebra theory was the algebraic description of the first cohomology group of the maximal ideal space, independently by R. Arens in **[3]** and H. Royden in **[58]**. They showed that for $\mathfrak{A}, \mathfrak{M}$ as above, $H^1(\mathfrak{M}, \mathbf{Z})$ is isomorphic to the quotient group of the group of units of $\mathfrak{A}$ by the subgroup of elements $\exp(y)$ with y in $\mathfrak{A}$.

3. Gleason's Program

Andrew Gleason launched the earliest attacks on the problem of analytic structure in the maximal ideal space of a uniform algebra.

In the case of the disk algebra $A(D)$ those maximal ideals m corresponding to an interior point of the disk, say the point a, have the algebraic property of being *simply generated*: every f in the ideal m can be written in the form: $f = g(z-a)$ with g in $A(D)$. Maximal ideals corresponding to boundary points of D are not simply generated. Gleason obtained the following striking result: *Let A be a uniform algebra and fix m in $\mathfrak{M}$. Suppose that the ideal m is finitely generated in the algebraic sense. Then some neighborhood U of m in $\mathfrak{M}$ can be given the structure of an analytic variety such that every h in A is analytic on U.*

He lectured on this result in the mid-fifties, and published it in **[29]**.

In another direction, Gleason observed the following: with $A, \mathfrak{M}$ as before, let m_1, m_2 be two points in $\mathfrak{M}$. Then $|f(m_1) - f(m_2)| \leq 2$ whenever f belongs to the unit ball of A. It may happen that there exists $k < 2$ such that $|f(m_1) - f(m_2)| \leq k$ whenever f belongs to this unit ball. In the case of the disk algebra, this occurs whenever m_1 and m_2 lie in the open unit disk. This suggests the following general definition: for m_1, m_2 in $\mathfrak{M}$, put $m_1 \sim m_2$ whenever $\exists$ such a $k < 2$, or, in other words, whenever the distance from m_1 to m_2 in the dual Banach space of A is less than 2. Gleason showed that $\sim$ is an equivalence relation on $\mathfrak{M}$. (Since $2+2=4$, the transitivity of the relation $\sim$ is not evident!) He called the equivalence classes under $\sim$ *the parts of* $\mathfrak{M}$.

For the case of the disk algebra $A(D)$, the open unit disk is one part and each point on the unit circle is a one-point part. For the case of the bi-disk algebra $A(D^2)$ which consists of all functions which are continuous on the closed bi-disk $D^2 = \{|z| \leq 1\} \times \{|w| \leq 1\}$ in $\mathbb{C}^2$ and analytic on the open bi-disk, the maximal ideal space $\mathfrak{M} = D^2$, and the parts are as follows: the open bi-disk is one part, each disk: $z = z_0, |w| < 1$ and each disk: $|z| < 1, w = w_0$ with $|z_0| = 1$ and $|w_0| = 1$ is a part; the remaining parts are the one-point parts on the distinguished boundary $\{|z| = 1\} \times \{|w| = 1\}$ of D^2. Thus the parts here are complex manifolds of dimensions 2, 1, and 0.

Gleason lectured on these ideas, [**28**], at the Conference on Analytic Functions at the Institute for Advanced Study in Princeton in September, 1957. This was a marvelous meeting. The people there interested in Banach algebras included R. Arens, R. C. Buck, L. Carleson, A. Gleason, K. Hoffman, S. Kakutani, Lee Rubel, H. Royden, I. Kaplansky, L. Waelbroeck, and myself. Many of the giants of function theory gave talks, both on one and several complex variables, and tolerated those of us who didn't know much about either one or several complex variables. The two weeks of the conference were for us enormously stimulating and provided the germ of much later work on Function Algebras.

Kakutani had studied H^∞ as a Banach algebra, and reported on his work in [**41**]. At the conference, he discussed the boundary behavior of a bounded analytic function in terms of normed ring theory, [**42**].

Earlier, Kakutani had raised the following basic question about H^∞ as a ring: the open unit disk is naturally embedded as an open subset Δ of the maximal ideal space $\mathfrak{M}$ of H^∞, and so its closure $\overline{\Delta}$ is contained in $\mathfrak{M}$. The set $\mathfrak{M}\backslash\overline{\Delta}$ was called the "Corona".

Is the Corona empty, i.e., is Δ dense in $\mathfrak{M}$? Suppose that the answer is "Yes" and consider an n-tuple of functions f_j in H^∞ with $\sum_{j=1}^n |f_j| \geq \delta$ on Δ, where δ is a positive constant. Then $\sum_{j=1}^n |f_j| \geq \delta$ on $\mathfrak{M}$ and so the f_j have no common zero on $\mathfrak{M}$. Hence the ideal generated by the f_j is contained in no maximal ideal of H^∞ and so is the whole ring. It follows that there exist g_j in H^∞, $j = 1, \dots, n$, such that

$$\sum_{j=1}^n f_j g_j = 1.$$

The problem of the existence of the g_j under the given assumption on the f_j turned out to be a very deep problem. This "Corona problem" was solved by Lennart Carleson in [**22**], and it follows that the Corona is indeed empty. Carleson's result and his method of proof has had a major impact on analysis. All this is treated in John Garnett's book mentioned in Section 7 below.

A breakthrough in the understanding of the maximal ideal space of H^∞ occurred at the conference, in the form of the birth of I. J. Schark, [**62**]. Schark's paper exhibited analytic structure in $\mathscr{M}\backslash\Delta$ for the first time. Schark never published again, since his name was put together from the initials of participants at the conference. So Schark did not perish; he vanished.

In his talk, Gleason formulated the following *Conjecture*: *Let m_1, m_2 be two points in the maximal ideal space $\mathfrak{M}$ of a uniform algebra. Then a necessary and sufficient condition for m_1 and m_2 to be in the same part of $\mathfrak{M}$ is that m_1 and m_2 can be connected by a finite chain of analytic images of the unit disk, contained in $\mathfrak{M}$.* A second idea Gleason introduced in [**28**] was the notion of a *Dirichlet Algebra.* The real parts of the functions belonging to a uniform algebra A on a space X can be viewed as "harmonic" on $\mathfrak{M}\backslash X$, as can uniform limits on $\mathfrak{M}$ of sequences of such functions. Gleason called A a *Dirichlet Algebra on X* if every real continuous function on X is the restriction to X of such a harmonic function, or, equivalently, if the real parts of functions in A form a uniformly dense subspace of the real continuous functions on X.

The disk algebra $A(D)$ may be viewed as a uniform algebra on the circle $|z| = 1$, with norm the supremum norm on the circle, rather than as a uniform algebra on the disk. $A(D)$ is a Dirichlet algebra on the circle.

Gleason wrote in [**28**] about Dirichlet algebras: "It appears that this class of algebras is of considerable importance and is amenable to analysis." It turned out subsequently that this preliminary judgment was right on target. At the time, in September 1957, Gleason's ideas were sufficiently strange and novel that I (and many of us, I imagine) did not fully grasp their significance.

4. The Summer of 1959 in Berkeley

In the summer of 1959 a lot of people working on Functional Analysis gathered, rather informally, in Berkeley. My wife Kerstin and I took our two boys, two and five years old, put them in our Chevy and drove across the country. It had been hot when we left the East Coast and got steadily hotter as we drove west until suddenly, as we came into Berkeley, a discontinuity occurred and we were in a cool and lush paradise, the sky blue, the air balmy, and all garden flowers blooming wildly.

I had along with me a recent paper by Henry Helson and David Lowdenslager, [**33**], in which they studied certain spaces of functions given by Fourier series on the torus. Earlier, Arens and Singer in [**6**], and Mackey in [**47**], had given a group-theoretic approach to analytic functions, based on the following observation: A Fourier series $f(x) = \sum_n c_n e^{inx}$ on the unit circle is the boundary function of a function analytic in the unit disk if and only if $c_n = 0$ for $n < 0$. Replacing the circle by the torus, one may consider Fourier series $f(\vartheta, \varphi) = \sum_{n,m} c_{nm} e^{in\theta} e^{im\varphi}$ in two variables. One specifies a half-plane S in the lattice $\mathbf{Z}^2$ and regards f to be "analytic", relative to S, if $c_{nm} = 0$ outside of S. An interesting example is obtained by taking S to be the set of points (n, m) in $\mathbf{Z}^2$ with $n + m\alpha \geq 0$, where α is a fixed irrational number. Helson and Lowdenslager showed in **[33]** that a series of classical boundary value theorems of function theory have counterparts for functions "analytic relative to S". Their results were dramatic and their proofs made elegant use of L^2-methods. Their paper stirred Solomon Bochner's interest, as he had looked at related questions at an earlier time. He showed that their proofs depended only on two properties: first, that for fixed S the class of S-analytic functions continuous on the torus is an algebra, and second, that the real parts of the functions in this algebra are dense in the real continuous functions on the torus. The group structure on the torus entered only through these properties. So Bochner, quite independently of Gleason, was led to the same Dirichlet algebras **[19]**. Thus it turned out that certain basic results about boundary-functions of analytic functions in the disk remain true, when properly stated, for an arbitrary Dirichlet algebra. How does this look?

For the case of the disk algebra, the measure $\frac{1}{2\pi}\, d\theta$ is the representing measure for the origin. For $p \geq 1$,the Hardy space H^p is defined as the closure of $A(D)$ in L^p on the circle with respect to this measure. Let now A, on X, be a Dirichlet algebra and fix m in $\mathfrak{M}$. Let μ be the unique representing measure for m on X, for the algebra A. We define $H^p(\mu)$ as the closure of A in $L^p(X, \mu)$. For f in $H^p(\mu)$, $f(m)$ is defined as $\int f\, d\mu$. One then has, for instance, the following:

THEOREM 1. *Let A, m, μ be as above. Fix a nonnegative function w on X which is summable with respect to μ. A necessary and sufficient condition for w to have a representation*

$$w(x) = |f(x)|^2 \text{ a.e.-}d\mu \text{ on } X$$

for some f in $H^2(\mu)$ with $f(m) \neq 0$ is that

$$\int \log w \cdot d\mu > -\infty.$$

THEOREM 2. *Let W be a closed subspace of $H^2(\mu)$ invariant under multiplication by elements of A, i.e., such that $f\varphi \in W$ whenever $\varphi \in W$ and $f \in A$. Assume also that 1 is not orthogonal to W. Then there exists a bounded function E_0 in W with $|E_0(x)| = 1$ a.e.-$d\mu$ such that*

$$W = \{E_0 g | g \in H^2(\mu)\}.$$

Theorems 1 and 2, in the case when A is the disk algebra, are classical results of, respectively, Szegö and Beurling.

When I realized, in Berkeley, how all these things fitted together I got quite excited. John Kelley and Errett Bishop had been studying Dirichlet algebras, and tutored me in the subject, and I also had the benefit of talking to Helson about his work with Lowdenslager. So I was able to prove the truth of Gleason's conjecture about parts, for the case of Dirichlet algebras, in the following form: *Let A be a Dirichlet algebra, $\mathfrak{M}$ its maximal ideal space and P a part of $\mathfrak{M}$. Then either P is a single point, or P is an analytic disk, i.e., P is the one-one image of the disk $|\lambda| < 1$ by a continuous map ψ such that $h \circ \psi$ is analytic on $|\lambda| < 1$ for each h in A* **[72]**.

When we left Berkeley to go home at the end of August, we ran into several people at gas stations and so on, whom we had met upon arriving, who had noted the Rhode Island plates on our car and had told us that they themselves came from the East. When they realized we were going back, they were amazed: "You've seen California and you're going back East!" they said. My five-year-old son said, "Let's go home to America!" (meaning Providence, Rhode Island).

5. Errett Bishop and the General Theory of Uniform Algebras

Dirichlet algebras were almost too good to be true. The general uniform algebra is much less tractable, largely due to the nonuniqueness of representing measures for fixed points m in $\mathfrak{M}$. However, a series of results about general uniform algebras was discovered, with important applications to many questions in analysis. In this general theory, the unquestioned leader was Errett Bishop. Bishop was on the faculty at Berkeley from 1954 to 1965 and then on the faculty of the University of California at San Diego until his untimely death in 1983.

He was one of the most remarkable people I have known. He was a mathematician of amazing insight and penetration, absolutely fearless and with a profound commitment to mathematics. In his last years he was somewhat isolated in the mathematical community, because of his absolute dedication to constructive methods in mathematics.

In the period about which I am writing, Bishop's work and personal contact with him was enormously stimulating to the rest of us, and led to much work by other people, both jointly with him and independently of him. There was the famous joint work by Bishop and Karel de Leeuw on the Choquet boundary and by Bishop and Phelps on Banach spaces. Stolzenberg and Bishop worked closely together on polynomially convex hulls, as did Rossi and Bishop on problems about complex manifolds. My own work on analytic structure in maximal ideal spaces, e.g. in the joint paper [7] with Aupetit, and work on the same problem by Gamelin in **[27]**, grew out of Bishop's rich paper **[15]**. And so on.

Here I can only mention a few of Bishop's contributions to the general theory of function algebras. The interested reader is referred to **[16]**, **[18]**,

[56] for more extensive discussions of his work. Further, his collected papers appear in **[76]**.

(i) *Peak points.* A point x_0 in X is called a *peak point* for the uniform algebra A on the space X if $\exists\ f$ in A with $f(x_0) = 1$ and $|f(x)| < 1$ on $X\backslash\{x_0\}$. In **[11]** Bishop showed that for X metrizable peak points exist and the set M of all peak points is a *minimal boundary* for A in the sense that for each g in A there is some point x in M with $g(x) = \|g\|$, and, of course, no proper subset of M has this property. It follows that M is a dense subset of the Šilov boundary $\check{S}$. Moreover, if m is a point of $\mathfrak{M}$ there exists a representing measure for m which lies on M.

The existence of peak points had earlier been observed by Gleason (unpublished).

Bishop was able to apply the notion of peak point to the problem of rational approximation. Let X be a compact subset of $\mathbb{C}$. As in Section 2 above, we write $R(X)$ for the uniform closure on X of those rational functions which are analytic on X. When does $R(X) = C(X)$? Clearly, for this to happen the interior of X must be empty. When X has connected complement in $\mathbb{C}$, Mergelyan's theorem shows that this is also sufficient. However, in general the condition is not sufficient, and to show this Mergelyan in 1952 in **[48]** constructed the following set S: remove from the closed unit disk $|z| \leq 1$ a countable family of disjoint open disks: $|z - a_j| < r_j$, $j = 1, 2, \ldots$ such that $\sum_j r_j < \infty$, and denote by S the closed set that remains. By Cauchy's theorem, the complex measure dz on the union of the circles $|z - a_j| = r_j$ together with the unit circle annihilates $R(S)$, and hence $R(S) \neq C(S)$. If A_j, r_j are chosen so that the interior of S is empty, we have the desired example. For obvious reasons, S is called a *Swiss Cheese.* It turned out that, in fact, Mergelyan had *rediscovered* the Swiss Cheese; in 1938 the Swiss mathematician Alice Roth had given such an example. The Swiss Cheese has been very useful to people constructing counterexamples in the study of Function Algebras. My colleague Bob Accola told me that Function Algebras *is* the study of the Swiss Cheese, but this is not strictly correct.

Let now X be an arbitrary compact subset of $\mathbb{C}$. Bishop showed the following: $R(X) = C(X)$ *if and only if each point x in X is a peak point for the algebra* $R(X)$. An extension of this result was found by Donald Wilken in **[74]**. A peak point is always a one-point part of the maximal ideal space. For $R(X)$ the maximal ideal space is precisely X. Wilken showed that *each part of X is either a one-point part, or has positive 2-dimensional Lebesgue measure.*

(ii) *The antisymmetric decomposition.* If A is a uniform algebra on X, a subset Y of X is called a *set of antisymmetry* if every function in A which is real-valued on Y is constant on Y. As example we may take X to be the solid cylinder $\{|z| \leq 1\} \times \{0 \leq t \leq 1\}$ and A to be the algebra of all continuous functions on X which are analytic on each slice: $t = t_0$, $|z| < 1$. Then each disk: $t = t_0$, $|z| \leq 1$ is a set of antisymmetry. For a general uniform algebra A on X Bishop showed in **[13]**: *Let* $\{Y_\alpha\}$ *be the family of all maximal sets of antisymmetry. Then the* Y_α *give a closed partition of* X *and a continuous function* f *on* X *belongs to* A *if and only if each restriction* $f|_{Y_\alpha}$ *belongs to the restriction* $A|_{Y_\alpha}$.

If the Y_α are the points of X, one recovers the Stone–Weierstrass theorem. A less complete result had been obtained earlier by Šilov, **[63]**. Bishop's result reduces the study of general uniform algebras to the study of *antisymmetric* such algebras, i.e., uniform algebras which contain no nonconstant real-valued function.

(iii) *Jensen measures.* The representing measure $\frac{d\theta}{2\pi}$ for the origin for the disk algebra $A(D)$ satisfies Jensen's inequality:

$$\log|f(0)| \leq \int_0^{2\pi} \log|f(e^{i\theta})| \frac{d\theta}{2\pi}$$

for each f in $A(D)$.

Let A be a uniform algebra and fix m in $\mathfrak{M}$. Can a representing measure μ be found for m which satisfies such an inequality? Arens and Singer had shown this to be true in certain cases. In **[15]** Bishop showed it in general: *Let* A *be a uniform algebra,* m *a point of* $\mathfrak{M}$. *There exists a representing measure* μ *for* m *such that*

$$\log|f(m)| \leq \int \log|f| \, d\mu$$

for each f *in* A.

Such a measure μ is called a *Jensen measure* for m. Jensen measures have turned out to be very useful.

6. Irving Glicksberg and Orthogonal Measures

Let A be a uniform algebra on the space X. A complex measure ν on X is called *orthogonal* to A if

$$\int f \, d\nu = 0 \text{ for every } f \text{ in } A.$$

We write $A^\perp$ for the family of all such measures. If we know $A^\perp$, then we can tell, using the Hahn–Banach theorem, whether a given function h in $C(X)$ belongs to A: $h \in A$ if and only if

$$\int h \, d\nu = 0 \text{ for each } \nu \text{ in } A^\perp.$$

The classical theorem of F. and M. Riesz identified all the orthogonal measures for the disk algebra. Frank Forelli's work in **[25]** gave a function-algebraic approach to this result.

In a series of papers **[30]**, **[31]**, **[32]**, the last jointly with me, Glicksberg analyzed measures orthogonal to a uniform algebra. He applied his results to obtain elegant new proofs of Bishop's results on general uniform algebras, as well as to problems in interpolation, approximation, and so forth.

In **[10]** and **[12]** Bishop considered a compact set X in $\mathbb{C}$ with $\mathbb{C}\backslash X$ connected and looked at the measures ν on X orthogonal to the algebra $P(X)$. He showed that such a measure ν always arises from a certain analytic differential $g(z)dz$ on the interior of X. In **[32]** Glicksberg and I adapted these ideas to Dirichlet algebras. Let A be a Dirichlet algebra on a space X. For each m in $\mathfrak{M}$, let λ be the representing measure for m and let $H^1(\lambda)$, as earlier, denote the closure of A in $L^1(X, \lambda)$. If $k \in H^1(\lambda)$ and $\int k \cdot d\lambda = 0$, then $k \cdot d\lambda$ is orthogonal to A, since if f is in A,

$$\int f(kd\lambda) = \left[\int f \cdot d\lambda\right]\left[\int k \cdot d\lambda\right] = 0.$$

Hence we get "obvious" orthogonal measures for A by forming convergent series

$$\sum_i k_i \cdot d\lambda_i$$

where each λ_i is a representing measure and $k_i \in H^1(\lambda_i)$ and $\int k_i d\lambda_i = 0$. We showed that every complex measure ν in $A^\perp$ has a representation

$$\nu = \sum_i k_i \cdot d\lambda_i + \sigma, \tag{1}$$

with k_i, λ_i as above and such that σ is orthogonal to A and is singular with respect to every representing measure for A.

As an application, we took a compact plane set X with connected complement and boundary ∂X, and took $A = P(X)$. By the classical Walsh–Lebesgue theorem, $P(X)$ is a Dirichlet algebra on ∂X. In this case, one can show that every measure σ appearing in (1) vanishes. (1) then quickly implies Mergelyan's theorem on polynomial approximation on X mentioned earlier.

Lennart Carleson in **[23]** gave an ingenious new proof of the Walsh–Lebesgue theorem, and went on to give a proof of Mergelyan's theorem, also based on Bishop's ideas.

Irv Glicksberg was an unusual person. He was ever cheerful, with unlimited enthusiasm and unfailing generosity. He enjoyed every bit of good mathematics that he met up with, and it usually stimulated new ideas in him. He was a delightful, indefatigable correspondent, a fanatic photographer, and fond of jaunty headgear. Politically, he was a staunch liberal, and so he found plenty to get mad about in the last twenty years. On most other questions he had a tolerant point of view.

I was planning to spend a year at the University of Washington with him in 1983, when I was shocked to hear of his death.

7. Function Algebras at MIT and at Brown

In the late fifties and early sixties, Iz Singer and Ken Hoffman presided over a very fruitful mathematical activity at MIT. Their students during that time included Andrew Browder, Hugo Rossi, and Gabriel Stolzenberg, each of whom made important contributions to the study of Function Algebras.

Hoffman and Singer jointly in **[38]** answered a series of questions on Function Algebras which had been posed by Gelfand. In **[37]** and **[39]** they studied *maximal* uniform algebras on a space X, i.e., algebras A such that if B is a closed subalgebra of $C(X)$ which contains A, then either $B = A$ or $B = C(X)$.

A major open problem at that time was to prove a *local maximum modulus principle* for function algebras. If z_0 is a point in the domain of analyticity of a function F and U is a neighborhood of z_0, then $|F(z_0)| \leq \max |F|$ taken over the boundary of U. The corresponding statement for a uniform algebra A with maximal ideal space $\mathfrak{M}$ should be this: fix m in $\mathfrak{M}$ and let U be a neighborhood of m whose closure lies in $\mathfrak{M}\backslash\check{S}$. Then $|f(m)| \leq \max |f|$ taken over the boundary of U, whenever $f \in A$. *Is this true*? We all tried to prove this, but, lacking insight into several complex variables, we had no luck. At last Hugo Rossi showed how to do it. I remember the excitement of a late evening phone call, Singer to Rossi when Rossi was in Princeton, where he told us about his proof. The secret was a clever use of the solution of the Cousin problem in n complex variables. Rossi's paper on this is **[55]**, in 1960. Much of what has been found about uniform algebras since then has depended on this local maximum modulus principle.

The local maximum modulus principle, as well as various examples of maximal ideal spaces which had been worked out in the meantime, as well as Gleason's conjecture about parts, all encouraged an effort to prove, in general, the existence of analytic structure in $\mathfrak{M}\backslash\check{S}$. One way to test this question was to look at examples in n complex variables. Let X be a compact set in $\mathbb{C}^n$ and let $P(X)$ be defined as in example (i) in Section 2 above. The maximal ideal space $\mathfrak{M}$ of the uniform algebra $P(X)$ has a natural identification with the so-called *polynomially convex hull* $\widehat{X}$ of X, which had come up in the 1930s in the work of K. Oka, **[51]** and **[52]**. $\widehat{X}$ consists of all points $z^0 = (z_1^0, \dots, z_n^0)$ in $\mathbb{C}^n$ such that

$$|Q(z^0)| \leq \max |Q| \text{ over } X$$

for every polynomial Q on $\mathbb{C}^n$.

If analytic structure exists on $\mathfrak{M}\backslash\check{S}$, then there must be complex analytic varieties contained in $\widehat{X}\backslash X$. Gabriel Stolzenberg in the winter of 1960–1961 constructed a set X on the boundary of the bi-disk: $|z_1| \leq 1$, $|z_2| \leq 1$ in $\mathbb{C}^2$ such that neither one of the coordinate projections $z_1(\widehat{X})$ and $z_2(\widehat{X})$ contains any open subset of the plane, while at the same time $\widehat{X}$ contains the point $(0, 0)$ and hence is larger than X. Then $\widehat{X}$ contains no analytic variety, for else $\widehat{X}$ would contain some proper analytic disk Δ and then either $z_1(\Delta)$ or $z_2(\Delta)$ would have nonvoid interior [**65**].

Thus the hope for analytic structure in $\mathfrak{M}\backslash\check{S}$ in the general case was gone forever. It was a heavy blow. From the perspective of today, almost thirty years later, I should say that Stolzenberg's example taught us that the story of polynomially convex hulls is much subtler than we had thought, but that some satisfactory understanding of these hulls is starting to emerge at the present time.

Stolzenberg himself made other incisive studies of polynomially convex hulls in the sixties, in [**66**] and [**67**].

In addition to the people just mentioned, MIT had in this period a number of junior faculty and academic visitors working on Function Algebras and related matters. These included Stephen Fisher, Ted Gamelin, John Garnett, Eva Kallin, and Donald Wilken. There was lively interaction between the Analysis Seminar at Brown, run by Andy Browder and myself, and these MIT people. Hoffman and Singer were good friends of mine, and much of my own work arose from conversations with them and others of the group.

Once, after Ken Hoffman and I had finished a particularly long-lasting and noisy mathematical conversation at my home in Providence, my three-and-a-half-year-old son came into the room, waving his arms and spouting a stream of nonsense syllables. "I am talking mathematics!" he told us.

Among the Ph.D. students working on Function Algebras who wrote their theses at Brown were Andy Browder and Robert McKissick, both borrowed from MIT; further Mike Voichick, John O'Connell, Bernie O'Neill, and Richard Basener (my students), Al Hallstrom, Jim Wang, and Kenny Preskenis (Browder's students), and Tony O'Farrell (Brian Cole's student). Stu Sidney (Gleason's student) and Lee Stout (Rudin's student) were part of this same mathematical generation, as were Mike Freeman and Laura Kodama, Bishop's students. H. S. Bear (John Kelley's student) is in this group, and Barney Weinstock (Hoffman's student) came somewhat later. Larry Zalcman was an MIT graduate student in this period. He wrote the volume *Analytic Capacity and Rational Approximation* [**75**], which gave a very valuable exposition in English of the recent work of Vituškin and his school on the algebra $R(X)$, example (iii) in Section 2 above.

Andy Browder joined the Brown department in 1961. One result of Browder's thesis concerned the topology of *polynomially convex sets*, i.e., sets X which coincide with their polynomially convex hull: *let X be such a compact set in $\mathbb{C}^n$.* Then the kth Čech cohomology of X with complex coefficients vanishes for $k \geq n$ [**20**]. It follows in particular that if Y is a compact orientable n-manifold in $\mathbb{C}^n$, then $\widehat{Y}$ is larger than Y. Identifying the set of "new points" $\widehat{Y}\backslash Y$ has turned out to be a difficult problem, only partially solved even for 2-manifolds in $\mathbb{C}^2$.

Eva Kallin joined the Brown faculty in 1965. B. Weinstock and G. Stolzenberg also taught at Brown for some years in the sixties. Kallin had written her thesis at Berkeley, with J. L. Kelley, and in it she had solved the following famous problem: *if a function belongs locally to a uniform algebra A, must it belong to A*? More precisely, if A is a uniform algebra and if a function f continuous on $\mathfrak{M}$ has the property that each point m in $\mathfrak{M}$ has a neighborhood U such that $f|_U = F|_U$, for some F in A, does then f belong to A? Kallin [**43**] gave a counterexample. G. Šilov who had earlier published an erroneous proof of the result, sent her a congratulatory postcard. Another result of Kallin's concerned the "n balls problem": consider n closed disjoint balls $B_1, \ldots, B_n$ in $\mathbb{C}^N$. Is their union polynomially convex? For $n = 1$ or 2 one sees at once that the answer is "Yes". For $n = 4$ the answer is unknown as of today. Kallin showed in [**44**] that the answer is "Yes" for $n = 3$.

One other major line of research at MIT at that time was Hoffman's work on the algebra H^∞ of bounded analytic functions on the unit disk. H^∞ is a uniform algebra. Its maximal ideal space, $\mathfrak{M}(H^\infty)$, is as mysterious a compact space as an analyst is likely to encounter. In his book *Banach Spaces of Analytic Functions*, which was published in 1962 by Prentice-Hall, Hoffman devoted Chapter 10 to H^∞ as a Banach algebra.

That book as a whole was a milestone. It showed to the world of classical analysts and to the world of functional analysts that they were brothers and sisters rather than strangers (as many had thought). One source of this recognition was for Hoffman, as it was for myself and many others, the towering figure of Arne Beurling who in his own work had combined classical and abstract analysis in essentially new ways.

One observation which Hoffman made was that H^∞ is *almost*, but not quite, a Dirichlet algebra on its Šilov boundary $\check{S}(H^\infty)$. For a uniform algebra A on a space X write $\log|A^{-1}|$ for the space of all functions $\log|f|$ such that f and f^{-1} both belong to A. Hoffman called A *logmodular* if $\log|A^{-1}|$ is uniformly dense in the real continuous functions on X. Dirichlet algebras are logmodular (trivially), but not conversely. Logmodular algebras still enjoy the property that representing measures for points in $\mathfrak{M}$ are unique. H^∞ is a logmodular algebra on $\check{S}(H^\infty)$. In [**35**] Hoffman developed the theory of logmodular algebras and showed that they enjoyed almost all the pleasant properties of Dirichlet algebras. In particular, their Gleason parts were either points or analytic disks. This last result raised the question of describing the Gleason parts of H^∞ explicitly. In the paper [**36**] Hoffman solved this very difficult problem, making use of the deep work of Lennart Carleson in [**21**] and Donald Newman in [**49**].

H^∞ as a Banach algebra and, in particular, as a subalgebra of L^∞ on the circle has, since Hoffman's work, been the subject of intensive investigation. This theory is closely connected with the modern theory of bounded linear operators on a Hilbert space. An exposition of the work on H^∞ from a function-theoretic point of view is found in John Garnett's book *Bounded Analytic Functions*, published by Academic Press in 1981. In particular, this book treats the Corona Problem, mentioned in Section 3 above, including the remarkable new solution of the problem by Tom Wolff.

A further development in the abstract direction came in the work of Lumer in **[46]**, where Lumer makes as his only hypothesis on a uniform algebra the uniqueness of the representing measures for the points of $\mathfrak{M}$. Other extensions of this theory are given by P. Ahern and D. Sarason in **[1]** and by K. Barbey and H. König in **[8]**.

8. Yale

I taught at Yale from 1951 to 1954 and at Brown after that. Yale provided a superb environment for a young analyst. The senior people in analysis, Rickart, Kakutani, Dunford, and Hille were very active, friendly and encouraging, and the junior people, Jack Schwartz, Henry Helson, Bill Bade, Bob Bartle, Frank Quigley, and myself had a very lively time in the analysis seminar. We all taught calculus, Math 12, out of Ed Begle's book, which is based on the axioms of the real number system. The combination of axioms, Yalies, and ourselves made a heady brew. Our wives were sick of conversations about Math 12 which went on at all department parties. The normal teaching evaluation which each of us got from our freshmen was, "While undoubtedly a brilliant mathematician, Mr. X just can't get it across".

Rickart early on saw the possibilities of an abstract function theory in his papers **[53]**, **[54]**, etc., and through the work of his students. Talking with him and with his student John Holladay got me to thinking about Function Algebras. One day in early 1953 Rickart showed to Kakutani and me a recent paper by the Russian mathematician Leibenson, **[45]**, in which Leibenson raised the following question: Let Γ denote the unit circle $|z| = 1$ and let A denote the disk algebra, viewed as a subalgebra of $C(\Gamma)$. Suppose φ is a function in $C(\Gamma)$ which is not in A. Is the closed algebra generated by φ and A then all of $C(\Gamma)$? He showed that it was if φ is real or if φ satisfies a Lipschitz condition.

Some months earlier I had heard about an intriguing recent result of Rudin: given an algebra of functions continuous in the closed disk $|z| \leq 1$. *Suppose every F in the algebra attains the maximum of its modulus on the boundary $|z| = 1$. Then if one schlicht function belongs to the algebra, every F in the algebra is analytic in $|z| < 1$.*

I did not then know Rudin's proof and spent a week of hard work, making up a proof of Rudin's theorem. My proof was function-algebraic in spirit and rather more complicated than Rudin's own, in **[60]**. When I saw Leibenson's question, I realized that I could use similar function-algebraic arguments to answer it. I showed that *every closed subalgebra of* $C(\Gamma)$ *which contains* A *either equals* A *or equals* $C(\Gamma)$.

I now asked myself what other closed subalgebras of $C(\Gamma)$ have this property of being "maximal" in $C(\Gamma)$. Let us consider a simple closed curve γ on a Riemann surface which is a torus, such that γ bounds a region Ω on this torus. Then the boundary functions on γ of all functions analytic on Ω and continuous on $\Omega \cup \gamma$ make up such a maximal subalgebra of $C(\gamma)$. Also $C(\gamma) \cong C(\Gamma)$. Since Ω need not be of the type of the disk, we have a new maximal subalgebra of $C(\Gamma)$. It was clear then that one should prove that if Σ is any finite Riemann surface with nice boundary $\gamma = \partial\Sigma$, then the algebra $A(\Sigma)$ of functions analytic on $\Sigma \backslash \partial\Sigma$ and continuous on Σ is a maximal subalgebra of $C(\gamma)$. With the kind help of Maurice Heins at Brown, I proved this for the case that $\partial\Sigma$ is a single contour in **[70]**. Hal Royden proved the general case in **[57]**.

The algebra considered by Leibenson, generated by φ and A on Γ, evidently is generated by the two functions: φ and z. Let now φ and ψ be any two functions continuous on Γ which together separate points on Γ, and denote by $[\varphi, \psi]$ the closed subalgebra of $C(\Gamma)$ which they generate. Of course, $[\varphi, \psi]$ may equal $C(\Gamma)$, e.g. if $\varphi = \bar{z}$, $\psi = z$. Suppose that $[\varphi, \psi]$ is a proper subalgebra of $C(\Gamma)$. Then we might expect that Γ lies embedded as the boundary curve $\partial\Sigma$ of some finite Riemann surface Σ such that φ and ψ extend analytically from $\partial\Sigma$ to Σ. In that case $[\varphi, \psi]$ would contain only boundary functions of functions analytic on Σ, and hence be a proper subalgebra of $C(\Gamma)$.

I badly wanted to prove that this is what happens. In the case that φ and ψ are real-analytic on Γ and hence can be viewed as defined and analytic in a little annulus containing Γ, I finally did prove it by the end of 1956, in **[71]**.

One can look at this question geometrically, by considering the image X of Γ in $\mathbb{C}^2$ under the map (φ, ψ). Then X is a simple closed curve in $\mathbb{C}^2$ and the hypothesis that $[\varphi, \psi] \neq C(\Gamma)$ is equivalent to the statement that $P(X) \neq C(X)$. The desired conclusion, the existence of a finite Riemann surface in which Γ is embedded, then becomes the existence of a finite Riemann surface Σ in $\mathbb{C}^2$ having X as its boundary.

In this language, and replacing $\mathbb{C}^2$ by $\mathbb{C}^n$, the problem is then as follows. *Given a simple closed curve* X *in* $\mathbb{C}^n$ *with* $P(X) \neq C(X)$. *Show that there exists a finite Riemann surface* Σ *in* $\mathbb{C}^n$ (*possibly admitting singular points*) *which has* X *as its boundary*. One expects that the finite Riemann surface Σ equals $\widehat{X}$, the polynomially convex hull of X. All this turned out to be true, as long as the curve X has some regularity. I proved it when X is a single real-analytic curve, Stolzenberg did the case when X is the union of finitely many differentiable closed curves **[67]**, and Herbert Alexander **[2]** did the case when X is merely rectifiable. Bishop's ideas in **[14]** and **[15]** played an important role in this work.

One application of this theory of function algebras on the circle came in the work of Royden in **[59]** on the maximum principle for bounded analytic functions on an open Riemann surface.

Suppose now that we replace the circle Γ by the unit interval I and study the closed subalgebras of $C(I)$ which are uniform algebras on I. The corresponding geometric problem in $\mathbb{C}^n$ is to identify the polynomially convex hull of a Jordan arc in $\mathbb{C}^n$. When J is a regular Jordan arc, satisfying the same smoothness conditions we imposed on the closed curve X above, it turned out that $\widehat{J} = J$, i.e., J is polynomially convex. We expect this, since intuitively we feel that "J cannot bound anything". Furthermore, when J is a regular Jordan arc, $P(J) = C(J)$, i.e., every continuous function on J is a uniform limit on J of polynomials in $z_1, \dots, z_n$. The proof uses both the fact that J is polynomially convex, and that J is smooth, and was given by H. Helson and J. Quigley, in greater generality, in **[34]**.

However, when J is merely topologically a Jordan arc, i.e., homeomorphic to the interval I, J may fail to be polynomially convex. Examples of this were given by me for $n = 3$, **[69]**, and by Rudin for $n = 2$, **[61]**.

The Peak Point Conjecture and Cole's Thesis. As we saw in Section 5 above, Bishop had shown, for an arbitrary compact plane set X, that $R(X) = C(X)$ whenever each point of X is a peak point of $R(X)$. The Peak Point Conjecture was the statement that if A is a uniform algebra on X and if every point of $\mathfrak{M}$ is a peak point of A (in which case, of course, $\mathfrak{M}$ and X coincide), then $A = C(X)$. A related conjecture, due to Gleason, was the statement that $C(X)$ is characterized as a uniform algebra on X by the fact that each part of its maximal ideal space is a single point.

During the 1960s many people tried to settle these conjectures without success. In his remarkable thesis at Yale in 1968, Brian Cole (Rickart's student) disproved both of these conjectures. His procedure was to make repeated adjunction of square roots to a given uniform algebra A so as to end up with an algebra $\widetilde{A}$ which is such that every function in $\widetilde{A}$ has a square root in $\widetilde{A}$. The proof given by Cole may be found in the appendix to A. Browder's book mentioned above in Section 1. Cole's thesis settled a series of other questions as well, and stimulated much further work.

In particular, Richard Basener at Brown was able to modify Cole's construction so as to obtain a compact set X lying on the sphere $|z_1|^2 + |z_2|^2 = 1$ in $\mathbb{C}^2$ such that $R(X)$ provides another counterexample to the peak point conjecture. Here $R(X)$ denotes the uniform closure on X of rational functions in z_1, z_2 which are analytic on X.

Brian Cole joined the Brown department in 1969.

9. Function Algebras on Smooth Manifolds

Let X be a compact set in $\mathbb{C}^n$. Under what conditions on X does $P(X) = C(X)$? Since the maximal ideal space of $C(X)$ is X and the maximal ideal space of $P(X)$ is $\widehat{X}$, a necessary condition for this is that X be polynomially convex.

Suppose now that X is a compact smooth manifold in $\mathbb{C}^n$ with or without boundary, and that X is polynomially convex. Does it follow that $P(X) = C(X)$? As we saw in Section 8, the answer is "Yes" in the case that X is a circle or an arc.

Let k denote the real dimension of X. For $k \geq 2$, it is clear that a new condition enters. If for instance Y is the 2-dimensional disk: $z_1 = \lambda$, $z_2 = \lambda$, $|\lambda| \leq 1$ in $\mathbb{C}^2$, then $\widehat{Y} = Y$ and $P(Y)$ contains exclusively functions analytic on Y. To rule out such a situation, one may consider the tangent space T_x to X for each point x in X. T_x is a k-dimensional real subspace of $\mathbb{C}^n$. If X is a complex-analytic manifold, as in the example, or if merely X contains a complex-analytic submanifold passing through x, this will show up by the presence of a *complex-linear subspace of* $\mathbb{C}^n$ in T_x.

We call such a subspace a *complex tangent* to X at x, and we call the manifold X *totally real* if it has no complex tangents. In 1968–1969 a breakthrough occurred. It was shown, under various conditions of smoothness on X, and arbitrary k, that *if* Σ *is a smooth totally real manifold in* $\mathbb{C}^n$ *and* X *is a compact and polynomially convex subset of* Σ, *then* $P(X) = C(X)$.

The real subspace $\mathbb{R}^n$ of $\mathbb{C}^n$, consisting of all points $(x_1, \dots, x_n)$ with all x_j real, is evidently a smooth totally real manifold, and every compact subset of $\mathbb{R}^n$ is polynomially convex. So one recovers the Weierstrass approximation theorem.

The above theorem was proved in R. Nirenberg and R. O. Wells, [**50**], L. Hörmander and J. Wermer, [**40**], and E. M. Čirka, [**24**], and the method of proof in these papers was based on Hörmander's solution of the $\overline{\partial}$-problem. Much further work on this problem, with weakened smoothness conditions and simpler, more elementary proofs, was done later on by Weinstock, Berndtsson, Harvey and Wells, and others.

I had earlier, in [**73**], proved the result for the case $k = 2$ when X is a 2-dimensional smooth disk, and M. Freeman, in [**26**], had settled the case of general smooth 2-manifolds. The method used by myself and by Freeman depended on the use of the Cauchy transform of a plane measure, and did not generalize to the case $k > 2$.

Bishop disks. Suppose that Σ is a smooth manifold which does have complex tangents. What then? If the dimension k of $\Sigma > n$, elementary linear algebra shows that Σ has complex tangents at every point. In his paper *Differentiable Manifolds in Complex Euclidean Space* in [**17**], Errett Bishop showed the following: *Assume $k > n$. Fix x in Σ and assume that the dimension of the largest complex-linear subspace of T_x is $k - n$. Then if U is any neighborhood of x on Σ, there exists an analytic disk in $\mathbb{C}^n$ whose boundary lies in U.*

Suppose now that X is a compact set lying on Σ which contains some open subset of Σ. It follows that $\widehat{X}$ contains a multitude of analytic disks whose boundaries lie in X. Every function in $P(X)$ is then analytic on each of these disks.

These "Bishop disks" have turned out to play an important role in the study of analytic continuation in several complex variables.

10. Hanover, N.H., Palo Alto, New Orleans, Yerevan

Many of us got together in the summers at a succession of meetings devoted at least in part to Function Algebras. The atmosphere was rather relaxed some of the time. The conference at Dartmouth College in Hanover was held in 1960. It was organized by Terry Mirkil et al. and was supported by the NSF. One weekday during the meeting Matt Gaffney showed up from Washington, representing the NSF, to see how the conference was going. At the Dartmouth math department he found *none* of the mathematicians, only one of the wives, looking for her husband. The rest of us were out in the lovely countryside. I myself was with Karel de Leeuw and Siggi Helgason on a sailboat on a lake. There were no dire consequences.

In 1961 there was a one-month conference in analysis at Stanford, under the auspices of the American Mathematical Society, and many of us were there and gave talks.

A conference fully devoted to Function Algebras was held at Tulane University in April 1965, organized by Frank Birtel et al. Most people interested in the subject attended, and a volume of the proceedings was published, *Function Algebras*, edited by F. Birtel, Scott-Foresman and Co. (1966).

In September, 1965 a number of us went to a big conference on analytic functions in the Soviet Union in Yerevan. We had a chance to meet and talk with many of the Russians who had similar interests, and I was very much struck by the warmth and friendliness of our hosts. The group around Shabat, Vitushkin, and Mergelyan was very active and doing fundamental work in approximation theory. It included E. Gorin, A. Gončar, E. M. Čirka, S. Melnikov, and E. P. Dolzhenko.

In those days the Russians were not party to an international copyright agreement, so they could freely translate foreign books into Russian. When an author came to the Soviet Union, he got his royalty for the translation in rubles. In this way Hoffman and I got some rubles. They had translated an article of mine so I got enough for one bottle of Armenian cognac and one fur hat. Hoffman's book had been translated, so he was amply supplied with rubles for cognac.

Gončar threw a party for many of us at his family home in Yerevan. The party was very high-spirited with singing, piano-playing and a greater density of liquor bottles on the table than I have ever seen. Of course many toasts were drunk. My toast to Mergelyan was "on your beautiful work which has inspired us all". V. P. Havin responded, with a toast to Mergelyan: "Your work has inspired not only the Americans."

Bibliography

[1] P. Ahern and D. Sarason, "The H^p spaces of a class of function algebras," *Acta. Math.* **117** (1967), 123–163.

[2] H. Alexander, "Polynomial approximation and hulls in sets of finite linear measure in $\mathbb{C}^n$," *Amer. J. Math.* **93** (1971), 65–74.

[3] R. Arens, "The group of invertible elements of a commutative Banach algebra," *Studia Math.* **1** (1963), 21–23.

[4] R. Arens and A. Calderon, "Analytic functions of several Banach algebra elements," *Ann. of Math.* (2) **62** (1955), 204–216.

[5] R. Arens and I. M. Singer, "Function values as boundary integrals," *Proc. Amer. Math. Soc.* **5** (1954), 735–745.

[6] ——, "Generalized analytic functions," *Trans. Amer. Math. Soc.* **81** (1956), 379–393.

[7] B. Aupetit and J. Wermer, "Capacity and uniform algebras," *J. Funct. Anal.* **28** (1978), 386–400.

[8] H. Barbey and H. König, *Abstract analytic function theory and Hardy algebras*, Lecture Notes in Math, no. 593, Springer-Verlag, Berlin and New York, 1977.

[9] F. Birtel (ed.), *Function algebras*, Scott, Foresman & Co., Glenview, IL, 1966.

[10] E. Bishop, "The structure of certain measures," *Duke Math. J.* **25** (1958), 283–289.

[11] ——, "A minimal boundary for function algebras," *Pacific J. Math.* **9** (1959), 629–642.

[12] ——, "Boundary measures of analytic differentials," *Duke Math. J.* **27** (1960), 331–340.

[13] ——, "A generalization of the Stone-Weierstrass theorem," *Pacific J. Math.* **11** (1961), 777–783.

[14] ——, "Analyticity in certain Banach algebras," *Trans. Amer. Math. Soc.* **102** (1962), 507–544.

[15] ——, "Holomorphic completions, analytic continuations, and the interpolation of seminorms," *Ann. of Math.* (2) **78** (1963), 468–500.

[16] ——, *Uniform Algebras*, Proc. Conf. on Complex Analysis, Minneapolis, 1964 (A. Aeppli et al., eds.), Springer, Berlin, 1965, pp. 272–281.

[17] ——, "Differentiable manifolds in complex Euclidean space," *Duke Math. J.* **32** (1965), 1–21.

[18] Murray Rosenblatt (ed.), *Errett Bishop: Reflections on him and his research*, Contemp. Math., Vol. **39**, 1985.

[19] S. Bochner, "Generalized conjugate and analytic functions without expansions," *Proc. Nat. Acad. Sci.* **45** (1959), 855–857.

[20] A. Browder, "Cohomology of maximal ideal spaces," *Bull. Amer. Math. Soc.* **67** (1961), 515–516.

[21] L. Carleson, "An interpolation problem for bounded analytic functions," *Amer. J. Math.* **80** (1958), 921–930.

[22] ——, "Interpolation by bounded analytic functions and the corona problem," *Ann. of Math.* (2) **76** (1962), 547–559.

[23] ——, "Mergelyan's theorem on uniform polynomial approximation," *Math. Scand.* **15** (1964), 167–175.

[24] E. M. Čirka, "Approximation by holomorphic functions on smooth manifolds in $\mathbb{C}^n$," *Mat. Sb. (N.S.)* **78** (120) (1969), 101–123.

[25] F. Forelli, "Analytic measures," *Pacific J. Math.* **13** (1963), 571–578.

[26] M. Freeman, *Some conditions for uniform approximation on a manifold*, Function Algebras (F. Birtel, ed.), Scott, Foresman & Co., 1966.

[27] T. W. Gamelin, *Polynomial approximation on thin sets*, Lecture Notes in Math., no. 184, Springer, Berlin, 1971, pp. 50–78.

[28] A. Gleason, *Function algebras*, Sem. Analytic Functions, Vol. II, Inst. Adv. Study, Princeton, N.J., 1957.

[29] ——, "Finitely generated ideals in Banach algebras," *J. Math. Mech.* **13** (1964), 125–132.

[30] I. Glicksberg, "Measures orthogonal to algebras and sets of antisymmetry," *Trans. Amer. Math. Soc.* **105** (1962), 415–435.

[31] ——, "The abstract F. and M. Riesz theorem," *J. Funct. Anal.* **1** (1967), 109–122.

[32] I. Glicksberg, and J. Wermer, "Measures orthogonal to Dirichlet algebras," *Duke Math. J.* **30** (1963), 661–666.

[33] H. Helson and D. Lowdenslager, "Prediction theory and Fourier series in several variables," *Acta. Math.* **99** (1958), 165–202.

[34] H. Helson and F. Quigley, "Existence of maximal ideals in algebras of continuous functions," *Proc. Amer. Math. Soc.* **8** (1957), 115–119.

[35] K. Hoffman, "Analytic functions and logmodular Banach algebras," *Acta. Math.* **108** (1962), 271–317.

[36] ——, "Bounded analytic functions and Gleason parts," *Ann. of Math.* (2) **86** (1967), 74–111.

[37] K. Hoffman, and I. M. Singer, "Maximal subalgebras of $C(\Gamma)$," *Amer. J. Math.* **79** (1957), 295–305.

[38] ——, "On some problems of Gelfand," *Uspekhi Mat. Nauk* **14** (1959), 99–144.

[39] ——, "Maximal algebras of continuous functions," *Acta. Math.* **103** (1960), 217–241.

[40] L. Hörmander and J. Wermer, "Uniform approximation on compact sets in $\mathbb{C}^n$," *Math. Scand.* **23** (1968), 5–21.

[41] S. Kakutani, *Rings of analytic functions*, Lectures on Functions of a Complex Variable (W. Kaplan, ed.), Ann Arbor, MI, Univ. Michigan Press, Ann Arbor, MI, 1955.

[42] ——, *On rings of bounded analytic functions*, Sem. Analytic Functions, Vol. II, Inst. Adv. Study, Princeton, N.J., 1957.

[43] E. Kallin, "A nonlocal function algebra," *Proc. Nat. Acad. Sci. U.S.A.* **49** (1963), 821–824.

[44] ——, *Polynomial convexity: The three spheres problem*, Proc. Conf. on Complex Analysis, Minneapolis, 1964 (A. Aeppli et al., eds.) Springer, Berlin, 1965, pp. 301–304.

[45] Z. L. Leĭbenzon, "On the ring of continuous functions on a circle," *Uspekhi. Mat. Nauk.* (*N.S.*) **7** (1952), 163–164.

[46] G. Lumer, "Analytic functions and Dirichlet problem," *Bull. Amer. Math. Soc.* **70** (1964), 98–104.

[47] G. Mackey, "The Laplace transform for locally compact Abelian groups," *Proc. Nat. Acad. Sci. U.S.A.* **34** (1948), 156–162.

[48] S. N. Mergelyan, *Uniform approximations to functions of a complex variable*, Amer. Math. Soc. Transl. No. **101**, 1954; reprint, Amer. Math. Soc. Transl. (1) **3** (1962), 294–391.

[49] D. J. Newman, "Some remarks on the maximal ideal structure of H^∞," *Ann. of Math.* (2) **70** (1959), 438–445.

[50] R. Nirenberg and R. O. Wells, "Approximation theorems on differentiable submanifolds of a complex manifold," *Trans. Amer. Math. Soc.* **142** (1969), 15–35.

[51] K. Oka, *Sur les fonctions analytiques de plusieurs variables.* I: *Domaines convexes par rapport aux fonctions rationnelles*, J. Sci. Hiroshima Univ. **6** (1936), 245–255.

[52] ——, *Sur les fonctions analytiques de plusieurs variables.* II: *Domaines d'holomorphie*, J. Sci. Hiroshima Univ. **7** (1937), 115–130.

[53] C. Rickart, "Analytic phenomena in general function algebras," *Pacific J. Math.* **18** (1966), 361–377.

[54] ——, "Holomorphic convexity for general function algebras," *Canad J. Math.* **20** (1968), 272–290.

[55] H. Rossi, "The local maximum modulus principle," *Ann. of Math.* (2) **72** (1960), 1–11.

[56] ——, Review of *Selected Papers of Errett Bishop*, Bull. Amer. Math. Soc. **19** (1988), 538–545.

[57] H. Royden, "The boundary values of analytic and harmonic functions," *Math. Z.* **78** (1962), 1–24.

[58] ——, "Function algebras," *Bull. Amer. Math. Soc.* **69** (1963), 281–298.

[59] ——, "Algebras of bounded analytic functions on Riemann surfaces," *Acta. Math.* **114** (1965), 113–142.

[60] W. Rudin, "Analyticity and the maximum modulus principle," *Duke Math. J.* **20** (1953), 449–457.

[61] ——, "Subalgebras of spaces of continuous functions," *Proc. Amer. Math. Soc.* **7** (1956), 825–830.

[62] I. J. Schark, "Maximal ideals in an algebra of bounded analytic functions," *J. Math. Mech.* **10** (1961), 735–746.

[63] G. E. Šilov, "On rings of functions with uniform convergence," *Ukrainian Math. J.* **3** (1951), 404–411.

[64] ——, "On the decomposition of a commutative normed ring in a direct sum of ideals," *Amer. Math. Soc. Transl.* **1** (1955), pp. 37–48.

[65] G. Stolzenberg, "A hull with no analytic structure," *J. Math. Mech.* **12** (1963), 103–111.

[66] ——, "Polynomially and rationally convex sets," *Acta. Math.* **109** (1963), 259–289.

[67] ——, "Uniform approximation on smooth curves," *Acta Math.* **115** (1966), 185–198.

[68] L. Waelbroeck, "Le calcul symbolique dans les algèbres commutatives," *J. Math. Pures Appl.* **33** (1954), 147–186.

[69] J. Wermer, "Polynomial approximation on an arc in $\mathbb{C}^3$," *Ann. of Math.* (2) **62** (1955), 269–270.

[70] ——, "Subalgebras of the algebra of all complex-valued continuous functions of the circle," *Amer. J. Math.* **78** (1956), 225–242.

[71] ——, "Function rings and Riemann surfaces," *Ann. of Math.* (2) **67** (1958), 45–71.

[72] ——, "Dirichlet algebras," *Duke Math. J.* **27** (1960), 373–381.

[73] ——, "Polynomially convex disks," *Math. Ann.* **158** (1965), 6–10.

[74] D. Wilken, "Lebesgue measure of parts of $R(X)$," *Proc. Amer. Math. Soc.* **18** (1967), 508–512.

[75] L. Zalcman, *Analytic capacity and rational approximation*, Lecture Notes in Math., no. 50, Springer-Verlag, Berlin and New York, 1968.

[76] *Selected Papers of Errett Bishop*; World Sci. Publ., Singapore, 1986.

Addendum: Concepts and Categories in Perspective

SAUNDERS MAC LANE

Doing mathematical research is known to be hard. Writing on the history of mathematics is not hard in the same way, but it is difficult. Part of the difficulty is that of picking the right things to bring out. As a small example of this sort of difficulty, I would like to observe now that my article in the first part of this series, under the title "Concepts and Categories in Perspective" seems to have missed some relevant perspectives.

My discussion of the role of "Theories" in §18 should have included the notion of a "sketch". A sketch S is a directed graph with a distinguished class of diagrams, as well as distinguished classes of cocones and cones (a cone consists of edges from a common vertex). A sketch S serves as a theory, if we take a model of S in a category A to be an assignment of an object of A for each vertex and an arrow of A for each edge, all so that all the distinguished diagrams become commutative in A, while the distinguished cones become limit cones and the cocones colimits in A. The notion of a Lawvere theory is a special case of a sketch. There are several variants of the definition, but the notion is due to Ehresmann (**MR 39** #278; A. Bastiani and C. Ehresmann **MR 48** #2211). At that time (1972) sketches did not attract much attention; Ehresmann published voluminously with many different elaborate definitions which were hard to sort out, especially because he and his students were hardly in touch with other workers in categories. His student C. Lair developed sketches further in his thesis (1970, **MR 53** #10888) and in later papers, such as Guitart and Lair (**MR 84h** #18012). The subsequent publication of Ehresmann's carefully annotated collected works (e.g., **MR 86i** #01059) has made his contributions more accessible. More recently the notion of a sketch was an essential part of the discussion of theories in the 1986 book by Barr–Wells (Toposes, Triples, and Theories, **MR 86f** #18001). By the efforts of John Gray and others, it now turns out that sketches are especially effective in aspects of computer science, partly because they are usually small and so are easy to put on a computer and more essentially because they readily handle many-sorted theories and their initial algebras, as used for data types.

This is a case in which a concept became relevant by way of its use in a different field.

My previous article reported the remarkable use of sheaves by Lawvere–Tierney to formulate Cohen's proof by forcing of the independence of the continuum hypothesis. I did not note that this is by no means the only case where toposes explicate forcing. Thus Marta Bunge showed (**MR 51** #2908) that this approach would also handle the famous Solovay–Tennenbaum proof by forcing (**MR 45** #3212) of the independence of the Suslin hypothesis. Then in 1980 Freyd used sheaves to get an amazingly simple new proof of the independence of the Axiom of Choice (**MR 82i** #03079). Subsequently, Scedrov has developed the connection of such forcing with classifying toposes in his memoir (**MR 86d** #03057). For some time I have thought that this connection with set theory calls for still further analysis of the relation between set-theoretic and sheaf-theoretic forcing, but I omitted to say so.

My article also tried to list the initial researches on categories in various countries—but I missed one such start (the first in the USSR?): An article by A. I. Mal'cev on "Defining relations in categories" (**MR 20** #3805). It was written shortly after his famous paper on the "Mal'cev operations", so evidently grew out of his interests in universal algebras.

This is by no means the only missed reference. My list of contributions from the "Swiss School" should certainly have included the work of F. Ulmer, as for example in his influential 1968 paper on "Properties of dense and relative adjoint functors" (**MR 36** #5190). Under the German school, I should add U. Oberst (1968), on the homology of categories (**MR 37** #1440). Walter Tholen has pointed out to me that I managed to hide the considerable activities in Rumania, where there was lively contact with mathematics in France. I had mentioned three Rumanians (I. Bucur, N. Popescu, and A. G. Radu) under the rubric "The Grothendieck School" and two more, M. Jurchescu and A. Lascu, under the heading "Eastern Europe". There should surely have been a separate title "Rumanian School", with these and additional entries such as

C. Bănică and N. Popescu (1963), Exactness of functions (**MR 33** #2700).

C. Năstăsescu and C. Nita (1965), Noetherian objects (**MR 34** #1374).

My most grievous omission is that of Yoneda. In the 1950s his work on homological algebra was right at the forefront, as I well knew from several contacts with him in Paris in 1955, and as in his basic 1954 paper on "Ext via long exact sequences" (**MR 16**, p. 947). Somewhere in this period, no one knows exactly where, he formulated the lemma stating that all the natural transformations from the hom functor $\hom(-, A)$ to a contravariant set-valued functor F are given by the elements of the set $F(A)$—a lemma which continues to play a basic role and which rightly bears his name. In these cases, as so often, the boundaries between fields (here between homological algebra and categories) are both indistinct and permeable.

History is difficult in part because the connections that matter are usually numerous, often hidden, and then subsequently neglected.

Lars Valerian Ahlfors was born in Helsinki in 1907 and obtained his Ph.D. from the University of Helsinki in 1930, having worked under the guidance of Ernst Lindelöf and Rolf Nevanlinna. His studies abroad included stays in Zürich and Paris. After a few years of teaching in Finland, he was invited in 1935 to Harvard University for a period of three years as visiting lecturer. In 1938, he returned to Helsinki as professor at the University. Meanwhile, in 1936, he had received one of the first two Fields Medals. In 1944, he left war-torn Finland to join the University of Zürich until accepting a permanent position at Harvard in 1946. His basic research has been in the area of complex analysis. In addition to a leading graduate text on this subject, his publications include books on Riemann surfaces (with L. Sario), quasiconformal mappings, and conformal invariants. Many of his papers were written jointly with Arne Beurling. He is a member of the National Academy of Sciences and a foreign member of the USSR Academy of Sciences.

The Joy of Function Theory

L. V. AHLFORS

It has become customary to write about the joy of everything, from the joy of cooking to the joy of sex, so why not the joy of function theory? It would perhaps be more proper to write about the great joy of all mathematics, but that would require volumes, and I shall limit myself to the joy of a single complex variable. I have participated both as an actor and as a spectator, and I shall use both aspects. On the other hand, it is only natural that this account will be partly autobiographical, but I shall try to keep my ego in check.

I shall begin with my years as an undergraduate at the University of Helsingfors (Helsinki). I was very lucky to have two great teachers: Ernst Lindelöf and Rolf Nevanlinna. As a freshman I should not have been allowed to take Nevanlinna's advanced calculus class, but I persuaded him to let me stay. Later, I found out that Lindelöf disapproved of this rash decision of his younger colleague. I was too young, and I am afraid I was a pest.

Lindelöf had retired from research, but he remained a devoted and excellent teacher. Advanced students were required to come to his home Saturdays at 8:00 a.m. to be either scolded or praised for the way they had handled his handwritten assignments. He had a private mathematics library from which he used to lend books, mostly in German or French. It never occurred to him to ask if the borrower could read the language. It was taken for granted. I remember vividly how he encouraged me to read the collected papers of Schwarz and also of Cantor, but he warned me not to become a logician, for which I am still grateful. Riemann was considered too difficult, and Lindelöf never quite approved of the Lebesgue integral.

Nevanlinna was a superb teacher. It was of course preordained that I should specialize in function theory. He was only twelve years older than I, and in due time he became not only my teacher and friend, but also my mentor. I was in great need of his help, for although I knew that I should learn as much mathematics as possible, it had not become clear to me that I was also expected to do original mathematics by myself.

When I had earned my first degree of "filosofie kandidat", an extraordinary piece of luck came my way. Lindelöf took me aside and told me that Nevanlinna had been invited to ETH, the Federal School of Technology in Zürich, to replace Hermann Weyl for the fall of 1929. You must find a way to go with him, he said, even if you have to go dog-class.

At that time Finland had not yet recovered from the civil war after the Russian Revolution. There was little money and no travel grants one could apply for. My father was a stickler for first class, and that was the way I finally traveled, undoubtedly at a great sacrifice.

The outside world had suddenly opened up to me, and I was thrilled. Never mind that I was on short rations and had to become a vegetarian to make ends meet. Never mind that the luxury around me was completely out of my reach. I had found what I wanted. For the first time I could listen to live lectures on contemporary mathematics.

Nevanlinna's course on meromorphic functions was an eye-opener, neither too elementary nor too advanced. From that point on I knew that I was a mathematician. What could have been more tantalizing than to hear of a problem that seemed approachable and had been open for twenty-one years, precisely my age?

Nevanlinna has spread the story that I suddenly disappeared from sight to return two weeks later with a full proof of the Denjoy conjecture. This was pure fiction, but obviously well meant. At that time I could still work nights and I had put in much thought on the problem, but I lacked the technique that was needed for a complete solution. In my desperation I showed my work to Nevanlinna, who saw at once that I was on the right track. He consulted with Pólya, and together these stars of the first magnitude showed me how to manipulate the inequalities that were needed for the punch line. They generously pressed me not to mention their names when the paper appeared;

I like to think that I would have done the same. In the end Pólya, who rightly did not trust my French, wrote my *Comptes Rendus* note. Whether I deserved it or not, my name became almost instantly known in the circle of leading analysts.

In retrospect, the problem by itself was hardly worthy of the hullabaloo it had caused, but it was of the kind that attracts talent, especially young talent. It is not unusual that the same mathematical idea will surface, independently, in several places, when the time is ripe. My habits at the time did not include regular checking of the periodicals, and I was not aware that H. Grötzsch had published papers based on ideas similar to mine, which he too could have used to prove the Denjoy conjecture. Neither could I have known that Arne Beurling had found a different proof in 1929 while hunting alligators in Panama. This proof was included in Beurling's famous thesis of 1933. The timing is of no consequence, but it is interesting that we all used essentially the same distortion theorem for conformal mapping.

Let me now break off this chronological review of events and discuss what I consider a major trend in the development of function theory. It has been said before, but is worth repeating, that analytic functions are by nature extremely rigid. It is impossible to change an analytic function at or near a single point without changing it everywhere. This crystallized structure is a thing of great beauty, and it plays a great role in much of nineteenth-century mathematics, such as elliptic functions, theta functions, modular functions, etc. On the other hand, it was also an obstacle, perhaps most strongly felt in what somewhat contemptuously was known as "Abschätzungsmathematik". Consciously or subconsciously there was a need to embed function theory in a more flexible medium. For instance, Perron used the larger class of subharmonic functions to study harmonic functions, and it had also been recognized, especially by Nevanlinna and Carleman, that harmonic functions are more malleable than analytic functions, and therefore a more useful tool.

The most active step in this direction was the introduction of quasiconformal mappings. I think everybody knows now that they were invented, independently and for different reasons, by Grötzsch and M. A. Lavrentiev. Over the years this important notion has changed the nature of function theory quite radically, and on many different levels. The ultimate miracle was performed by O. Teichmüller, a completely unbelievable phenomenon for better and for worse. He managed to show that a simple extremal problem, which deals with quasiconformal mappings of a Riemann surface, but does not involve any analyticity, has its solution in terms of a special class of analytic functions, the quadratic differentials. Over the years since Teichmüller's demise his legacy has mushroomed to a new branch of mathematics, known as Teichmüller theory, whose connection with function theory is now almost unrecognizable.

World War II had a strong dampening effect on mathematics. International communications between mathematicians stopped almost completely, and almost all experienced important and sometimes radical changes in their lives. For this reason, when dealing with the development of mathematics in this century it is unavoidable to separate the antebellum and postbellum eras.

My personal life has been strongly influenced by my work in mathematics. It was my mathematics that caused Harvard, presumably on the advice of Carathéodory, to offer me a position in the Harvard mathematics department. I did not think I was mature enough to accept anything on a permanent basis, but I was persuaded by W. Graustein, the chairman, to come for a trial period of three years. My stay lasted from 1935 to 1938, and I am grateful for the chance it gave me to broaden my mathematical knowledge. I was made a full professor at Helsinki University in 1938, but after a happy year the war engulfed Finland, and the forced separation from my family put an end to the good days.

In the summer of 1944 I was offered a position at the University of Zürich. I accepted, but had no idea how to get there. In September Finland signed an armistice with Russia, and became automatically an enemy of Germany. Fortunately, I was allowed to join my family in Sweden. After a long wait we were able to continue our travel through wartorn Britain and France to Zürich. My stay in Sweden would not have been possible without the Fields Medal and the hospitality and financial help of Arne Beurling. Switzerland was a haven, but I was relieved when in 1947 I was asked to rejoin the Harvard faculty.

There were many refugees in Cambridge, Massachusetts at that time, and it was said that the language in Harvard Square was English with a Viennese accent. Among prominent mathematicians were R. von Mises, M. Schiffer, Stefan Bergman, and Z. Nehari. Before my return to Harvard I was hardly aware of Schiffer's earlier work, but when I learned a little more I was enormously impressed by his idea of interior variation and its connection with the coefficient problems and, at that time rather unexpectedly, with quadratic differentials. Even if the proof of the Bieberbach conjecture would ultimately follow a different but equally fascinating path, Schiffer's part in the development of pure function theory remains one of the high points. I would be remiss if I did not also mention H. Grunsky and his famous inequalities. During the war he was one of the standard-bearers of German mathematics who did not succumb under the poisonous atmosphere.

At Harvard I had many brilliant students. From the prewar period I remember R. Boas who taught me how to pronounce mathematical formulas, and H. Robbins who tried hard to be an *enfant terrible*. M. Heins has remained a close friend through all these years. I regret that I had to leave Harvard before he finished his Ph.D. under Joe Walsh. All three have become renowned mathematicians.

After the war there was a younger and much bigger crop. I cannot forget J. Jenkins with his photographic memory and his uncanny zest for perfection. He has become justly famous for having put many vaguely expressed ideas in univalent functions and extremal problems on a sound footing. P. Garabedian learned more from S. Bergman than from me, but I still count him as my student. He has spread his talents over many fields, equally at home in all. Among my prize students were H. Royden, C. Earle and A. Marden. They are busily working on the cutting edge of function theory in a very general sense, some of it beyond what I can follow. R. Osserman has surprised everybody with his startling developments in minimal surfaces. There were many more, but at this point I take refuge in my failing memory.

James A. Donaldson is Professor and Chairman of the Mathematics Department at Howard University. After growing up in Madison County Florida, he graduated from Lincoln University (Pennsylvania) where he majored in mathematics. He received his Ph.D. at the University of Illinois (Champaign-Urbana) in 1965 with a thesis under the guidance of R. G. Langebartel. Following appointments at Howard University, the University of Illinois at Chicago, and the University of New Mexico, he returned to Howard in 1971. He has served as a member of the AMS Council, in CUPM, and as a consultant to NSF, the Sloan Foundation, and several State Boards of Education. His research publications include numerous papers on differential equations and applied mathematics.

Black Americans in Mathematics

JAMES A. DONALDSON

INTRODUCTION

Mathematics, as much as any other great discipline, belongs to all mankind. Seminal contributions to this discipline have been made by different groups, geographical and racial, in various times and eras. This is important because the continuing evolution and development of society throughout the world depends heavily upon scientific and technological advances, many of which are made possible through the application of important mathematical discoveries. To be sure, many factors — some economic, some social and political — have affected the contributions of different groups at various times and therefore the development of mathematics and science within the group during a given period.

To accomplish the objective of chronicling some of the participation and involvement of Black Americans in mathematics, we give here a brief description of circumstances confronting Black people in America. The signing of the Emancipation Proclamation in 1863, twenty-five years before the founding of the American Mathematical Society, was the landmark event in the granting of freedom to all Black people living in America and marked the beginning of the healing of the terrible wounds resulting from slavery, the cruel American institution that had shackled both the enslaved and the free.

After termination of the U.S. Civil War in 1865, free Black people and newly freed Black people, fortified by hope and quiet determination, struggled to prepare themselves in every way for full membership in society. Education then, as now, was viewed as the key to realizing this desire. Consequently this period, shortly before and after 1865, saw the founding of several educational institutions (Lincoln University in Pennsylvania, Wilberforce, Howard, Shaw, Johnson C. Smith and others which will be called traditionally Black institutions or TBIs) with the goal of providing higher educational opportunities for newly freed Black people and other people of African descent.

These institutions (TBIs) were of essentially two types: those that offered classical higher education in the arts and sciences, and those that offered industrial education. Although a number of predominantly white educational institutions during this period permitted Black people to matriculate, the salient fact is that the overwhelming majority of Black people attended the newly established TBIs. This pattern continued until well into the twentieth century.

After termination of reconstruction with the election of Rutherford Hayes to the presidency of the United States in 1876, many economic, political, and educational advances won by the newly freed Black Americans were arrested, and concerted efforts were launched to return Black people to bondage. These efforts included the legal implementation of segregation as a way of life in many sections of the country.

It should come as no surprise that the mere existence of the TBIs, especially those offering classical higher education, came increasingly under attack. As reaction in the country strengthened and Black people were exposed to most vicious attacks, a few good people were unswerving in their support of offering the highest classical education to Black people. On the other hand, the proponents of industrial education found the notion of classical education (Latin, mathematics, Greek, philosophy, psychology) for Black people ridiculous, and proclaimed vociferously that Black people did not need instruction in these subjects, but rather needed instruction in "how to work." Despite these attacks and many betrayals of solemn promises, Black people kept their hopes and aspirations alive. Their thirst for knowledge was unabated, and they continued to view education as an avenue for the determination and the realization of their dreams.

EARLY BLACK MATHEMATICIANS AND SCIENTISTS

Examples of Black people's achievements in intellectual activities during this period and to the present day have been of enormous importance both for them and their supporters. Mathematics[1] and science were disciplines held in the highest regard for this purpose.

Well before the signing of the Emancipation Proclamation, Benjamin Banneker (1731–1806), a free Black, had demonstrated to the world the ability of Black people to excel in mathematics and science. Also a surveyor, Banneker assisted Andrew Ellicot in laying out the District of Columbia [**3**, **6**]. He exchanged letters with the leading scientists of his day, and also with Thomas Jefferson, then a prominent American figure in science, and Secretary of State of the United States of America. An evaluation of the mathematical accomplishments of Benjamin Banneker is provided by Jefferson who forwarded Banneker's ephemeris (astronomical almanac) to de Condorcet (a leading mathematician, philosopher, statesman, revolutionary, and member of the French Academy), Secretary of the Academy of Sciences at Paris, with a cover letter[2] which contained:

> I am happy to be able to inform you that we have now in the United States a negro, the son of a Black man born in Africa, who is a very *respectable* (emphasis added) mathematician. I procured him to be employed under one of our chief directors in laying out the new federal city on the Potowmac, & in the intervals of his leisure, while on that work, he made an Almanac for the next year, which he sent me in his own hand writing, & which I inclose to you. I have seen very elegant solutions of Geometrical problems by him...

Charles L. Reason[3] (1818–1893) was a brilliant student whose achievements in mathematics earned for him the position of assistant instructor in his high school before he reached the age of fourteen. He became the first Black person to receive a faculty position in mathematics at a predominantly white institution when he was appointed to chairs in belles lettres, French and mathematics at Central College, Cortland County, New York in 1849. Central College was an institution founded for higher education, without distinction as to race. Reason resigned this position after one year, and several years later he was appointed Principal of the Institute for Colored Youth in Philadelphia, Pennsylvania where he remained for three years until beginning his duties as Principal of school no. 6 of New York City in 1856. He held that position for 37 years until his death in 1893. Reason was a master teacher, an abolitionist, a protector of Black people fleeing from bondage, and an avid defender of the public school system of New York.

---

[1]The reader should note that in this paper we include in our designation *mathematics* all that is now encompassed by the designation *mathematics and the mathematical sciences*.

[2]Letter from Thomas Jefferson to the Marquis de Condorcet, dated August 30, 1791, Ford, ed., *The Works of Thomas Jefferson*, volume VI, pp. 310–312.

[3]Anthony R. Mayo, *Charles Lewis Reason—a brief sketch of his life*, commemorating the fiftieth anniversary of his death, August 16, 1893, n.d., twelve-page pamphlet.

Elbert F. Cox
ca. 1925

When Edward Alexander Bouchet (1852–1918) earned a doctorate in physics from Yale University in 1876, he became the first Black and one of the first Americans to earn a Ph.D. The first such degree had been conferred in the United States only fifteen years earlier, also by Yale University. (In 1862 Yale University became the first American institution to confer the Ph.D. in mathematics[4].) Bouchet, unlike his contemporaries, was unable to find employment at a higher level than in a high school. For the next eighty-five years, this was not an uncommon circumstance for most Black scientists.

From 1876–1902, Bouchet taught at the Institute for Colored Youth in Philadelpha, the same high school from which Reason had resigned twenty years earlier. Subsequent academic positions were also in high schools. Although there is no doubt that Bouchet made important contributions in these positions by serving as an inspiration and model for a great many young people, one can only wonder how much poorer American science was during this period because he had no opportunity to develop a career at the level of his academic attainment.

Kelly Miller, the son of a slave, pursued the Ph.D. in mathematics and astronomy at Johns Hopkins University around the time the American Mathematical Society was founded **[16]**. Miller's studies were aborted because of his inability to obtain funds for tuition and other fees. It is quite likely that if financial assistance had been forthcoming, he would have been the first Black to earn a Ph.D. in mathematics. He joined the faculty of Howard University where he rose through the ranks to professor of mathematics.

The first Black person to earn the Ph.D. in mathematics was Elbert F. Cox. Cornell University conferred this degree upon him in 1925. He had received his undergraduate degree from Indiana University in 1917. Before joining the faculty of Howard University in 1929, Cox had taught at several other TBIs. During his tenure at Howard University he directed the research of many students for the master's degree. He retired in 1961 and died in 1969.

When Dudley W. Woodard was awarded a Ph.D. in mathematics by the University of Pennsylvania in 1928, he became only the second Black person to earn that degree in mathematics and the first of several Black Americans (William W. S. Claytor, George H. Butcher, Jr., Jesse P. Clay, Sr., and Orville Keane) to earn a Ph.D. in mathematics from that venerable institution. After joining the mathematics faculty at Howard University, he moved through the ranks to professor, served a term as Dean of the College of Liberal Arts from 1920–1929, and developed the M.S. degree program in mathematics at Howard University in 1929. Professor Woodard, who received his undergraduate degree from Wilberforce University in 1903, was the first graduate of a TBI to earn the Ph.D. in mathematics.

---

[4]See Part II of *A Century of Mathematics in America*, p. 88 ff.

The third Black person to earn a Ph.D. in mathematics was William W. S. Claytor (University of Pennsylvania, 1933). Claytor worked on embedding problems. These problems had received the attention of many well-known mathematicians (S. Mac Lane, H. Whitney, Kuratowski, and others). Claytor's solution to the embedding problem of Kuratowski attracted considerable attention throughout the topological community, and is regarded by many mathematicians as one of the high points of classical point set topology theory. His scientific results were published in the *Annals of Mathematics* [4, 5]. After earning his doctorate, Claytor in 1933, not unlike Bouchet before him in 1876, was unable, because of unyielding racial restrictions, to secure an academic position permitting him to develop a career at the level of his academic attainments. It is now acknowledged that the racial practices in this country had a negative impact on the full contribution of his mathematical research genius and that mathematics and science are the poorer by far because Black scientists of Claytor's caliber were not given the opportunity to contribute according to their talent.

In 1980 when the National Association of Mathematicians inaugurated its Williams W. S. Claytor Lecture Series, several well-known mathematicians were invited to share with the organization some of their reminiscences of Claytor and his work. We include here excerpts from some of their letters.[5]

R. L. Wilder wrote

> Toward the end of his stay at Michigan (Claytor was a postdoctoral fellow at the University of Michigan around 1937), the question of where he could get a "job" came up. We topologists concluded he should join the University of Michigan faculty. Today I'm confident, there would be no hesitancy about this on the part of the Michigan administration, but that was about thirty years ago. To our surprise, Harry Carver (the late statistician and founder of the *Annals of Mathematical Statistics*), who, to our minds, was very conservative and reactionary, took the matter into his hands. He had cultivated a great liking and respect for Bill, and decided to use his influence (which was considerable) to get Bill appointed. But it was to no avail; the administration was simply afraid (I am sure this was the case more than racial prejudice). I finally wrote to Oswald Veblen, head of the School of Mathematics at the Institute for Advanced Study, who quickly replied that he'd find a place for Bill at the Institute. However, when I told Bill this, he shook his head and replied, "There's never been a Black at Princeton, and I'm not going to be a guinea pig." I've always felt this was the turning point in Bill's life, and a great mistake on his part. I knew

[5]See *Proc. of the Eleventh Annual Meeting of the National Association of Mathematicians* (1981), pp. 11–17.

how he felt and argued with him, but he was adamant. I am sure that if he had accepted, he would have found lots of friends at the Institute, and that his future would have been quite different.

Samuel Eilenberg wrote

> His mathematical talent never came to full fruition, for reasons which neither he nor many of his friends had any way of controlling. I very deeply felt the tragedy of the situation.

David Blackwell wrote

> And I remember a lecture he gave on the Jordan Curve Theorem. It was a model of beauty, elegance, clarity, and simplicity, like the man himself. He was a great mathematician, and a great man.

And Gail S. Young wrote

> I have always regarded Bill Claytor's life as an outstanding example of the penalties individuals have had to pay for segregation. Another individual of the same vintage, regarded as an equally promising young mathematician, was on the Michigan faculty and then at Princeton; however, in that period, it was impossible for Claytor to get a position in any strong mathematics department. In fact, it was not until about 1955 that the University of Michigan, the school in which he studied as a postdoctoral student, gave a Black a teaching assistantship. I remember the long discussions preceding that decision and the fears that were expressed about students' reactions. Of course, there was not the slightest problem.... The two papers on which his reputation rests are brilliant. That he could not get a job in any research-oriented department was tragic. It would not happen now, I think. We do seem to have made some progress, however slow, in the past 40 years; but it all came too late to do him any good.

The tragic circumstance of the career of Claytor and others may be summed up by a statement of Professor Frank R. Lillie, in the obituary of his student, the great biological scientist, Ernest E. Just [**17**].

> An element of tragedy ran through all Just's scientific career due to the limitations imposed by being a Negro in America. He felt this as a social stigma, and hence unjust to a scientist of his recognized standing. That a man of his ability, scientific devotion, and of such strong loyalties as he gave and received, should have been warped in the land of his birth must remain a matter of regret.

In 1934 Walter Talbot earned the Ph.D. in mathematics from the University of Pittsburgh. Heavy teaching and administrative duties at Lincoln University in Missouri for many years prevented him from pursuing his research interests. After joining the mathematics faculty of Morgan State University, he played a leading role in increasing Black participation in some of the professional mathematical societies during the late 1960s.

Reuben McDaniel and Joseph Pierce earned their Ph.D. degrees in mathematics in 1938 from Cornell University and the University of Michigan, respectively.

David Blackwell completed his graduate studies at the University of Illinois (Urbana-Champaign) in 1941. He taught on the faculties of Southern University, Clark College, and Howard University, where he was chairman of the mathematics department for several years before joining the faculty of the University of California (Berkeley) in 1954. David Blackwell has contributed to several areas of mathematics: set theory, measure theory, probability theory, statistics, game theory, and dynamic programming. His name is attached to a theorem in statistics, the Rao–Blackwell theorem, which is important in estimation theory and tests of hypotheses. Blackwell, a former vice-president of the American Mathematical Society and a former president of the Institute of Mathematical Statistics, is currently the only Black person among some 1500 members of the National Academy of Sciences. He is the recipient of numerous honors and awards, including the R. A. Fisher Award and the TIMS/ORSA John von Neumann Theory Prize.

J. Ernest Wilkins, Jr. had not yet reached the age of nineteen when the University of Chicago conferred upon him the Ph.D. in mathematics in 1942. He has enjoyed distinguished careers in both academia and industry, and has published numerous papers in the areas of projective differential geometry, calculus of variations, special functions, optics, and probability theory. Wilkins, a member of the National Academy of Engineering, is a former member of the Council of the American Mathematical Society and a former president of the American Nuclear Society.

In 1943 Harvard University conferred upon Jeremiah Certaine a Ph.D. in mathematics, and in 1944 the Ph.D. in mathematics was conferred upon Wade Ellis, Sr., Warren H. Brothers, and Clarence Stephens by the University of Michigan, and upon Joseph J. Dennis by Northwestern University. When Oberlin College appointed Wade Ellis, Sr. to its faculty in 1949, he became, I believe, the second Black mathematician to teach at a predominantly white institution. Later, he was Associate Dean of the Graduate School at the University of Michigan.

These names exhaust the list of those Black mathematicians that have been identified as receiving their doctorates in mathematics during the twenty-year period commencing with the year that the degree was conferred upon

Elbert Cox. It is noteworthy that many of these mathematicians received their undergraduate degrees at TBIs. The table in Appendix A shows that the TBIs continued to play a primary role in producing graduates who later received doctorates in mathematics.

We turn our attention to Black women in mathematics who have not yet appeared in our account. The first two Black American women to earn Ph.D. degrees in mathematics were Marjorie Lee Browne and Evelyn Boyd Granville who earned their degrees in 1949 from the University of Michigan and Yale University, respectively. Why hadn't any Black American women earned mathematical doctorates before then? Some reasons are contained in Vivienne M. Mayes' article [**11**]. Of special relevance to mathematics is her statement:

> The idea of encouraging Blacks, and especially females, to prepare for academic careers was unheard of. Since Black colleges were so few in number, it was not economically sound to plan on teaching appointments or even to pursue advanced academic degrees. In those days we were counseled to prepare for health professions, the ministry, or public schoolteaching, the few careers which offered an opportunity for livelihood.

Majorie Lee Browne was born in Tennessee and received her B.S. degree in mathematics from Howard University. After several years of employment in a private secondary school and a small private college, she enrolled in the mathematics graduate program at the University of Michigan. Initially she was able to attend only during the summer months. But later a fellowship made possible her attendance during the entire year. After completing her doctorate in mathematics she joined the faculty of North Carolina Central University where she prepared numerous students for careers in mathematics and related areas. She built the undergraduate and graduate programs at that institution.

Evelyn Boyd Granville graduated from Dunbar High School in Washington, D. C. and completed her undergraduate degree in mathematics at Smith College. After writing her doctoral dissertation under the supervision of Einar Hille, she completed a postdoctoral fellowship at New York University in 1950. Despite this superior training and research experience, her application for a position at one institution with working conditions which would permit the continuation of her research was met with laughter [**9**]. During her career she held several positions in government and industry, and academic positions at Fisk University and California State University at Los Angeles where she retired.

Evelyn Boyd Granville
1989
(Photograph courtesy of Marvin T. Jones & Associates, Washington, DC.)

The hiatus of thirteen years before the next Black American woman earned the Ph.D. degree in mathematics was broken when the University of Washington conferred upon Gloria Hewitt, a Fisk University graduate, the Ph.D. degree in mathematics in 1962. Argelia Velez-Rodriguez, however, a naturalized American citizen, had earned the doctorate in mathematics from the University of Havana in 1960. In 1965 Ohio State University conferred the Ph.D. upon Thyrsa Frazier Svager. One year later, 1966, Eleanor Green Dawley Jones received her Ph.D. in mathematics from Syracuse University; Vivienne M. Mayes received the degree from the University of Texas at Austin, and Shirley Mathis McBay received hers from the University of Georgia.

Syracuse University conferred the Ph.D. degree in mathematics upon Geraldine Darden in 1967; St. Louis University conferred the Ph.D. in mathematics upon Mary Sylvester DeConge in 1968; Emory University conferred the Ph.D. in mathematics upon Etta Zuber Falconer in 1969; and Deloris Spikes earned hers at Louisiana State University in 1971. This completes the list of Black American women who earned doctorates in mathematics during the period 1949–1971. For additional material on this topic one should read Patricia Kenschaft's fascinating article in [**8**].

We conclude this section with some mathematicians who do not fit nicely into any of the categories mentioned above, but whose contributions to research and administration deserve mention. Charles B. Bell obtained his undergraduate degree from Xavier University (New Orleans) and his doctorate in mathematics from the University of Notre Dame. Currently on the mathematics faculty of San Diego State University, Bell has written papers in the area of nonparametric statistics. He has been on the faculties of Case Western Reserve University, University of Michigan, Tulane University and the University of Washington. Albert T. Bharucha–Reid had conferred upon him only two degrees: a bachelor of science degree from Iowa State University, and an honorary doctorate from Syracuse University. He taught on the faculties of Oregon State University, Wayne State University, Georgia Institute of Technology and Atlanta University. During his career Bharucha–Reid published many papers in probability, random equations and statistics and wrote and edited several books. His career was cut short by his untimely death in 1985.

A continuing serious problem has been the small pool of people capable of providing effective leadership in traditionally Black institutions. Because of the great need for effective administrators, some Black mathematicians felt compelled to assume administrative positions in these institutions. This list includes Luna Mishoe, former President of Delaware State College; Deloris Spikes, President of Southern University; William L. Lester, Provost of Tuskegee University; Jesse C. Lewis, Vice President for Academic Affairs at Norfolk State University; and L. Clarkson, Vice President of Academic Affairs at Texas Southern University.

Participation in the Professional Mathematical Organizations

One readily observes that Black mathematicians participate in programs and activities of the Mathematical Association of America (MAA) in far greater numbers than in those of other professional mathematical organizations. This may be due partially to the fact that many Black mathematicians are employed at educational institutions where there are few opportunities for research and partially to the laissez-faire attitude of many of the other organizations. Since the late sixties MAA has expressed more concern and interest in problems of Black mathematicians than the other mathematical organizations. Indeed, several officers and members of the MAA (George Pedrick, Gail Young, Israel Herstein, Creighton Buck, R. D. Anderson, et al.) played a key role, through their assistance to Walter Talbot in 1969–1970, in bringing large numbers of Black mathematicians, especially those from nonwhite universities, into the full range of activities and programs of the organization.

Before 1970 circumstances were markedly different. When meetings of both the MAA and the American Mathematical Society (AMS) were held in the south, rank discrimination was practiced against Black participants. These participants were not permitted to eat or stay in the same establishments as their white counterparts and consequently were unable to benefit from much of the informal but important discourse which occurs outside the formal sessions. It was this situation in the early 1950s which compelled the Fisk University mathematics department, while Lee Lorch was its chairman, to beseech the governors of the MAA to insert an antidiscrimination clause in the organization's bylaws.

The MAA inserted an antidiscrimination clause in its bylaws after much consideration; but, because of its organizational structure, the organization had some difficulty with its enforcement. To be sure, in the early sixties there is a vivid account of the walk-out of Dr. A. Shabazz (then Lonnie Cross) of Atlanta University from a MAA sectional meeting in South Carolina when he discovered the discriminatory housing and eating arrangements that were to be in effect.

I began attending meetings of the AMS, the leading professional mathematical organization in America, in 1966. At one of these meetings I was told by a very prominent mathematician that the organization had been indifferent, insensitive, and silent far too long on far too many issues of importance in opening the profession to all people with talent in mathematics. He cited names of several Black mathematicians who had been unable to find suitable positions because of racial discrimination and stated that he knew of nothing that the AMS had done to help them! Apparently many influential members of the organization, through their public actions, considered efforts to provide access to career opportunities for all mathematicians, irrespective of race, as counterproductive and political.

Even though David Blackwell, William Claytor, Albert Bharucha-Reid and others had read papers before the society fairly early, it was not until the sixties that significant numbers of Black mathematicians began reading papers before this organization. David Blackwell, Albert Bharucha-Reid and Floyd Williams have given invited hour addresses before the society, and many others have contributed papers to special sessions and the general program of the Society. Several Black mathematicians have published in the Society's journals.

David Blackwell was a vice-president of the AMS; David Blackwell, J. Ernest Wilkins, Jr., James Donaldson, and Albert Bharucha-Reid were members of the AMS Council. Walter Talbot, Gloria Gilmer, Eleanor Jones, and Rogers Newman have served on the Board of Governors of the Mathematical Association of America, and Sylvia Bozeman is serving currently on its board. Several Black mathematicians have served on committees of the American Mathematical Society, the Mathematical Association of America, and the Society for Industrial and Applied Mathematicians.

In the early seventies, controversy arose over efforts of the AMS Council to establish a reciprocity agreement between AMS and the South African Mathematical Society (SAMS). The AMS Council was requested in unambiguous terms to choose between the reciprocity agreement with SAMS and the continuing membership of Black American mathematicians in the organization. We are grateful the AMS Council chose to keep its Black members. However, many AMS members were very disappointed with a subsequent move of the AMS Council to terminate all reciprocity agreements in an aftermath to the controversy over the proposed agreement with SAMS. The AMS membership soundly defeated this move and thereby restored some hope that the Society would take no steps to inhibit continuing efforts for full membership for Black mathematicians.

Many mathematicians came eventually to the conclusion that the special problems and concerns confronting Black mathematicians were not being addressed by any of the professional mathematical societies. This assessment served as a catalyst for founding the National Association of Mathematics (NAM) in 1969.

The organization's annual meeting, which coincides with annual winter meetings of AMS-MAA, features a program for the general scientific and mathematics communities, and special activities treating problems of concern for mathematicians teaching at educational institutions with large Black student enrollment. The organization's Claytor–Woodard Lectures have been delivered by numerous mathematicians: David Blackwell, J. Ernest Wilkins,

Jr., Albert Bharucha-Reid, Wade Ellis, Jr., James Joseph, Jr., Raymond Johnson, J. Robinson, Japeth Hall, et al.

Attracting Minority Students for Careers in Mathematics

Segregation was rampant when I attended public school in the south in the 1940s and 1950s. In my home state, Florida, our Black teachers attempted to compensate for the gross material inequities of this monstrous system by instilling in us self-confidence, self-worth, and a greater appreciation of our ability to achieve in mathematics and science. Teachers from this school system, such as Lennie Collins, Alma Jean McKinney, and Juanita Miller, directed many students to pursue careers in science and mathematics. Unfortunately, in the school systems of today, many Black students are steered, either intentionally or unintentionally, from programs that would prepare them to pursue a college or university degree in any area, not to mention pursuing a career in mathematics or science [7].

I, not unlike many students from disadvantaged groups, entered the university with very little or no perception of what a person with a degree in mathematics could do outside of teaching. There were no examples known to me at that time of Black people working in non-teaching positions in mathematics. This is true for many students from minority groups today. Increasingly now, having knowledge of members of their group with teaching positions in mathematics is not part of their experience. Thus, it falls upon faculty members to encourage and motivate these students, some already victimized by misguided well-intended efforts to help by those teachers employing a variant of the "missionary" approach, and others frustrated and discouraged by mathematics teachers displaying cultural imperialism in their interaction with students.

The teacher plays a primary role in attracting students to a discipline. For this reason early interactions of a teacher with the students are very important. The teacher must establish quickly a nonthreatening environment in which all students feel free to participate without the fear of being humiliated or embarrassed, must provide constant positive reinforcement, and must refrain from prejudging the ability of students to learn.

The effectiveness of a teacher in getting a student interested in mathematics or any subject is determined as much by the teacher's sincere, persistent, and visible involvement in the subject himself/herself and in the struggle to realize the aspirations of the students as by the teacher's enthusiasm in presenting the subject. The teacher must have confidence that students can learn, must demonstrate through action the belief that students can learn and the expectation that they will learn. The teacher must be aware that in the case of minority students, as well as for other students from underrepresented groups in mathematics and science, there are other very important interests which cannot be placed on hold — for example, the struggle by minorities for full participation in all aspects of society.

An example of how not to motivate students to consider careers in mathematics may be found in Barry Beckham's description of the experiences of Blacks on white campuses [2]:

> Another black graduate, now an executive with a national financial institution, related to me how he had been discouraged from majoring in math because of his low SAT score. Driven to succeed anyway, he registered for and passed all but one of the required courses for the concentration by his senior year. When he reported his surreptitious achievement to the department chairman, that faculty member tried to block the student's registering for the final, necessary course. After graduating from..., the student matriculated at the University of Michigan and earned his M.B.A.

Accordingly, we need to look at aspects of successful programs which have produced numerous undergraduate mathematics majors from minority groups, and have prepared many for graduate work which led eventually to a Ph.D. in mathematics or related areas. In these programs faculty members:

(1) have had confidence that students can learn and master mathematical concepts and have shown by their action that they believe that students can so learn and that they have expected them to learn,
(2) have insisted upon high standards in the students' mathematical work,
(3) have built and have continually reinforced students' confidence in their ability to do mathematics by introducing some material beyond that covered in the normal curriculum,
(4) have involved students in various aspects of mathematical life in the department,
(5) have presented mathematical ideas and concepts with great clarity and patience to students, and
(6) have displayed some identification with the students and their problems.

Most, if not all, of these elements were part of the successful programs of Professors Claude Dansby (Morehouse College), James Frankowsky (Lincoln University, PA), Lee Lorch (Fisk University and Philander Smith College), Abdulalim Shabazz (Tuskegee University and Clark-Atlanta University), and Clarence Stephens (Morgan State University, State University of New York, Geneseo, and State University of New York, Potsdam). Oral testimonies of the effectiveness of their methods and programs abound. In the literature

[10, 16] and in personal communications are found accounts of the methods of Lorch, Stephens, and Shabazz.

What Prospects for the Future?

The growing national need for highly skilled scientists coupled with the inescapable demographic fact that one-third of the high school population in the 1990s will be minority students make solving the problem of attracting and retaining minorities into science increasingly urgent. Self-interest may serve to focus universities' attention on recruiting minority scientists at all levels in ways that the simple demand for justice did not.

It is indeed true that most minority students need financial support to defray college and university expenses. This support, usually in the form of loans which must be repaid, weighs heavily upon career choices of students. To students who must obtain large burdensome loans to defray their college expenses, careers in professional fields which pay lucrative salaries appear more attractive than careers in academia or research where salaries are less lucrative. To neutralize this phenomenon, more innovative and creative financial aid programs must be made available to minority students, especially in the sciences and mathematics.

I offer the following recommendations for increasing the participation in mathematics and science by members of minority groups.

(1) The underrepresentation of members of minority groups in mathematics should be given a top priority as an important and fundamental problem to be solved by this nation.

(2) Public and private institutions, civic, social, and religious organizations, professional societies, and government and industry of this nation must accept a greater commitment to ensuring full participation in mathematics and science by members of minority groups.

(3) Increased financial support should be made available to minority students interested in pursuing careers in mathematics and science.

(4) Counselors must exercise healthy amounts of scepticism in relying upon test scores in advising minority students about possible programs of study or future career choices.

(5) In order to attract, retain and encourage minority students who are to pursue careers in mathematics and science, teachers and instructors should raise their expectations of what students can do, and be generous with remarks that raise student self-esteem and strengthen student confidence.

(6) The number of innovative support programs should be multiplied at all levels.

There are several indications that the nation can expect to see increased numbers of Blacks and other minorities in mathematics and science. Innovative and creative programs such as the MESA Program in California and several western states, the McKnight Black Doctoral Fellowship Program and the Minority Junior Faculty Development Fellowship Program in Florida, the Ford Foundation Minority Fellowship Program, and the Minority Fellowship Program of the Council for Institutional Cooperation (a consortium of the "big ten" universities and the the University of Chicago) are providing improved opportunities to minorities for careers in mathematics and science as well as in other areas. Various professional organizations, which include the American Association for the Advancement of Science, the Mathematical Association of America, the American Council on Education, the Education Commission of the States, have issued reports that focus partially or wholly on this problem. Of particular note is the recently issued report of the MAA *Task Force on Minorities*, and the document *One-Third of a Nation* issued jointly by the American Council on Education and the Education Commission of the States [**13**].

Another hopeful sign pointing to an increase in minority mathematicians was the decision in 1976 by Howard University to establish a Ph.D. program in mathematics. Despite limited funding, the program has produced seven graduates since 1984 and currently has nine students enrolled. Degrees have been conferred in areas ranging from set theory to applied mathematics.

ACKNOWLEDGMENTS

I thank Dr. Elinor D. Sinette, acting director of the Howard University Moorland-Spingarn Research Center, and her staff for making resources of the center available to me; Raymond Johnson, Lee Lorch, J. Arthur Jones, J. Ernest Wilkins, Jr., George H. Butcher, Jr., and Abdulalim Shabazz for sharing personal recollections and private papers; Creighton Buck for calling my attention to Charles Reason; Ms. Shirley Heppell of the Cortland County (New York) Historical Society for providing information about Charles Reason and Central College; Ms. Georgette Fowler for reading critically an earlier draft and offering suggestions for its improvement; and Ms. L. Thurgood of the Office of Scientific and Engineering Personnel of the National Research Council for providing the data contained in Appendix C below. Also, special thanks for many useful suggestions go to the editor and the editorial committee of *A Century of Mathematics in America*.

APPENDIX A

The table below contains names of some institutions where Black Ph.D. mathematicians have earned their undergraduate degrees. Those institutions with an asterisk are TBIs.

Undergraduate Institutions

Alabama A. and M. University*
Alabama State College*
Alcorn A. and M. University*
Allen University*
Arkansas A. and M. University*
Atlanta University*
Bishop College
Bryn Mawr College
Clark College*
Fisk University*
Florida A. and M. University*
Grinnell College
Hampton University*
Howard University*
Indiana University
Jackson State University*
Johnson C. Smith University*
Langston University*
Lehigh University
LeMoyne College*
Lincoln University, MO*
Lincoln University, PA*
MIT
McNeese State College
Mississippi State University
Morehouse College*
Morgan State University*
North Carolina Central*
Northwestern University
Ohio State University
Ottawa University
Paine College*
Prairie View A. and M. University*
Princeton University
Rust College*
Rutgers
Savannah State College*
Seton Hill
Smith College
South Carolina State*
Southern University*
Spelman College*
St. Augustine*
Talladega College*
Temple University
Texas Southern University*
Tougaloo College*
Tuskegee University*
UCLA
University of California (Berkeley)
University of Chicago
University of Illinois
University of Michigan
University of Pittsburgh
University of Texas
Wayne State University
Wilberforce University*
Xavier University (New Orleans)*
Yale University

Appendix B

Some academic institutions that have granted doctorates in mathematics to Black Americans are contained in the table below.

Doctoral Institutions

Auburn University
Brown University
Carnegie-Mellon University
Catholic University
City University of New York
Columbia University
Cornell University
Emory University
Florida State University
George Washington University
Harvard University
Howard University
Indiana University
Johns Hopkins University
Louisiana State University
MIT
New York Universit
Northeastern University
Northwestern University
Notre Dame University
Ohio State University
Oklahoma State University
Pennsylvania State University
Purdue University
Rensselaer Polytechnic Institute
Rice University
Rutgers University
Stanford University
Stevens Institute of Technology
Syracuse University
Tulane University
University of Alabama
University of California, Berkeley
University of California, Irvine
University of California, LA
University of Chicago
University of Georgia
University of Houston
University of Illinois
University of Iowa
University of Maryland
University of Miami
University of Michigan
University of Mississippi
University of New Mexico
University of Pennsylvania
University of South Carolina
University of Texas
University of Washington
Vanderbilt University
Washington University
Wayne State University
Yale University

APPENDIX C

The table below contains data about doctorates awarded in mathematics in the United States since 1975 according to ethnicity.

Doctorates in Mathematics
conferred upon U.S. Citizens[6], 1975–1987

Year	Total	White	Black	Asian	Hispanic	Native American
1975	923	796	11	62	8	3
1976	803	698	5	49	9	0
1977	744	649	10	42	10	1
1978	666	563	13	43	5	1
1979	603	505	11	46	10	0
1980	582	496	12	42	5	0
1981	525	448	9	40	5	1
1982	499	437	6	32	6	1
1983	457	395	3	34	7	0
1984	443	380	4	30	11	3
1985	418	350	7	33	12	0
1986	402	343	6	28	12	1
1987	397	319	11	41	11	0

REFERENCES

1. D. J. Albers, "David Blackwell," in *Mathematical People*, Birkhauser, Boston, 1985.

2. W. Barry Beckham, "Strangers in a Strange Land: The Experience of Blacks on White Campuses," *Educational Record* **68**, **69** (1988), pp. 74–78.

3. Silvio A. Bedini, *The Life of Benjamin Banneker*, Scribner, New York, 1971.

4. W. W. S. Claytor, "Topological immersion of Peanian continua in a spherical surface," *Ann. of Math.* **35** (1934), 809–835.

5. ——, "Peanian continua not imbeddable in a spherical surface," *Ann. of Math.* **38** (1937), 631–646.

6. Shirley Graham, *Your Humble Servant*, Messner, New York, 1949.

7. J. Arthur Jones, "Blacks in Science: A Growing National Crisis," *Proc. of the Eleventh Annual Meeting of the National Association of Mathematicians* (1981), 20–24.

8. P. C. Kenschaft, "Black Women in mathematics in the United States," *Amer. Math. Montly* **88** (1981), 592–604.

9. ——, "Black men and women in mathematical research," *Journal of Black Studies* **18** (1987), 170–190.

10. L. Lorch, "Blacks on the council," *Notices of the Amer. Math. Society* **30** (1983), 401–402.

11. V. M. Mayes, "Lee Lorch at Fisk: a tribute," *Amer. Math. Monthly* **83** (1976), 708–711.

---

[6]Holders of U.S. permanent residence visas are included here, also.

12. Julia Boublitz Morgan, "Son of a slave," *John Hopkins Magazine* (1981), 20–26.

13. *One-Third of a Nation, a Report of the Commission on Minority Participation in Education and American Life*, American Council on Education and Education Commission of the States, Washington, D.C., 1988.

14. National Research Council, Office of Scientific and Engineering Personnel, *Doctorate Records File*, Private Communication, 1989.

15. Virginia K. Newell, Joella H. Gipson, L. Waldo Rich, and Beauregard Stubblefield, *Black Mathematicians and their Work*, Dorrance, Ardmore, Pa., 1980.

16. John Poland, "A modern fairy tale," *Amer. Math. Monthly* **94** (1987), 291–295.

17. Frank Lillie, "Obituary: Ernest Everett Just," *Science* n.s. **95** (1942), 11.

J. L. Kelley received his Ph.D. in 1940 from the University of Virginia, studying under G. T. Whyburn. He taught at Notre Dame, then served during the war as a mathematician at the Ballistic Research Laboratory, Aberdeen Proving Ground. Following an appointment at the University of Chicago, he moved to the University of California, Berkeley in 1947. In 1950 he refused to sign the loyalty oath imposed by the University Regents, was dismissed from his tenured position, and taught at Tulane and Kansas until the oath was declared unconstitutional. He then returned to Berkeley and later served two terms as chairman. His books include Exterior Ballistics *(with E. J. McShane and F. V. Reno),* General Topology, Linear Topological Spaces *(with I. Namioka and others), and* Measure and Integral *(with T. P. Srinivasan). Concerned with problems of mathematical education, he wrote elementary texts and lectured on Continental Classroom (NBC-TV) in 1960. He retired in 1985.*

Once Over Lightly

J. L. KELLEY

Peter Duren to J. L. Kelley, 10/2/87, *for the AMS Committee on History of Mathematics*:

"*... We invite you to write some kind of autobiographically oriented historical article for inclusion in a centennial volume. We rely on you to make an appropriate choice of topic.*"

J. L. Kelley to Peter Duren, 10/28/87:

"*I'm pleased and honored by your invitation to write an article for one of the Society's centennial volumes... I want to write about the mathematics that most interested me and about the changes in mathematics and mathematical education during my time. I also want to write about universities and about mathematics and politics in war and in peace, and about students.*

...All of this is too much on too many topics, so I propose to try a sketchy autobiography, touching lightly on these matters and full of gossip and name dropping..."

I am a member of a threatened species. For the first thirteen years of my life my family was not urban, nor suburban, but just country. We lived in small towns, the largest with fewer than 2500 inhabitants; the roads were unpaved, we had no radio and television hadn't been invented. I was born in my family's house (there was no hospital in town) and about the only hint of modernity at my birth was that I was an accident, the result of a contraceptive failure. But I was a genuine, twenty-four-carat country boy, a vanishing breed in these United States.

My schooling began in Meno, Oklahoma, which was then a village of a few hundred people, two churches, one general store, a blacksmith's and a one-room school. There was no electricity and the town center was marked by a couple of hundred feet of boardwalk on one side of the road. I went to school at a very early age because my mother was the school teacher and there weren't any babysitters. I remember the first day of school; I got spanked.

There were about thirty students in the school, spread over the first eight grades. Most of the time was devoted to oral recitation, reading aloud, spelling, and arithmetic drill, with various groups performing in turn. We were supposed to study or do written work while other groups were reciting, but listening wasn't forbidden and we often learned from other recitations (simian curiosity is not a bad teacher). The first couple of years of arithmetic were almost entirely oral, quite independent of reading. We recited the "ands" and the "takeaways", as in "seven and five is" and "eleven take away three is", and we counted on our fingers. Eventually, we got so we could do elementary computations without moving our lips, but it was a strain.

The arithmetic I was taught by my mother during the two years in Meno, and thereafter by a half dozen different teachers in four or five other small towns, was mostly calculation. Compared with today's programs: there was more oral work then, and less written; the textbooks then were unabashedly problem lists with a minimum of explanatory prose and they weren't in color, but then and now not very many students read what prose there was; the textbooks then were much shorter. Then and now, most teachers assumed that boys were better at arithmetic, especially after the third or fourth grade; and the end result, then and now, of the first half dozen or so years of arithmetic classes was the ability to duplicate some of the simpler answers from a five-dollar hand calculator. Of course we didn't have hand calculators so this seemed much more important than it does now.

Perhaps it's worth recounting that the mathematics program I was taught in the first six or eight years differed from that taught my father. Somewhere about the seventh or eighth grade there used to be a course called "mental arithmetic", which was problem solving without pencil and paper or, in my father's time, without slate and crayon. He also studied "practical arithmetic" where they learned about liquid measure and bulk measure, liquid ounces and ounces avoirdupois, bushels and pecks, furlongs and fortnights, gallons and pints and gills, interest and discount, and other esoteric matters. Some of these subjects still appear in the late elementary math curriculum, but even though the French Revolution did not overrun England, its system of measurements is conquering the world.

But the mental arithmetic course has apparently vanished from our schools. I regret its demise. A modest competence in mental arithmetic and a five-dollar calculator would, I think, ensure arithmetic competence as measured by the usual standard tests, as well as saving an enormous amount of student and teacher time.

But to return to my own schooling. After arithmetic and a rather muddled study of measurement, I entered high school and an algebra class. The former experience was frightening; the latter devastating. I didn't understand why letters at the beginning of the alphabet were called constants and those at the end were variables; it seemed odd to me that a variable could take on a value, or several different values if it wanted to; I didn't know what a function was, and why a string of symbols should be called an identity some of the time and an equation, or a conditional equation, at other times; and disastrously, I decided that our teacher, who was inexperienced, did not understand these things either. This was quite unfair although it was comforting and the real difficulty was probably my own pattern of being literal-minded (or perhaps simple-minded) in times of insecurity. But the mathematics was abominably organized, and the quantifiers "for every" and "there exists", weren't mentioned, so no one without prior information or divine inspiration *could* tell an equation from an identity. At any rate, my teacher indicated by her grading that she agreed with my assessment of my understanding of the course.

The following year I took my last high school mathematics course, geometry. It was a traditional course, very near to Euclid; it talked about axioms and postulates, defined lines and points in utterly confusing ways. The woman who taught us had a chancy disposition and she had been known to throw erasers at inattentive students. It was the loveliest course, the most beautiful stuff that I've ever seen. I thought so then; I think so now.

One would suppose that I, having fallen in love with geometry, would immediately have pursued mathematics passionately, and one would be wrong. The mathematics course that, then and now, follows euclidean geometry is algebra again. In my junior year in high school I decided to be an artist (we had a sensational art teacher that year) and in my senior year I decided to be a physicist (I had a sensational physics teacher).

It is time to pause a bit, with me proudly graduating from high school, to explain what was going on with the rest of the world. We had moved to California in 1930 along with the rest of the "okies" and so my last high school year was in a downtown high school in Los Angeles. Times were hard. One-third of the men in LA County were out of work and no one counted how many women needed work. But women weren't neglected. There was considerable rumbling about women taking jobs away from men that needed them and, for example, the state legislature in Colorado passed an act denying teaching jobs to married women (this was one of the reasons we emigrated from Colorado); but women had not yet advanced to the dignity of unemployment statistics.

We were poor and it was not a good time to be poor. One summer a couple of years later I worked with my father trucking oranges from the LA basin up to the central valley and peddling them, buying potatoes and fruit in the central valley and peddling it in LA. I remember the Los Angeles basin with stacks of oranges a hundred and fifty feet long with purple dye poured over them so people couldn't steal them to eat or sell; and I remember the camp outside Shafter where hundreds and hundreds of "okie" families lived and everyone, including children of four, picked up potatoes and sacked them following the potato digging machine. There was food rotting, and people hungry, and my view of the glories of unrestrained capitalism became and remains a trifle jaundiced.

But I digress.

One of California's truly great educational innovations was tuition-free junior colleges. I entered Los Angeles Junior College in 1931, at the bottom of the depression, faced only with a three-dollar student activity fee and a block-long line to see a dean for permission to pay the fee with four bits down and four bits a month. But the fee included admission to LAJC's little theater productions, football games and many other goodies and my sister worked in the bookstore and got books for me, so I really had it made.

Besides four semesters of physics (I was still going to be the great physicist) I took of necessity Intermediate Algebra, College Algebra, Trigonometry, Analytic Geometry (even the words have archaic significance) and finally a year and a half of calculus. Calculus was almost as nice as geometry (analytic geometry wasn't really geometry, since Descartes muddled over what Apollonius discovered). And experimental equipment displayed a distinct antipathy towards me. So I entered UCLA, well-trained by very good teachers at LAJC, wanting to be a mathematics major and wondering just how a mathematics major made a living.

As far as I could find out, there was very little market for mathematically trained people. Teaching, actuarial work, and a very few jobs at places like Bell Labs, seemed to be the size of it. I had no money so graduate school seemed unlikely, and high school teaching looked like the best bet. Consequently I undertook three courses in education in my first three semesters at UCLA in order to prepare for a secondary credential. The courses were pretty bad and besides, the grading was unfair, e.g., I wrote a term paper for Philosophy of Education and got a B on it; my friend Wes Hicks, whose handwriting was better than mine, copied the paper the next term and got a B+, and our friend Dick Gorman *typed* the paper the following term and got an A.

Of course teaching is a low prestige field in this country. The prestige of a field of study is apparently a direct function of the technical complexity of the surrounding society. Engineering, and especially civil engineering, seems to be the prestige field at a relatively early developmental stage (e.g., pre-World War I U.S., pre-World War II India), to be overtaken by chemistry and chemical engineering as technology develops (World War I was a chemist's war), followed in turn by electrical engineering and physics (World War II, radar and nuclear weapons). It has been said that the last war will be a mathematician's war, so mathematics is now deadly and hence reasonably prestigious.

Fortunately for history, the precise time that mathematics acquired prestige among students at Berkeley is recorded. My student Eva Kallin explained to me that during her first couple of years at Berkeley she suppressed the fact she was a math major when talking to an interesting new man; later it was OK to be a math major, and a little later it was a *very* definite plus.

Back to UCLA. Los Angeles itself was then a gaggle of small towns held together by a water company, and UCLA was on a new campus, plopped down on the west side of town in the middle of an expensive real estate development. Too expensive for most of us students, so we drove, hitchhiked, car pooled or bussed from our homes to the school. There were about 4500 students and the math department was on the top floor of one wing of the chemistry building. It was definitely not a prestigious location. But mathematicians were usually viewed with an uneasy mixture of awe and contempt like, say, minor prophets. Our prophetic powers were used: math courses were prerequisites for courses in other fields, and math grades were often used to section physics classes into fast and slow groups. But mathematics was scorned as being irrelevant to the "real" world.

E. R. Hedrick, of Goursat-Hedrick *Cours d'analyse*, chaired the department—he later became chancellor. I enrolled in the last term of calculus, won the departmental prize for a calculus exam ($10), then blew the final on my calculus course and got a B (Wes Hicks said they should have offered a fifty-cent prize). I got shifted from my part-time job in a school parking lot to a part-time job in the math department office keeping time sheets for readers, recording grades, and whatever. I took all of the courses in geometry, mostly from P. H. Daus, admired Hedrick's flamboyant lecturing style, conceived quite a fancy for my own mathematical ability, and quit taking education courses, thus abandoning a career as a high school teacher. (I could *always* go back and get a teaching credential if I had to.)

In midyear 1935–1936 I graduated and was given a teaching assistantship in the department at $55 a month, which was enough to live on, and so became one of the multitude feeding at the public trough at the taxpayers' expense. Of course I could never have gone to college except at a public school—I could barely manage to cope with UCLA's $27 per semester fee—so I *like* public schools, and the public trough is just fine. The fall of 1935 was notable for another event: I received my first college scholarship. It paid $30.

During my last year at UCLA I began to learn how to teach (I was terrified) and I was first exposed to the R. L. Moore method of instruction, which was fascinating (more on this method anon). W. M. Whyburn, who took his degree at Texas, introduced me to the Moore methodology in a real variable course, told me I had to leave to get a Ph.D. (I didn't even realize that UCLA had no mathematics doctoral program), and arranged a teaching assistantship at the University of Virginia for me. In 1937 I was granted an M.A. and headed for Virginia.

I crossed the Mississippi river for the first time that September, carried in a brakeless old Packard 120 by a maniac who had advertised in an LA paper for riders going east. He dropped me off in Knoxville, I took the train to Charlottesville and enrolled in the university.

I didn't know what to expect. I'd consulted the U. Va. catalogue about requirements and it stated that "The requirements for graduate degrees in mathematics are the province of the School of Mathematics", which is not very informative although it's a classy way to go. (Consider the number of deans and faculty committees that are bypassed! But wait until I get to Witold Hurewicz' theory of deanology.)

As it turned out, I didn't need to know what the requirements for a degree were. G. T. Whyburn, E. J. McShane, and G. A. Hedlund *told* me what to do and I did it. That first year I took Point Set Topology from Whyburn and Calculus of Variations in the Large from McShane. The C of V was horribly difficult for me in spite of valiant attempts by A. D. Wallace, George Scheigert, and B. J. Pettis to teach me enough algebraic topology to understand the lectures. But the topology course was *geometry*, and she was my friend. Here are some results that we proved in the course, to give the flavor of the material.

Suppose that X is a separable topological space whose topology has a countable base, and that each neighborhood of a point contains a closed neighborhood of the point (i.e., X is regular). Then X is normal, and in fact metrizable. If X is locally connected, then it is the continuous image of a closed interval (it is a Peano space) and it is itself arcwise connected. Moreover, if two distinct points of a Peano space are not separated by some cut point x, i.e., don't lie in distinct components of $X\backslash\{x\}$, then the two points both lie on some simple closed curve.

The course on point set topology contained beautiful mathematics and it was done in a fascinating way. Whyburn stated theorems, drew pictures, gave examples, and we were left to find proofs. Each day he listened with enormous patience to our clumsy presentations of proofs of previously announced results. If no one of us had a proof of a result and we all gave up on it, he presented a proof himself. Otherwise he just listed more results, all chopped up into lemmas and propositions that we might be able to prove. It was often brutally difficult and it was always enormous fun. It gave us great self-confidence and a really deep understanding of a body of material.

By the end of the year I'd written a couple of papers and considered myself a mathematician. Indeed, mathematics has been my pleasure and my support since then, and it sure beats working for a living. Of course there is some drudgery. The last two years before my Ph.D. I taught thirteen hours a week (the same course at 8:30, 9:30, and 11:30—Whyburn didn't believe in having his students do too many different preps because it took too much of their time). But I had an assistant, Truman Botts (later the Director of the Conference Board of Mathematical Sciences), who tried to teach me to fence and, pounding out the Revolutionary Etude on a beat-up old piano in the gym, explained to me that composing was certainly a better idea then returning from France to Poland to fight.

My self-satisfaction after the first year at Virginia knew no bounds, and so it was probably just as well that during my second year I was taken down a notch. I tried to solve a problem of K. Menger: is it possible to construct a metric for a Peano continuum X so that X is (metrically) convex? I spent months on the problem, couldn't do it, and was abashed when both Ed Moise and R. H. Bing, independently, established the conjecture.

Before leaving the lovely lawns of Mr. Jefferson's University let me mention two more notable facts, the first about the mathematics that was being done in this period, and the second about the university students' honor system and why it worked. First, during my second year there I was taught J. Alexander's duality theorem about the relation of the homology of a nice subset of n space and that of its complement. It was a major turn-on for me, and so I read Pontrjagin's beautiful proof of the duality theorem for compact subsets of n space, but then I didn't know how the necessary duality theorem for locally compact groups was proved so I had to read that, and to straighten this out, I went through Emma Lehmer's translation of Pontrjagin's book on topological groups, and so (it was a year or two after my Ph.D. by now) I wandered into functional analysis.

At the University of Virginia the honor system worked. Partly this was because it was a university of reasonable size (four or five thousand), rather than a megaversity, but most importantly because the faculty and administration stayed out of it. The only possible penalties, if guilt was established, were resignation from the university or dismissal, and dismissal showed on one's record. To the best of my judgement, this worked better and with fewer injustices than faculty or administration systems. I think the governing principle is that students are better at this sort of problem than professors.

Let me describe, with some nostalgia, what being a mathematician was like in the decade or so after 1938. First, there weren't so many of us. About 100 Ph.D.s a year were granted except for the war years, and even the Christmas meeting of the Society drew only four or five hundred people. Society meetings were always held on college campuses, virtually all of the participants lived in the college dormitories and ate in the cafeteria, and almost everyone knew everyone else. Irving Kaplansky could say with only mild exaggeration that he knew every mathematician in the United States. It was a smaller world.

The mathematicians then were like mathematicians now, only more so. John Wehausen, an early editor of the *Mathematical Reviews*, once told me that mathematics was one of the psychologically hazardous professions. "Every mathematician, for most of his early life, is the brightest person he knows, and it's a great shock when he finds there are people that can do easily things that are very hard for him" according to John. I think that this is true, and that within every mathematician, more or less suppressed or laughed at, is an arrogant little know-it-all, and simultaneously a stricken child who has been found wanting. Johnny von Neumann has said that he will be forgotten while Kurt Gödel is remembered with Pythagoras, but the rest of us viewed Johnny with awe.

Arrogance in good graduate students is much admired though one usually hopes that they will grow out of it. I remember Murphy Goldberger's description of a physics student: "He understands everything, he knows everything, he's incredibly quick, he can barely contain his contempt for the rest of us." Students in theoretical physics are much like math students, although Feynman insisted that the difference between math and physics is the difference between masturbation and sexual intercourse.

Perhaps a few anecdotes about mathematicians will help characterize the breed. Paul Erdös was one of the characters. For many years Erdös wandered about the world in almost periodic fashion with a long list of mathematical problems in his head and the rest of his possessions in two suitcases. He must hold the record for writing the most joint papers with the most different authors. He had an elaborate code: "epsilons" were children and very young epsilons with that profound look knew all of mathematics, but couldn't talk; "bosses" were wives, and "slaves" were husbands. He enjoyed being an eccentric, and was a charming but absent-minded house guest.

Witold Hurewicz, for a brief period, expounded a theory of "deanology". It began like this: Let S be the set of frustrated scholars, let B be the set of frustrated businessmen, and let D be the set of deans. *Axiom*: $D = S \cap B$. And so on. The fascinating part of the theory was the method of reproduction. Sons of deans are not deans, but potential deans marry deans' daughters. He expounded this theory once at a rather formal dinner given by a rather pompous host, and his hostess said, "But I'm a dean's daughter", and that stopped the conversation. Afterward Witold, looking mischievously penitent, said to me, "But what could I do? It's exactly what I meant." Witold was a gentle, elfin man, incredibly insightful and inventive, and he wrote mathematics like poetry.

No list of eccentric mathematicians would be complete without Norbert Wiener. Many mathematicians like to show off, a sort of delayed "show and tell" syndrome, but Wiener really *demanded* attention. He was short, a bit plump, and had a neat pointed beard that he wore pointed up in the air. It was rumored, and it was quite possibly true, that he wore his bifocals upside down. He feared that his students called him "Wienie" (they called him Norbie). His standard ploy when attending a lecture was to walk in late, walk down to the front row, take out a magazine, read ostentatiously, then sleep ostentatiously, wake abruptly at the end of the lecture to ask a pointed question, or sometimes to make a little mini-lecture of his own. For awhile he had a game of asking others for a list of the ten finest American mathematicians. At one math meeting (Duke, sometime in 1938–1940) a number of people concocted a response. They would run briskly through a list of nine mathematicians, omitting Wiener's name, and then look thoughtful and puzzled about the tenth until Wiener's squirming was unbearable. It sounds cruel, but I suspect Wiener knew what was going on and enjoyed the attention.

But to return to the autobiographical business. In 1940 I wrote a thesis, Whyburn made me revise it, McShane made me revise it again, and Hedlund said *he'd* make me revise it except it was too late in the year. So it was accepted and then Sammy Eilenberg spent a couple of weeks revising and making me revise. This training, with a post-graduate bit from Paul Halmos a few years later, is how I learned to write mathematics.

On a rainy day in Charlottesville in June 1940, I was granted a Ph.D. degree. But this remarkable occurrence was overshadowed by the commencement address. Italy had just entered the War and Franklin Roosevelt said, "... The hand that held the dagger has plunged it into his neighbor's back... ". It seemed pretty clear that the war that had begun in Spain in 1937 would now engage us, and within a year and a half it did. So I'll be getting on to the bit about how I won the war.

The Christmas meeting of the Society in 1941 was held in Chicago, and was titillated by the news that three aliens, two of them enemy aliens, had been caught taking pictures near a radar station on Long Island. They offered the unlikely story that they were on their way from Princeton to Chicago. Further details were soon available. It turned out that their names were Paul Erdös, S. Kakutani, and Arthur Stone. Oswald Veblen of the Institute for Advanced Study, also known by his irreverent young admirers as his Grey Eminence or the Great White Father, finally got them out of stony lonesome.

Oswald Veblen, Jimmy Alexander, and Gilbert Ames Bliss were at the Ballistics Research Laboratory at Aberdeen Proving Ground in World War I, and they redid exterior ballistics following the methods of computational astronomy. In the second war Veblen acted as recruiting agent for mathematical types for Aberdeen. There was already a mathematics unit at Aberdeen under Franklin V. Reno, who was trained in astronomy. He had set up a system of cameras obscura to obtain ballistic data on bombs, and he devised the standard method for constructing bombing tables. He was meticulous; at the laboratory the smallest known unit of measurement was called the Reno. It was defined as the width of a milli-frog's hair.

Veblen talked to Reno and asked if he needed help. Reno said yes, but he wanted someone he could boss around or else someone who would boss him around, and Veblen got Jimmy McShane to be the big boss. A little later McShane wanted me. I was teaching at Notre Dame; they didn't want to let me go. Veblen sent his assistant, Gerhard Kalisch, who was an alien and couldn't work at Aberdeen, to teach in my place, and I went to Aberdeen. Veblen had persuasive ways.

Our group at Aberdeen, known at various times as the math unit, the math section, and the theory section, set up new computational procedures for exterior ballistics and did troubleshooting on all sorts of projects. The construction of artillery firing tables had long since been turned over to the computing branch, as had the tables for level bombing a few years before. Theory section projects during the war included such exotica as: tables of Fresnel integrals, as well as of various statistical variables; reduction procedures to obtain aerodynamic constants from spark range (shadowgraph) data; ballistics for dive bombing; ballistics for the Draper-Davis lead computing gunsight; construction of ballistic theory and procedure for range firing and making tables for rocket air-to-ground firing; taxonomic work on known aerodynamic data for bomb shapes; measurement of aerodynamic constants for some bomb shapes; and emergency work on a variety of fouled-up ordnance and projects.

Let me try to sketch the path of a single continuing problem.

Suppose an arrow moving through the air is yawing; i.e., the axis of symmetry is at an angle to the velocity vector. Then there are forces of drag and lift and, if the yaw is small, one can measure them in dimensionless constants (or functions, if the velocity varies over much of a range) and these can be used to predict, after a fashion, the behavior of the projectile. But what if the arrow is spinning? Of course there are inertial effects, but are there nontrivial aerodynamic effects?

Here is an experiment devised by Bob Kent, head of interior ballistics at Aberdeen. He constructed a "bomb", a wooden cylinder with a couple of lugs at the side and a weight in the front. When fired from a "smooth bore" shotgun (no spin) it wiggled its tail a bit and then flew like an arrow, stable as can be. When fired from a "rifled shotgun" so that it rotated, it started out well, then developed a flat spin and tumbled. This worked *every* damn time, and the most reasonable explanation is that the aerodynamic Magnus force and couple can cause instability.

On the other hand, the British had a high-accuracy bomb (I think it was called the Tall Boy) which they deliberately spun, to average out asymmetries, so not every spinning bomb is unstable. We (the theory section and Alex Charters of the spark range group) measured the aerodynamic coefficients for real bomb shapes in the twenty-foot wind tunnel at Wright Field (at night because 35,000 h.p. takes more power than the City of Dayton can spare during the daytime). Tare effects (effects of the suspension system) messed up results on the small standard practice bomb but the results for the general purpose bombs were consistent and useful. "Statically" stable bombs can be dynamically unstable, and increasing static stability can remedy matters.

A stability problem of just this sort came up very late in the war. After the Allied invasion of Europe a minor scandal erupted. Alongside the roads of Normandy a lot of American 2000-pound bombs were sprinkled; the fusing wasn't designed for every possible landing position and reports said that the bombs went into flat spins. The problem wound up at Aberdeen.

Of course one could redesign the whole thing, but that's very expensive and very slow and so a simpler fix seemed very important. A hydraulic engineer from Cal Tech, Bob Knox, suggested running water channel tests on an (interval $\times$ bomb body shadow) and try various tail patterns, all this on the basis of an analogy (with the wrong γ) between rotationally symmetric flow and two-dimensional flow. A couple of GI's and I tried this in Bob's lab at Cal Tech. (I ran into a friend, Hans Albert Einstein, there and introduced him to Bob, who turned out to be his colleague.) The experiments suggested adding to each flat plate of the tail a plate so the cross section was a line interval with a triangular form on the rear third of the plate (making the cross section a rough cusp, point forward). So we designed a fix, had it made, and it worked.

There was a little fuss at a conference later involving some high brass (civilian and military) about who deserved credit for this remarkable wing design and this flattered me. Bob Kent told me later that he'd looked through my notebooks and he thought that the design was a pretty wild guess, if it was based on *that* data. But a bit of Irish luck never hurts.

Jimmy McShane had a health problem and had to go back to Charlottesville, Reno's health was not good, and so I ran the section for the last year or so. The astronomer Edwin (Red Shift) Hubble was my boss. He was a pleasant well-spoken man, still very much influenced by his Rhodes scholarship. He talked of "shedules" and "leftenants" and such and his irreverent underlings spoke behind his back of "that skit about the shedule" and so on. Hubble was rather reserved and we saw nothing of him outside of office hours, but we understood he read Horace with the commanding general. Bob Kent, who never entirely grew up, was known to remark, "Dr. Hubble, known to his intimates as Dr. Hubble,... ", but in fact we all worked together very well.

The war ended for the theory section, not with a whimper, but a bang. The European war had dribbled out in daily rumors of new coups, new crises, and new German governments, so we were unprepared for the end. But for the Pacific War *we were prepared.* We'd hoarded ration tickets for liquor and the entire section, mathematicians, secretaries and computers (people who used desk calculators) had a historic party at Tony Morse's house in Aberdeen, and within weeks we began to drift away, out of town.

I wanted to get back to mathematics, get the rust out of the tubes. For three years, except for some conversations with Herb Federer and Tony Morse about set theory and a bit with Chuck (C. B., Jr.) Morrey on area, I'd only thought about useful (i.e., potentially murderous) mathematics. I asked Veblen for help and he helped. He arranged that my new boss, the University of Chicago, and the Institute for Advanced Study split my salary for a year and I went to Princeton.

At that time the Institute was mathematics heaven, the place all good mathematicians wanted to go, and it really was heavenly. It was the first time I'd had no responsibility save mathematics, and the fabled characters of my time drifted in and around the Princetitute. Veblen, Alexander, von Neumann, Weyl, Lefschetz, Eilenberg, Montgomery. The words make a litany.

The social life and the social knife at Princeton were a revelation to me. "The Veblens live a very simple life. I think it must be very expensive to live so simply", said Dolly Schoenberg, married to Iso and daughter of Landau, who married the daughter of Ehrlich, whose wife was related in some fashion to one of the Minkowski brothers. (I've probably mixed up a lot of this—my memory isn't too good.) "You know he's a son of a bitch, but you have to like him because he's so sincere about it," said one anonymous friend of mine about another ditto.

Something I like to remember. My father-in-law was the physician for Hans Albert Einstein's family in Greenville, S.C., and we knew Hans and his wife and made acquaintance with his aunt Mrs. Winteler, who took a liking to my young son. While Mrs. Winteler was visiting her brother (Hans Albert's father) Albert Einstein in Princeton they invited my son, my wife and me to tea at his house on Mercer Street. I'd known Albert Einstein to speak to (the Institute wasn't crowded that year) but this was the first time we'd actually had a conversation. He was gentle, he was thoughtful, he talked about mathematics and physics and me, and I remember his saying, "Your job is easier than mine. What you do only has to be correct, but what I do has to be both correct and right." He was absolutely without pretension, without condescension, and he impressed the bloody hell out of me.

There was only one other famous person who, in person, so surpassed my expectations. The first professionally produced play I saw was Lillian Hellman's "The Children's Hour" and I was enormously impressed, and later I liked her plays, her other writing, and her politics. So in 1960, somewhat embarrassed with myself, I got my Tulane philosopher friend Jimmy Feibleman to take me to lunch with her. She was great. I think I've read everything she wrote, as well as some of the snide stuff that was written about her after.

But I digress, and so back to mathematics. In 1946–1947 there was a lovely seminar at Chicago. It started out with functions of positive type and carried on through works of M. H. Stone, Gelfand, Raikov, Shilov, Tannaka, and others. Seymour Sherman, Paul Halmos, Irving Kaplansky, Will Karush, Al Putnam, Marshall Stone sometimes, and I took part. In a certain sense I at last began to understand the role of linearity, and the wobbly path that led from Alexander's duality theorem to the Fourier integral became clear.

The Chicago seminar had a decisive effect on the direction of my work. In Berkeley, in 1948–1949, I was booked to teach algebraic topology, and I asked if I could do topological algebra instead. I got an absent-minded approval, which is what I'd hoped. It was sort of a topics course, not yet approved for the catalogue, and neither algebraic topology nor topological algebra had ever been taught at Berkeley, and I doubt that my question was really understood.

But back to the real world for a little. The hot war was over and the cold war had begun. Our intelligence services imported and/or protected a most unattractive batch of German and Japanese war criminals on the grounds that we needed their expertise. Allen Dulles' amateur spooks were legitimized as the CIA, and the domestic spook front also brightened up as a massive "security" program was put in place to harass the citizenry and provide program music for that thrilling melodrama, "China is getting lost, or the battle against monolithic, atheistic, godless communism". In particular, a bill was passed that denied federal employment to members of the Communist Party and required federal employees to sign a statement as to whether they were or had been members of the party. All federal employees had to sign, at least if they wanted to remain federal employees. All of this seems pretty routine now, but it did take a little time for our gallant ally Russia to become the evil empire Russia.

At any rate all the employees of the Laboratory signed a statement that they weren't and hadn't been communists. Then, sometime in the years 1946–1948, McShane and Everett Pitcher and I were called before a Federal Grand Jury in New York and questioned about Frank Reno. It turned out that Frank had been a communist, that he had known Whittaker Chambers, and that Chambers had denounced him. There was also some talk about Reno giving documents to Chambers, but no charge was ever made. However, Frank had signed a statement saying he had never been a communist, and so his federal employment was over permanently.

Frank tried to get all sorts of jobs without much luck. Jimmy McShane and I both recommended him, explaining why he could not have a federal position, to a number of people including Abe Taub at his computing laboratory at the University of Illinois, the only computing lab in the country that was not on federal money. As a result Abe Taub was hauled before a loyalty board under threat of losing his own security clearance.

Reno was never again able to use his very considerable talents as a practical astronomer, statistician, and applied mathematician. The FBI kept potential employers informed as to his past and they pressured him to register as a foreign agent. On the basis of his signed denial of earlier membership in the Communist Party, he was charged with fraud (accepting his salary falsely) under an act that was passed because of Anaconda Copper misdeeds. He was convicted and sentenced to two years imprisonment and, with time off for good behavior, he served the sentence. When I visited him in Leavenworth sometime in 1952–1953 he said it wasn't too bad, that one had to be careful of psychotics and never to settle a bet even though the inmates called him "doc" and appealed to him as an authority. He said most inmates were wild kids who'd driven stolen cars across state lines which made it a federal beef.

I saw Reno just once after that visit in Leavenworth. My family and I were car camping around Boulder, I picked up Frank in or near Denver, he camped with us for two or three days, and we talked a bit. He told of his mining engineering father; of violence in Leadville, not far from where we camped; of his graduate school days in the observatory at the University of Virginia. After the University of Virginia he got a job as a statistician in the agriculture department in Washington—he was probably a bearcat at civil service exams—just about at the bottom of the depression. He was recruited into the Communist Party (by Steve Nelson I think), was active in the Party and knew Whittaker Chambers. He told me that Chambers used to demand, cajole, and threaten in order to get money from him and other party members. Reno left the Party when he got the Aberdeen job, and he heard no more from Chambers.

I digress to recall that Frank did all of the ballistics for horizontal bombing, adopting the drag data of the Gâvre commission as a guess at the drag of bombs (a reasonably good guess), making the necessary ballistic tables, designing and supervising the construction of instrumentation for range bombing, and establishing procedures for making bombing tables. It was a first-class job and he was decorated for exceeding his authority in doing it. The pickle barrel into which our bombers could drop their bombs under ideal conditions, from 20,000 feet, had a radius (probable circular error) of about 140 feet, which was better than that attained by any other air force.

Frank was very much part of the last hundred years of American applied mathematics, and he and his family are very much a part of American history. His grandfather was the Major Reno who fought under Custer at the Little Big Horn and later became a general and had a fort named for himself. Frank told me in detail of Custer, of West Point and the battle of Bull Run, and of the Indian Wars; of Custer's last battle and of Reno's fight—30% casualties in twenty minutes; of the Sioux, of the Dakotahs and the Hunkpapas and the other subtribes; of statesman-sachem Sitting Bull, of Crazy Horse and Gall and the other two war chiefs.

All of this was related in the high mountains, under the stars, before sleeping. The last night he explained, to ears unbelieving of such jury-rigged Rube Goldberg gimmickry, how the astronomical scale of distance was established.

The next day Frank rode with us on our journey for fifty or a hundred miles, reluctant to part. I did not see him again, and except for a few letters, that is all. I mourn him and the way the country treated him, and that only a poor man's Horatio speaks for him.

I arrived in Berkeley in 1947, just in time to observe the death of Joe College. He was done in by the returning war veterans who entered the University on the G.I. Bill of Rights. It was too much to expect a new freshman with thirty missions over the Burma-China hump to stay off the senior bench, or to wear a freshman beanie, and hazing was definitely out of the question. So, in spite of occasional revivals of fraternity rituals, Joe College died; the University blossomed.

It is easy to describe the Berkeley math department of that period: very strong in analysis, statistics, set theory and the foundations of mathematics, and not strong in other areas. It was a harmonious group, although there was a bit of jealousy of the statisticians because they could get consulting money and were generally a little more prosperous than the rest of us (something like the computing science people today). But this was temporary; statistics emigrated to become a separate department sometime in 1949. There was also occasionally a little nervous hostility toward the work in foundations, accompanied by a shaky lack of confidence that we understood the foundations of our own field. This hostility has now pretty well vanished, and unfortunately the intimacy and convenience of a small department has also vanished.

There was one curious action in the early 1950s that distinguished our department amongst other departments. In late 1949 or early 1950 we agreed that if any of us were dismissed, for any reason whatsoever, then each of the others would contribute up to ten percent of his yearly salary to support the dismissed person or persons. This agreement was called "Mathfund" and there was a reason for its existence. There was a peculiarly virulent outbreak of anti-Communist fervor in Sacramento and one of the University vice-presidents had a brilliant idea: Let's stop attacks on the University by getting the faculty and other employees to sign a loyalty oath denying membership in the Communist Party. Of course the state constitution already required an oath of office, a promise to support the constitution of the U.S. and of the state of California, and forbade any other oath or test, but that sort of detail didn't bother our administrative executive types.

The faculty got upset. The Academic Senate had a great deal of power at that time, because it won an argument with the University president in the early 1920s and was not yet being choked by sheer numbers, excessive structure, and a statewide superstructure designed to suggest that all the University's campuses are like Berkeley.

The Senate held interminable meetings, a group of "non-signers" emerged, the Korean War began and a good many of the non-signers breathed a sigh of relief and signed on, the scared Senate passed a resolution that membership in the Communist Party was inconsistent with membership in the University, and a "compromise" was arranged. The Senate's Committee on Privilege and Tenure resigned, a new blue ribbon committee was appointed, and each non-signer had the privilege of appearing before the committee.

In the spring of 1950 the various non-signers appeared before the committee, and the committee brought in its report in April or May. The committee argued sturdily for all the non-signers except five, and these it "could not recommend for continued employment" although there was no evidence of membership in the CP. So much for tenure.

If memory serves, two of the five people thus unceremoniously dumped were women, Elizabeth Hungerland of the Philosophy Department and Margaret O'Hagan of Decorative Art. The three men nominated for firing were Nevitt Sanford, Harold Winkler, and me. Sanford was a professor of psychology, a psychiatrist and author, and later founded the Wright Institute. Hal Winkler was in the Political Science Department and was later the first president of Pacifica, the mother foundation for the public radio stations,

KPFA, WPFW, WBAI, WPFW and KPFK. And I was associate professor of mathematics, John Kelley.

I hit the panic button and wrote Veblen, Whyburn, McShane, Lefschetz, and a couple of others. It was June, I had a wife and three children and just two months' salary in sight. Then Bill Duren called me from Tulane, told me that S. T. Hu was going to the Institute for a couple of years, and in that courtly southern way he gravely said that he understood I might be free to accept an appointment. *Jeez*!

Later that summer the Regents rehired the five, gave everyone thirty days to sign and then fired *all* the non-signers. Hans Lewy, Pauline Sperry, and I were fired from math, Charles Stein and Paul Garabedian left in disgust, R. C. James left soon after, S. Kakutani refused to accept a position because of the treatment of his mathematical colleagues, another of our department went on a self-imposed exile for three years, and I heard that preliminary talks about bringing the Courant group to Berkeley ceased abruptly. Chandler Davis and Henry Helson declined to take positions at UCLA because of the oath. (I only learned that this past summer.) This was a fair amount of carnage in just one field, and it's hard to say how much the oath damaged the University. Postscript: The next fall Monroe Deutsch, former Provost of the Berkeley campus, and the faculty group called "Friends of the Non-signers" (chaired by Milton Chernin, with Frank Newman, later a justice of the California Supreme Court, as treasurer) took political command of the Senate and sent the Committee on P and T back to do its homework again, and they did. But we were long gone and the Regents' edict was unchanged. Quite a few of the non-signers returned to Berkeley three years later under some sort of amnesty but our complete legal vindication by the California Supreme Court waited until 1956.

A last word about our famous loyalty oath. The Regents' problem with us non-signers wasn't communism; it was insubordination. For example, in my case: at that time I did consulting work for Aberdeen Proving Ground, Redstone Arsenal, Sandia Corporation, and Los Alamos and was cleared for highly classified material. I see no way that the Berkeley administration and the Committee on P and T could have failed to know this; the problem with me was that I wouldn't *say* that I wasn't a communist.

But back to Tulane; it was lovely. Bill Duren, Don Wallace, B. J. Pettis, Paul Conrad, and Don Morrison were there, the graduate school was vigorous though not large, and the food was magnificent. Gumbo, oysters, shrimp, and crab; crayfish bisque! I drool to think of it.

I taught two courses; formally I was on half-time and the rest of my salary was covered by Mina Rees' invention, an ONR grant. An unpleasant incident: a colonel wrote Bill and/or the ONR to complain that a known communist or at least an associate of a known communist was feeding at the Navy trough, but nothing came of it. It's the sort of thing one expects of colonels. They're always starting revolutions, or committing coups, or whatever; it's part of their midlife crisis, the syndrome that we called "bucking for B. G." at Aberdeen. If you are a colonel and haven't taken the precaution of marrying a Senators' daughter or finding a communist or otherwise displaying political acumen, you're at the end of your line. That step from colonel to B. G. is *the* biggie.

There was a more upsetting occurrence. In 1951–1952 I was called to Albuquerque (or was it Los Alamos?) for a Loyalty Board Hearing. There were three charges: (1) I hadn't signed the U. C. loyalty oath, (2) I continued to associate with a known communist, Frank Reno, and (3) I was careless in handling classified material and uncooperative with an FBI agent in Berkeley. Certainly (1) was true, (2) was true except that Reno wasn't a communist and hadn't been since I'd known him, and (3) contains a good bit of truth. I had some stuff from Aberdeen, some of it was marked "Restricted", most of which I'd written myself, I had no private office and the stuff was kept in a couple of cartons in a non-private office. I was also rude to an FBI agent. Later, under the Freedom of Information Act, I read his letter stating that since I was fired I would undoubtedly want to work again at Aberdeen and he recommended that I not be hired there. I wish I'd been ruder.

The Loyalty Board seemed very reasonable, they recommended clearance for me, the district manager concurred, Eisenhower was elected, the general manager appealed, and clearance was denied. For me no appeal, no witnesses, no hearing, no nothing. It ended my work for the AEC. A patent or two was taken out in my name (or my name plus Charlie Runyan's) and an FBI agent in New Orleans got my signature and gave me one dollar in the coin of the realm. Some years later I got a notice that some patent was being released, but I don't know what. Not sure I'm cleared to know.

I did retain security clearance, through "Confidential" at least, after the loyalty hearing and I continued to do some consulting for Redstone and for Aberdeen up to the time I returned to Berkeley. I wanted out of classified work entirely, but I was barely employable in the crazy freaked-out atmosphere of the witch hunt, and I hung on to security clearance as a possible protection, a security blanket.

The Tulane appointment was for two years. Rochester needed a mathematician and I'd thought an appointment was arranged there; but John Randolph wrote that his dean turned it down because it could make getting federal grants difficult—and it might easily have done so. Lee Lorch offered me a job at Fisk (he was fired from Fisk at the end of the following year) but Baley Price had just offered me a place at the University of Kansas. (It was May or June of the year again and I was jobless.) Baley had rescued Nach Aronszajn and Ainsley Diamond when Oklahoma State freaked out. Nach ran an excellent seminar, I made some good friends, and it was a good year. The following spring some signs of normalcy appeared at Berkeley and the University even agreed to put non-signers who returned on the payroll, and so I went back. I really didn't expect to stay more than a year or so because I was still outraged by the University's behavior. I had dreams of taking off my sandals, shaking off the dust and stalking out, but I never got around to it.

But before we relax in Berkeley's ivory towers, let me announce a profound truth made clear by my eleven years experience and observation (1942–1953) of soldiers and spooks: The military, with assistance from security spooks, is deadly effective against both research and development. Here are some more or less current examples of what I modestly think of as the Kelley principle.

The September 1988 *Scientific American* carries a fascinating article on Halley's Comet and its meeting with five spacecraft that obtained data to analyze the gases and dust in the vicinity of the comet and photographed the nucleus, the tiny solid body in the comet's head. The space probes were the Sakigake, Suisei, Vega 1, Vega 2, and Giotto. Two were launched by Japan, two by Russia and one by Europe. U.S. science? Well, the Air Force is in charge. Their Challenger, a monument to the Wild Blue Yonder syndrome, is a press agents' dream when it works, but it is obviously not a comet chaser. Not to worry: Halley's Comet will be back in eighty years.

Another example: "Star Wars", otherwise known as Pie in the Sky, is based more on fantasy than on science according to many scientists. And the project has already led to suppression of scientific dissent and dissemination of incorrect data (as revealed by the Woodruff affair at U. C.'s Lawrence Livermore Laboratory).

Here is a last picturesque example: The Stealth bomber stole onto the front pages of our local newspaper recently, after a long, well-announced development that began no later than Jimmy Carter's presidency. It's a swoopy looking machine, sort of an up-to-date Batmobile, but in spite of its long and public history, no prototype has yet flown (according to our local press). If this is indeed a military research and development project, for what war is it being prepared?

But we're all tired of soldiers and spooks and so let's go back to Berkeley, a quiet place in the mid 1950s. The students did not riot, except for panty raids, and a dog named Wazu narrowly escaped being elected president of the student body. Students were not very much involved in politics, and the university administration encouraged political lethargy. It was, for example, forbidden to invite a candidate for public office to speak on campus, and a firm foundation for the Free Speech Movement of 1964–1965 was under construction.

Toward the end of the decade the students showed a bit more initiative. They began to publish "Slate", an evaluation of courses and teachers that appeared at the beginning of each registration period. It caused a flutter in the dovecote—not that students haven't always evaluated faculty, but it's not usually been systematic, with comments on lectures, exams, and grading patterns.

But the Berkeley students of the fifties were not confrontational, although one could say that some finished their political activities with a splash. You see, in those benighted days a sort of dog-and-pony show called the House Unamerican Activities Committee (HUAC) roamed the countryside in search of headlines, and in 1959–1960 they were booked into a hearing room in the City Hall in San Francisco. A bunch of U. C. students tried to attend the hearing, found themselves unwelcome, sat down outside the hearing room and were presently washed down the stairs by fire hoses manned by the San Francisco Fire Department. Unfortunately the cameras weren't ready and almost none of the students could be identified, though a rough idea of HUAC tactics could be deduced. No one was hurt, no one was convicted of anything, and the general popularity of the dog-and-pony show took a satisfying drop.

But back to mathematics. There was a major development in the math ed business in the last half of the fifties. The war had focused a lot of attention on scientific training, and especially on mathematical training. A super-committee, the Commission on Mathematics, was formed (by ETS, AMS, and MAA if my memory is right) to investigate the situation and make such recommendations as seemed needed. The super-committee was set up in 1956 and found that the wrong math was being taught and often taught badly, and recommended a major effort to improve matters. A serious effort was begun in 1957 and then Sputnik was launched. It was like striking oil.

Sputnik raised enormous questions, and our own experts and newspaper pundits responded with something like panic. Was it possible that the Russkies were ahead of us on something? What was wrong with our own program (see the Kelley Principle)? Couldn't an astronaut just throw nuclear bombs over his shoulder at us? All of a sudden there was a *lot* of money around for space flight and for technical training—enough money for technical training that some was even available for mathematics.

A long-term program improvement project, The School Mathematics Study Group (SMSG), was set up under E. G. Begle, first at Yale University in 1957 and then later at Stanford. The group labored for more than a score of years, with impressive results. Every high school program today shows improvements that began with SMSG, every university program is changed because of changes in the high school programs, and Begle's bunch of Ph.D. students remain outstanding in the math ed biz. SMSG, the Madison project, the Ball State project, Minnemath, and many others changed the character of pre-college mathematics instruction. All of this and a theme song, *New Math*, by Tom Lehrer, to boot.

I got involved with the math ed biz because of an over-developed sense of outrage. I attended a conference in the 1950s that was re-examining the requirements for a California state secondary teaching credential, and neither the old nor the newly-proposed credential required, for example, that a teacher of ninth grade algebra had passed ninth grade algebra. At the time there was a tremendous shortage of math teachers, many high schools did not even offer four years of mathematics to their students, and now there was to be an emphasis on mathematics training! It sometimes seemed that requiring a math minor from physical education majors was the most constructive action possible, since athletic coaches often taught math on their sports' off term.

But not all was lost. The California State Bureau of Secondary Education was headed by a sharp-tongued classical scholar named Frank Lindsay ("State buildings don't have to be cheap, they just have to look cheap.") who used the state textbook adoption system to upgrade the mathematical curriculum. E. G. Begle, who had moved to Stanford by then, served as adviser and a whole bevy of district math specialists, administrators, and pre-collegiate and collegiate math teachers became involved in the California program. The California Mathematics Council played a truly professional role and the statewise math curriculum, the teaching of pre-college math, and the preparation of teachers were all improved.

Of course nothing stays fixed without a lot of continuing attention. Thus, for example, an intern system of training teachers was set up successfully at Berkeley at the end of the 1950s by Clark Robinson, but as the pressure for schoolteachers slackened and the outside financing ended, the program was junked. Another example: The math department offered a Math for Teachers major (Harley Flanders and I set it up) that lasted for years. It was dropped only recently, in honor of my retirement and the current shortage of high school math teachers.

But let us look at Berkeley, and examine briefly the University itself during the decade of the 1960s. (See *Education at Berkeley, Report of the Select Committee on Education*, Univ. of Calif. Berkeley, Academic Senate, March 1966 for a detailed point of view.) Berkeley was the *big U*; its 27,000 students overcrowded the classrooms, jammed the libraries, and overwhelmed the faculty. It was very different from its pre-war counterpart—at least very different from UCLA a quarter century earlier, and it didn't match the movies nor the stories of college. It was a new kind of animal, a megaversity, a maverick, a supermarket of ideas, but self-service only.

The customers at the big U were better off financially than pre-war students. It was possible, and quite common, for reasonably vigorous, reasonably able students to be entirely self-supporting, and this encouraged independence and self-confidence. And the increasing graduate enrollment maintained a reasonable level of intellectual and political sophistication on the campus.

In 1964–1965 the students demonstrated against restrictions on free speech on the campus and against over mechanization of the teaching process. (Do not roll, spindle, or mutilate me!) Several hundred were arrested for taking part in civil disobedience, the faculty was deeply concerned, a free speech policy was established, and something like a new kind of university seemed to be coming into existence (see *Education at Berkeley*, loc. cit.). This frightened the Regents, the newspaper reporters, and the voters, and before you could say 1968 Ronald Reagan was Governor, and at his first Regents' meeting President Clark Kerr left his position as he had entered it, fired with enthusiasm. (I stole that last line from Clark Kerr.)

The University had no monopoly on turmoil. Voting rights, desegregation of schools, free speech, and above all ending the Vietnam War, made the 1968 Democratic Convention a noisy showplace for democracy. The antiwar movement, the war against the war, became the focus of American political activity. The Resistance, Stop the Draft Week, the War Resister's League, Draft Counseling, the Vietnam Day Commencement, the Peace Brigade, the march to Kezar—these and many more events, organizations, points of view, became a single stream of protest, and finally, at long last, we stopped the war. Not when we wanted to, not the way we wanted to, but for the very first time the American people stopped a war. *We won*! (Read Mark Twain's writing on the Philippine War, or U. S. Grant on the Mexican War. There have been unjust American wars opposed by strong, articulate people, but this was, I think, the first such to be stopped by the American public.)

Many of us have some unpleasant memories from the anti-war movement (be careful of shirt-sleeved policemen who wear black gloves) but we have good memories too. (My son announced his parents' brief imprisonment with an engraved card.) But I think we all know, even the most burned-out of us, that what we did was important, perhaps the most important thing we have ever done.

Here is a last story to add here. It is a painful story because it concerns friends, acquaintances, and colleagues rather than anonymous administrators, politicians and officials.

In 1981 I accepted an invitation to lecture at Birzeit University on the Israeli-occupied West Bank. I gave a two-week series of lectures at Birzeit and a couple of talks at the University of Bethlehem. I lived on the West Bank at Ramallah, used public transportation, gossiped with local mathematicians and observed a visit of the Israeli army to the University.

In 1982 the Human Rights Committee of the AMS recommended that the Council of the AMS protest the continuing violations of academic privileges of Birzeit faculty by occupying Israeli authorities. The Council refused to take action and later, despite the representations of a distinguished former AMS president, refused to reconsider. I resigned from the Society in protest.

I do not believe that there was or is reasonable doubt as to the circumstances at Birzeit, and I think the Council has quite properly deplored repression in less severe cases. But the problems of our colleagues at Birzeit and the other Palestinian universities remain, and reproach us.

Saunders Mac Lane studied at the University of Göttingen, where he received his Ph.D. in 1934 under the supervision of Paul Bernays and Hermann Weyl. After early positions at Harvard University, he moved to the University of Chicago in 1947. His research has ranged through algebra, logic, algebraic topology, and category theory. Among his books are Homology *and (with Garrett Birkhoff) the influential text* A Survey of Modern Algebra. *His numerous honors include a Chauvenet Prize and a Distinguished Service Award from the MAA and a Steele Prize from the AMS.*

The Applied Mathematics Group at Columbia in World War II

SAUNDERS MAC LANE

In articles in the first part of this series, Mina Rees and Barkley Rosser have each given effective summaries of the research work of American mathematicians during WWII; each of these articles gives a good description of the activities of various applied mathematics groups, including the one at Columbia. The present article, by concentrating attention on just this one group, will try to give some feel as to "how it really was". That try cannot really succeed, but I will depend not only on my memory. Luckily, I have a copy of the final report [2] which I wrote about the activities of the Applied Mathematics Group at Columbia (AMG-C) in airborne fire control. This report was originally classified "Confidential" and then "Restricted". After it was finally declassified, on June 4, 1958, my friend John Coleman, then Executive Officer of the National Research Council (NRC), procured a copy for me.

1. Background

The NRC had been started in WWI, and then continued, by executive order of President Wilson, as a subsidiary of the National Academy of Sciences. The NRC was not enough for WWII, so civilian war research in 1942 was organized under the National Defense Research Committee (NDRC), headed by James Bryant Conant, then President of Harvard University. Conant, as president there, had tightened up the appointment policy at Harvard—six years up or out, with special committees to examine proposed appointments from outside. I have the impression that his policies have been widely copied, so that today every University aims to be as good as Harvard and by the same methods. At that time, Marshall Stone, in faculty meetings at Harvard, disagreed sharply with Conant about these appointment policies. Conant was perhaps a bit of an autocrat, at Harvard and with the management of the NDRC; with the priorities of war-time, this may have been necessary. This whole story may suggest that Conant was not too sympathetic to involving mathematicians in the work of the NDRC.

In 1942 many mathematicians were lobbying to get more involvement of mathematics in the war effort. Dean R. G. D. Richardson at Brown University, eager to develop applied mathematics, had appointed (from Germany via Turkey), William Prager, an expert on plasticity. Brown then organized sessions to help the war effort by retreading many pure mathematicians as applied ones. I was a retreadee, but it did not take with me; the applications of 19th-century style elasticity to problems of plasticity did not catch my real interest.

On a larger stage, the NDRC was reorganized in late 1942 and acquired an "Applied Mathematics Panel" (AMP) headed by Warren Weaver (Rockefeller Foundation), with vice-head Thornton C. Frey (Bell Labs). The intent was to establish Applied Mathematics Groups at various universities, to give them contracts to study suitable projects—those formulated by the Panel in response to requests for help from the military services or their contractors. The work on the contracts was to be supervised by government employees of the AMP, called Technical Aides (in our case Edward Paxson and later Mina Rees). This was apparently like the method the government used to supervise industrial contracts. It often did not actually work that way. My final report cites several cases where AMP established the approved study only after the report on that study had been written.

The Applied Mathematics Group at Columbia (AMG-C) was established in March, 1943 with Professor E. J. Moulton, an applied mathematician from Northwestern, as its director. At about that time, Warren Weaver himself asked me to join. When I accepted, he wrote me (at Harvard) a long letter to get me thinking about a problem in which various gases and liquids circulate in an elaborate arrangement of tubes and pipes. I did not get anywhere with this question; only much later, after the Smyth report on the development of the Atomic Bomb, did I guess that it had to do with the gaseous diffusion process for separating isotopes of uranium. It then seemed to me sad that so few mathematicians were involved in that Manhattan project; the veterans of that project dominated science policy in this country for thirty years. (To the best of my knowledge, the only mathematicians involved in Manhattan were John von Neumann, Stan Ulam, C. J. Everett, and Jack W. Calkin.)

However, it is clear that the secrecy about the problem was such that nothing about it would have been delegated to a bunch of mathematicians in a project housed in a converted apartment building on Morningside Heights next to Columbia University. It also happened that in the late spring of 1945 Paul Erdös, a Hungarian mathematician well-known to many of us, came to visit AMG-C. He told us what the Manhattan project was up to. Of course, he had no clearances, but he did get around.

There was also a rumor that, at Los Alamos, Everett, by training an algebraist, had made a slide rule calculation of a constant required for the H-bomb which differed significantly from Teller's. Everett's was correct. Also, D. C. Lewis, Jr. recalls that in 1942, when he was collaborating with G. D. Birkhoff about chromatic polynomials, the discussion shifted to bellicose mathematics. Birkhoff was then enthusiastically working on the effect of introducing a given amount of energy in a confined portion of space (ideally, just at a single point). He of course, gave no hint that this concerned an atomic bomb, but in retrospect, this seems likely.

In my view, it was a loss to the war effort that more mathematicians were not earlier involved in war research. At an AMS meeting in 1944, Marshall Stone, then President of the AMS, criticized Warren Weaver for this delay. I am not at all sure that Weaver is responsible; in 1941, nobody would have thought that mathematics would be of help in problems that seemed to belong to physics, say, or to engineering.

As it was, AMG-C came into being so late that there was no question of designing new devices which hardly could have been ready in time. The center of interest was the study of how to use the gadgets which had been designed and were on hand.

2. How to Use the Given Gadgets

The original intent was that AMG-C would tackle any sort of military problem which required some use of mathematics; there were such efforts involving classical applied mathematics, as for example in work of J. J. Stoker. He was first at AMG-C, then transferred in 1944 to the group at NYU, where his work concerned the properties of water waves on sloping beaches, with possible reference to islands in the Pacific.

What subsequently happened was that AMG-C became a group of people specializing in all the varieties of airborne fire control—how best to make use of the various lead computing sights which were on hand. I estimate that this came about, first, because Dr. Weaver had expert knowledge of these matters and, second, because such fire control had very high military priority for the bombing raids over Germany and later, with B-29s, over Japan.

To the first point: The British and C. S. Draper, at MIT, had designed gyroscopic lead computing sights. The gunner on a bomber tracks an approaching fighter, a gyroscope on his gun measures the rate of change of angle and multiplies this by "time of flight" (of the bullet) to determine the angle by which the gun direction should "lead" the fighter in order to score a hit. This summary is vastly oversimplified. When I actually joined AMG-C (living in a dismal rented room and commuting sometimes back to my home in Cambridge) I found that Dr. Weaver had written an analysis of such lead computing sights, and that his description needed to be expanded. I then wrote a longish report: "An introduction to the analysis of the performance of lead computing sights (Mark 18)". This was then put together in a good binding and seems to have been considerably used at the time. It may indicate that there is a tendency in an emergency to do research on those things that we had known before; in my case, Garrett Birkhoff and I had recently written *A Survey of Modern Algebra* and I thought that I knew how to prepare a good exposition. But that report on the Mark 18 clearly depended on the prior work of Weaver and of engineers such as Dr. Draper. At the time, I did think that the engineering design was very much on the quick and dirty side, with too little prior mathematical analysis of the possibilities. Some of my later experience suggests that this may then have been characteristic of much of engineering design at that time, in which the mathematical input was on the intellectual level of the widely used Granville's Calculus. I cannot now further document this opinion.

The second reason for the concentration on fire control was the military situation. Thus Mina Rees [**3**], reporting a summary by Warren Weaver, quotes a letter from Brigadier General Harper, head of a training command. "The problems connected with flexible gunnery are probably the most critical being faced by the Air Force today". The letter went on to ask the AMP to train competent mathematicians for practical service in operations research sections in the various theaters which had flexible gunnery problems. This task was assigned to AMG-C, which did then find 10 and later 8 willing mathematicians. They were then exposed to our knowledge for a couple of months (including, I think, that report of mine) and then sent off to the theatres. The general effect on AMG-C was a concentration on the many questions involved in fire control.

3. The Mathematics Which Was Needed

My final report summarises the many questions of fire control. The mathematics needed was by and large elementary. We spoke of the pitch, roll, and yaw of a fighter plane, and soon we had a good command of spherical trigonometry. There was a constant flow of classified documents from all sorts of other agencies, in particular many from Great Britain, where there was an evident interest in these matters. The documents were circulated through all the mathematicians at AMG-C, but of course stored overnight in a suitable safe. Then, as now, the literature was too extensive to master it all. I recall one British report which reduced an important problem to a trigonometric formula. The formula was obscure, but the real use demanded numbers, so the report went on to compute a table:

0°	10°	20°	30°	...	90°
2.00	2.01	1.98	1.97	...	2.01

(This is just my recollection; the real figures may still be classified.) Some one of us became curious, studied the formula and found that it was

$$2 \cos^2 x(1 + \tan^2 x).$$

This may illustrate the fact that a knowledge of high school trigonometry was useful in war research—and that most of our problems involved chiefly elementary mathematics. As evidence, I include excerpts from recent letters to me from some of my colleagues at AMG-C:

E. R. Lorch writes about "those exciting but not exhilarating days when we worked in the dingy apartment on 118th street. In my own case the problems involved trigonometry (spherical when the going was rough) and differential calculus (but not beyond the second derivative). The problems were tough, annoying, and without lustre. Of course they were connected to situations of life and death".

D. C. Lewis, Jr. writes: "Most of my own work was concerned with earth-bound fire control for anti-aircraft weapons. At one point, I was given the job of calculating the probability of hitting an aeroplane flying a straight line course, using the then existing anti-aircraft equipment—later, I had the job of revising existing equipment so as to better take care of cases when the target is taking evasive action—a rather futile endeavor as far as actual application to World War II is concerned—some of the theoretical results were published under the title 'Polynomial least square Approximations' (*Amer. J. Math.* **69** (1948), 273–278)."

Daniel Zelinsky writes: "What I remember best is my contact with the Laredo Air Force Base, where they were trying to use some scaled down training exercises to assess the accuracy of some of the gunsights. My contribution was to convince them that the system was not linear—if you divide everything by 2 (all speeds, bullet speeds, etc.) a mechanical sight will probably become totally inaccurate, even if at full speeds it could work well."

George Piranian reports: "One of my first assignments at AMG-C concerned the scattering of electromagnetic waves by a cloud of spheres of uniform size. Using a classical formula, the computing staff had determined the degree of scattering for various wavelengths of the radiation. A laboratory group elsewhere in New York City had found that cigarette smoke is a reasonable substitute for a uniform fog, and had tried to obtain empirical verification of the results obtained by computation. The discrepancy between the two sets of results was unacceptably large, and Walter Leighton instructed me to join forces with Leon Brillouin to find the error.

"I was helpless, but Brillouin declared, 'We must study the formula; I will look at it tonight'. The next day, he held victory between his teeth. A big shot who had derived the formula applied it to an extreme case (perhaps that of a single sphere of large radius or many ridiculously small spheres), found that his formula erred by a factor of $1/2$, and remedied the defect by throwing in a fudge factor of 2. Said Brillouin: 'The factor is not 2, but a number between 1 and 2; its value depends on the ratio between radius and wavelength'. The moral: For reliable results, engage a competent worker."

4. Aerodynamics

To compute leads for machine guns, one also needed to study the courses followed by fighter planes: A pursuit course. The simplest example in 2 dimensions is the course followed by one point moving at a given speed so as to be directed always at another point moving in a straight line at constant speed. This results in a simple differential equation. With aerodynamic effects, it is more complicated. Stimulated by Dr. E. W. Paxson, the Brown University Group prepared a report on "Aerodynamic pursuit curves"; their equations worked well in the vertical plane or in some other "plane of action". But then Paxson discovered that when it isn't vertical there is no single plane of action; the problem is really three-dimensional, and there the equations in the Brown report don't allow successive approximations. The pursuit curve problem was then considered at AMG-C. With considerable stimulus from John Tukey (from Princeton), Leon Cohen (a topologist working at AMG-C) found more manageable equations; from these a battery of young women working by hand on the desk top Marchant computers then available, computed some 33 such courses. (How different it would be today.) Others at AMG-C, such as Walter Leighton, George Piranian, and Daniel Zelinsky, contributed other items. That report by Leon Cohen is still alive; at any rate someone interested in these matters recently asked me for a copy, which I was able to get for him, and it is reported that a Ph.D. thesis at MIT was based in part on this work. My final report says (p. 7), "Dr. Cohen presents the equations in the form in which step-by-step computation of such courses is possible. In the opinion of the author the success of Dr. Cohen (a topologist) demonstrates that in war work applied problems are not necessarily solved most effectively by people bearing the trade labels of applied mathematicians". This comment now seems to me needlessly snide. Fortunately on a prior page I had also noted that "the unsatisfactory conclusion (of the Brown study) is, in the opinion of the author, primarily due to lack of liaison.

After the study was set up, there was little attempt to explain to those working on it which gunnery problems really required the theory of pursuit curves". I quote this now because I suspect that the same lack of liaison applied in many other cases of wartime studies.

These pursuit curve studies were completed after various changes at AMG-C. Late in 1943, I recall that I was dissatisfied with some of the management arrangements, so for a period I worked there only part time. In the summer of 1944, Walter Leighton left to set up another Applied Mathematics Group at Northwestern University, with the active participation of Adrian Albert, who did not wish to leave the Chicago location; that group was also concerned with fire control. Then in August 1944 Professor E. J. Moulton left AMG-C and I became director, with Magnus Hestenes and later Irving Kaplansky as associate directors. I used my acquaintance with the mathematical community to bring in a number of able mathematicians, including Leon Cohen, Samuel Eilenberg, Irving Kaplansky, George Mackey, Harry Pollard, and Daniel Zelinsky. As director, I often found myself in disagreement with Warren Weaver. However, the contract administration at Columbia was in the hands of Dean Pegram. As a young man in North Carolina he had dated Isabel Elias, who later married Virgil L. Jones and whose daughter Dorothy was my wife. Dean Pegram still admired Isabel; he and I got along famously. After a day of war work, Eilenberg and I would often adjourn to discuss the relation between the homology and homotopy of topological spaces.

5. Calibration of Gunsights

As already noted, AMG-C had many different studies about fire control. Some bomber guns had no computing sights, but only metal ring sights. The gunners were given various rules for their use—position firing and zone firing. AMG-C tried to compare and improve these rules, and attempted to consider what would be different if an attacking fighter had offset guns, not firing along the nose direction. My final report says (p. 31) "In the initial design of lead computing sights the idea had been that the lead was obtained by multiplying the angular rate by the actual present time of flight of the bullet, as obtained essentially from ballistic tables. Misguided early enthusiasts (including the author) went to considerable extents trying to justify this particular approximation. The essential result of study was the observation that the multiplier used in computing kinematic lead has no reason to be exactly the present time of flight. Leighton discovered that the use of 90% of present time of flight would be more effective." This result led to considerable efforts to "calibrate" sights by hopefully finding the optimal percentage good for this or that circumstance. This is a clear illustration of the point that our efforts were directed at making do as best one could with the gadgets at hand.

We tried to compare "true lead" with the lead actually computed by the sight. True lead consists of ballistic lead plus kinematic lead; the first of these was found from ballistic tables. We tried various formulas; my report says, "In this connection we see the importance of using real mathematicians on problems not involving technical mathematical knowledge beyond the undergraduate level, for the real mathematician endeavors to avoid mere horsepower. Dr. M. R. Hestenes in this sense did real mathematics on this problem. He appealed to basic ballistic theory (the differential equations) rather than to the derivative ballistic tables." His results came out in a report AMG-C #247, revised. They were extensively used at AMG-C and elsewhere in 'calibrating' sights.

I note that Hestenes had been a student of G. A. Bliss at Chicago. Bliss had worked effectively in the first World War on ballistics.

AMG-C carried on certain "assessments", showing that the mark 18 sight had "substantially smaller class B errors than the K-3 sight". My report says, "It is difficult to measure the extent to which these results may have had influence. The general conclusion was presented on numerous occasions to (military) officers. Both the Army and the Navy, toward the end of the war, did carry on programs emphasizing the procurement of this (the mark 18) sight and it is possible that AMP recommendations had a real part in these decisions." This is a characteristic of such war research; it is almost impossible to know then or now what it may have really contributed. We may have originally thought that the purpose was primarily to produce reports, but we soon learned that this was not it—though we did go on to produce many reports.

My final report (p. 79) puts it this way: "In the early stages of AMG-C there were only infrequent contacts between members of the group and service officers. Such contacts as there were came with related sections of NDRC. Only belatedly did we learn the great importance of direct contact with Army and Navy agencies. By virtue of such contact it was possible to get authentic information as to the needs and interests of the services and it was also possible to present effectively the recommendations, results, and suggestions which were obtained in the scientific work at AMG-C. In a sense, the accomplishments in such personal relations were greater than any achieved by the mere circulation of documents."

This would seem to support the case that mathematicians were not brought into war research early enough.

6. Accomplishments

At Harvard, I had developed a considerable admiration for the ability and imagination of my colleague Hassler Whitney, whose extensive contributions to topology and geometry are noted in his article (Moscow, 1935): "Topology moving toward America" in the first part of this series. In October, 1943, I recommended that the AMG-C enlist his services. He agreed. As best I recall, he did not stay often at Columbia, but instead visited many service facilities, with very effective results.

George Piranian recalls it for me as follows: "Immediately after lunch on a gray day in the fall of 1943, the entire scientific staff of AMG-C gathered to witness your induction and indoctrination of Hassler Whitney. You described the difficulties with the mark 18 gunsight, and Hassler's quick perception and active engagement were spectacular. I believe that immediately after the assembly's dispersal, Hassler withdrew to his office and began writing a scientific report. A few days later, there was a question whether Hassler should be permitted to see his own report. The paper was classified, and Hassler's security clearance was held up."

The clearance was eventually cleared. When AMG-C in November 1945 received a Naval Ordnance Development Award, Whitney received the first individual citation, which for his case read in part:

"a. Suggestions for the design of naval types of sights not actually used in this war, may be of future interest.

"b. Fundamental study of tracking problems for sights. Whitney early recognized the importance of a thorough-going analysis of the nature and limitations of the tracking problem with a view both to the design of future gunsights and to the optimum utilization of existing sights.

"c. Adaptation of the mark 23 gunsight for rockets in the U.S. Early in 1945 the British method of adapting this gunsight for firing rockets from fighter planes reached this country. Whitney immediately saw the importance and initiated calculations, consulted with members of the Bureau of Ordnance on this and had taken an active interest in the training program.

"d. General study of rocket sights for Naval fighter planes.

"e. Skid. Whitney was one of the first scientific workers to recognize the importance of the errors caused by skid of a fighter airplane in attack" (Skid = plane not banked correctly for the intended turn)."

There were also individual citations for:

Irving Kaplansky, adaptation of Gunsight mark 23 for rockets;

Magnus Hestenes, for fundamental deflection formulas, as noted above, and for work on the "stabilized" S-3 sights; one of his investigations led to a modification in the design of this sight;

Walter Leighton, for the calibrations of the mark 18 sight at a naval ordnance plant, based on Leighton's data, and for his administrative initiatives at AMG-C and AMG-N;

D. P. Ling, for studies of the dome type control of the mark 18 type gyroscopic gunsight, and for detached service, for example with training officers at the Naval Air Station at Inyokern;

Saunders Mac Lane, for consultations at the Naval Air Station, Patuxent and at the Naval Air Station at Jacksonville, on training of gunners, for administration at AMG-C, and for serving as Vice-chairman of the Army-Navy-NDRC Airborne Fire Control Committee. (Apparently, my job was to prepare the minutes for the committee; my first ever trip on an airplane was taken to get to that conference at Jacksonville.)

This listing may indicate what were considered, late in 1945, as the chief results of all those numerous studies.

7. Operations Research

Operations research had proved very effective in locating enemy submarines. When AMG-C was requested to train mathematicians for related work with the Air Force, we did get 10 and later 8 men to spend two months at Columbia to learn what we thought we knew about airborne fire control. I do not have a complete list of them but among them were W. L. Duren, Jr., P. W. Ketchum (later at Urbana), John W. Odle, R. H. Bing, R. V. Churchhill, W. L. Ayres, V. W. Adkisson, from Arkansas, and Edwin Hewitt, just graduated at Harvard. Hewitt was without doubt the most flamboyant. The following comes from my final report (p. 11ff) starting with this discussion of "qualitative rules for the use of gunsights":

"The fighter with its guns bearing will be moving more or less directly toward the bomber. The bomber, meanwhile, is moving forward so that from the viewpoint of the bomber the fighter will appear to drift astern. Hence the important conclusion that against pursuit attacks the bomber's gun should be aimed on the side of the fighter toward the bomber's tail.

"The difficulty found with gunners in this respect is well illustrated by the following story due to Edwin Hewitt. In the early days of the eighth air force, Hewitt argued with a certain nose gunner trying to convince him to 'aim toward his tail'. The nose gunner swore up and down that he should rather aim toward his nose, and Hewitt and the gunner parted in violent disagreement. On the next mission this same nose gunner espied two ME190s making pursuit attacks off the port bow. The nose gunner drew a careful bead on the outside plane, aiming inside this plane according to his ideas. He gave him a good burst, said, 'There, I got the bastard', looked up and was amazed to see a wing falling off the inside plane. His aim had been toward the tail of the inside plane. He came back convinced that Hewitt and other 'Feather merchants' might have something on the ball."

I can't now guarantee all the details in this story, but it does serve to illustrate well why operations analysts with some mathematical know-how could be effective in the combat theatre.

W. L. Duren, Jr. has reminded me of two other AMG-C trainees in the second group who had influence in postwar developments. First, the late George Nicholson was later prominent in the formation of ORSA, the Operations Research Society of America, and continued as postwar advisor to the Air Force. Second, Stanley J. Lawwill had a Ph.D. from Northwestern in mathematics and electrical engineering. He became General LeMay's postwar head of operations analysis in the strategic bomber command and went on to found ANSER, a nonprofit think tank that serves the Pentagon with weapons systems analysis.

Duren also recalls that he worked on the adjustments of a possible vector sight for the B-29, using calculations made under his instructions at AMG-C. Subsequent tests in New Mexico showed that this sight gave 30 times the hit probability of some other sights, but it was not adopted in preference to a sight designed by a contractor. It later appeared that tests based on pursuit curves had little relevance to the Pacific theatre to counter strafing, non-pursuit attacks by fighter planes. For this and other reasons, the flexible machine guns were often simply removed from bombers there. What one calculates may not fit reality.

At AMG-C there was some study of the tactical employment of the B-29. Under this head my report states:

"The very great interest in the effectiveness of the B-29 bomber led to the project AC-92 set up in the summer of 1944... under the general direction of Warren Weaver... a number of different activities. (A group at the University of New Mexico) carried out extensive and realistic air experiments to study the efficiency of the fire control system. The general conclusion of this work tended to suggest that the CFC fire control system was inadequate to the defense of the bomber. This conclusion was not one fully justified by the data actually found in the experiments; in some sense the experiments were done in order to establish a conclusion anticipated in advance.[1]

"One of the objectives of AC-92 was that of considering the strategic employment of the B-29 'in the large'. The idea was that mathematical consideration of the whole problem of aerial warfare might lead to effective results. It is the opinion of the author that this attempt was couched and carried out in such general form as to be meaningless. In particular, one famous document purported to study the relative merits of different operations by methods of mathematical economics used to compute the relative loss in manpower to

[1]These statements by the author are based on a quite incomplete knowledge of this work, and so may be subject to correction or revision.

the enemy and to us. It is the author's opinion that this particular document represents the height of nonsense in war work."

There were other parts of this study AC-92 which had to do more with development and were probably more successful. But for the B-29 the proposed central fire control system (CFC) "was described in the appropriate technical order in great detail, but the principles of which were not always clearly set forth". This led to Study 143 at AMG-C, "carried on by Dr. D. C. Lewis, with the active assistance of Dr. M. R. Hestenes and Dr. F. J. Murray. Modifications of the CFC were under continuous discussion. The contracting company themselves developed two modifications, one of which was known as the 'press bang':

"The press-bang system was essentially a mechanization of some of the apparent-speed methods of eye shooting. The early AMG-C analysis of these methods gave reason to doubt that such a formula would be as efficient as the more classical time of flight times angular rate method. The press-bang system was also briefly tested on the Texas machine. Reports have it that even though the system was carefully and explicitly adjusted to give optimum results on two of the canned courses for this machine, its efficiency was not too great. Nevertheless the company was ready to transfer a large portion of its production (of the CFC) to production of the press-bang. On the basis of such incomplete information it appears to the author that this is a case in which a commercial company unduly and unwisely pushed one of its fire control projects more vigorously than prior theory and subsequent tests indicated was appropriate."

My criticisms quoted here from my report may well be totally wrong; I no longer have the documents to check them. I include them to indicate that mathematical input on engineering design is a complex issue. These cases may also indicate that the problems with the Military-Industrial complex may have started back then. The later overeager manufacture of "World models" as, for example, by the Club of Rome, may have had an origin in the hurried work of WWII on operations analysis, and this may also apply to the ambitious "Global Systems Analysis" sometimes favored at the International Institute of Applied Systems Analysis (IIASA) in Vienna.

8. AMG-C Dispersed

With the Japanese surrender, questions about the fire control for the B-29 became moot. During the last two weeks of August 1945 I hastily wrote up that final report. As the quotations above may indicate, I included at several points an indication of my frustration at things which I thought had been done wrong. The report with these indications was promptly classified; until now, it is likely that no one except my diligent secretary, Betty Amitin, has ever read them. By the first of September, the mathematicians at AMG-C were all dispersed, though Arthur Sard faithfully stayed on as director to wind up the final report. The technical aides had enjoyed the work; E. W. Paxson went on to do operations research at the Rand Corporation, while Mina Rees did pioneering work as the first program director for mathematics for the Office of Naval Research. This was the first federal agency with the statutory authority to support basic research; its methods were later a pattern for the NSF.

From AMG-C, Dr. Ling went to the Bell Labs. Most of the other mathematicians returned to their university work. From Churchill Eisenhart and from Allen Wallis and Milton Friedman at the Statistical Research Group at Columbia I had learned the importance of statistics; back at Harvard, I taught an undergraduate course on statistics. Also, when I taught sophomores about ordinary differential equations, I had a much better understanding, which I hope I transmitted to the students. My experience with engineers stood me in good stead when I later served as consultant to the Dean of Engineering at Purdue. Also, I had learned that the best typewritten "reports" could involve arrant scientific nonsense. This was of surpassing value to me much later, when for 8 years I was Chairman of the Report Review Committee of the NAS. Others at AMG-C were enriched by their experience. Walter Leighton, for example, later served for many years as a scientific adviser for AFOSR (The Air Force Office of Scientific Research). He was alert to a subsequent call for scientists to get clearance for classified work when later emergencies threatened.

At NYU (the Courant Institute) and at Brown University under the guidance of R. G. D. Richardson the wartime Applied Mathematics Groups were an important step to the development, with government support, of excellent new centers of applied mathematics. There was not such a result at Columbia. This is not surprising. By and large, at Columbia we did not do "applied mathematics" but just applications, sometimes of elementary mathematics. Our results do indicate that, in a time of emergency, imaginative "pure" mathematicians, such as Hassler Whitney, can make vital contributions to applications. The earlier articles by Peter Hilton, Mina Rees, and Barkley Rosser in the first part of this series give other such examples, while the study of the Manhattan project (Rhodes [**4**]) indicates that this happens with other sciences.

This use of "pure" science for practical emergencies is one fundamental reason for the government support of scientific research which developed so generously in this country after WWII. This is especially important now, when the terms of this support are being reconsidered. In particular, the prospect of setting up large centers under the NSF should give us pause. Such centers tend to concentrate power in the hands of people with administrative skills but with little personal knowledge of the able younger mathematicians. Many of the troubles which I think I identified above seem to me due to this example of excessive centralization of the conduct of mathematical research during the war.

It is not the center, but the individual who really counts in discovery or application.

My final report for AMG-C ends thus:

"Scientific war research, like other scientific activities, is not automatically immune from nonsense; especially because of the pressure of the work it is possible to set up problems which look superficially sensible, but which turn out to be either hopeless of solution or meaningless in application. This tendency is especially strong when the problem comes to the scientist through a long chain of channels."

9. Aspects of Applied Mathematics in the USA before WWII

At the start of her article [**4**], Mina Rees quotes from William Prager: "In the early thirties, American applied mathematics could, without much exaggeration, be described as that part of mathematics whose active development was in the hands of physicists and engineers rather than professional mathematicians. This is not to imply that there were (in the early thirties) no professional mathematicians interested in the applications, but that their number was extremely small. Moreover, with a few notable exceptions, they were not held in high professional esteem by their colleagues in pure mathematics, because of a widespread belief that you turned to applied mathematics if you found the going too hard in pure mathematics."

Professor Prager arrived in this country in 1941. He was then a mature mathematician, age 38. I deem it remarkable that so many mature mathematicians from Europe have been able to develop so well in this country; it must be difficult to adapt to a foreign culture. Professor Prager made great contributions at Brown, but I doubt that he ever was really assimilated to the American mathematical community and he surely had little direct knowledge of the American scene in applied mathematics before his arrival in this country. Here is my own summary of how it then seemed to me:

At Harvard, E. V. Huntington held a part-time chair which had some applied designation. He did teach an (idiosyncratic) course on statistics but taught no applied math, and was really interested chiefly in axiomatics. Harvard had appointed J. H. Van Vleck, half in mathematics and half in physics, in order to right the applied balance. As best I could see, his real interests (and his subsequent Nobel Prize) were all in physics. George Birkhoff had made decisive contributions to dynamical systems which are still influential today, but he also worked on many other topics, and would have viewed himself as simply a mathematician, not as an applied mathematician.

At Yale, E. W. Brown, an expert on celestial mechanics and a past president of the AMS, taught a course on mechanics, which I took (1928–1929), to my great subsequent profit. His lectures on Hamiltonian mechanics were delivered from badly yellowed notes. They were clear, but at the time, I noted that new things were going on in topology, in logic and in algebra, but apparently not in mechanics. A course there in theoretical physics with Leigh Page and his then new text never mentioned the new developments in quantum mechanics. The remarkable legacy of J. W. Gibbs at Yale seemed dissipated; his student Irving Fisher had gone into economics and his last student Edwin B. Wilson into Public Health (at Harvard).

At Princeton, Oswald Veblen's projective relativity theory did not really succeed, while H. P. Robertson was not very active.

At Chicago, there had been an active group in mathematical astronomy. F. R. Moulton was effective in this group, but when he left about 1928, the remaining applied mathematicians there were discouraged. This is described in more detail in my article about Chicago in Part II of this series.

At Wisconsin, Max Mason and Warren Weaver did applied mathematics. They published together in the 1920s a monograph *The Electromagnetic Field.* The preface raises "such a searching question as the reconciliation of quantum ideas on energy interchanges with general theory. The great scientific task of the next fifty years is the development of a new 'electromagnetic' theory." However, their students at Wisconsin were not encouraged to take part in that scientific task, because both Mason and Weaver left soon for administrative positions: Mason to be president at the University of Chicago and then to the Rockefeller Foundation, where he was soon joined by Weaver.

MIT was not notable for applied mathematics until Norbert Wiener, under wartime stimulation, shifted his interests from Brownian motion and Tauberian theorems to stationary time series, as in his notable monograph [7] (**MR 11**, p. 118) on this subject; I understand that this study was stimulated by work at the wartime Radiation Lab at MIT.

As Prager suggests, some applied mathematics may have been done by engineers in this period, but the general plan of study at universities with engineering departments called for two years of calculus with an exposure to a list of tricks for solving ordinary differential equations. This may have been what was then usually needed for engineering design, but work at AMG-C suggests that the designs could often have been improved with just the same sort of (not very exciting) mathematics.

Other examples would support the case that applied mathematics as then taught at universities in the USA did not then present real challenges. The number of applied mathematicians in the thirties was indeed, as Prager says, extremely small, but what was the cause? There had been native American applied mathematicians, but by 1925 many were inactive and their courses had become dull, with little indication that there were new things to be discovered. At a time when pure mathematics abounded with excitement, this inevitably meant that lively students of mathematics were not likely to go on in applied directions.

10. Optimal Control

A general conclusion of this article might seem to be that little subsequent mathematics, beyond statistics and operations analysis, came from the wartime work at AMG-C. I did know that Magnus Hestenes had done decisive work on optimal control, but I did not know the background until he wrote me recently as follows:

> "I was very interested in the aerial flights L. W. Cohen made on paper. He matched actual flights which were photographed at Patuxent. I found out much later that the work at Patuxent was directed by my brother Arnold. However, there was a "spin off" of this work that you might find interesting. When the war was over Paxson went to RAND. Sometime later, I came to UCLA. Paxson got in tough with me and asked me to determine time-optimal flights for a fighter plane. Using what I had learned at Columbia about flights of airplanes, I set out to formulate this problem as a variational problem. I found that the usual variational formulation did not fit very well. It was too clumsy. And so I reformulated the Problem of Bolza so that it could be applied easily to the timeoptimal problem at hand. It turns out that I had formulated what is now known as the general optimal control problem. I wrote it up as a RAND report and it was widely circulated among engineers. I had intended to rewrite the results for publication elsewhere and did so about 15 years later. It was delayed because I became chairman at UCLA and had many obligations with regard to the Institute for Numerical Analysis. This was almost 10 years prior to the publication by Pontryagin et al. on this subject. You may be interested to know that their work also was an outgrowth of studies of aerial combat, a study that was requested by the Russian government. Thus, the theory of Optimal Control, both here and in Russia, was developed in response to studies of aerial combat.
>
> "As for applied mathematics at the University of Chicago, Bliss insisted that all Ph.D. students take courses in applied mathematics. Unfortunately, the only applied courses were mechanics, potential theory, and differential equations. A student could take courses in physics also. When Bliss retired this requirement disappeared.

"I found that my war experience gave me a broader outlook on the role of mathematics in our society. It also gave me a greater appreciation of research mathematicians. A good researcher, no matter what field, would tackle a problem with vim, vigor and imagination. He needed no guidance. This was not true for some mathematicians who were not researchers although they did good work with guidance.

"In my latter years I asked myself what branch of mathematics was most useful in applications. I came to the conclusion that it was algebra. Accordingly, algebra can be viewed to be the basic course in applied mathematics. I doubt if many algebraists view algebra as applied mathematics."

11. A Roster of People

The work of an organization depends fundamentally on the people involved. Fortunately, I have a list of all those at AMG-C who received a lapel emblem with the ONR award in 1945. I will list first the mathematicians in the research staff; whenever possible, I add a subsequent position or two, plus the citation of just one characteristic paper or, when possible, a book. This may suggest the wide variety of interests of the research staff. For the computing and secretarial staff, the best I can do is just the list of names, but these people were equally vital for the urgent studies we made at that distant time.

Research Staff, AMG-C

Leon Brillouin (born 1889)
Adjunct Professor, Columbia University.
Science and Information Theory, 2nd ed., New York: Academic Press, 1962.

Eleazer Bromberg (1913–)
Professor and Assistant Director, Courant Inst., NYU.
Buckling of a very thin rectangular block, Comm. Pure Appl. Math. **23** (1970), 511–528.

Leon W. Cohen (1903–)
Program Director, National Science Foundation, 1953–1958.
On Differentiation, Acta Sci. Math. (Szeged) **38** (1976), 239–251.

Samuel Eilenberg (1913–)
Professor and Chairman, Columbia University.
Automata, Languages and Machines, Vols. A and B, New York: Academic Press, 1974 and 1976.

Churchill Eisenhart (1913–)
Chief Scientist, Eng. Lab., National Bureau of Standards.
Some canons of sound experimentation, Bull. Inst. Internat. Statist. **37** (1960), 339–350.

Holly C. Fryer (born 1908)
Professor of Statistics and Chairman, Kansas State University.
Concepts and Methods of Experimental Statistics, Boston: Allyn and Bacon, Inc., 1966.

Gustav A. Hedlund (1904–)
Professor, University of Virginia and Yale University.
(with W. H. Gottschalk) *Topological Dynamics*, Providence, R.I.: Amer. Math. Soc. Colloquium Publ. **36** (1955).

Magnus R. Hestenes (1906–)
Professor and Chairman, UCLA.
Calculus of Variations and Optimal Control Theory, New York: John Wiley & Sons, Inc., 1966, 405 pp.

L. Charles Hutchinson (born 1914)
Associate Professor, Northeastern University.

Irving Kaplansky (1917–)
George Herbert Mead Distinguished Service Professor, The University of Chicago; Director, Mathematical Sciences Research Institute, Berkeley, Ca.
Commutative Rings, Boston: Allyn and Bacon, Inc., 1970, 180 pp.

Walter Leighton (1907–1988)
Professor and Chairman, Western Reserve University and the University of Missouri.
On self-adjoint differential equations of second order, J. London Math. Soc. **27** (1952), 37–47.

Daniel C. Lewis (1904–)
Professor, Johns Hopkins University.
Autosynartetic solutions of differential equations, Amer. J. Math. **83** (1961), 1–32.

John L. Lewis
Director, Cinemath Tech. Animation Studio, New York.

Donald P. Ling (1912–
Research Mathematician, Bell Telephone Laboratories.
Geodesics on surfaces of revolution, Amer. Math. Soc. Trans. **59** (1961), 415–429.

Edgar L. Lorch (1907–)
Professor, Columbia University.

Compactification, Baire functions, and Daniell Integration, Acta Sci. Math. (Szeged) **24** (1963), 204–218.

George W. Mackey (1916–)
Professor, Harvard University.
Induced Representations of Groups and Quantum Mechanics, New York: W. A. Benjamin, Inc., 1968, 167 pp.

Saunders Mac Lane (1909–)
Max Mason Distinguished Service Professor, The University of Chicago.
Categories for the Working Mathematician, New York: Springer-Verlag, 1971, 262 pp.

E. J. Moulton (born 1887)
Professor, Northwestern University.

Francis J. Murray (1911–)
Professor, Columbia University; Duke University.
(with J. von Neumann) *On Rings of Operators IV*, Annals of Math. (2) **44** (1943), 718–808.

George Piranian (1914–)
Professor, The University of Michigan.
(with D. M. Campbell) *Normal analytic functions and a question of M. L. Cartwright*, J. London Math. Soc. (2) **20** (1979), 467–471.

Harry Pollard (1919–1971)
Professor, Purdue University.
Celestial Mechanics, Carus Math. Monographs, No. 18, Washington, D.C.: Math. Assoc. of America, 1976, 134 pp.

Arthur Sard (1909–1980)
Professor, Queens College, NYC; Research Assoc., University of California, San Diego.
The measure of the critical values of differentiable maps. (*Sard's Theorem*), Bull. Amer. Math. Soc. **48** (1942), 883–890.

Paul A. Smith (1900–1980)
Professor, Columbia University.
Fixed-point theorems for periodic transformations, Amer. J. Math. **63** (1941), 1–8.

Herbert Solomon(1919–
Head, Statistics Section, ONR; Professor of Statistics, Stanford University.
Geometric Probability, CBMS Regional Conference Series No. 28, Philadelphia, Pa: SIAM (1978).

J. J. Stoker (1905–)
Professor, Courant Institute, NYU.
Water waves, The mathematical theory with applications, New York: Interscience (1957), 567 pp.

Robert L. Swain (1913–1962)
Ohio State University, Professor, State Teachers College New Paltz, NY.
Approximate isometries in bundle spaces, Proc. Amer. Math. Soc. **2** (1951), 727–729.

Robert M. Thrall (1914–)
Professor, University of Michigan and Rice University.
(with W. Allen Spivey) *Linear Optimization*, New York: Holt, Rinehart and Winston, Inc., 1970, 530 pp.

Hassler Whitney (1907–)
Professor, Institute for Advanced Study.
Geometric Integration Theory, Princeton, NJ: Princeton Univ. Press, 1957, 387 pp.

Daniel Zelinsky (1922–)
Professor, Northwestern University.
(with O. E. Villamajor) *Galois theory for rings with finitely many idempotents*, Nagoya Math J. **27** (1966), 721–731.

Computing Staff

Georganne Beazley
Reba Beller
Eloise Buikstra
Claire Cohen
Deborah Davidson
Evelyn Garbe
Frances Gelbart
Lucy LaSala
Grace Lesser
Mary J. Lewis
Anna Merjos
Phyllis Monderer
Mary Anne Moore
Virginia K. Osburn
Angela Pelliciari
Mrs. Joe L. Piranian
Mae Reiner
Ellen Swanson
Irene Wiener
Marion S. Wolff

Secretarial and Administrative Staff

Phyllis Ackerman
Betty Amitin
Pearl Anton(ofsky)
Martha S. Brisbane
Betty Campagna
Frances Galloway
Sara S. Gelof
Marion Harris
Ruth Jackson
Juanita Lewandowski
Clara B. Lipsius
L. Veretta Olton
Rose O'Rourke
Helene M. Pastrone
Charlotte D. Pattee
Ruth P. Russell
Lenora Salmon
Pearl Sklar
Frances Stitt
Anna K. Stroh
Dorothy R. Townsend

References

[1] Peter Hilton, Reminiscences of Bletchley Park, 1942–1945, *A Century of Mathematics in America, Part* I (Amer. Math. Soc., Providence, RI, 1988), 291–301.

[2] Saunders Mac Lane, *Final Report of the Applied Mathematics Group, Division of War Research, Columbia University.* Consisting of four parts; Part 2, *Aerial Gunnery Problems.* October 31, 1945 (Confidential; subsequently declassified).

[3] Mina Rees, The mathematical sciences and World War II, *A Century of Mathematics in America, Part* I (Amer. Math. Soc., Providence, RI, 1988), 275–289. (First published in Amer. Math. Monthly, vol. **87** (1980), 607–621.)

[4] Richard Rhodes, *The Making of the Atomic Bomb.* New York: Simon and Schuster, Inc., 1956. 886 pp.

[5] J. Barkley Rosser, Mathematics and mathematicians in World War II, *Notices Amer. Math. Soc.* **29** (1982), 40–42; reprinted in *A Century of Mathematics in America, Part* I (Amer. Math. Soc., Providence, RI, 1988), 303–309.

[6] Smyth, Henry de Wolf. *Atomic Energy for Military Purposes*, Princeton: Princeton University Press, 1945, 266 pp.

[7] Norbert Wiener, *Extrapolation, Interpolation, and Smoothing of Stationary Time Series. With Engineering Applications*, Cambridge: The Technology Press of the Massachusetts Institute of Technology, 1949. ix + 163 pp.

The Education of Ph.D.s in Mathematics

SAUNDERS MAC LANE

1. Introduction

All these historical articles about the development of major departments of mathematics in the United States might perhaps be supplemented by some direct attention to one of the major objectives of such departments: the training of the next generation of research mathematicians. This article will be a first try; inevitably, most of the examples will be drawn from experience at the University of Chicago, but the issues raised may well apply elsewhere. I will also add some small adjustments to my prior article (in Part II of this series) about the department of mathematics at Chicago.

2. Requirements

Marshall Stone has observed that when he came to Chicago in 1946 he soon managed to reduce the formal requirements for the Ph.D. to their essentials. Previously, students had been required to take 27 quarter courses; after finishing a prospective thesis, they underwent an oral examination covering a considerable selection of those 27.

The effect is best described by a sad example. One student had gone away to teach, finally wound up a thesis, and returned to face that examination. One of the courses covered was on algebraic topology. Successive questions about homology groups, cohomology groups, and fundamental groups elicited no answer. Finally, an examiner brought up the subject of covering spaces. The student brightened, but did not have the definition. Then came a helpful question: "Please give an example?" The response was immediate: "The circle and the line." At that point, some cruel professor asked: "Which covers which?"

After this tragic denouement the department gave up that full list of 27 courses and that final exam on them all. The new system required a master's degree on an explicit list of basic courses, followed by written and oral exams on the list. For the Ph.D. there were then two topics, chosen by the student to match his expected interests, with successive oral exams on each, plus inevitably that thesis, with a final oral exam on the subject of the thesis. In this way, that last exam has a real and realistic target. I have the impression that this plan, or its analogues, is now in place in many graduate departments.

3. Direction of Theses

For this, there are no rules. Some professors, such as L. E. Dickson or Irving Kaplansky at Chicago, attract and hold many students (more than 60 in each of these cases). Some professors are active, but not so prolific. At a given period, one faculty member may be remarkably effective; in a ten-year period at Chicago, Irving Segal directed theses by I. M. Singer, Richard Kadison, Edward Nelson, Brian Abrahamson, and Bert Kostant (and perhaps others). What a list! Some faculty members set high standards. André Weil knew what was top quality mathematics and expected his students to live up to that standard. As a result, some fell by the wayside, some had trouble finishing a thesis when the results did not seem to meet the intended standard, others did manage to finish but stopped then and there (this happens to nearly every thesis director). But some of Weil's students really went on; I think for example of Arnold Shapiro, who accomplished much before his untimely death.

There are different procedures for guiding thesis students. Some faculty (I was one) try to see each student once a week. Some, perhaps more wisely, wait until the student has something to say. Some look for the main results in the thesis. Others read it word for word, with ample blue pencil, in the hopes of training the student for clearer writing later. I know one active mathematician who claimed that his Doctor-father never did read his thesis.

And how does the student or the teacher choose a problem? I recall one eager student to whom I assigned a specific topological problem which would seem to require some substantial algebraic computation. In those weekly sessions the student never presented more than a half-page of computation, if that. Eventually the student gave up. Sadly, I thought that perhaps I had ruined a promising career by assigning an impossible problem. Then two things happened. For other reasons, I finally needed the solution of that problem. With 20 pages of straightforward calculations by methods which had been at hand for the student, I had the solution and used it. At about the same time, that student came back to Chicago and, wisely, consulted not me but one of my colleagues. However, that second attempted thesis did not get finished. So, in this case, I may not have really contributed to ruination.

There are other cases of real regret. In 1946, Eilenberg and Mac Lane had described the homotopy type of a space with just two nonvanishing homotopy groups; it involved a cohomology class. At that time Joseph Zilber, a graduate student at Harvard, came to me to say that he could do the same for any number of homotopy groups. I did not encourage him. The idea was later developed in the USSR (**MR 13**, pp. 374, 375), and so is now known as a Postnikov system; it might have been a Zilber system.

There are cases where the professor does not really see what was accomplished. At Harvard, in 1945, I suggested to Roger Lyndon that he compute some cohomology groups of suitable groups. He did, and I checked all the details and found them correct. I did not really understand what he had done; a year later, I heard about spectral sequences from Leray, but I still did not understand that Lyndon's thesis had really developed a spectral sequence (**MR 86d** #20060).

There are more cheerful cases. About 1955, Anil Nerode asked me to guide his thesis in logic. I agreed, but soon left for a sabbatical in Paris. Nerode proceeded to complete a good thesis on his own. This may indicate that one need not always assign a problem to a student. However, if I had really understood his thesis, I would have seen that he was trying to invent cartesian closed categories (which were developed only much later). I take solace in the fact that he went on to learn much more logic than I ever could have taught him.

Others will know different examples; I draw the conclusion that guiding a thesis is a rewarding but very difficult task. It does not at all fit the cynical statement that a thesis is a paper written by the professor under unusually difficult conditions, but not signed by him.

4. The Academic Parent

The usual relation "A is a student of X" seems to me highly ambiguous, because X may be undefined, empty, or multiple.

There are many examples of the empty set of thesis directors. Nerode is almost an example. Another at Chicago is William Howard. I seem to recall that he had attempted to write a thesis first with one and then with another faculty member. Nothing quite succeeded. I knew him only casually. One day he came to me with a rather brief manuscript in logic, proposed as his thesis. I read it and earnestly advised him to consider also certain related questions. He simply refused. I then accepted that brief manuscript as a thesis. I have never regretted the outcome; Howard has gone on to do significant research in proof theory and intuitionistic logic. I am proud to count him as one of my students.

There are theses directed at university U for submission at university V. I know one example where U is Harvard and V is Chicago, and other examples with U = Chicago, for students who were non-tenure faculty at Chicago. Such examples, when they work out, can be very satisfying.

There are real orphans. Once, in the early 1950s, I visited LSU, where Professor H. L. Smith (of the well-known Moore-Smith limits) had just died, leaving a student with a half-finished thesis, understood by none of the local faculty members. I talked with him, made recommendations to him, read the results, and finally wrote the department that it was an adequate thesis. He went on to an effective teaching career.

There are theses with multiple faculty direction. In earlier days at Chicago, Professors Bliss and Graves had several common students. In the Stone age, Kadison was a joint student of Stone and Segal, while Murray Gerstenhaber was joint with Albert and Weil. For my own part, there was Michael Morley. In the late 1950s, he came to me with a result in logic using the compactness theorem, hoping that it would be his thesis. Instead, I said, in effect: "Mike, applications of the compactness theorem are a dime a dozen. Go do something better." I no longer know why I thought he could do something better, but I knew he was interested in model theory, and I encouraged him to go to spend time at Berkeley, where there were real experts on model theory. He went there, talked with Professor R. L. Vaught, and came back with a proof of a theorem about categoricity in power, now well known as Morley's theorem. So he had two thesis directors, and both of us can be happy with the result, as well with its many extensions by logicians.

There are surely other cases of multiple thesis directors. For example, Walter Feit has described to me the development of his interest in finite group theory. First, while studying for a master's degree at Chicago (in 1951), he became fascinated with Galois groups. He then moved to the University of Michigan, where he learned much about group representations from Richard Brauer (1951–1952). Then Brauer left for Harvard, so Feit worked with R. M. Thrall, learned about Chevalley finite simple groups from Dieudonné, a visiting professor, and continued to consult Brauer in finishing his thesis (**MR 17**, p. 1051; **20**, #65).

From these and other examples, it seems in any event clear that it often helps a starting mathematician to learn from more than one mentor and so to fit the different ideas together.

5. Colloquia

Graduate students also can learn much from visiting mathematicians, who give talks at colloquia and/or seminars. One notable example has to do with Galois theory. Since the late 1890s, this theory had been a major part of algebra courses, as a natural continuation of the theory of equations so central to algebra. It was a substantial part of Heinrich Weber's influential

"Lehrbuch der Algebra". At the first AMS colloquium, James Pierpont from Yale lectured on Galois theory. A text written at Columbia in 1928 covered just Galois theory. I recall studying it hard at that time, but it and the other presentations were complicated, full of excessive use of permutation groups in a noninvariant way, and so very difficult to understand. Finally in 1931, the influential van der Waerden *Moderne Algebra* gave a really clear presentation of the Galois group, described invariantly as the group of those automorphisms of the root field which leave the base field pointwise fixed. This book was based on lectures by Emil Artin and Emmy Noether.

Artin continued to think about Galois theory, to find a new and still more conceptual presentation making effective use of linear independence arguments. In the year 1937–1938 he gave a (two-hour) colloquium lecture at Chicago with it all there (as later written up in his booklet with Milgram, **MR 4**, p. 66; **5**, p. 225). It was a spectacular lecture; I think the students learned much; I know that I did and that Birkhoff and I subsequently incorporated the results in *Survey of Modern Algebra.*

In an article in Part II of this series, Gian-Carlo Rota asserts that Artin overemphasized Galois theory. I disagree. The emphasis was there before Artin. What Artin did, in two stages, was to make it clear and perspicuous, as it had not been. If there was emphasis, this emphasis has led to results, as in the well-known Galois theory of Grothendieck (**MR 36** #179ab).

6. Group Theory, Decline and Revival

This is an instance of the effects of graduate work. In the years 1900–1920, group theory was very active in the United States. At Chicago, E. H. Moore was interested in the axiomatics for groups and also found a set of generators and relations for the symmetric group; these relations play a role today in braid groups and conformal field theory. There was much interest in listing all groups of a given order. There was an influential book by Miller, Blichfeldt, and Dickson. At Stanford, Manning worked on permutation groups. Then the interest died down. In 1928, Oystein Ore came to Yale, where he gave courses on group theory. I learned from one of these courses, as did Marshall Hall. Then Ore's interest turned to lattice theory, as for example in the study of the lattice of subgroups.

Subsequent activity in group theory was for a period somewhat limited. Reinhold Baer at Urbana published extensively on group theory, finite and infinite. Marshall Hall studied group extensions, word problems and free groups; by 1959 he had organized his interests in a text (**MR 21** #1966). Hans Zassenhaus at Hamburg had written an influential text on group theory (with perhaps too much emphasis on the elaborate proof of the Schreier–Zassenhaus theorem (**MR 11**, p. 77)); he later taught at Notre Dame and at Ohio State. Combinatorial group theory was not much pursued. In Europe

Graham Higman, Weilandt, Specht, Kuros, and others were active; Philip Hall was quietly influential, as in his work on the classification of p-groups (**MR 2**, p. 211). Richard Brauer taught at Toronto, then at the University of Michigan and Harvard, quietly but steadily developing methods of representation theory (**MR 13**, p. 530; **17**, p. 824). His interests were clearly directed at finite simple groups, as in his 1945 paper with H. F. Tuan: "On simple groups of finite order, I" (**MR 7**, p. 371) and in the influential 1955 paper with his student K. A. Fowler: "On groups of even order" (**MR 17**, p. 580). Michio Suzuki, who later constructed some finite simple groups, also had effective contacts with Brauer. However in the 1950s, finite group theory was not in the center of mathematical interest in the USA.

In 1955–1956 I spent a sabbatical year in Paris, working on homological algebra. I tired of it. When I came back, I thought (for reasons not now clear to me) that group theory was due for revival. I taught a two-quarter course on both finite and infinite groups; at the end of the second quarter, I was at the end of my knowledge. Joseph Rotman was one of my students; he later wrote an effective text on group theory (**MR 34** #4338).

John Thompson was also a student in that course; at the end he said that he wanted to work in group theory. I encouraged him, and in his case I usually spent two or more hours a week talking with him, or more accurately, listening to what he had done. At first he tried to solve the Burnside problem, to show that a finitely-generated group in which every element has order dividing the fixed integer k must be finite. He did not succeed (and it later took Novikov and Adian years to prove the contrary). After talking the problem over with me, Thompson turned to other questions. With some understanding of his potential I took care to invite to Chicago a number of group theorists, including Reinhold Baer and Marshall Hall. Thompson soon attacked and then solved the following problem: A group with an automorphism of prime order fixing only the identity is necessarily nilpotent. In this work he made astute use of results of a paper of Philip Hall and Graham Higman on p-solvable groups (**MR 17**, p. 344).

On page 370 of Part I of this series, Marshall Hall says that he gave Thompson this problem; this is doubtless correct. Hall also says that, "From then on John Thompson came to Columbus [where Hall then taught] to work with me on a Ph.D. topic." This sentence seems to me a considerable exaggeration. Clearly, Thompson learned from several of us. John also resisted other advice. At an early stage one of Weil's students remarked to John that finite group theory was not it; it would be better to consider the (then just discovered) Chevalley groups. Thompson wisely persisted in his own direction.

A major aspect of thesis direction seems to me that of finding a topic which lies in the natural interests and talents of the student.

Group theory continued. Adrian Albert had spent a sabbatical year 1957–1958 at Yale, where he set up an NSF grant for a "special year" on algebras. He returned to Chicago, convinced of the merit of special years; in addition he had acquired from his mentor Dickson a real interest in group theory. So he soon organized a special year on group theory at Chicago. Those present included John Thompson and Walter Feit. From that special year came the Feit–Thompson odd-order paper: Every finite simple group is cyclic or of even order (**MR 29** #3538).

The subsequent development of finite group theory is detailed by Daniel Gorenstein in Part I of this series. The account above gives some elements of the background, and indicates that the revival came from many different contributions. I am happy to have played a small part in a field where my chief related work has been in the cohomology of groups.

7. A Diploma Mill?

In my article on Chicago in Part II of this series, I allowed that Chicago under the chairmanship of Gilbert Bliss had become a "diploma mill". If one views the department as a competitor for the accolade "Best Department" this may be a correct assessment. But this is a strictly internal view; adequate history must also consider external forces, as to the fit and tension between a research department and the rest of society. (See in this connection the article by William L. Duren, Jr. in Part II.) Now at that time, many universities in the Midwest and South were building up and so desperately needed trained faculty. This need must have been especially apparent at Chicago, for many of the previous Chicago students were already located at these growing schools; they must have told Bliss and others at Chicago of their needs. For these reasons, there could be a reasoned policy of taking in many students, equipping them with good knowledge and an adequate thesis, and finding subjects such as number theory and the calculus of variations which would allow for handy thesis topics. This appears to have been the policy at Chicago, and it was not intended to block the education of exceptional students. It did involve real faculty interest in the life of the students (for example, there were bridge parties in the Eckhart common room with students and faculty). One of my fellow students at Chicago in 1930–1931, Julia Bower, read my earlier article and then wrote me hoping that I might succeed "in expressing the special combination of discipline and joy that characterized that mathematical community".

Arthur Everett Pitcher was born in 1912 *and took his Ph.D. at Harvard in* 1935 *under the direction of Marston Morse. He was assistant to Morse at the Institute for Advanced Study during the following year and then Benjamin Peirce Instructor at Harvard for two years. He held faculty positions at Lehigh University from* 1938 *until his retirement in* 1978. *His first paper of about twenty was a precursor to his thesis and was written jointly with Morse. It appeared in the* Proceedings of the National Academy *in* 1934 *entitled* "On Certain Invariants of Closed Extremals." *He gave an invited address to the American Mathematical Society in* 1955, *published in the* Bulletin *of the Society under the title* "Inequalities of Critical Point Theory." *During World War II, he served in the army at the Ballistics Research Laboratory at Aberdeen Proving Ground and later in scientific intelligence in the European Theatre. He was an associate secretary of the American Mathematical Society from* 1959 *to* 1966 *and secretary from* 1967 *to* 1988. *At the time of the AMS centennial, he wrote* A History of the Second Fifty Years of the American Mathematical Society, 1939–1988, *thus updating R. C. Archibald's semicentennial history of* 1938.

Off the Record

EVERETT PITCHER

Most of this talk[1] is in the style of personal reminiscences. I was aware of the American Mathematical Society (AMS) for as long as I can remember. My father was a mathematician, a student of E. H. Moore. His thesis was concerned with the complete independence of Moore's axioms for general analysis. His subsequent work, a good deal of it joint with E. W. Chittenden, was in the field of topological spaces. I still have much of his library, including books that I used as a student:

Bôcher, *Introduction to higher algebra*
Burnside and Panton, *Theory of equations*
Carathéodory, *Vorlesungen über reelle Funktionen*
Gibbs, *Vector analysis*

[1]The author was the principal speaker at the first banquet to honor individuals who have been members of the American Mathematical Society (AMS) for twenty-five years or more. The banquet was held at the Annual Meeting in Phoenix on 14 January 1989. The speech has been edited for publication.

Goursat, *Cours d'analyse, v. II*
Grace and Young, *Algebra of invariants*
Hausdorff, *Grundzüge der Mengenlehre* (1914 ed.)
E. H. Moore, *Introduction to a form of general analysis, New Haven Colloquium* (autographed presentation copy)
F. Riesz, *Les systèmes d'équations linéaires à une infinité d'inconnues*
de la Vallée Poussin, *Cours d'analyse*
Veblen, *Analysis Situs, Cambridge Colloquium*

From the time I was three until his premature death when I was eleven, my father was head of the mathematics department at Adelbert College of Western Reserve University in Cleveland. He was a member of the Council at the time of his death, a fact of which I became aware on systematic reading of minutes of the Council when I became secretary of the Society.

My father used to be absent between Christmas and New Year's, for that was the time of the Society meetings. These were usually in New York at Columbia University. Travel was by overnight train. I used to get the registration badge when he returned, a handsome metal frame rather than the ubiquitous plastic of today. The existence of such a substantial badge is consistent with the fact that the annual meeting was held jointly or concurrently with the annual meeting of the American Association for the Advancement of Science (AAAS).

Attendance was fewer than 100 members. Of course the total membership in 1920 was 770 persons. The year 1916 at Columbia was the first exception, with 131 persons in attendance. By 1926, attendance at annual meetings usually exceeded 200.

I did my undergraduate work at Western Reserve. Toward the end of my high school years, I had waivered among bacteriology, chemistry, and mathematics but settled on mathematics as I began college. I was somewhat raised by hand in a small department (three professors, with two more in other colleges of the university). My next contact with the AMS was at the end of my sophomore year. I was just eighteen. There was an assistant professor whom I knew well, M. G. Boyce, who was a product of the Bliss school of the calculus of variations. He asked me if I would like to go to the Summer Meeting of the AMS with him. He had a new car, his first, and was driving. Of course I agreed. This was the Summer Meeting of 1930 in Providence on the Brown campus. I note that there were 218 members present, 78 papers, Colloquium Lectures by Solomon Lefschetz, and invited addresses by T. H. Hildebrandt and J. D. Tamarkin. Society membership in 1930 was 1,926.

One should not be misled by that number of 78 papers. Thirty-one of the 78 were on the supplementary program. That is, they were classified and listed but were not presented in person.

I think that there was no registration fee. I can remember explaining that I was only an undergraduate student, but being made welcome. One of the Brown graduate students, I think it was H. L. Krall, showed me around the campus. It was a great revelation to me to learn that graduate students at Brown had offices. There were five desks in a garret where my guide was quartered.

The Mathematical Association of America (MAA) met on Monday, 8 September and on Tuesday morning. The group picture was taken Tuesday noon. A group picture was a compulsory event through 1948. The Society met Tuesday afternoon through Friday morning.

Rooms were at Pembroke College, $1.50 per day for singles, $1.00 per person for doubles. Meals in the cafeteria cost $2.00 per day.

Although there was no registration fee, there was a special fee of $2.00 for the Colloquium Lectures, which consisted of four lectures each of one and one quarter hours. The Society was keeping its books on the basis of separate funds at the time. All income and expenditure was allocated. There was no such entity as overhead. This fee went into the fund to pay for the publication of the lectures. I did not pay the $2.00, but I did hear about fifteen minutes of one of the lectures.

The entertainment included a clambake, my first contact with clams and lobster. A half-day outing was a feature of summer meetings for many years.

The first meeting I recall attending seriously was the Annual Meeting of 1935; i.e., 30 December 1935 to 3 January 1936. I was assistant to Marston Morse at the Institute for Advanced Study (IAS) then and was in the employment market. As some of you will remember, this was at a time when one did not apply for a position. One depended on the efforts of an adviser or mentor for recommendations and was approached by a potential employer, but one attended meetings to be visible by contributing a paper and to be available for interview.

This was a joint meeting with the MAA and AAAS and included a section meeting jointly with the Econometric Society (ES) and the Institute of Mathematical Statistics (IMS) and a dinner of the National Council of Teachers of Mathematics. At the dinner, R. C. Archibald spoke on Babylonian mathematics with special reference to recent discoveries.

The Gibbs Lecturer was Vannevar Bush, who spoke on "Mechanical Analysis." I became very familiar with this version of mechanical analysis not long after at Aberdeen Proving Ground, when I was called to active duty in the Army. The Ballistics Research Laboratory had a companion to Bush's MIT differential analyzer in Aberdeen and used a second such machine at the Moore School of the University of Pennsylvania.

The differential analyzer was an enormous analog machine with shafts and gears connecting the principal elements, which were mechanical integrators used in the manner proposed by Lord Kelvin. It was used in Aberdeen to calculate trajectories of artillery projectiles at widely spaced intervals of initial variables. The calculation of a single trajectory could take the better part of an hour of running time, this after extensive mechanical preparation. Firing tables were then calculated by high order inverse interpolation, using hand computation with the assistance of a motor driven (or hand cranked) desk calculator. It was a treat to observe skilled operators of a motor driven calculator, for they depended on listening to the spinning dials rather than reading them in multiplication.

The meeting was concurrent with a meeting of AAAS in St. Louis. The Society used the Coronado Hotel as headquarters and most of the sessions were on the campus of St. Louis University. Parenthetically, the hotel overbooked and had no room for me. When I stood my ground about the confirmed reservation, they gave me the Queen Marie Suite of five rooms, named after the famous Queen of Rumania, who had recently occupied it in the course of a highly publicized tour.

I do not find any reference to a registration fee in the program. Since it was a joint meeting with AAAS, I would expect that there was one. R. D. Carmichael gave his retiring address as vice president of AAAS and chairman of Section A.

The program included thirty-nine contributed papers delivered in person, with twenty-seven more on the supplementary program. There was a joint AMS-MAA lecture by J. L. Synge on "Tensorial Methods in Dynamics." There was also a lecture by G. Szegö on "Some Recent Investigations Concerning Sections of Trigonometric and Related Series" and one by Thomas Rawles on "Mathematical Theory of Index Numbers" in the section with ES and IMS.

When I first went to Lehigh in 1938, I used to go quite regularly to fall and spring meetings in New York. These were ordinarily held at Columbia. It was a pleasant trip, for this was a time when there was good passenger train service: two hours by train with breakfast on the train into New York and a subway to Columbia, return by the reverse route in the bar and dining car.

I should (or perhaps I should not) tell you of another trip to a New York meeting. I believe it took place in 1946. I was out of the Army but was at the IAS, having not yet returned to Lehigh, so I was not a participant in the madness that I am about to recount. In any vita of André Weil, there is a hiatus of two years. He does not mention the time that he spent at Lehigh. I think he was quite unhappy there—and no wonder, for this was his first position in the new world and it was the time of the GI Bill with the flood of students in introductory calculus, at a level of teaching with which he was unfamiliar. At any rate, he was one of a group who drove to a New York meeting. On the old highway US 22, this was a three-hour trip, but not the way this group chose to do it. Someone, though I do not wish to attribute this to Weil, drew a straight line on the map from Bethlehem to New York, a straight line being well known to yield the shortest distance between two points. Then they followed it as closely as possible on the roads on the map for about six hours. This may remind you of a familiar example in the calculus of variations.

My first national meeting as associate secretary was the Annual Meeting of 1961 at the old Hotel Willard in Washington, D.C. The associate secretary arranges the program, that is, screens the abstracts of contributed papers and assigns times and space within certain constraints, the hotel being previously visited to assure that the space is adequate. Nevertheless, one is faced with rooms of varying size and suitability and must guess which sections of contributed papers will draw the larger audiences. Too few people in a room is uncomfortable and too many can be a disaster.

There were two sessions of contributed papers in applied mathematics and probability, each with eight papers. In the second session on the morning of the last day of the meeting, there was a paper by Edward O. Thorp, then at MIT, with the title "Fortune's Formula: The Game of Blackjack." The abstract concluded with a summary calculation that in the casino game (with one deck) a player with capital of $3,200, making large bets when the remainder of the deck is in his favor and nominal bets otherwise, can win about $10 per hour (characterized as a living wage), with probability of ruin of less than 1%. The abstract promised the strategy for so doing but did not state it. As I recall, I had assigned this session to one of the larger rooms. It filled completely. Every seat was taken. The aisles were full of people standing and sitting. Not only that, I did not recognize many in the audience.

The Society has tried sporadically to publicize its meetings and was engaged in such an effort at this time. In so doing, we had made known the existence of this paper. A lot of outsiders attended, including administrative assistants of a number of congressmen, for whom attendance was an assignment.

The system involved memorizing some or all of the cards as played in order to learn the composition of the unseen partial deck from which the play proceeded. It further required knowing what constituted a favorable partial deck. Subsequent use of the system by earnest practitioners, known as "card counters," has resulted in their being barred from casinos when detected and in the introduction by the casinos of multiple decks. The alternative of shuffling the deck after each hand is not acceptable to the management because it uses time.

The launching of Sputnik in 1957 had many effects, some apparent even in meetings. There was a perceived shortage of mathematicians that was followed by an increased number of contributed papers presented at meetings as interest in mathematics was heightened and more Ph.D.'s appeared. Inasmuch as space for presentation of papers at meetings had already been booked at hotels and threatened to be inadequate, there was a brief period during which is was necessary to limit the number of contributed papers at annual meetings to 200. This may have discouraged the submission of papers at a meeting or two but almost no papers were turned away or postponed to another meeting.

The largest meeting of the Society was the joint meeting with MAA in New Orleans in January of 1969. There were 4,175 registered mathematicians in attendance. In the academic year 1968–1969, there were at least 1,156 new Ph.D.'s in the mathematical sciences, awarded by 130 universities in the U.S. and Canada. This was a result of a combination of circumstances. Because of the perceived shortage of mathematicians, opportunities for support of education and research had appeared, not the least of which were the NDEA fellowships offered through the National Defense Education Act. By 1969, this effort may have overshot the mark. In any event, there was a scramble for jobs in January of 1969. There were 722 papers at the meeting in New Orleans. Almost all of them were contributed ten-minute papers, many by aspiring job seekers. The employment register had no fewer than 182 prospective employers and 465 applicants for positions. This meeting should be mentioned very cautiously in the presence of Hope Daly, now head of the Meetings Department in the Providence office. She was then new to the Society and was assigned to the register.

The meeting of January 1989 produced a different kind of record attendance. I had thought that the joint AMS-MAA lecture by Paul Halmos, titled "Matrices I have met," in January 1986, with attendance of 1,260, had set a record that was only just now broken. However, the joint AMS-MAA lecture by John Kemeny and the first of the Colloquium Lectures of Victor Guillemin each had larger attendance. But the new attendance record of 1,863 was set by the AMS-MAA lecture by Stephen Smale titled "Story of the higher dimensional Poincaré conjecture (what actually happened on the beaches of Rio de Janeiro)."

At the meeting of December 1941 at Lehigh, Tomlinson Fort was the Chairman of the Committee on Arrangements and I was a member. Department members were co-opted to work. The student contract on dormitory rooms was a September-to-June contract. I well remember that prior to the meeting Malcolm Smiley and Bill Transue and I tramped from room to room in the dormitories arranging with students to sublet their rooms during the meeting. I found the accounting for the meeting recently. We rented rooms to mathematicians for a total of 585 nights at $1.25 a night, from which $0.75 per night was paid to the student. We sold 250 banquet tickets at $1.85. Income of $1,193.75 exceeded expenses, which included a tea, by

$73.89, which was donated to the local chapter of the Red Cross. There is, of course, no accounting for university space, janitor service, etc.

The program at that meeting had a session of six papers on numerical computation devices, forty-nine contributed papers, a symposium on applied mathematics consisting of two one-hour addresses, the award of the Cole Prize to Claude Chevalley, and an hour address by Oscar Zariski of The Johns Hopkins University.

Incidentally, the system once used of putting meeting participants in student rooms during the school year had possibilities for the students. There was a meeting long ago, at Cornell I think, where a resident woman learned that mathematicians would occupy rooms during vacation. She left her calculus assignment on the desk with a "pretty please" on it. It is reported that the temporary occupant was Hassler Whitney, who dutifully wrote out the assignment.

There used to be another kind of burden on the local faculty at the site of a meeting. There was a time when host institutions received no financial support from the AMS-MAA. If university funds were not available or if the university support was not swallowed by university overhead, department members sometimes found themselves contributing personally in order to be gracious hosts. This factor also increased the unwillingness of institutions to host meetings. It was as recently as 1953 that reimbursement of expenses was authorized of $25 for a one-day meeting, $50 for a two-day meeting, and $100 for a Summer or Annual Meeting, with an annual ceiling of $600.

The Society used to be more loosely organized than it is now. The Chicago section was organized in 1897 and from then through 1923 had an independent set of officers and meetings. The bylaws then provided for the authorization of sections. The San Francisco section first met in 1902 and maintained its autonomy through 1929. The Southwestern section functioned from 1906 through 1928. It was really a "south central section." The first meeting was in Columbia, Missouri. After a time, the meetings of the sections were designated as regular meetings of the Society and were incorporated into the system of numbering of meetings. The division into sections and their relative independence was a natural consequence of distances and travel time.

There was a letter in the Notices this past October from Hugh J. Miser, telling of a meeting attended by his father, Wilson L. Miser in Eckhart Hall about 1910 when he was a graduate student. When the meeting was over in mid-afternoon, the assembled group adjourned to a softball game, for which they had just enough persons for two teams and an umpire.

Membership in the Society is accomplished by the process of application, election, and payment of dues. For many years, an application required the signatures of two supporting members. So far as I know, these served no real purpose. When an application was received without them, the secretary and

American Mathematical Society, Chicago Section, meeting held at Northwestern University, Evanston, January 1902. (1) E. H. Moore, (2) Thomas F. Holgate, (3) Henry S. White, (4) Alexander Ziwet, (5) Jacob Westlund, (6) unidentified, (7) Oskar Bolza, (8) J. W. Glover, (9) Ida M. Schottenfels, (10) C. A. Waldo, (11) Frank L. Griffin, (12) unidentified, (13) E. J. Townsend, (14) W. J. Davidson, (15) George T. Sellew, (16) unidentified, (17) J. B. Shaw, (18) R. J. Aley, (19) C. L. Bouton, (20) T. Proctor Hall, (21) D. F. Campbell, (22) unidentified, (23) Ellery W. Davis, (24) John H. Mc Donald, (25) unidentified, (26) W. L. Risley, (27) F. R. Moulton, (28) G. W. Meyers, (29) Oswald Veblen, (30) R. W. E. Summerville, (31) W. H. Short, (32) L. W. Dowling, (33) M. J. Newell, (34) H. G. Keppel, (35) unidentified, (36) unidentified.

an accommodating person down the hall supplied the signatures. It was not until 1973 that the requirement was dropped.

Election was once quite formal. Names of applicants were accumulated by the secretary, presented to the Council at a meeting, and listed in the minutes. At the next meeting, applicants were elected.

The system has been mechanized, institutionalized, and delegated to the point that it is now handled by the Membership and Sales Department, with record vote delegated to the secretary and the associate secretaries, who conduct a ballot by mail.

One must not suppose that election was automatic. I found two examples of persons who were proposed for membership and not elected. One was a routine application. It was noted that the person was simultaneously to be a nominee of an institution, so the application was rejected. The other was a specific denial for no stated reason. This was long enough ago that the application is not available, so I have no information except the name of the applicant, which I do not recognize or identify. It was a woman, but that is surely not the reason for rejection since women were members almost from the beginning of the Society.

Another application for membership that was twice denied came from Nicolas Bourbaki. The first appearance was in 1948 as an application for individual membership, and the second was in 1950 as an application for membership by reciprocity with the Société Mathématique de France. I have recounted the circumstances, the furor that was engendered, and the response to the second denial at length in my book *A History of the Second Fifty Years, American Mathematical Society, 1939–1988.*

Meetings on a university campus depend on an invitation from a host. Attitudes toward issuing such invitations vary with place and time. The ideal campus for a meeting from the point of view of the AMS and MAA is an institution that is easily accessible, very large, and academically superior. It will have put forth many Ph.D.'s and have sheltered many visitors over the years. The meeting is a return home or the reliving of a pleasant experience for many people, with large meeting attendance as a consequence.

On the other hand, a meeting may make a lot of work for local faculty. One likes to see one's friends but not too often. So the desirable institutions do not necessarily welcome repeat business.

There is another requisite that I should mention. From the point of view of the Society, the institution should be inexpensive. Universities are capable of burying many of the costs of a meeting in general university overhead. With tighter budgets, they are reluctant to do this, and at publicly supported institutions, there may be a mandate against it. Because of either the financial stringency or the legal requirement, institutions set up conference bureaus with staff, whose charges are added to direct space charges and itemized service charges. Sometimes the additional supervision is valuable and worth the cost. Note, however, that institutions come to regard meetings as a profit center. Some of the groups holding meetings are commercially financed trade associations. One must continually remind university administrations that an academic society is not itself a profit center from which the university should properly carve its slice. One must withdraw from negotiations when one is classified with trade associations as to charges.

Lesser institutions that would like to become better known sometimes seek meetings eagerly. They are less likely to overcharge for space and services. If their facilities are unsatisfactory or if they are not easily accessible to travelers, this gives rise to the awkwardness of rejecting an invitation after a general solicitation for invitations.

The burden on the faculty of an institution used to be relatively greater than it is now. Local volunteers drew a much larger fraction of the work when the total staff supplied from the Society was the secretary to the secretary of the AMS and perhaps one other person.

The Society had no staff at all until 1914. F. N. Cole was both secretary from 1896 to 1920 and associate editor of the *Bulletin* in 1897–1898 and then an editor until 1920. But he had no regular assistant until the appointment of Dr. Caroline E. Seeley in 1914 to the position called clerk, which she held until 1935. Despite the lowly title, which was on the books until her departure, she became associate editor of the *Bulletin*, 1925–1934, and of the *Transactions*, 1924–1936.

By the end of the 1930s, that is, up to the beginning of *Mathematical Reviews*, the staff had grown to only four persons, housed in space belonging to Columbia University. While each of the four had a nominally assigned area of activity, all were available and were called upon for whatever needed to be done.

Meetings initially were operated entirely by impressed volunteers. As the staff grew beyond the lone Miss Seeley, one or more members of the headquarters staff could be made available for the registration desk at the Annual or Summer Meeting. The individuals who drew this duty most frequently over the longest period were Margaret Kellar, who was office manager for a number of years, and Muriel Scribean, who was the accountant.

The first person employed with the specific title of manager of meetings was Marian Leigh, who served in that capacity in 1960–1965. Since that time, the Meetings Department has grown with the size of the task until now Hope Daly, who took over the job in 1975, is the head of a department of ten.

The growth in personnel reflects the increase in size of the task. There used to be an Annual Meeting and a Summer Meeting, each of a few hundred people, within the purview of the Meetings Department. Now it is responsible for the management of those two meetings of from one to three thousand people, one with an extensive Employment Register, a spring Council meeting, a Summer Conference, a Summer Symposium in Applied Mathematics, a set of Summer Research Conferences, two or three shorter symposia per year, and meetings of the Agenda and Budget Committee and of the Executive Committee and Board of Trustees.

MATHEMATICAL STATISTICS IN THE EARLY STATES[1]

By Stephen M. Stigler

University of Wisconsin, Madison

The history of mathematical statistics in the United States prior to 1885 is reviewed, with emphasis upon the works of Robert Adrain, Benjamin and Charles Peirce, Simon Newcomb, and Erastus De Forest. While the period before 1850 produced little of substance, the years from 1850 to 1885 saw such innovations as an outlier rejection procedure, randomized design of experiments, elicitation of personal probabilities, kernel estimation of density functions, an anticipation of sufficiency, a runs test for fit, a Monte Carlo study, optimal linear smoothing, and the fitting of gamma distributions by the method of moments. Reasons for the rapid acceleration in the growth of the field are explored.

1. Introduction. In 1799 Thomas Jefferson received a letter from a young man asking which branches of mathematics it would be most useful for him to study. Jefferson's reply praised Euclid and Archimedes as useful sources, and stated that trigonometry "... is most valuable to every man. There is scarcely a day in which he will not resort to it for some of the purposes of common life; the science of calculation also is indispensable as far as the extraction of the square and cube roots, Algebra as far as the quadratic equation and the use of logarithms are often of value in ordinary cases: but all beyond these is but a luxury; a delicious luxury indeed; but not to be indulged in by one who is to have a profession to follow for his subsistence" (Smith and Ginsburg, 1934, page 62). Jefferson listed "Algebraical operations beyond the 2nd dimension" and calculus as among these luxuries, and doubtless would have added probability if he had been asked.

What follows is a report on an investigation into the question: Did Jefferson's contemporaries and their 19th century descendants follow his advice? How, by whom, and why was the study of the "delicious luxury" of probability and mathematical statistics pursued in the United States, in the first century of its existence? There were good reasons for expecting that little, if any, American work would be found. American science generally did not reach advanced stages of development before the latter half of the nineteenth century, as American capital and genius had found other pursuits more rewarding. (See de Tocqueville,

Received November 1976; revised June 1977.

[1] This research was supported by the National Science Foundation under Grant No. SOC 75-02922, and was presented as a Special Invited Paper at the Annual Meeting of the Institute of Mathematical Statistics, New Haven, Conn., on August 18, 1976.

AMS 1970 *subject classifications.* Primary 62-03; Secondary 01A55.

Key words and phrases. History of statistics, randomization, density estimation, sufficiency, runs test, Monte Carlo, gamma distribution, smoothing.

Reprinted from *The Annals of Statistics*, Vol. 6, No. 2, pp. 239–265, with permission of the Institute of Mathematical Statistics.

1840, and, for a modern assessment and recent references, Reingold, 1972.) Nonetheless, this investigation was undertaken with a cautious optimism. Although standard histories of statistics have little to say about American work of this period, the years from 1770 to 1850 had been ones of great interest in this subject in Europe: perhaps some Americans would have been inspired to join in its study, and contribute to its development. But, at least in the years before 1800 this was not the case. Upon closer examination it appears that the founding fathers' most significant contribution to mathematical statistics was their decision to leave the Federalist Papers unsigned! (Mosteller and Wallace, 1964.)

Fortunately as the 19th century progressed some signs of American interest in this field appeared. In what follows we shall review the early development of mathematical statistics in America, from its beginnings until 1885. The choice of 1885 as a cutoff date had been made for several reasons. First, by 1885 the field (if we may be so anachronistic as to call nineteenth century mathematical statistics a "field") has achieved a relative maturity, both in Europe and America. Those institutions most responsible for the early development of statistical techniques, the geodetic surveys, the observatories, and the insurance companies, all had reached advanced stages of growth, although the spread of these techniques to other areas of application was still only tentative. Second, 1885 marks the beginning of the major statistical works of Galton and Edgeworth, works that led directly to that of Karl Pearson, R. A. Fisher, and the twentieth century explosion of interest in the field. And third, 1885 marks the culmination of the work of one of the major American figures to contribute to the early states of statistics, Erastus De Forest.

In Section 2 the situation before 1850 will be surveyed, and the few oases in this statistical desert discussed. Section 3 will consider the Peirces, a family closely associated with the emergence of mathematical research in America, and Simon Newcomb. Section 4 will discuss early work at Yale, and the remarkable achievements of Erastus De Forest. Finally, Section 5 will consider the reasons for the late development of mathematical statistics in America, and attempt an assessment of these early efforts.

2. Before 1850. Prior to about 1850, the level of attainment in mathematical statistics in America—indeed the level of attainment in American mathematics generally—was quite primitive. In fact, the most important American mathematical publication to appear before 1850 was Nathaniel Bowditch's 1829–1834 translation of Laplace's *Mécanique Céleste*, and the only widely circulated American work on probability I have found to appear before 1850 was an anonymous book review! (Anonymous, 1832.)

At first this dismal assessment of early American work may seem incredible. After all, Harvard could boast of a professorship of mathematics as early as 1727, the year of Newton's death. But upon closer inspection it appears that this early professor was no fit companion of Newton: his only publication was

Edward Wigglesworth (1732–1794)

an arithmetic, and, we are told, he was removed from his chair in 1738 as "guilty of many acts of gross intemperance, to the dishonor of God and the great hurt and reproach of society" (Cajori, 1890, page 24). It is true that Thomas Jefferson's *Notes on the State of Virginia* (1785) and Benjamin Franklin's *Observations Concerning the Increase of Mankind* (1755) made important contributions to non-mathematical statistics, but the closest I have found to a contribution to our subject is the publication in 1789 of the first American life table, by a Harvard professor of divinity, Edward Wigglesworth (1732–1794) (O'Donnell, 1936, page 371). His publication contained no mathematics, and his dour expression is adequate commentary on the fortunes of statistics in America before 1800.

Nor did the beginning of the 19th century bring early relief from this drought of mathematical and statistical research. We may recall that the first quarter of the 19th century had produced exciting work in Europe: Gauss, Laplace, and a host of lesser workers had written books on probability and statistics. The greatest of these was Laplace's *Théorie analytique des probabilités*, first published in 1812. What, we might ask, was happening in American mathematics in 1812? You may complain that preoccupation with the war of 1812 may make such a comparison unfair, but recall that 1812 also marked the climax of the Napoleonic

Robert Adrain (1775–1843)

wars in Europe. Well, in 1812 twenty-five books on mathematics were published in the United States (Karpinski, 1940). Twenty-two these (including all those written by Americans) were books of tables or elementary texts on arithmetic or geometry. Two of the 25 were reprintings of English works on surveying, and one was a reprinting of an English work on fluxions (incidentally, this was the *first* text on calculus to be printed in the new world).

This situation persisted until about 1850. There was, however, one minor but interesting exception to this assessment, one brief but early spark, that hinted at the latent Yankee ingenuity that would erupt in the latter half of the century. That spark was the Irish-American Robert Adrain (1775–1843). For in 1809, Adrain published two derivations of the normal probability distribution, derivations that were published independently of and nearly simultaneously with Gauss's *Theoria Motus*.

Adrain had been born in Ireland in 1775, been well-trained in mathematics there, and emigrated to America after being badly wounded while an officer with the insurgent forces in the Irish uprising of 1798. The move was a wise one in many respects; he not only escaped the English gallows, he also switched from being an Irish mathematician, of which there were many, to being an American mathematician, of which there were few. He had taught mathematics in Ireland and continued to do so in America: by his death in 1843 he had taught at several Academies, Rutgers, Columbia, and the University of Pennsylvania (Coolidge, 1926; Babb, 1926).

His sole contribution to our field (and his sole original contribution to mathematics) appeared in a mathematical magazine he started in 1808. The magazine, called the *Analyst*, was one of several dedicated to problems and recreations that appeared in the first half of the 19th century. Adrain's paper, "Research concerning the probabilities of the errors which happen in making observations," was presented as a solution to a problem in surveying that had been posed as a Prize Question in the second number of the magazine. In the fourth number Adrain began his solution by presenting two derivations of the normal distribution, both of which were more wishful thinking than proofs. Both have been analyzed in modern notation by Coolidge (1926); the first was reprinted in the original notation by Abbe (1871) and the second by Merriman (1877, page 140). Essentially the first began by supposing probabilities of errors of observation would be proportional to the quantity measured, and by an obscure argument arrived at a differential equation of which the "simplest" solution was the normal density. The second derivation argued that a bivariate error distribution should be symmetrically distributed with respect to either axis, and the chance of an error should decrease in all directions from the origin, and must have continuous contours. The contour curve "must be the simplest possible having all the preceding conditions, and must consequently be the circumference of a circle" (Adrain, 1809, page 97); with independence of coordinates this leads to the bivariate normal distribution. Having derived the normal distribution of errors, Adrain went on the deduce the "most probable" solutions to several estimation problems by maximizing the likelihood function; that is, he found the least squares solutions. The problems he considered were those of determining the most probable position of a point in space (i.e., estimating a mean), correcting the dead reckoning at sea (i.e., reconciling an observed latitude with the recorded times and directions sailed), and correcting a survey (i.e., reconciling a system of inconsistent survey measurements).

Adrain's work remained nearly totally obscure until it was rediscovered by

Abbe in 1871 (Abbe, 1871). Only two references by other writers seem to exist. His rule for correcting dead reckoning was incorporated, with a reference to Adrain, in the third edition of Bowditch's *New American Practical Navigator* (1811, page 208), and his method of correcting a survey was cited by Gummere in his treatise on that subject (1817, page 116), but neither author mentions the connection with probability or with the general method of least squares.

Adrain, in his original paper (1809) and in two later applications (1818), does not mention Legendre's earlier work on least squares, and many writers have concluded that his discovery of the method was independent of Legendre's. However, Babb (1926) tells us that Adrain had an original copy of Legendre's 1805 work in his library, and Coolidge (1926) documents an instance where Adrain borrowed from a contemporary without citation.

Also, Adrain's formulae deriving the method of least squares are quite similar to Legendre's, and his groping toward the normal distribution would be more easily explained if Adrain had Legendre's work in hand and was working toward the method of least squares. On the other hand, there is no reason to doubt that his derivation of the normal distribution was done in ignorance of Gauss's work. The publication of the number of Adrain's magazine containing his paper was evidently delayed; notwithstanding the nominal year of publication 1808, internal evidence (such as a May 1809 date in a problem on page 110) suggests a spring 1809 publication as more likely. But Adrain's *manuscript* was dated 1808 (Abbe, 1871), and Gauss's book *Theoria Motus* did not reach even Paris until May 1809 (Plackett 1972, page 243), the preface being dated March 28, 1809. These facts and the total dissimilarity of their derivations of the normal density make it clear that Adrain must be counted an independent discoverer of this density, although his work had no apparent impact on the development of statistics.

3. The Peirces and Simon Newcomb. The emergence of American mathematical research was closely linked to Benjamin Peirce (1809–1880). Peirce was born in 1809 and graduated from Harvard in 1828, a classmate of Oliver Wendell Holmes. His most influential teacher was the self-taught sea captain and translator of Laplace, Nathaniel Bowditch (1773–1838). Peirce revised and proofread Bowditch's translation, and wrote several elementary books before being appointed to Harvard's Perkins professorship of Astronomy and Mathematics in 1842, a chair he held until his death in 1880. Incidentally, his first published work was the solution to a problem in one of Adrain's mathematical magazines in 1825 (Archibald, 1925).

Benjamin Peirce's main contributions to our field were three. The first is of a general nature, as a teacher and researcher in mathematics. He is generally regarded as the first American professor of mathematics for whom research was more than a hobby. With Peirce the character of mathematics in American universities changed from that of a service department to that of a major field

Benjamin Peirce (1809–1880)

of research. Although only his *Linear Associative Algebra* (1870) is considered today as a genuinely important piece of research, his texts on all areas of mathematics published from 1835 on and his many papers on mathematical astronomy marked a change in the level of work in American mathematics. Peirce is also reported to have been an inspiring and stimulating teacher, although descriptions of his technique make one wonder. One of his students was Charles Eliot, later President of Harvard, who wrote: "His method was that of the lecture or monologue, his students never being invited to become active themselves in the lecture room. He would stand on a platform raised two steps above the floor of the room, and chalk in hand cover the slates which filled the whole side of the room with figures, as he slowly passed along the platform; but his scanty talk was hardly addressed to the students who sat below trying to take notes… No question ever went out to the class, the majority of whom apprehended imperfectly what Professor Peirce was saying" (Eliot, 1925). Another student, later a famous mathematician himself, wrote: "Although we could rarely follow him, we certainly sat up and took notice. I can see him now at the blackboard, chalk in one hand and rubber in the other, writing rapidly and erasing recklessly, pausing every few minutes to face the class and comment earnestly, perhaps on the results of an elaborate calculation, perhaps on the greatness of the Creator" (Byerly, 1925).

Peirce's second contribution to statistics was specific. In 1852 he published the first significance test designed to tell an investigator whether an outlier should be rejected (Peirce, 1852, 1878). The test, based on a likelihood ratio type of argument, had the distinction of producing an international debate on the wisdom of such actions (Anscombe, 1960, Rider, 1933, Stigler, 1973a). While the debate was never satisfactorily resolved, Peirce was the victor in one sense: From 1852 to 1867 he served as director of the longitude determinations of the U. S. Coast Survey, and from 1867 to 1874 as superintendent of the Survey. During these years his test was consistently employed by all the clerks of this, the most active and mathematically inclined statistical organization of the era. Few statisticians have such an opportunity to put their test into routine use!

Benjamin Peirce's third major contribution to our field was his son, Charles Sanders Peirce (1839–1914). It is to his son and other workers of the period after 1860 that I shall shortly turn.

Now, during the first half of the 19th century America was largely preoccupied with territorial expansion and was primarily an agricultural nation. The English divine and wit Sidney Smith spoke more in truth than jest when he wrote early in the century: "Why should the Americans write books, when a six weeks' passage brings them, in their own tongue, our sense, science and genius, in bales and hogsheads? Prairies, steamboats, grist-mills, are their natural objects for centuries to come. Then, when they have got to the Pacific Ocean—epic poems, plays, pleasures of memory, and all the elegant gratifications of an ancient people who have tamed the wild earth, and set down to amuse themselves" (Smith, 1818). Sidney Smith's forecast was accurate, but his timing was off. With respect to science and mathematics, the period after mid-century saw rapid advancement. For mathematical statisticians, the three most important evidences of this growth were the rapid expansion of the U. S. Coast (later Coast and Geodetic) Survey, which was charged with measuring and mapping the new land; the founding and staffing of new observatories; and the growth of higher education, with its increasing emphasis on research. Charles Sanders Peirce was a product of this changing, intellectually charged atmosphere.

C. S. Peirce, son of Benjamin Peirce, was born in 1839 and like many young men of the era, went into the family business after he finished his schooling. Only in his case, "schooling" meant Harvard University and the "family business" was the newly expanded U. S. Coast Survey. Charles Peirce is best known today as a philosopher and logician—in fact there is today a C. S. Peirce Society which publishes a journal largely devoted to his thought. But for nearly thirty years he was an assistant at the Coast Survey, and a major portion of his life's work was tied to physical science and mathematics (Weiss, 1934; Eisele, 1974).

C. S. Peirce (1839–1914)

While his time with the Coast Survey included the years his father was superintendent, no charge of nepotism seems to have been leveled; in fact the son is generally conceded to have been the intellectual superior of the two.

Charles Peirce's interests covered an enormous range, and probability and statistics formed an integral part of both his philosophical views and his scientific method. Probability was basic to his view of scientific logic, and in one passage he defined "induction" to be "reasoning from a sample taken at random to the whole lot sampled" (Peirce, 1957, page 217). Indeed, Peirce's work contains one of the earliest explicit endorsements of mathematical randomization as a basis for inference of which I am aware (Peirce, 1957, pages 216–219).

What was perhaps Peirce's most influential statistical work came in the field of psychophysics, or experimental psychology. In 1884 Peirce and a student, Joseph Jastrow, performed an experiment to test the existence of a least perceptible difference in sensations. Gustav Fechner, in an important 1860 book (*Elemente der Psychophysik*) had argued that for each sense there was a nonzero threshold, such that if two sensations differed by less than the threshold they could not be distinguished. Peirce and Jastrow performed a large scale experiment involving the sensation of pressure, with themselves as subjects, and they effectively refuted the existence of such a threshold. Their methodology is of particular interest.

Peirce and Jastrow's (1885) report would be considered as a good example of a well-planned and well-documented experiment today; as a nineteenth century experiment it was unexcelled. (Incredibly, Peirce later described the precautions he took as "more careful and studied and elaborate than the memoir states" (Eisele, 1957).) They sought to refute the notion of a least perceptible difference by performing what we would now call a quantal response experiment using probit analysis, and showing that the results were inconsistent with Fechner's theory. Two slightly different known weights would be presented sequentially to the subjects, and they would state (or guess) in which of two possible orders they had been presented. In addition, the subject would estimate the confidence he had in his judgment on a scale from 0 to 3. The frequency of correct guesses for differing combinations of weights was then used to fit a probit response curve. Experiments of this general type were familiar to psychophysicists; this was called the method of "right and wrong cases." The authors gave few details on the method of fitting the probit response curve. They took "dosage" to be the ratio of the two weights used, so it could be assumed that a dosage of 1.0 led to a 0.5 probability of a correct guess, and there remained only one parameter, the scale parameter, to be determined. They apparently determined this separately for each dosage level d (by $\hat{\sigma}_d = (d - 1)/\Phi^{-1}(\hat{p}_d)$) and averaged the results. But they were painstaking in their description of the experimental procedures followed, and two aspects of these were strikingly original.

The first novel point was the way in which the estimates of confidence were used. Peirce and Jastrow used them to fit a relationship of the form $m = c \log (p/(1 - p))$ for each subject where m was the estimate of confidence, c an "index of confidence" peculiar to each subject, and p the "true" probability of a correct guess, estimated by the observed relative frequency. Peirce's conception of probability was that of an objective frequentist, but his work here shows he was also one of the first individuals (perhaps the very first) to experimentally elicit subjective or personal probabilities, determining that these probabilities varied approximately linearly with the log odds.

The second departure from tradition was the manner in which the order of presentation of the weights was determined. The Peirce–Jastrow experiment is the first of which I am aware where the experimentation was performed according to a precise, mathematically sound randomization scheme! The assignment was done in blocks of twenty-five (to achieve balance) by using, alternately, two well-shuffled packs of 25 cards, one with 12 red and 13 black cards, and one with 13 red and 12 black cards. They wrote "A slight disadvantage in this mode of proceeding arises from the long runs of one particular kind of change, which would occasionally be produced by chance and would tend to confuse the mind of the subject. But it seems clear that this disadvantage was less than that which would have been occasioned by his knowing that there would be no such long runs if any means had been taken to prevent them." Jastrow's later development and advocacy of this methodology had an important influence on later psychological research (Jastrow, 1888), although randomization in the design of experiments did not become part of the mainstream of statistical thought until R. A. Fisher's book on the subject appeared, a half-century later.

The breadth of Peirce's interests and the statistical turn of his mind are illustrated by a paper he read to the Philosophical Society of Washington in 1872. He, the abstract says, "called attention to the striking resemblance between the map showing the distribution of illiteracy ··· in the United States, given in the Report of the Census of 1870, and the map showing the distribution of rainfall during the three winter months published [by the Smithsonian]. Mr. Peirce suggested as a possible explanation for the resemblance, that the copious winter rains would produce agricultural plenty, which in its turn would favor indolence" (Peirce, 1872). I expect that farmers of his day would have taken offense at his informal path analysis, if they could have read his paper.

In another work the question whether or not meteorologists could successfully predict tornados led Peirce in 1884 to derive a latent structure measure of association for 2×2 tables (Peirce, 1884, Goodman and Kruskal, 1959). But his work for the Coast Survey is probably of more immediate interest to modern mathematical statisticians. In one 1879 paper on the "Economy of Research" (Peirce, 1879), he provided a rigorous mathematical analysis of the problem of optimally allocating experimental observations between competing experiments, under a model with two components of variance. This paper attacked the allocation problem from the point of view of a quite general utility theory, and contains an early and possibly independent formulation of a basic result in what economists call marginal utility theory.

Probably Peirce's best known statistical investigation was an 1873 paper "On the theory of errors of observations" (Peirce, 1873). At the close of this paper he presented the results of an extensive empirical investigation into the nature of laws of error. He hired an untrained 18 year old boy to react on a telegraph key to signals received. Five hundred measurements a day were recorded for 24 days, and Peirce sought to determine the distribution, that is, the density, of the reaction times for each day. The manner in which he estimated this density is interesting, particularly in view of recent work. He did *not* just present a histogram. Rather, "The curve has, however, not been plotted directly from the observations, but after they have been smoothed off by the addition of adjacent numbers in the table eight times over, so as to diminish the irregularities of the curve. The smoother curve on the figures is a mean curve for every day drawn by eye so as to eliminate the irregularities entirely." What he had done was to employ a form of repeated adjustment that was similar to techniques then in use for the interpolation and smoothing of mortality tables. What his technique did was to replace each ordinate of the histogram by a binomially

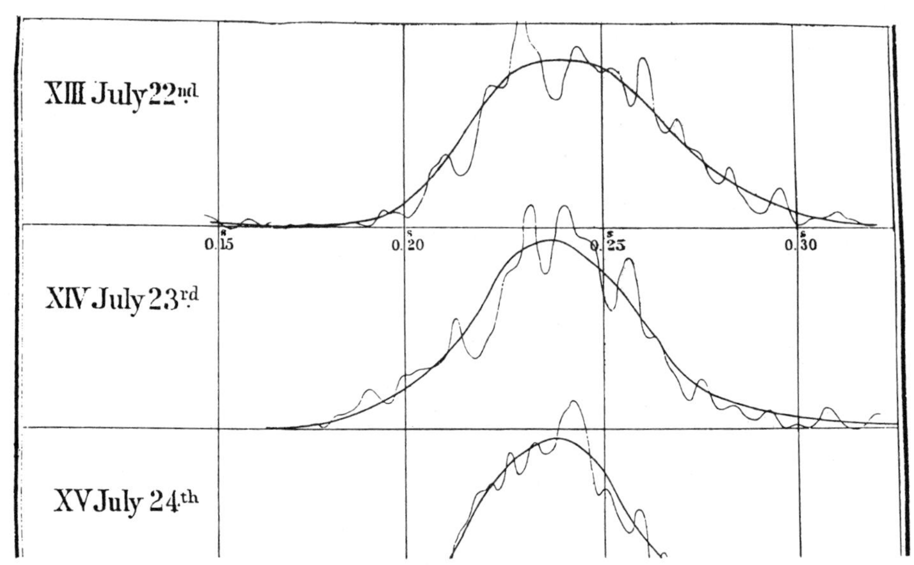

FIG. 1. C. S. Peirce's graph of his estimated probability density, by the kernel method and by a freehand sketch (smoother curves).

weighted average of nine consecutive ordinates. This is what we would now call a "kernel-type" estimate of the density using a binomial kernel that produces essentially the same effect as would a normal density kernel, although this kernel estimate produces a curve slightly out of phase with Peirce's. Fifty-five years later E. B. Wilson and Margaret Hilferty (1929) reanalyzed these data and concluded that Peirce's qualitative conclusion that the distribution differed little from the normal was not supported under closer scrutiny. In particular, data set 14 (Figure 1) was found to have a skewness of $\mu_3/\sigma^3 = 5.74$ and kurtosis of $\beta_2 - 3 = 63.6$. But Peirce had provided experimental evidence that human reaction times exhibited a regularity and distribution at least qualitatively similar to the normal curve's bell shape, and he had contributed substantially to American work on a major line of development in statistical thought, the concept of a distribution (Stigler, 1975).

Two years after Peirce's paper appeared, the British Astronomer Royal G. B. Airy added an appendix to the second edition of his text on the theory of errors (Airy, 1875) presenting an essentially similar example, apparently inspired by Peirce (although no citation was given). Unlike Peirce, however, Airy omitted the raw data and published only the smoothed frequency counts, and in this form Peirce's innovation was later to draw Karl Pearson's scorn in his famous paper on chi-square: "$\cdots$that Appendix really tells us *absolutely nothing* as to the goodness of fit of his 636 observations $\cdots$ to a normal curve. [We] find that he has *thrice smoothed* his observation frequency distribution before he allows us to examine it. It is accordingly impossible to say whether it really does or does not represent a random set of deviations from a normal frequency curve" (Pearson, 1900).

Simon Newcomb (1835–1909)

In his 1879 paper Charles Peirce had argued that in scientific research, the expected marginal utility of further investigation decreases as experimentation continues. He based this argument on the fact that the probable errors of estimated quantities are convex functions of sample size, and felt that the principle held more generally. "All the sciences exhibit the same phenomenon, and so does the course of life. At first we learn very easily, and the interest of experience is very great; but it becomes harder and harder, and less and less worth while, until we are glad to sleep in death" (Peirce, 1879). As if to confirm this as prophecy, Peirce became increasingly withdrawn in later life and died in isolation on a farm, in 1914.

Another, purely intellectual, descendent of Benjamin Peirce was the astronomer-statistician-economist Simon Newcomb (1835–1909). Newcomb was a student of Peirce's at Harvard's Lawrence Scientific School, where he graduated in 1858. But that cold statement of fact fails to capture the flavor of Newcomb's youth, which was like that of a plot from a Horatio Alger novel. He was born in Nova Scotia where his father, a teacher, lived a nomadic life. At 16 Simon Newcomb began an apprenticeship to a doctor that was to last 5 years. But his autobiography (Newcomb, 1903) tells a dramatic tale of how after 2 years his prospects for a career in medicine faded as the doctor proved to be a total fraud, a quack herbalist, and he was forced to flee in the middle of the night from what had become onerous servitude, to seek his way in the world with little more than his wits to support him. At 18 he made a living as a teacher, relying on what he had taught himself in odd hours. At 21 he obtained employment he Cambridge with the Nautical Almanac office. At 23 he had a Harvard degree, and at 24 he read a paper to the American Association for the Advancement of Science on the untenability of the hypothesis that the asteroids had a common origin. He went on to become America's most honored scientist in the 19th century (Campbell, 1924).

Probability and statistical thinking played a major role in Simon Newcomb's lifework. His early work in robust estimation has received attention recently with the resurgence of interest in the subject (Stigler, 1973a), and he was the first probabilist to present the logarithmic distribution of leading digits (Newcomb, 1881). But the large volume of his published work—at least 541 notes, papers, and books (Archibald, 1924), much of this at least tangential to statistics—makes it impossible to do him justice in a short space. Rather, I will only briefly mention one minor, but interesting passage he published in some "Notes on probability" at the age of 25.

This paper appeared in the *Mathematical Monthly*, one of the better of the numerous magazines which fanned the growing interest in mathematics in America before the *Analyst* (1874–1883) approached, and the *American Journal of Mathematics* (cofounded by Newcomb in 1878) finally achieved international respectability. Among other topics, Newcomb considered the problem of estimating the number of serially numbered tickets in a bag, based on the numbers observed on s tickets drawn, with replacement. That is, based on a random sample of size s from a discrete distribution, uniform from 1 to n, estimate n. Before he went on to present a sound and well-explained Bayesian analysis of this problem, he gave the clearest statement of the idea of sufficiency I have encountered before Fisher. (See also Stigler 1973b.) Simon Newcomb wrote, "Let i be the largest number drawn in the s drawings. The number of Tickets, then, cannot be less than i. We need not know any of the drawn numbers except the largest, because after we know this, every combination of smaller numbers will be equally probable on every admissible hypothesis, and will therefore be of no assistance in judging of these hypotheses" (Newcomb, 1860–1861). Of course Newcomb did not abstract the concept, but at the least, this early statement is evidence of a clear mind capable of quickly reaching to the essentials of a problem, a signal of the brilliant career that was to come.

4. Work at Yale: E. L. De Forest. The major figures I have discussed so far—the two Peirces, Simon Newcomb, even Wigglesworth—all attended Harvard and continued their careers at Harvard, in Washington, or both. This hints at a Boston-Washington scientific axis that did indeed exist. For example, Admiral Charles Henry Davis was instrumental in obtaining government appropriations needed to start the Nautical Almanac in 1849, and he played a key role in locating its headquarters at Harvard with Benjamin Peirce in charge. Admiral Davis was also Peirce's brother-in-law. But it would be a mistake to suppose that the *only* scientific activity in America was localized in these two centers.

Another center of learning where early and important contributions to mathematical statistics were made was Yale University. It was there in the Sheffield Scientific School that the first doctorate in America was awarded for a thesis in mathematical statistics. The thesis was on the method of least squares, and it was awarded to Mansfield Merriman in 1876, the nation's centennial. Merriman went on to become the first American statistician to capture the market for elementary statistics textbooks, with his *A Textbook on the Method of Least Squares* (First Edition, 1884; Eighth Edition, 1907), and his extensive "List of writings relating to the method of least squares" remains the best bibliography of this subject (Merriman, 1877a).

From 1871 until his death in 1903, the major intellectual force in science at Yale was J. Willard Gibbs (Wheeler, 1951). Shortly after Gibbs died, Lord Kelvin visited Yale and forecast that "by the year 2000 Yale will be best known to the world for having produced J. Willard Gibbs" (Fisher, 1930), and while some modern Yale professors might question that prediction, no one of them who is familiar with Gibbs' work could take it as an insult. Gibbs himself published nothing in statistics, although he taught least squares (Wilson, 1931) and his work on statistical mechanics relied heavily on probabilistic concepts. Gibbs, through his development of vector analysis and of statistical mechanics, may indirectly have had a more profound influence on 20th century work in mathematical statistics than any other man I have mentioned, but his more obvious impact was as a teacher. Two of his students, Irving Fisher and E. B. Wilson, served as presidents of the American Statistical Association, and a third, E. L. Dodd, had a significant impact on mathematical statistics, partly as a teacher of Sam Wilks.

But the Yale man I most wish to discuss here was not a student of Gibbs, although he was Gibbs' contemporary and his work shows some Gibbsian influences. I wish to turn to the remarkable work of Erastus Lyman De Forest. De Forest's name is not widely recognized today, but his name was well known to Edgeworth and Karl Pearson, who respected and cited his work. Between 1870 and 1885 De Forest published a series of over 20 papers which cover such topics as a runs test for the residuals from a regression; a Monte Carlo determination of the variance of a statistic; the gamma distribution for one, two, and three dimensions; a measure of skewness; and an analysis of the bivariate normal distribution.

De Forest was born in 1834, the son of a Yale graduate, and he received two degrees himself at Yale, a B.A. in 1854 and Ph. B. in 1856. I suspect he was

De Forest (1834–1888)

viewed by the Yale administration as an ideal student: he did well in his studies (Gibbs is said to have called him "one of the most brilliant and promising" of Yale's students (Wolfenden, 1968)) and he was both independently wealthy and generous. Yale's Erastus De Forest Professorship of Mathematics was endowed by him in 1888 (Anderson, 1896; Wolfenden 1925, 1968).

Shortly after he graduated, De Forest surprised his family (and possibly himself) by vanishing from sight while on a visit to New York, leaving his suitcase behind and no forwarding address. His family feared the worst, and the search concentrated on New York's East River, but two years later he turned up in Australia, teaching in Melbourne. Nowadays we would probably say he had been "finding himself." The reports of his trip are contradictory, but he must have eventually decided that he preferred bulldogs to kangaroos, for he returned to New Haven and after 1865 seldom ventured further than New York.

The direction of most of De Forest's work seems to have been determined by a project he undertook in 1867–1868 for his uncle, the president of Knickerbocker

Life Insurance Company in New York. In the process of determining the company's policy liabilities, De Forest encountered the problem of smoothing mortality or life tables.

Let $u_1, u_2, \cdots, u_n$ be a sequence of numbers; the problem is to "adjust" or smooth the sequence in the hope that a better estimate of an underlying functional relation is thus obtained. We have already encountered one example in C. S. Peirce's density estimate; in the primary application that motivated most early work on the subject, the u_i's would be empirically determined estimates of the probabilities that an individual of age i in the class under study would die in the next year. A plot of u_i vs. i would be an empirically determined "mortality curve" that would give the chance of death in the next year for individuals of all ages. However, if the u_i's are simple crude death rates or relative frequencies of deaths in a sample population, as may well be the case, the plot of u_i vs. i will show marked irregularities, in contradiction to the smooth relation believed to hold.

Long before De Forest, actuaries had grappled with this problem, employing a variety of parametric models and averaging schemes. De Forest's early work centered on that species of averaging which Sheppard later named "linear compounding." That is, replace each u_i by a symmetric linear function of surrounding values, say

$$v_i = l_0 u_i + l_1(u_{i+1} + u_{i-1}) + \cdots + l_m(u_{i+m} + u_{i-m}) .$$

Of course a different rule would be needed near the extremes of the series, and an asymmetric rule could be used, too. Many schemes equivalent to ones of this type, such as the one Peirce applied, had been considered before De Forest, but they were (with one exception) ad hoc in nature. De Forest's first innovation was the introduction, in two papers in the *Smithsonian Reports* for 1871 and 1873, of optimality criteria into this problem (De Forest, 1873, 1874).

De Forest supposed that the observed u_i differed from underlying values U_i by small errors "of an accidental nature" which he supposed independent, with equal variances (we will use σ^2 for the variance; De Forest used ε for the "probable error": $\varepsilon = .6745\sigma$). He then assumed that the U_i sequence was "smooth" in the sense that any $2m + 1$ U_i's differed little from a polynomial of degree j in i; that is, given $2m + 1$ U_i's, a polynomial $g(x)$ of degree j could be found such that $U_i = g(i)$, approximately. In his 1873 paper (De Forest, 1874) he carried out much of his investigation for the case $m = 2$, $j = 3$ and so we too shall specialize to this case. Thus he supposed that any 5 consecutive U_i's could be represented "very nearly" by a cubic in i. By making his assumption of smoothness a local one and relying on local weights, a great deal of flexibility was retained over assuming U_i cubic for *all* i, or assuming a particular parametric model.

One approach De Forest considered was to determine l_0, l_1, l_2 by least squares: if a cubic function of the index or year is fit to $u_{i-2}, u_{i-1}, u_i, u_{i+1}, u_{i+2}$ by least squares, ignoring all other u_k's, the ordinate of the fitted cubic at i will be the required $v_i = l_0 u_i + l_1(u_{i-1} + u_{i+1}) + l_2(u_{i-2} + u_{i+2})$ and will give the minimum mean square estimate of U_i under the cubic assumption and the restriction to estimates linear in the local $u_{i-2}, \cdots, u_{i+2}$. An alternative (and equivalent) formulation is, since $\text{Var}(v_i) = (l_0^2 + 2(l_1^2 + l_2^2))\sigma^2$, to minimize $l_0^2 + 2(l_1^2 + l_2^2)$ subject to the condition

$$U_i = l_0 U_i + l_1(U_{i-1} + U_{i+1}) + l_2(U_{i-2} + U_{i+2})$$

for every cubic U_k in k. De Forest solved this problem and found the l's (which of course do not depend on the u's), but as he noted (De Forest, 1873, page 335) he was thus far anticipated by 1867 work of the Italian astronomer Schiaparelli, of which he had at first been unaware. But De Forest continued, and broke new ground when he noticed that the minimum mean square error criterion applied to each five u_i's separately need not produce a very smooth relation globally, notwithstanding the assumption the function was cubic locally. As an alternative to the criterion "minimize $l_0^2 + 2(l_1^2 + l_2^2)$ subject to the constraint $U_i = l_0 U_i + l_1(U_{i-1} + U_{i+1}) + l_2(U_{i-2} + U_{i+2})$ for all cubics U_i" he proposed a criterion based on smoothness: minimize the probable error of the fourth difference of the smoothed series v_i, or equivalently, minimize $E(\Delta^4 v_i)^2$, subject to the same constraint. He solved this problem for several different cases.

As a contribution to nineteenth century work on smoothing or adjustment, De Forest's introduction of this measure of smoothness as an optimality criterion was well ahead of its time, and his work was not generally appreciated until Wolfenden (1925) rediscovered it in the 1920's. By then, others had come upon his main techniques independently. Variants of De Forest's and others' criteria are currently enjoying great popularity in the related field of spline interpolation.

While De Forest's introduction of optimality criteria into interpolation and smoothing problems was a major, if unappreciated, advance at the time, modern statisticians are liable to be more interested in his evaluations of the fit of the smoothed to the observed series. De Forest was acutely aware of the contradictory combination of the desires for a close fit to the observed series and for smoothness, and he devised several goodness-of-fit tests to determine whether or not the series had been over or under-smoothed.

The first tests he discussed (De Forest, 1874, 1876, 1877) were of the nature of large sample significance tests based on the magnitude of the residuals. In the first place, if independent (possibly theoretical) estimates of the variances of the errors were available, then these could be compared with the residuals. For example, if the u_i were relative frequencies based on given numbers of trials, then a binomial model using the fitted values to estimate the probabilities would provide estimates of variances to compare with the corresponding residuals. To actually perform this test he dropped the "equal variances" assumption and took, for each year, the ratio of the squared residual over the estimate of variance for that year, and averaged these ratios over all years. He then compared the difference between this average and 1.0, with the calculated sample variance of the ratios, a sort of large sample two-tailed t-test. In suggesting that a difference of $\frac{3}{2}$ or 2 times the estimated probable error be considered large (De Forest, 1876, page 12), he seems to have had a significance level of 0.31 or 0.18 in mind. De Forest noted that this test was similar to one proposed in 1871 by Thiele, although he criticized Thiele's choice of n-m as a divisor in calculating the average ratio.

If no separate or theoretically based estimate of variance was available, De Forest suggested that the residuals be compared with the fourth differences of the original series. In a privately printed pamphlet, he proposed as a statistic, the average (over the series) of $\log (r_i/d_i)$ (our notation), where r_i and d_i are the absolute values of the ith residual $u_i - v_i$ and a constant multiple of the corresponding fourth difference of the u_i's. The multiple was chosen so that $E(d_i^2) = \sigma^2$. The basic idea was that if the U_i's were locally cubic, fourth differencing would eliminate their effect leaving only variation due to random errors; then by choosing the constant multiplier so that $E(d_i^2) = \sigma^2$, an estimate of σ^2 not based on the residuals could be obtained. Similar procedures were later rediscovered in the ballistics literature; see Von Neumann et al. (1941).

This was an interesting and novel idea, although its execution was flawed by his neglecting the autocorrelation of the d_i's, among other types of correlation. But the manner in which he sought to determine the asymptotic variance of this second statistic may be of more interest than the statistic itself. He began by making a quick determination of the standard deviation of $\log (r/d)$, using a first order differential approximation, based on the supposition that r and d are independent absolute values of normal random variables. Today we might recognize this as half the logarithm of a random variable with an F-distribution with 1 and 1 degrees of freedom, but De Forest's work shows little feeling for exact sampling distributions.

De Forest did not, however, have full confidence in his derivation. He wrote: "The demonstration [of the formula for the asymptotic standard deviation is] not a strictly rigorous one, and it has been thought desirable to test the accuracy of the formula by trials made on a sufficiently large scale, in the following manner" (De Forest, 1876, page 23). De Forest's "following manner" was a Monte Carlo study! From a table of the normal distribution, he found 100 percentiles for the absolute value of a normal deviate ranging from the 0.005th to the 0.995th, in steps of 0.01. These numbers he "inscribed upon 100 bits of card-board of equal size, which were shaken up in a box and all drawn out one by one, and entered in a column in the order in which they came" (De Forest, 1876, page 23). He then repeated this procedure to get four columns in all, and, considering them in pairs, took ratios, then logs. These he squared and averaged. He found close agreement with his formula, which he then adopted as "trustworthy."

Of course, we can now suggest more efficient methods of proceeding, and the correlation structures of both De Forest's Monte Carlo experiment and his analysis were not the same as that of his intended application. Nonetheless, his appeal to a Monte Carlo experiment for verification of his analysis was a remarkable innovation in the study of sampling distributions.

Another of De Forest's innovations was first mentioned in this same privately printed pamphlet, and more fully developed in several papers in the *Analyst* (De Forest 1876, 1877, 1878a, 1878b), a journal with international circulation and impact printed in Des Moines, Iowa. This is the idea of testing fit by analyzing the grouping of signs of the residuals. Step by step he was led to a runs test.

Unknown to De Forest, Quetelet had employed one type of runs test in 1852, with a different aim (Stigler, 1975). Quetelet had examined the distribution of lengths of runs of days of rainfall to test independence against the alternative of Markov dependence; De Forest sought to examine the signs of successive residuals to determine whether or not the small number of runs would provide evidence of too much smoothing. Actually, he began by considering the distribution of the number of runs of each of several given lengths, and comparing the numbers of runs actually observed with the numbers expected under the hypothesis that the fitted curve was the actual curve (plus or minus a probable error) (De Forest, 1876, pages 29 ff). In a later paper (De Forest, 1878b), though, he approached the modern version of the test when he proposed counting the number of "permanences" (non-changes) of signs, which equals the number of terms in the series less the number of runs.

In suggesting a test of fit based on the number of permanences of signs in the residuals, De Forest provided no exact distribution theory; he did not attempt a combinatorial theory of runs. Rather he provided approximations to the mean and probable error of his statistic, based on an unproved assumption that asymptotically the number of permanences among the residual's sign behaved as would a like statistic based on tosses of a fair coin. A little reflection shows that with the mode of fitting he employed this is *not* the case. Since the fitting is accomplished by local averaging, the signs of consecutive residuals will show a strong negative association rather than be approximately independent. A test based on the latter assumption would be severely biased, as the number of runs will tend to greatly exceed what would be expected under the null hypothesis, even with an adequate fit.

This problem did not escape De Forest's notice, and he provided an approximate rule to deal with this dependence. He reasoned that if n_1 terms were averaged in a simple arithmetic average, then one would expect the signs of the n_1 residuals, on average, to be evenly divided (De Forest, 1878a). Thus given one residual is positive, the probability the succeeding one is positive is $(\frac{1}{2}n_1 - 1)/(n_1 - 1) = (1 - 2r^2)/(2 - 2r^2) = q$, where $r = n_1^{-\frac{1}{2}}$, the ratio of the probable error of the mean to that of a single term. Then q is the chance two successive residuals form a permanence. As he did not use simple means but weighted means, he took r to be the corresponding ratio, $(l_0^2 + 2(l_1^2 + l_2^2))^{\frac{1}{2}}$ in our example, so n_1 becomes a sort of effective sample size. De Forest then, by appealing to a binomial model with probability of success q, provided an approximate means of correcting the distribution of the number of permanences for this dependence (De Forest, 1878b). He felt this correction should be adequate when $r < \frac{1}{2}$. De Forest's model for the behavior of successive signs is equivalent to a two state Markov chain with transition matrix

$$\begin{bmatrix} q & 1-q \\ 1-q & q \end{bmatrix}.$$

In a later series of papers in the *Analyst* De Forest was led by a series of steps to the consideration of some families of probability densities. He began by considering "repeated adjustments"; that is, iterated smoothing of a series by the same linear smoothing scheme (De Forest, 1878c, and following papers). An iterated linear smoothing scheme is itself a linear smoothing scheme whose coefficients are derivable from the original coefficients by convolution; Peirce's density estimate is one example. De Forest employed generating functions and differential equation approximations to difference equations to determine the character of the limiting curve of coefficients. For recent contributions to this problem see Greville (1966, 1974).

One limiting curve De Forest was led to was of course the normal (De Forest, 1879), but when he considered the limiting properties of unsymmetric adjustment schemes, he was led to something new. By considering differential equation approximation to the coefficients of binomial distribution, he derived the gamma distribution, which he called the "gamma curve" (De Forest, 1882–1883, page 140). The gamma distribution had appeared as a sampling distribution earlier than 1882 (see Lancaster, 1966), but apparently not outside of that sampling context. Pearson's and Edgeworth's work on asymmetric curves was yet to come, and the English school appears not to have noticed De Forest's work before about 1895, when Edgeworth called Pearson's attention to it. Pearson's own derivation of the gamma (or Type III) curve had then appeared (Pearson, 1895a), but he graciously acknowledged De Forest's priority as respects this type (saying De Forest's deduction had "the advantage of greater generality" and praising "the excellency of his work," Pearson, 1895b). Actually, De Forest's anticipation of Pearson went beyond the simple gamma curve. De Forest also showed how this density could be fit by the method of moments, and explored the third moment as a measure of skewness (he called it "cubic mean inequality") that was useful for distinguishing between the gamma and its limiting form, the normal. This work of De Forest has been commented on by Edgeworth (1896, 1902), Pearson (1895), Hatai (1910), McEwen (1921), Walker (1929) and Wolfenden (1925, 1942).

Another area in which De Forest worked was that of multivariate densities. Starting with the problem of smoothing two and three dimensional arrays, he derived differential equations for the limiting curve of coefficients after repeated adjustments, and was led to consider multidimensional normal (De Forest,

1881a, 1881b, 1882) and gamma distributions (De Forest, 1884). In the normal case he did not restrict attention to the independent case as he did in the gamma, although he noted that a simple rotation of axes was sufficient to reduce the general case to the independent (De Forest, 1881a). In this he was preceded by Bravais, to whom he referred (Walker, 1929, page 96). He added little new to the study of the bivariate normal, although in his final paper in 1885, after fitting a general bivariate normal distribution to target data, his thoughtful check for marginal asymmetry with respect to the transformed axes was a refreshing change from European work of that period.

In 1885 De Forest's health began to fail, and he ceased mathematical work. De Forest died in 1888; his work spanned two decades and was wholly on topics in mathematical statistics. It was widely circulated and extensively abstracted in the German *Jahrbuch über die Fortschritte der Mathematik* (Garver, 1932), but its impact was diminished by his failure to develop his methods much beyond the limited class of problems in adjustment which had suggested them in the beginning.

5. Additional work and conclusions. I have surveyed a major portion of American work in mathematical statistics before 1885, but the survey has by no means been complete. I have omitted discussion of early work on the errors-in-the-variables problem by a Monmouth, Illinois attorney (Adcock, 1877–1878) and by an assistant with the U.S. Lake Survey (Kummell, 1879), published in 1877–1879. I have skipped an early use of the range as a short-cut estimate of a standard deviation (Wright, 1882), and countless computational algorithms designed to simplify the calculation of least squares estimates, including the aptly named Doolittle method (Doolittle, 1881). I have included no discussion of work on the design of experiments, including an 1885 note which suggested that an experiment designed so that the factors would be orthogonal would "secure the maximum precision with the minimum of computation," after which the discussants allowed that they had known that all along (Woodward, 1885).

Also hidden from view are the gaffes, blunders, and absurdities that have sometimes crept into our forefathers' work. I have spared you their promiscuous use of dx and ∞ as positive real numbers, and their petty disputes over ill-posed problems in probability. But lest the picture seem totally one-sided, it may be worth noting as evidence that American understanding of concepts did have limits, that just a year after getting his degree, Mansfield Merriman, Ph. D. Yale 1876, wrote of Gauss's elegant demonstration of the "Gauss–Markov" theorem that "The proof is entirely untenable" (Merriman, 1877b). In charity to Merriman it might be added that no less a mathematician than Poincaré also misconstrued the nature of Gauss's result (Poincaré, 1912, page 188).

Despite these omissions, it should be clear that by the latter part of the last century, the United States had produced a quantity and variety of work in statistics that, while not the equal of European efforts, at least permits a respectable comparison. It had taken Americans quite a while to show an interest in mathematical statistics. I think the major reason for this was not a lack of talent, but the fact that the United States was quite late in undertaking a systematic and large scale measurement of its land, and equally late in founding observatories and beginning extensive astronomical observation. In Europe the major impetus to the development of mathematical statistics in the eighteenth and early nineteenth centuries had come from astronomy and surveying. The concepts of linear models, least squares and similar methods, had been developed between 1750 and 1820 in Europe, primarily for the reduction and analysis of astronomical observations. These techniques had then received further refinement when applied in the major geodetical surveys, for example in the survey of Britain from 1783 on.

In both spheres of activity the U.S. lagged. When President John Quincy Adams proposed a program for the construction of observatories in 1825, his phrase "lighthouses of the sky" was derisively trumpeted in the press, and funding was denied (Shepherd, 1975, page 285). The Harvard observatory was not operational before 1839, the U.S. Naval Observatory began observation in 1845. While the U.S. Coast Survey was founded in 1807 with a Swiss in charge (he was Ferdinand Hassler, Simon Newcomb's grandfather-in-law), work was begun only in 1816, and the survey did not really get on the ground on a large scale before the middle of the century, over 50 years after the British survey had reached a similar state (Cajori, 1890, pages 286 ff).

When America finally did commit itself to astronomy and land survey, it moved rapidly and energetically, and work in statistics progressed accordingly. Of the men I have discussed, only Wigglesworth and De Forest had no direct tie with astronomy or the Coast Survey. Even the European-educated Adrain's stumbling upon the normal distribution was inspired by an attempt to apply Legendre's methods for analyzing astronomical data to a problem in surveying.

De Forest's work belongs to another, separately developing tradition, that of actuarial mathematics. Here too the British led, as the major American insurance companies only reached full development in the mid-nineteenth century.

While early American work has not received much attention from historians, it did make some international impact. Peirce's outlier technique stimulated a debate which at one point involved the British Astronomer Royal. Simon Newcomb's work on robust estimation influenced the direction of Edgeworth's work. De Forest's precursor to the chi-square test and his anticipation of the gamma distribution and the method of moments may have played a role in Karl Pearson's later work on these subjects, as may also have been true of American work on the errors-in-the-variables problem, although I know of no direct evidence on this latter point. But at the least, the burst of activity in statistics after 1850, some at a remarkably high level, signaled that the talents available in America were second to none. Statistics in the early States had remained largely a dormant field; the second century would tell a different story.

Acknowledgments. I am grateful to James C. Hickman for comments and references, to T. N. E. Greville for access to Wolfenden's unpublished manuscript on De Forest, to Librarians S. Hunchar at the University of Pennsylvania, R. J. Mulligan at Rutgers, J. A. Schiff and C. M. Hanson at Yale, and C. J. Radmacher at the Warren County Library, Monmouth, Illinois; and, for comments on the first draft, to Persi Diaconis, Churchill Eisenhart, Charles C. Gillispie, Frederick Mosteller, Robin Plackett, Oscar Sheynin, George J. Stigler, John W. Tukey, and two referees.

REFERENCES

ABBE, C. (1871). A historical note on the method of least squares. *Amer. J. of Science and Arts* 101 (Third series v. 1), 411-415.

ADCOCK, R. J. (1877-1878). Note on the method of least squares. *Analyst* **4** 183-184 (1877); **5** 21-22, 53-54 (1878).

ADRAIN, R. (1809). Research concerning the probabilities of the errors which happen in making observations, etc. *The Analyst; or Mathematical Museum* **1** (No. 4), 93-109. Available on microfilm as part of the American Periodical Series.

ADRAIN, R. (1818). Investigation of the figure of the earth, and of the gravity in different latitudes. *Trans. Amer. Philos. Soc.* **1** (New Series) 119-135.

ADRAIN, R. (1818). Research concerning the mean diameter of the earth. *Trans. Amer. Philos. Soc.* **1** (New Series) 353-366.

AIRY, G. B. (1875). *On the Algebraical and Numerical Theory of Errors of Observations*, 2nd ed. Macmillan, London.

ANDERSON, JOSEPH (Ed.) (1896). E. L. De Forest. *The Town and City of Waterbury Connecticut* **3** 895-897. Price and Lee, New Haven.

ANONYMOUS (1832). Doctrine of Probabilities. *Amer. Quarterly Rev.* **11** 473-494.

ANSCOMBE, F. J. (1960). Rejection of outliers. *Technometrics* **2** 123-147.

ARCHIBALD, R. C. (1924). Simon Newcomb 1835-1909, Bibliography of his life and work. *Memoirs of the National Academy of Science* **17** 19-69.

ARCHIBALD, R. C. (1925). Benjamin Peirce: Biographical sketch and the writings of Peirce. *Amer. Math. Monthly* **32** 8-30.

BABB, M. J. (1926). Robert Adrain—Man and Mathematician. *The General Magazine and Historical Chronicle* **28** 272-284.

BOWDITCH, N. (1811). *The New American Practical Navigator*, 3rd ed. E. M. Blunt, New York.

BYERLY, W. E. (1925). Reminiscences. *Amer. Math. Monthly* **32** 5-7.

CAJORI, F. (1890). *The Teaching and History of Mathematics in the United States.* Bureau of Education, Circular of Information No. 3, 1890 (Whole Number 167). G.P.O., Washington.

CAMPBELL, W. W. (1924). Simon Newcomb. *Memoirs of the National Academy of Science* **17** 1-18.

COOLIDGE, J. L. (1926). Robert Adrain, and the beginnings of American mathematics. *Amer. Math. Monthly* **33** 61-76.

DE FOREST, E. L. (1873). On some methods of interpolation applicable to the graduation of irregular series, such as tables of mortality, etc. *Annual Report of the Board of Regents of the Smithsonian Institution for 1871*, 275-339.

DE FOREST, E. L. (1874). Additions to a memoir on methods of interpolation applicable to the graduation of irregular series. *Annual Report of the Board of Regents of the Smithsonian Institution for 1873*, 319-353.

DE FOREST, E. L. (1876). *Interpolation and Adjustment of Series.* Tuttle, Morehouse and Taylor, New Haven.

DE FOREST, E. L. (1877). On adjustment formulas. *Analyst* **4** 79-86, 107-113.

De Forest, E. L. (1878a). On the grouping of signs of residuals. *Analyst* **5** 1–6.

De Forest, E. L. (1878b). On repeated adjustments, and on signs of residuals. *Analyst* **5** 65–72.

De Forest, E. L. (1878c). On the limit of repeated adjustments. *Analyst* **5** 129–141.

De Forest, E. L. (1879). On the development of $[p + (1 - p)]^{\infty}$. *Analyst* **6** 65–73.

De Forest, E. L. (1879–1880). On unsymmetrical adjustments, and their limits. *Analyst* **6** 140–148, 161–170 (1879); **7** 1–9 (1880).

De Forest, E. L. (1880a). On some properties of polynomials. *Analyst* **7** 39–46, 73–82, 105–115.

De Forest, E. L. (1880b). On a theorem in probability. *Analyst* **7** 169–176.

De Forest, E. L. (1881a). Law of facility of errors in two dimensions. *Analyst* **8** 3–9, 41–48, 73–82.

De Forest, E. L. (1881b). On the elementary theory of errors. *Analyst* **8** 137–148.

De Forest, E. L. (1882). Law of error in the position of a point in space. *Analyst* **9** 33–40, 65–74.

De Forest, E. L. (1882–1883). On an unsymmetrical probability curve. *Analyst* **9** 135–142, 161–168 (1882); **10** 1–7, 67–74 (1883).

De Forest, E. L. (1883). A method of demonstrating certain properties of polynomials. *Analyst* **10** 97–105.

De Forest, E. L. (1884). On an unsymmetrical law of error in the position of a point in space. *Trans. Connecticut Academy of Arts and Sciences* **6** 122–138.

De Forest, E. L. (1885). On the law of error in target-shooting. *Trans. Connecticut Academy of Arts and Sciences* **7** 1–8.

Doolittle, M. H. (1881). Paper No. 3. *Report of the Superintendent of the U.S. Coast Survey for 1878.* Appendix 8, 115–120. G. P. O. Washington.

Edgeworth, F. Y. (1896). The asymmetrical probability curve. *Philosophical Magazine*, Fifth series v. **41** 90–99.

Edgeworth, F. Y. (1902). Article "Error, law of." In *Encyclopedia Britannica*, Vol. 28 of 9th and 10th Editions Combined, Vol. 4 of Supplement to 9th Edition.

Eisele, C. (1957). The Charles S. Peirce–Simon Newcomb Correspondence. *Proc. Amer. Philos. Soc.* **101** 409–433.

Eisele, C. (1974). Charles Sanders Peirce. In *Dictionary of Scientific Biography* **10** 482–488 (C. C. Gillispie, ed.). Scribner, New York.

Eliot, C. W. (1925). Reminiscences of Peirce. *Amer. Math. Monthly* **32** 1–4.

Fisher, I. (1930). The application of mathematics to the social sciences. *Bull. Amer. Math. Soc.* **36** 225–243.

Franklin, B. (1755). Observations concerning the increase of mankind, peopling of countries, etc. In *The Writings of Benjamin Franklin* **3** 63–73 (Albert Henry Smyth, ed.), 1905. Macmillan, New York.

Garver, R. (1932). The Analyst, 1874–1883. *Scripta Mathematica* **1** 247–251, 322–326.

Goodman, L. A. and Kruskal, W. H. (1959). Measures of association for cross classifications. II: Further discussion and references. *J. Amer. Statist. Assoc.* **54** 123–163.

Greville, T. N. E. (1966). On stability of linear smoothing formulas. *SIAM J. Numer. Anal.* **3** 157–170.

Greville, T. N. E. (1974). On a problem of E. L. De Forest in iterated smoothing. *SIAM J. Math. Anal.* **5** 376–398.

Gummere, J. (1817). *A Treatise on Surveying*, 2nd ed. (First Edition published in 1814.) Richardson, Philadelphia.

Hatai, S. (1910). De Forest's formula for an unsymmetrical probability curve. *Anatomical Record* **4** 281–290.

Jastrow, J. (1888). A critique of psycho-physic methods. *Amer. J. Psychology* **1** 271–309.

Jefferson, T. (1785). *Notes on the State of Virginia.* Reprinted 1955 by Univ. of North Carolina Press, Chapel Hill.

Karpinski, L. C. (1940). *Bibliography of Mathematical Works Printed in America through 1850.* Univ. of Michigan Press, Ann. Arbor. Supplements in *Scripta Mathematica* **8** 233–236; **11** 173–177; **20** 197–202.

KUMMELL, C. H. (1879). Reduction of observation equations which contain more than one observed quantity. *Analyst* **6** 97–105.

LANCASTER, H. O. (1966). Forerunners of the Pearson χ^2. *Austral. J. Statist.* **8** 117–126.

MCEWEN, G. F. (1921). Rapid methods of approximating to terms in a binomial expansion. *J. Amer. Statist. Assoc.* **17** 606–621.

MERRIMAN, M. (1877a). A list of writings relating to the method of least squares, with historical and critical notes. *Trans. Conn. Acad. of Arts and Sciences* **4** 151–232.

MERRIMAN, M. (1877b). On the history of the method of least squares. *Analyst* **4** 33–36, 140–143.

MERRIMAN, M. (1907). *A Text-Book on the Method of Least Squares*, 8th ed. Wiley, New York. First Edition 1884. Another text, *Elements of the Method of Least Squares*, published 1877.

MOSTELLER, F. and WALLACE, D. (1964). *Inference and Disputed Authorship: The Federalist.* Addison-Wesley, Reading, Mass.

NEWCOMB, S. (1860–1861). Notes on probability. *The Mathematical Monthly* (ed. by J. D. Runkle) **3** 119–125.

NEWCOMB, S. (1881). Note on the frequency of use of the different digits in natural numbers. *Amer. J. Math.* **4** 39–40.

NEWCOMB, S. (1903). *The Reminiscences of an Astronomer.* Harper, New York.

O'DONNELL, T. (1936). *History of Life Insurance.* American Conservation Co., Chicago.

PEARSON, K. (1895a). Contributions to the mathematical theory of evolution. II. Skew variation in homogeneous material. *Phil. Trans. Roy. Soc. Lond. A* **186** 343–414. Reprinted in *Karl Pearson's Early Statistical Papers* (pages 41–112), Cambridge Univ. Press (1956).

PEARSON, K. (1895b). On skew probability curves. *Nature* **52** 317 (Aug. 1, 1895).

PEARSON, K. (1900). On the criterion that a given system of deviations from the probable in the case of a correlated system of variables is such that it can be reasonably supposed to have arisen from random sampling. *Philosophical Magazine*, fifth series, **50** 157–175. Reprinted in *Karl Pearson's Early Statistical Papers* (pages 339–357). Cambridge Univ. Press (1956).

PEIRCE, B. (1852). Criterion for the rejection of doubtful observations. *Astronomical Journal* **2** No. 21, 161–163.

PEIRCE, B. (1878). On Peirce's criterion. *Proc. Amer. Acad. Arts and Sciences*, **13** (v. 5 N.S.), May 1877—May 1878, 348–349.

PEIRCE, C. S. (1872). On the coincidence of the geographical distribution of rainfall and of illiteracy, as shown by the statistical maps of the ninth census reports (Abstract). *Bull. Philos. Soc. of Washington* **1** 68.

PEIRCE, C. S. (1873). On the theory of errors of observations. Appendix No. 21 (pages 200–224 and plate 27) of *Report of the Superintendent of the U.S. Coast Survey for the year ending June 1870.* G.P.O., Washington. Reprinted in *The New Elements of Mathematics* by C. S. Peirce, ed. by C. Eisele, Humanities Press, Atlantic Highlands, N.J., 1976, Vol. 3, pt. 1, pp. 639–676.

PEIRCE, C. S. (1879). Note on the theory of the economy of research. Appendix No. 14 (pages 197–201) of *Report of the Superintendent of the U.S. Coast Survey for the year ending June 1876.* G.P.O., Washington. (Also in *Collected Papers of Charles Sanders Peirce*, vol. 7. See pages 84–88 for some "Later reflections" on this subject.) Reprinted in *Operations Res.* **15** 643–648 (1967).

PEIRCE, C. S. (1884). The numerical measure of the success of predictions. *Science* **4** 453–454. Reprinted in *The New Elements of Mathematics* by C. S. Peirce, ed. by C. Eisele, Humanities Press, Atlantic Highlands, N.J., 1976, vol. 3, pt. 1, pp. 682–683.

PEIRCE, C. S. and JASTROW, J. (1885). On small differences of sensation. *Memoirs of the National Academy of Sciences for 1884* **3** 75–83.

PEIRCE, C. S. (1957). *Essays in the Philosophy of Science.* (V. Tomas, ed.) Bobbs-Merrill, Indianapolis. The material on pages 195–234, "Lessons from the History of Science," was written c. 1896, originally published in *Collected Papers of Charles Sanders Peirce* (C. Hartshorne and P. Weiss, eds.). **1** 19–49. Harvard Univ. Press (1931).

POINCARÉ, H. (1912). *Calcul des probabilités*, 2nd ed. Gauthier-Villars, Paris.

REINGOLD, N. (1972). American Indifference to Basic Research: A Reappraisal. *Nineteenth-Century American Science, A Reappraisal*, 38-62 (George H. Daniels, ed.). Northwestern Univ. Press.

RIDER, P. R. (1933). Criteria for rejection of observations. *Washington University Studies-New Series, Science and Technology*, No. 8.

SHEPHERD, J. (1975). *The Adams Chronicles*. Little, Brown, Boston.

SMITH, D. E. and GINSBURG, J. (1934). *A History of Mathematics in America Before 1900*. Carus Mathematical Monograph No. 5.

SMITH, SYDNEY (1818). Travellers in America. *Edinburgh Review* **31** 132-150. Reprinted in *The Works of the Rev. Sydney Smith*, D. Appleton, New York (1860, pages 107-115).

STIGLER, S. M. (1973a). Simon Newcomb, Percy Daniell, and the history of robust estimation 1885-1920. *J. Amer. Statist. Assoc.* **68** 872-879.

STIGLER, S. M. (1973b). Laplace, Fisher, and the discovery of the concept of sufficiency. *Biometrika* **60** 439-445.

STIGLER, S. M. (1975). The transition from point to distribution estimation. 40th session of the I.S.I., Warsaw, Poland.

TOCQUEVILLE, A. DE (1840). The example of the Americans does not prove that a democratic people can have no aptitude and no taste for science, literature, or art. *Democracy in America*, Vol. 2, Book 1, Chapter 9; pages 35-40 of the 1945 edition. Knopf, New York.

VON NEUMANN, J., KENT, R. H., BELLINSON, H. R. and HART, B. I. (1941). The mean square successive difference. *Ann. Math. Statist.* **12** 153-162.

WALKER, HELEN M. (1929). *Studies in the History of Statistical Method*. William and Wilkins, Baltimore. Reprinted Arno Press (1975), New York.

WEISS, PAUL (1934). Charles Sanders Peirce. In *Dictionary of American Biography* **14** 398-403 (D. Malone, ed.). Scribner, New York.

WHEELER, L. P. (1951). *Josiah Willard Gibbs*. Yale Univ. Press.

WILSON, E. B. and HILFERTY, M. M. (1929). Note on C. S. Peirce's experimental discussion of the law of errors. *Proc. National Acad. Sci.* **15** 120-125.

WILSON, E. B. (1931). Reminiscences of Gibbs by a student and colleague. *Scientific Monthly* 210-227.

WOLFENDEN, H. H. (1925). On the development of formulae for graduation by linear compounding, with special reference to the work of Erastus L. De Forest. *Trans. Actuarial Soc. of America* **26** 81-121.

WOLFENDEN, H. H. (1942). *The Fundamental Principles of Mathematical Statistics*. Macmillan, Toronto.

WOLFENDEN, H. H. (1968). Biography of Erastus L. De Forest. Unpublished, c. 1968.

WOODWARD, R. S. (1885). Some practical features of a field time determination with a meridian transit (Abstract). *Bull. Philos. Soc. of Washington* **8** 55-58.

WRIGHT, T. W. (1882). On the computation of probable error. *Analyst* **9** 74-78.

SUPPLEMENTARY REFERENCES[2]

AIRY, G. B. (1856). Letter from Professor Airy, Astronomer Royal, to the editor. *Astronomical Journal* **4** 137–138.

NEWCOMB, S. (1859–1861). Notes on the Theory of Probabilities, *Mathematical Monthly* (Runkle, ed.), Vol. 1, pp. 136–139, 233–235, 331–335; Vol. 2, pp. 134–140, 272–275; Vol. 3, pp. 119–125, 343–349.

[2]Added to the original article by the author for this volume.

NEWCOMB, S. (1886). A generalized theory of the combination of observations so as to obtain the best result. *American Journal of Mathematics*, Vol. 8, pp. 343–366.

PLACKETT, R. L. (1972). The discovery of the method of least squares. *Biometrika* **59** 239–251. Reprinted 1977 in *Studies in the History of Statistics and Probability*, Vol. 2 (M. G. Kendall and R. L. Packett, eds.) (London: Griffen).

WIGGLESWORTH, E. (1793). A table showing the probability of the duration, the decrement and the expectation of life, in the states of Massachusetts and New Hampshire formed from sixty-two bills of mortality... in the year 1789. *Memoirs of the American Academy*, Vol. 2, part 1, Boston, 1793.

WINLOCK, J. (1856). On Professor Airy's Objections to Peirce's Criterion. *Astronomical Journal* **4** 145–47.

DEPARTMENT OF STATISTICS
UNIVERSITY OF WISCONSIN
MADISON, WISCONSIN 53706

WILLIAM FELLER AND TWENTIETH CENTURY PROBABILITY

1. Twentieth Century Probability

When William Feller was born in 1906, Lebesgue measure had just been invented, and Fréchet was to introduce measure on an abstract space about ten years later. Thus, the technical basis of modern mathematical probability was developed about the time of Feller's early childhood. Since that time the subject has been transformed, by no one more than by Feller himself, into an essential part of mathematics, contributing to other parts as well as drawing from them.

In the first part of our century, few probabilists felt comfortable about the basis of their subject, either as an applied or as a purely mathematical subject. In fact, it was commonly judged that there was no specific mathematical subject "probability," but only a physical phenomenon and a collection of mathematical problems suggested by this phenomenon. A probabilist joked that probability was "a number between 0 and 1 about which nothing else is known." In the discussions of the foundations of probability, there was no clear distinction made between the mathematical and the real. For example, one influential theory was that of von Mises, based on the concept of a "collective," which was defined as a sequence of observations with certain properties. Since "observation" is not a mathematical concept and since the properties were properties which no mathematical sequence could have, the theory could survive in its original form only by an affirmation that it was not a formal mathematical theory but an attempt at a direct description of reality. Instead, the theory was restricted to remove the mathematical objection, unfortunately losing in intuitive content what it gained in mathematical significance. The fate of the theory was an inevitable result of the increasing demand of mathematicians for exact definitions and formal rigor. The present formal correctness of mathematical probability only helps indirectly in analyzing real probabilistic phenomena. It is unnecessary to stress to statisticians that the relation between mathematics and these phenomena is still obscure. Or if not obscure it is clear to many but in mutually contradictory ways.

Formalizations of mathematical probability by Steinhaus in 1923 and Fréchet in 1930 were too incomplete to have much influence. The first acceptable formalization was by Kolmogorov in his 1933 monograph. Of course before that, and in fact for at least two centuries before that, there had been mathematicians who made correct and valuable contributions to mathematical probability. Mathematicians could manipulate equations inspired by events and expectations before these concepts were formalized mathematically as measurable sets and integrals. But deeper and subtler investigations had to wait until the blessing and curse of direct physical significance had been replaced by the bleak reliability of abstract mathematics.

Reprinted from *Proceedings of the Sixth Berkeley Symposium on Mathematical Statistics and Probability*, Vol. 2, pp. xv–xx, with permission of the University of California Press.

William F. Feller

Some probabilists have scorned the measure theory, functional analysis invasion of their subject, thinking it could do no more good than the discovery by Molière's character that he had been talking prose all his life. But in fact this invasion, to which Feller contributed so much, enriched the subject enormously in bringing it into the framework of modern mathematics, providing it with the possibility of undreamed of contacts with seemingly quite different mathematical fields. The definitive acceptance of mathematical probability as mathematics was, however, quite unnecessary for a large part of probabilistic research. For example, much of the distribution theory of sums of independent random variables can be considered an analysis of the convolutions of distribution functions. Random variables need never be mentioned. But even such researches are interesting largely because of their probabilistic significance. Many would not have been thought of and many more would have not been carried out, even if thought of, without this significance. Thus, the acceptance of probability as mathematics influenced research that could have been written without the probabilistic context. Even now this acceptance is not complete. In fact, many mathematics students are unaware of the place of probability in their subject. This situation is preserved by the special flavor given to probability by its linguistic heritage. Terms like "random variable" are here to stay and to continue misleading students on the state of probability theory, although "random variable" has a purely mathematical meaning whereas other familiar terms like "inclined plane" do not.

But the situation was even more confused forty years ago. A student can hardly visualize the difficulty of working in a field without a formal basis, without any sophisticated textbooks, in which it was respectable to have a serious discussion on what "really happens" when one tosses a coin infinitely often. The first sophisticated book was Lévy's remarkable 1937 book which was not written as a textbook and which yielded its treasures only to readers willing to make extreme efforts. A distinguished statistician in the early 1930's when asked how probability was taught at his university expressed surprise at the idea of teaching probability as a separate subject—it would be a "pointless *tour de force*." At that time random variables still were so mysterious that another distinguished mathematical statistician stated in a lecture that it was not known whether two random variables which were uncorrelated had to be independent. At that time the idea that a random variable was just (mathematically) a function was still so unfamiliar that it did not occur to the speaker to consider the sine and cosine functions on $(0, 2\pi)$, with the uniform distribution on that interval, as random variables, trivially uncorrelated and not independent.

But it is true that probability has lost some of its glamour along with its mystery. Luckily, the subject still has its basic physical background to draw on, still a source of ideas and problems. A further present feature is the interplay between mathematical probability and other parts of mathematics, for example, partial differential equations and potential theory.

It was a wonderful thing to be entering the field of probability when Feller did, in the early thirties. To one with his classical background, the field was obviously full of unsolved problems. Of course, it was not obvious at the time that the field was as rich as it has turned out to be, but it was clear that the subject was new in the sense that it had been barely touched by modern techniques. For example, discrete and continuous parameter Markov processes were just beginning to be studied in a nontrivial way. The multiplicity of classical type problems suggested by probability was such that it was not surprising that many probability papers indulged in probabilistic slang only long enough to reach the safe territory of integral equations or some other respectable established topic. Thereafter both writer and reader could relax, knowing that the introductory slang was as unessential as it was unexplained.

Wiener's work on Brownian motion (1924) was an exception. Although Wiener even later never used or knew the slang or even many of the elementary results of probability theory, his Brownian motion analysis was quite rigorous. In fact, an early problem in stochastic processes was to create a general theory which would include his approach to Brownian motion!

Since the thirties, mathematical probability has exploded. Several journals are devoted to it in its pure form and some other journals, for example the *Annals of Mathematical Statistics*, are barely distinguishable from probability journals. Sophisticated text books and specialized books are appearing all the time and there is even talk that the subject has reached or passed its peak. Feller, who was one of the researchers who brought the field to its present state, liked to relax in his advanced research by playing with elementary problems, polishing their known solutions. Let us hope that the new crop of researchers will be able to continue both his research and his purifying of old results.

2. William Feller

Feller was born in Zagreb, Yugoslavia on July 6, 1906, the ninth of twelve children of the well to do owner of a chemical factory. He attended the University of Zagreb (1923–1925), where he received the equivalent of an M.S. degree, and the University of Göttingen, where he received his Ph.D. in 1926 and remained until 1928. In 1928, he left Göttingen for the University of Kiel, where he worked as Privatdozent until in 1933 he refused to sign a Nazi oath and was forced to leave. It was at Kiel that he did his first work in probability. After a year (1933–1934) in Copenhagen, he went to Stockholm where he spent the next five years at the University and (July 27, 1938) married Clara Nielsen who had been his student in Kiel.

In 1939 the Fellers emigrated to Providence, where he became associate professor at Brown University and the first executive editor of *Mathematical Reviews*, founded that year. The only current mathematics review journal was then becoming corrupted by Nazi ideas. Much of the success of *Mathematical*

Reviews has been due to the policies initiated by Feller. *Mathematical Reviews* was founded in a less frantic scientific age when it was reasonable to have critical reviews, before the age of speed and preprints. There is now some opinion that traditional reviewing is an obsolete luxury, like peaceful universities. But even if this is true, the very speed of mathematical development that has made it true is in part a tribute to the success of *Mathematical Reviews* in furthering research.

In 1945, Feller accepted a professorship at Cornell University and remained there until 1950 when he moved to Princeton University as Eugene Higgins Professor of Mathematics. He held this position until his death (January 14, 1970), but in addition was a Permanent Visiting Professor at The Rockefeller University where he spent the academic years 1965–1966 and 1967–1968. A great attraction at The Rockefeller University was the opportunity to talk to geneticists.

Feller's first probability paper (1935) was on the central limit theorem, and in fact the properties of normalized sums of independent random variables were the subject of much of his later research, both from the point of view of distribution theory and from that of asymptotic bounds of the sums. Some of his deepest analytical work was in connection with the latter, work related to the general forms of the iterated logarithm law. It was in the context of distribution functions and their convolutions, not of random variables, and thus did not need the mathematical formalization of probability provided by Kolmogorov only a few years before. The central limit theorem paper gave necessary and sufficient conditions for convergence to a Gaussian limit.

In 1906, Markov did the first work on the sequences of random variables with the property that now bears his name. Progress was slow at first and some of Markov's work was repeatedly rediscovered. Kolmogorov's 1931 paper on continuous parameter Markov processes was a turning point, the first systematic investigation of these processes including the processes of diffusion. Feller wrote his first paper on these processes in 1936, going considerably beyond Kolmogorov and proving the appropriate existence and uniqueness theorems for the integrodifferential equations governing the transition probabilities. The main interest of both authors was in these equations. The stochastic processes themselves were secondary, although they inspired the analysis, and it is not surprising in view of the general historical remarks made above in Section 1 and the state of the subject at the time that Kolmogorov defined Markov processes incorrectly and Feller added an incorrect characterization (independent increments) to Kolmogorov's definition. All they needed was the Chapman-Kolmogorov equations, and the process giving rise to them was almost irrelevant. For Feller, as distinguished, say, from Lévy, it was usually the differential or integral equations or the semigroups arising in a probability context that interested him, rather than sample properties. On the other hand he kept these properties in mind, and although he usually did not treat them specifically he had a sure feeling for them and they inspired much of his analysis.

Feller completely transformed the subject of Markov processes. Going beyond his 1935 paper, he put the analysis into a modern framework, applying semigroup theory to the semigroups generated by these processes. He observed that the appropriate boundary conditions for the parabolic differential equations governing the transition probabilities correspond on the one hand to the specification of the domains of the infinitesimal generators of the semigroups and on the other hand to the conduct of the process trajectories at the boundaries of the process state spaces. In particular, he found a beautiful perspicuous canonical form for the infinitesimal generator of a one dimensional diffusion. In this work, he was a pioneer yet frequently obtained definitive results.

Feller is best known outside the specialists in his field for his two volume work *An Introduction to Probability Theory and Its Applications*. He never tired of revising this book and took particular pleasure in finding new approaches, new applications, new examples, to improve it. The book is extraordinary for the almost bewildering multiplicity of its points of view and applications inside and outside pure mathematics. No other book even remotely resembles it in its combination of the purest mathematics together with a dazzling virtuosity of techniques and applications, all written in a style which displays the enthusiasm of the author. This style has made the book unexpectedly popular with non-specialists, just as its elegance and breadth, not to mention its originality, has made it an inspiration for specialists. Feller had planned two more volumes, and it would have been fascinating to see if his excitement in his subject could have brightened the usual dull measure theoretic details which would inevitably have had to appear in later volumes. Perhaps his unequalled classical background could have diluted, and made more palatable with applications and examples the concentrated dosage of preliminaries other mathematicians find necessary before studying Markov processes.

Feller was never heavily involved in statistics, although he was interested in it. He was not afraid of dirtying his fingers with numbers and in fact at one time he liked to work out least squares problems on hand computers as relaxation! He was president of the Institute of Mathematical Statistics in 1946. His attitude towards applications was unusual. On the one hand, his research was almost entirely in pure mathematics. On the other hand, he had far more than an amateur's interest in and knowledge of several applied fields, including statistics and genetics. He wrote a paper on extra sensory perception, and he wrote several papers applying the sophisticated ideas of a modern probabilist to genetics. He took an excited delight in applications of pure theory and nothing pleased him more than finding new ones. On the other hand he had a low boiling point for poor thinking, and nothing made him more excited than what he considered improper scientific thinking whether he favored or opposed the conclusion. Thus, he had great contempt for those who buttressed insufficient statistics on lung cancer and cigarettes with emotionalism or those who adduced uninformed arguments against Velikovsky's theories.

Feller was a member of the U.S. National Academy of Sciences, the Royal Danish Academy of Sciences, and the Yugoslav Academy of Sciences, as well as a member of the American Academy of Arts and Sciences and the American Philosophical Society. His wife accepted the National Medal of Science for him shortly after his death. But apart from his mathematics those who knew him personally will remember Feller most for his gusto, the pleasure with which he met life, the excitement with which he drew on his endless fund of anecdotes about life and its absurdities, particularly the absurdities involving mathematics and mathematicians. To listen to him deliver a mathematics lecture was a unique experience. No one else could generate in himself as well as in his auditors so much intense excitement. In losing him, the world of mathematics has lost one of its strongest personalities as well as one of its strongest researchers.

J. L. DOOB

Early Days in Statistics at Michigan

CECIL C. CRAIG

For me this period began in 1922 when I arrived in Ann Arbor with an M.S. degree intending to take courses in Actuarial Science. Professor J. W. Glover, who set up the actuarial program in Michigan, which still flourishes, conceived the idea in about 1910, that such a curriculum should include courses in mathematical statistics. In 1916 he brought back to Michigan a recent graduate, Harry C. Carver, to develop courses in that subject. In 1922 there were only two schools in the country, the State University of Iowa and the University of Michigan, where courses in mathematical statistics were offered. Carver's first course, Mathematics 49 and 50, each for 2 hours credit, ran throughout the year at a precalculus level. A second more mathematical course was given by Professor R. W. Barnard, who later taught pure mathematics at the University of Chicago. I took this course and learned some mathematics but not much statistics. I began teaching an advanced course after I got my doctor's degree which was a result of a year in Lund, Sweden, working under Professor S. D. Wicksell.

In those days the *Journal of the American Statistical Association* was well established, but manuscripts with any mathematical content had little chance of being published by the *Journal*. I heard Professor Carver say on more than one occasion that there ought to be a place in this country where a paper in mathematical statistics could appear. I have always thought that the trigger for the founding of the *Annals of Mathematical Statistics* was a paper of mine that was rejected by the *Journal* because it was too mathematical. Carver reacted strongly to this and shortly afterward he joined with a friend, J. W. Edwards, who was trying out a new lithoprinting process, in putting out the first issue of the *Annals of Mathematical Statistics* in 1930. Carver

Cecil C. Craig (April 14, 1898–June 16, 1985) was Emeritus Professor of Mathematics and Director of the Statistical Research Laboratory, University of Michigan. This memoir was written shortly before his death on the occasion of the 50th anniversary of the establishment of the Institute of Mathematical Statistics.

Reprinted from *Statistical Science*, Vol. 1, No. 2, pp. 292–293, with permission of the Institute of Mathematical Statistics.

assumed the financial responsibility for the new journal and with the aid of two assistants and his friend's support he served as its editor until 1935 when he turned the *Annals* over to the newly formed Institute of Mathematical Statistics. There was a sufficient supply of scholarly papers offered for publication but the supply of funds to meet the bills was not enough to avoid severe strains. At times toward the end of World War II the *Annals* came close to going broke. I don't know if Carver ever told anybody the cost in dollars of his devotion to statistics but I doubt if he knew closely. Fortunately, the publishers of the *Annals* and the officers of the Institute allowed a really large inventory of back numbers to accumulate during the second World War. Once the war was over, it turned out that there was a healthy market for those back numbers. The faithful industry of Paul Dwyer and Carl Fischer handled the sale of this merchandise. Only their friends knew how hard they worked, but enough money came in to put the *Annals* on a sound financial footing.

The remainder of the 1920s and the first of the 1930s were marked by a steady growth in this country in the number of people whose principal interest lay in mathematical statistics. By living and working in the city where the new *Annals* were edited and by regular attendance at the national meetings, it was easy for me to become widely acquainted with the members of the new group. I spent the year 1930–31 in Stanford University where Harold Hotelling was beginning a career in statistics. When I left Stanford to return to Ann Arbor, Hotelling also left to accept an appointment at Columbia University. On my way back across the country I stopped for a few days in Iowa City where Egon Pearson was lecturing. There a rather remarkable group of students was working with H. L. Rietz, who deserved to be known as the dean of American mathematical statisticians. These students were S. S. Wilks, A. T. Craig, Selby Robinson, and Carl Fischer. They all earned doctorates under Rietz, and I made friends with all of them. Only Fischer, who recently retired from Michigan, is still alive. When I left Iowa City, I went to Minneapolis where I spent five weeks listening to my first series of lectures by R. A. Fisher. Some time in the next few years I became well acquainted with B. H. Camp of Wesleyan.

In 1935, the summer meetings of the mathematics societies were held in Ann Arbor. The attendees included enough people interested in mathematical statistics to fill the reception room in the Betsy Barbour dormitory on this campus. They were convened to discuss a proposed organization of a new society devoted to mathematical statistics. I do not recall all of the thirty to forty people who were present, but I do remember Rietz, Wilks, A. T. Craig, Carl Fischer, Selby Robinson, and Paul Rider from Iowa and H. C. Carver, C. C. Craig, T. E. Raiford, and A. L. O'Toole from Michigan. Others whom I do not recall from there were Hotelling, Camp, Gertrude Cox, and W. A. Shewhart.

I know that previously Carver's idea of the proper form to be assumed by an organization of mathematical statisticians was that of the actuaries, with qualifying examinations for different grades of membership. But at the actual organization meeting this form of a society was not seriously proposed and a form very close to what we have today was adopted with only brief discussion. The Institute of Mathematical Statistics was created on September 12, 1935 in Ann Arbor, Michigan, with the following elected officers: President, H. L. Rietz; Vice President, W. A. Shewhart; Secretary-Treasurer, A. T. Craig. The five-year-old *Annals of Mathematical Statistics* was adopted as the official journal of the new society.

From his joining of the faculty of the University of Michigan until his retirement in 1960 the dominant figure in statistics in Michigan was Harry Carver. He was a native of Waterbury, Connecticut, and he took a B.S. degree from Michigan in 1915. He had a spare well-muscled figure more than 6 feet tall, a sandy complexion, and the coordination of a natural athlete. He won an "M" in track and worked out for years with the cross country team. He was good enough at pocket billiards to have made his living at that game. He could beat ordinary golfers using only a five iron. He greatly enjoyed bridge and belonged to a group which regularly met for poker. I do not know that he was exceptionally good at card games.

As a high school student, he was known for repairing and riding motorcycles. Later, as a student, he enjoyed rebuilding second-hand automobiles, making them better than new. He became known for his fast driving. He made a practice of leaving Ann Arbor at the same time as the train carrying the track team and arriving first into Chicago. Sometime later he discovered California and with his second wife he more than once drove there nonstop; one driving while the other slept in the back seat.

But soon he became interested in airplanes and became a qualified pilot. He and a friend acquired a small plane and he enjoyed taking acquaintances for rides. He quickly became aware of the problems in navigation encountered by the pilots of the long-range, high-speed planes being supplied to the Air Force. He cultivated friends among officers in the Air Force. He enrolled in and completed the training course being given to United States Air Force cadets. He applied his quick mind and mathematical ability to improving navigational methods then in use. He showed how to use small calculating machines to get numerical results in navigation problems more quickly and accurately. As he neared retirement age he spent much time in Air Force bases in Texas and California.

After retirement he made a study of climatic data for the United States and selected Santa Barbara, California, as the best place to live. He rented an apartment and moved there for several years. At age 80 he quit driving an automobile "while he was ahead" as he put it. His health deteriorated and he moved back to Ann Arbor and ended his days at age 87 in the Michigan Union.

Carver had a very quick mind and he had a warm and sympathetic manner. Taking a course with him was an experience his students did not forget. He directed the work of ten doctoral students. He bordered on the eccentric; his diet seemed to consist largely of crackers and milk. He made a practice of offering to buy a class a dinner if it could beat him at one of five indoor sports—card games or billiards or pool—or at one of five outdoor sports such as running or putting the shot. He never lost.

Churchill Eisenhart majored in mathematical physics at Princeton University, receiving an A.B. degree in 1934. Continuing at Princeton the following year, he studied under S. S. Wilks and received an A. M. in mathematics in 1935. He then moved to University College London, where he received a Ph.D. in 1937 under the supervision of Jerzy Neyman. Later he taught at the University of Wisconsin, served as an applied mathematician during World War II, then in 1946 joined the staff of the National Bureau of Standards, where he established and headed (1947–1963) the Statistical Engineering Laboratory. He was then an NBS Senior Research Fellow until his retirement in 1983. The recipient of many honors for his work in mathematical statistics and its applications, Dr. Eisenhart is currently a Guest Scientist in the Center for Computing and Applied Mathematics of the National Institute of Standards and Technology (successor to the NBS).

S. S. Wilks' Princeton Appointment, And Statistics At Princeton Before Wilks*

CHURCHILL EISENHART

The following paragraphs provide a detailed account of the circumstances of the appointment of Samuel Stanley Wilks (1906–1964) to a position in the Department of Mathematics, Princeton University, in 1933; the state of statistics at Princeton then and in prior years; and an explanation of the three-year delay before Wilks taught his first course in statistics at Princeton.

The key figure in Wilks' appointment was my father, Luther Pfahler Eisenhart (1876–1965), who, in the spring of 1933, was not only willing, but, as Chairman of the Department of Mathematics (1928–1945), Dean of the Faculty (1925–1933), and Chairman of the University Committee on Scientific Research (1930–1945), was also able to effect Wilks' appointment to an

*Adapted from "Samuel S. Wilks and the Army Experiment Design Conference Series," in *Proceedings of the Twentieth Conference on the Design of Experiments in Army Research Development and Testing held at the U.S. Army Engineer Center, Fort Belvoir, Virginia, 23–25 October 1974* (ARO Report 75-2), Research Triangle Park, North Carolina: U.S. Army Research Office, June 1975.

instructorship in Mathematics on a more or less emergency basis over the opposition of almost every member of his department.

A few words are in order on how my father became interested in, and partial to statistics: My father's primary mathematical interest was differential geometry, and his research was exclusively in that area. Exactly when he began to take an "outside" interest in mathematical statistics I do not know. It may have been as early as 1913, when he corresponded with Edward L. Dodd on various aspects of the latter's paper entitled "The probability of the arithmetic mean compared with that of certain other functions of measurements", which was published in the *Annals of Mathematics* (Vol. **14**, pp. 186–198, June 1913), of which my father was then an editor. At any rate, thereafter Dodd sent my father reprints of many of his subsequent papers on functional and statistical properties of various types of "means", which my father kept and ultimately turned over to me when I became interested in such matters in the early 1930s.

Early in 1924, "at the request of the Commission on New Types of Examination of the College Entrance Examination Board", my father "formed a committee of mathematicians to examine critically certain statistical methods used in the investigations of the Commission" (*American Mathematical Monthly*, Vol. **31**, No. 4 (April 1924), p. 209). The "mathematicians" of the Committee included the economic statisticians W. Randolph Burgess and W. L. Crum (1894–1967) of the Federal Reserve System and of the Economics Department, Harvard, respectively; the mathematicians E. V. Huntington (1874–1952) and J. H. M. Wedderburn (1882–1948), of Harvard and Princeton, respectively; and the mathematical statistician, H. L. Rietz, of the University of Iowa.

The findings of this Committee, my father's continued advisory relations with the higher-ups of the College Entrance Examination Board (CEEB), and Wilks' contributions at Iowa (and under Hotelling at Columbia) to the solution of statistical problems arising in educational testing, made it possible for my father to arrange a part-time appointment with the CEEB concurrent with his initial University appointment —a relationship with the Board, and its successor, the Educational Testing Service, that continued until Wilks' death.

Harold Hotelling (1895–1973) took steps to assure that my father was kept informed of the Student-Fisher revolution in statistical theory and practice. He had gone to Princeton University as a J. S. K. Fellow in mathematics, 1921–1922, after receiving his A. B. (1919) and an M.S. (1921) from the University of Washington, in Seattle. His interest in statistics predated his going to Princeton in the fall of 1921. He had hoped to find some work in probability theory and the mathematics of statistics going on there in the Mathematics Department. Finding none, he undertook instead a program of study and research in topology (then called "analysis situs") and differential geometry, under the direction of Professor Oswald Veblen (1880–1960) and my father, Luther Pfahler Eisenhart. He stayed on at Princeton, 1922–1924, as an Instructor in Mathematics and received his Ph.D. from Princeton University in June 1924, his doctoral dissertation being on "Three-dimensional manifolds of states in motion." His paper "An application of analysis situs to statistics" (*Bulletin of the American Mathematical Society*, Vol. **33**, (1927), pp. 467–476), had to do with topological aspects of serial and multiple correlations.

Following receipt of his Ph.D., Hotelling returned to the West Coast, to Stanford University, where he was a Junior Research Associate and then Research Associate (1925–1927), in the Food Research Institute; and finally, an Associate Professor of Mathematics (1927–1931), in the Department of Mathematics. In 1931, he was called to Columbia University, in New York City, as Professor of Economics to develop further the existing work there in mathematical economics, and to initiate a program in mathematical statistics.

Hotelling's paper on "The distribution of correlation ratios calculated from random data", in *Proceedings of the National Academy of Sciences*, **11**, no. 10 (October 1925), pp. 657–662, made him the first person in the United States to respond *in kind* to R. A. Fisher's signal contributions to the theory of small samples—his derivation employed the same kind of geometrical reasoning in terms of Euclidean N-dimensional space that Fisher had used so effectively. This paper carries a footnote that I've always considered to be very significant. I believe it affords an explanation of why so many American mathematicians had difficulty following Fisher's geometrical proofs. Anyone who attempts to duplicate Fisher's geometrical reasoning soon discovers that a crucial step is the correct evaluation of the relevant *element of volume*. Hotelling, at this juncture in his paper, gives a general expression for the relevant element of volume, which he numbers "(17)", and then remarks in a footnote:

"This important expression for the volume element has been used in lectures [at Princeton University] by Professors O. Veblen and L. P. Eisenhart. I do not find it in any of the treatises on calculus, analysis or differential geometry, save for the special case in which the manifold of integration is a surface. It may readily be proved by showing first that (17) is a relative invariant under arbitrary transformations of the parameters; and second, that if the parameters of the hypersurface are orthogonal at a point, (17) becomes at this point the simple expression for the volume element in cartesian coordinates."

During his years at Stanford, Hotelling wrote and published a stream of important original contributions to statistical theory and mathematical economics; reviews of American and English books on statistical methods, (e.g., of *Statistical Analysis* by Edmund E. Day (New York: The Macmillan Company, 1925), in *Journal of the American Statistical Association*, Vol. **21**, No. 155 (Sept. 1926), pp. 360–363), in which he deplored the obsoleteness of teaching and research in statistics in the United States and placed the blame squarely on the doorsteps of departments of mathematics; and expository articles on "British statistics and statisticians today" (*Journal of the American Statistical Association*, Vol. **25**, No.170 (June 1930) pp. 186–190), "Recent improvements in statistical inference" (*Journal of the American Statistical Associates*, Vol. 26, March 1931 *Supplement*, pp. 79–87; discussion, pp. 87–89) in which he did his very best to acquaint American readers with the "new look" in statistics. He regularly sent reprints of all of these to my father. When my father gave them to me in the fall of 1932, as I was reading up on "Student-Fisher statistics", it was quite clear that my father had more than a superficial knowledge of the papers on statistical theory, and had "got the message" of Hotelling's book reviews and expository articles.

An event that was to be instrumental in bringing both mathematical economics and modern statistical theory and methodology to the Princeton campus was the arrival of Charles F. Roos (1901–1958) as a National Research Fellow in Mathematics for the academic year 1927–1928. Roos had received his Ph.D. in theoretical economics at the Rice Institute in 1926 under Professor G. C. Evans (1887–1973), who at that time was developing a new mathematical theory of economic phenomena termed "economic dynamics", and had spent 1926–1927 at the University of Chicago working with Professor Henry Schultz (1893–1938) who at that time was deeply engaged in his epochal research on statistical laws of demand and supply as one facet of his life's work on the theory and measurement of demand. Roos came to Princeton primarily to broaden and sharpen his knowledge of mathematics as a basis for making further contributions to Professor Evans' new "economic dynamics". While there he succeeded in convincing some members of the Department of Economics and Social Institutions that the Department could not afford to continue to neglect much longer the advances in economic theory and methods pioneered by Evans and Schultz.

In 1928 my father became the chairman of the Mathematics Department. One of his early acts in this capacity was to arrange for the loan by the Bell Telephone Laboratories, Inc. of a member of its Technical Staff, Dr. Thornton C. Fry, author of *Probability and Its Engineering Uses* (D. Van Nostrand, 1928), to give a course at Princeton on "Methods of Mathematical Physics" as a Visiting Lecturer in Mathematics during the first semester 1929–1930. One result of Fry's visit to Princeton was that a course in probability, taught by H. P. Robertson (1903–1964), Associate Professor of Mathematical Physics, using Fry's book as the text, was offered by the Mathematics Department during the second semester of my sophomore year (1931–1932). It was this course that first interested me in probability and mathematical statistics and started me on my career.

In 1931 steps were taken that led to a course in "modern statistical theory" being offered for the first time at Princeton by the Department of Economics and Social Institutions during the second semester of my senior year (1933–1934). What happened was this: Professor Frank D. Graham (1890–1949) of this department approached my father in his capacity as Dean of the Faculty, and suggested that one way to overcome lack of competence in his department with respect to the latest developments in mathematical and statistical methods in economics would be to send one of the young instructors in his department to study with Professor Henry Schultz at the University of Chicago. (The possibility of hiring a new staff member from the outside to this end had been considered earlier but put aside—the Depression was in full swing, and there was a freeze on new university appointments.) My father was favorable to this proposition, subject to an additional provision: that the individual concerned also study the modern theory of statistical inference with Harold Hotelling for the purpose of initiating a course in this subject on his return. The "victim" that Professor Graham had in mind was Acheson J. Duncan; and this is how it came to pass that Duncan, with financial assistance from the International Finance Section of Princeton University, spent the first half of the academic year 1931–1932 studying with Professor Henry Schultz at the University of Chicago; and the second half with Professor Hotelling at Columbia.

This assignment was very disruptive to Duncan at that time. When asked to undertake it he was already at work on his doctoral dissertation on "South African gold and international trade"; and his acceptance of it delayed until 1936 his completion of the requirements for his Ph.D. in economics. He also lost out on one of the features that "sweetened" the proposition, an opportunity to visit the West Coast—when the plans were made, Hotelling was at Stanford University, but had moved on to Columbia University before the time arrived for Duncan to study under him. This assignment was to be instrumental in changing the direction of Duncan's subsequent career. (After serving as Assistant Professor of Economics and Statistics at Princeton until 1945, he became a Professor of Statistics at Johns Hopkins University, retiring as Professor Emeritus in 1971. Since 1960 he has been a member and Chairman (1972–1979) of the prestigious Committee E-11 (statistical methods) of the American Society of Testing and Materials.)

Duncan returned to Princeton in the fall of 1932 to resume his duties as Instructor in the Department of Economics and Social Institutions, and to begin to ready himself to teach his projected new courses, unaware—as was also my father—that before his course in "modern statistical theory" would get under way, Wilks would have joined the Princeton University faculty.

The program worked out for Duncan on his return to Princeton was this: He would participate as an assistant in the course, "Elementary Statistics", taught by Professor James G. Smith (1897–1946) in the Department of Economics and Social Institutions, scheduled for the fall semester in 1933, serving as instructor in charge of the "laboratory" or "workshop" sessions in which the students gained practical experience in graphical and tabular presentation, and in the computation of descriptive statistics, index numbers, moving averages, link relatives, etc. Then, as a sequel to this course, Duncan's new course on "modern statistical theory" would be offered by the same department during the second semester of the academic year 1933–1934.

I took these two courses in the fall of 1933 and spring of 1934, respectively. In Smith's course we used as text *Principles and Methods of Statistics* by Robert E. Chaddock (1879–1940), published by the Houghton Mifflin Company in 1925, but the scope, nature, and mode of presentation is more accurately reflected by Professor Smith's *Elementary Statistics. An Introduction to the Principles of Scientific Methods*, published the following year (New York: Henry Holt and Company, 1934). Some of R. A. Fisher's contributions to statistical methodology were alluded to, but only very briefly, as tips on recent developments that would warrant looking into, not as integral parts of the course. In Duncan's course, on the other hand, built as it was around Hotelling's lectures, and the then available mimeographed chapters of Hotelling's never published book, *Statistical Inference*, the contributions of "Student" (William Sealy Gosset) and R. A. Fisher occupied the center of the stage a large part of the time.

In the spring of 1933 a crisis developed of which I was totally unaware at the time, and the particulars of which I was not to learn until some years later. Wilks was at Cambridge University working with Wishart on the last lap of his two-year fellowship program and would be needing a permanent post, or at least a new source of income, by fall. He had sent résumés of his professional career to the universities in the United States known to have programs in probability and mathematical statistics, indicating that he was in need of an instructorship or other full-time position beginning with the academic year 1933–1934. The replies that he received were all negative—the United States was in the depth of the Depression, colleges and universities were having to make do with dramatically reduced income from endowment and other sources, and all, it seemed, were tightening the belt, and none were planning to take on additional personnel. With an exceptional training in mathematical statistics, with four substantial research papers, and two research notes already published, one joint research paper accepted for publication, and two research papers nearly ready for publication, he was one of the most promising young men in mathematical statistics and applied mathematics generally, yet he had no prospect of a job. Wilks' situation seemed hopeless and was rapidly becoming desperate. Here he was in England with his wife and son; his fellowship funds, which were never really adequate for married people, or couples with children, were about to run out; and no prospect of employment.

Hotelling, knowing full well of my father's desire to build up a program in probability and mathematical statistics at Princeton and of the need of the College Entrance Examination Board for assistance from someone of Wilks' caliber on multivariate sampling distribution problems arising in educational testing, appealed directly to my father to take Wilks on at Princeton, stressing the long-term advantages to Princeton and the at-the-moment desperateness of Wilks' situation. Thus it came to pass late in the spring of 1933 that my father, as Chairman of the Mathematics Department, offered Wilks an instructorship in the Department of Mathematics for the academic year 1933–1934, and advised him of a tentative arrangement that he had made with Professor Carl C. Brigham of the Department of Psychology and Associate Secretary of the College Entrance Examination Board (the central office of which had been at Princeton for some years) to work part-time also with the Board on problems arising in the scaling of achievement tests. It was not until many years later that I learned from my father that he had brought off this coup over the opposition of almost every member of his department. I have often wondered whether he would have been able to bring it off a year or even six months later because, although he continued as Chairman of the Mathematics Department until 1945, in mid-1933 he gave up his post as Dean of the Faculty to become Dean of the Graduate School.

Wilks arrived in Princeton in September 1933. As a new instructor in the Department of Mathematics, he found himself teaching the usual undergraduate courses in analytic geometry, calculus, and so forth during the academic year 1933–34. In addition to such teaching that first year, Sam continued his research, primarily in multivariate analysis; gave me helpful guidance in the preparation of my senior thesis on "The accuracy of computations involving quantities known only to a given degree of approximation"; and spent the remainder of his "spare time" on his "second job" with Professor Brigham and the College Entrance Examination Board. The following year, 1934–1935, Sam's program was much the same, except that he now guided my post-graduate reading and study in probability and statistical theory and methodology in preparation for my becoming a doctoral candidate in statistics under J. Neyman and E. S. Pearson at University College, London, 1935–1937.

Wilks taught his first statistics course at the University of Pennsylvania, in Philadelphia, during 1935–1936. (Dr. George Gailey Chambers, Professor of Mathematics, University of Pennsylvania, had died on 24 October 1935, shortly after his graduate course "Modern Theory of Statistical Analysis" had gotten under way. Sam was commissioned to complete the teaching of this course.) During the same period Sam gave an informal course—i.e., not listed in the official university course catalog—to three Princeton seniors, Walter W. Merrill, John O. Rohm, and William C. Shelton, on much the same material; and supervised Shelton's senior thesis on "Regression and analysis of variance". (Shelton continued in statistics, rising to become Special Assistant to the Commissioner of Labor Statistics. Merrill and Rohm took up accounting and law, respectively.)

Wilks was promoted to an assistant professorship in 1936; and in 1936–1937 taught his first statistics courses at Princeton: a graduate course during the fall term and an undergraduate course during the spring term. A Princeton senior that year who took the graduate course, Irving E. Segal (later a Professor of Mathematics at Chicago and at MIT), wrote a senior thesis under Sam's supervision that was subsequently published in the *Proceedings of the Cambridge Philosophical Society* **34**, pt 1 (Jan. 1938), pp. 41–47.

The publication, in the January 1973 issue of the IMS *Bulletin*, of Professor Harry C. Carver's letter of 14 April 1972 to Professor William Jackson Hall on the "beginnings of the *Annals*" prompts me to correct a mistaken conjecture contained therein on why Sam Wilks was not permitted to teach a course in mathematical statistics during his first few years as an instructor in the Mathematics Department there. Professor Carver wrote:

> "... one day I asked [Wilks] how it was that he was not teaching a course in mathematical statistics at Princeton. He replied that he had tried to start such a course there, but his superiors turned down his request each time,—probably because mathematical statistics and probability had not yet rung a bell in the staid Eastern Colleges."

The fact of the matter is that mathematical statistics and probability *already had* "rung a bell" at Princeton: two years before Wilks' arrival, Acheson J. Duncan had been sent off at university expense to study with Professors Henry Schultz and Harold Hotelling for the express purpose of readying himself to initiate courses in "mathematical economics" and "modern statistical theory" on his return. It was this prior arrangement and commitment, not lack of appreciation of the importance of mathematical statistics and probability—or of Wilks' exceptional qualifications—that constituted the primary obstacle to Wilks' offering an undergraduate course in mathematical statistics during his first three years as a member of the Mathematics Department of Princeton University. Duncan's course on "modern statistical theory" had been scheduled to be offered for the first time during the spring term of 1934 before the possibility of Wilks' coming to Princeton had even been considered. In view of the expense that the university had incurred in underwriting Duncan's year of training in preparation for the offering of this course, and the sacrifice that Duncan had made in postponing work on his doctoral dissertation in order to acquire the requisite training at the university's request, it would have been very improper and cruel to have shelved

Duncan's course and let Wilks start one instead. I am sure that Wilks recognized this; and was also cognizant of the other factors that delayed his getting a course of his own in the Mathematics Department.

The three-year delay between Sam's arrival at Princeton and his first officially recognized course in statistics under the auspices of the Mathematics Department was the result of at least four factors.

First, there was the priority that circumstances had accorded to Duncan's course in the Department of Economics and Social Institutions. Furthermore, that department had taken the initiative in the matter, and was desirous of modernizing its outlook and course offerings with respect to mathematical economics and statistics. This department was aiming to improve its own offerings in statistics for economics students by integrating and updating the Smith-Duncan sequence of courses within that department. The extent to which this aim was achieved is evidenced by the two volumes *Fundamentals of the Theory of Statistics*: *Vol.* 1, *Elementary Statistics and Applications*; *Vol.* 2, *Sampling Statistics and Applications*, authored jointly by Professors Smith and Duncan and published by the McGraw-Hill Book Company, Inc., in 1944, 1945, respectively.)

Second, under the circumstances, any course on "mathematical statistics", "statistical analysis", "statistical inference", or whatever, to be offered by Wilks in the Mathematics Department would have to be an additional new course, and would require the approval of the all-powerful Course of Study Committee of the Faculty. A new course at Princeton had to be described in detail by the department proposing to offer it. Faculty approval gave the department the right to teach the described subject matter. I am not sure that this was an exclusive right, but I doubt that the Course of Study Committee would have approved teaching essentially the same material in two departments. Hence a major obstacle to Sam's teaching an undergraduate course in statistics was the historical fact that statistics had been the province of the Department of Economics and Social Institutions.

Third, until Sam was promoted to an assistant professorship in 1936, he was only an instructor; and in a department having the stature, nationally and internationally, of Princeton's Mathematics Department it was definitely not customary for an undergraduate, much less a graduate course, to be initiated by and be the sole responsibility of an individual with the rank of instructor.

A fourth, and very inhibiting factor was the unfavorable mathematical "climate" that prevailed in Fine Hall, the home of Princeton's Mathematics Department, during Sam's early years at Princeton. Geometry had occupied the center of the stage in this department, for over a quarter of a century, with algebra and analysis accorded much less exalted roles. Then, in 1932, the new Institute for Advanced Study, an institution completely distinct from Princeton University, had come into being, and the members of its School of Mathematics were granted office space in the Mathematics Department's Fine Hall until the completion of their first building, Fuld Hall, in 1939. Albert Einstein (1879–1955) arrived to take up his post in the Institute during the Winter of 1933, and Hermann Weyl (1885–1955) arrived a few months earlier. John Von Neumann (1903–1957) was already there (Lecturer, 1930–1931, Princeton, then Professor of Mathematical Physics, 1931–1933; Professor of Mathematics, Institute for Advanced Study, 1933–1957); as were also E. U. Condon (1903–1961; Assistant Professor of Mathematical Physics, Princeton, 1928–1931; Associate Professor, 1931–1938, Professor, 1938–1947), and E. P. Wigner (Lecturer in Mathematical Physics, Princeton, 1930; Professor, 1930–1936; 1938–1971). With this galaxy of mathematical physicists all together in one place for the first time, the mathematical theory of relativity and quantum mechanics were definitely the fashion of the day in Fine Hall—a difficult "climate" in which to initiate a program in mathematical statistics.

By 1936–1937, the division of territory between the Department of Mathematics and the Department of Economics and Social Institutions had been resolved. The latter would be restricted to instruction in statistical theory and methods pertinent to the economic and social sciences; and the basic general undergraduate course (s) in statistical theory and methodology, and the graduate courses in advanced mathematical statistics would be the province of the Mathematics Department. As we have already said, Wilks taught his first statistics course at Princeton in the fall of 1936, the graduate course leading to his lithographed lecture notes on *Statistical Inference*—(1937); and in the spring of 1937, a sophomore course with calculus as prerequisite, quite possibly the first carefully formulated college underclass course in mathematical statistics at this level. It was offered thereafter for a number of years to students in all fields in the second half of the sophomore year. The material presented in this course, extended and polished, became generally available a decade later in his "blue book", *Elementary Statistical Analysis* (1948). A third course, also one semester in length, was added in 1939–1940. It was an upperclass course for students who wanted to specialize in statistics, and consisted of a rather thorough mathematical treatment of statistical theory in the classroom plus a laboratory section devoted to applications and computations. This course was taken also by beginning graduate students. Wilks' first doctoral student, Joseph F. Daly received his Ph.D. in 1939. George W. Brown and Alexander M. Mood followed in 1940. World War II demolished his plans for sabbatical leave to lecture in South America and accept an offered exchange professorship for one semester at the National University in Santiago, Chile. As World War II progressed, Sam became ever more deeply involved in war research and in due course was released from academic duties entirely. Helped by two of his graduate students, T. W. Anderson and

D. F. Votaw, Jr., and Henry Scheffé, he succeeded in seeing through to lithoprinted publication the graduate level text, *Mathematical Statistics* (1943), before becoming totally involved in war work. This was the forerunner of his polished comprehensive treatment bearing the same title published as a typeset book in 1962.

In keeping with my father's policy of promotions as soon as merited without regard to leave of absence, Sam was promoted to a full Professor of Mathematics in 1944, effective on his return to academic duties; and plans were laid for a Section of Mathematical Statistics within the Department of Mathematics. Following the war there was a steady flow of able graduate students and postdoctoral research associates, some of whom, like Robert Hooke and Henry Scheffé, were changing from mathematics to statistics. By the time of Sam's death (1964), Princeton had granted Ph.D.s to approximately forty students in mathematical statistics and probability, all of whom had studied to some extent with Wilks, and the dissertations of about half had been supervised by him.

A Conversation with David Blackwell

MORRIS H. DEGROOT

David Blackwell was born on April 24, 1919, in Centralia, Illinois. He entered the University of Illinois in 1935, and received his A.B. in 1938, his A.M. in 1939, and his Ph.D. in 1941, all in mathematics. He was a member of the faculty at Howard University from 1944 to 1954, and has been a Professor of Statistics at the University of California, Berkeley, since that time. He was President of the Institute of Mathematical Statistics in 1955. He has also been Vice President of the American Statistical Association, the International Statistical Institute, and the American Mathematical Society, and President of the Bernoulli Society. He is an Honorary Fellow of the Royal Statistical Society and was awarded the von Neumann Theory Prize by the Operations Research Society of America and the Institute of Management Sciences in 1979. He has received honorary degrees from the University of Illinois, Michigan State University, Southern Illinois University, and Carnegie-Mellon University.

The following conversation took place in his office at Berkeley one morning in October 1984.

"I EXPECTED TO BE AN ELEMENTARY SCHOOL TEACHER"

DeGroot: How did you originally get interested in statistics and probability?

Blackwell: I think I have been interested in the concept of probability ever since I was an undergraduate at Illinois, although there wasn't very much probability or statistics around. Doob was there but he didn't teach probability. All the probability and statistics were taught by a very nice old gentleman named Crathorne. You probably never heard of him. But he was a very good friend of Henry Rietz and, in fact, they collaborated on a college

Reprinted from *Statistical Science*, Vol. 1, No. 1, pp. 40–53, with permission of the Institute of Mathematical Statistics.

algebra book. I think I took all the courses that Crathorne taught: two undergraduate courses and one first-year graduate course. Anyway, I have been interested in the subject for a long time, but after I got my Ph.D. I didn't expect to get professionally interested in statistics.

DeGroot: But did you always intend to go on to graduate school?

Blackwell: No. When I started out in college I expected to be an elementary school teacher. But somehow I kept postponing taking those education courses. [Laughs] So I ended up getting a master's degree and then I got a fellowship to continue my work there at Illinois.

DeGroot: So your graduate work wasn't particularly in the area of statistics or probability?

Blackwell: No, except of course that I wrote my thesis under Doob in probability.

DeGroot: What was the subject of your thesis?

Blackwell: Markov chains. There wasn't very much original in it. There was one beautiful idea, which was Doob's idea and which he gave to me. The thesis was never published as such.

DeGroot: But your first couple of papers pertained to Markov chains.

Blackwell: The first couple of papers came out of my thesis, that's right.

DeGroot: So after you got your degree...

Blackwell: After I got my degree, I sort of expected to work in probability, real variables, measure theory, and such things.

DeGroot: And you *have* done a good deal of that.

Blackwell: Yes, a fair amount. But it was Abe Girshick who got me interested in statistics.

DeGroot: In Washington?

Blackwell: Yes. I was teaching at Howard and the mathematics environment was not really very stimulating, so I had to look around beyond the university just for whatever was going on in Washington that was interesting mathematically.

DeGroot: Not just statistically, but mathematically?

Blackwell: I was just looking for anything interesting in mathematics that was going on in Washington.

DeGroot: About what year would this be?

Blackwell: I went to Howard in 1944. So this would have been during the year 1944–1945.

DeGroot: Girshick was at the Department of Agriculture?

Blackwell: That's right. And I heard him give a lecture sponsored by the Washington Chapter of the American Statistical Association. That's a pretty lively chapter. I first met George Dantzig when he gave a lecture there around that same time. His lecture had nothing to do with linear programming, by the way. In fact, I first became acquainted with the idea of a randomized test by hearing Dantzig talk about it. I think that he was the guy who invented a test function, instead of having just a rejection region that is a subset of the sample space. At one of those meetings Abe Girshick spoke on sequential analysis. Among other things, he mentioned Wald's equation.

DeGroot: That's the equation that the expectation of a sum of random variables is $E(N)$ times the expectation of an individual variable?

Blackwell: Yes. That was just such a remarkable equation that I didn't believe it. So I went home and thought I had constructed a counterexample. I mailed it to Abe, and I'm sure that he discovered the error. But he didn't write back and tell me it was an error; he just called me up and said let's talk about it. So we met for lunch and that was the start of a long and beautiful association that I had with him.

DeGroot: Would you regard the Blackwell and Girshick book (*Theory of Games and Statistical Decisions.* New York, John Wiley & Sons, 1954) as the culmination of that association?

Blackwell: Oh, that was a natural outgrowth of the association. I learned a great deal from him.

DeGroot: Were you together at any time at Stanford?

Blackwell: Yes, I spent a year at Stanford. I think it was 1950–1951. But he and I were also together at other times. We spent several months together at Rand. So we worked together in Washington, and then at Rand, and then at Stanford.

"I Wrote 105 Letters of Application"

DeGroot: Tell me a little about the years between your Ph.D. from Illinois in 1941 and your arrival at Howard in 1944. You were at a few other schools in between.

Blackwell: Yes. I spent my first postdoctoral year at the Institute for Advanced Study. Again, I continued to show my interest in statistics. I sat in on Sam Wilks' course in Princeton during that year. Henry Scheffé was also sitting in on that class. He had just completed his Ph.D. at Wisconsin. Jimmie Savage was at the Institute for that year. He was at some of Wilks' lectures, too. There were a lot of statisticians about our age around Princeton at that time. Alex Mood was there. George Brown was there. Ted Anderson was there. He was in Wilks' class that year.

DeGroot: He was a graduate student?

Blackwell: He was a graduate student, just completing his Ph.D. So that was my first postdoctoral year. Also, I had a chance to meet von Neumann that year. He was a most impressive man. Of course, everybody knows that. Let me tell you a little story about him.

When I first went to the Institute, he greeted me, and we were talking, and he invited me to come around and tell him about my thesis. Well, of course, I thought that was just his way of making a new young visitor feel at home, and I had no intention of telling him about my thesis. He was a big, busy, important man. But then a couple of months later, I saw him at tea and he said, "When are you coming around to tell me about your thesis? Go in and make an appointment with my secretary." So I did, and later I went in and started telling him about my thesis. He listened for about ten minutes and asked me a couple of questions, and then he started telling *me* about *my* thesis. What you have really done is this, and probably this is true, and you could have done it in a somewhat simpler way, and so on. He was a really remarkable man. He listened to me talk about this rather obscure subject and in ten minutes he knew more about it than I did. He was extremely quick. I think he may have wasted a certain amount of time, by the way, because he was so willing to listen to second- or third-rate people and think about their problems. I saw him do that on many occasions.

DeGroot: So, from the Institute you went where?

Blackwell: I went to Southern University in Baton Rouge, Louisiana. That's a state school and at that time it was *the* state university in Louisiana for blacks. I stayed there just one year. Then the next year, I went to Clark College in Atlanta, also a black school. I stayed there for one year. Then I went to Howard University in Washington and stayed there for ten years.

DeGroot: Was Howard at a different level intellectually from these other schools?

Blackwell: Oh yes. It was the ambition of every black scholar in those days to get a job at Howard University. That was the best job you could hope for.

DeGroot: How large was the math department there in terms of faculty?

Blackwell: Let's see. There were just four regular people in the math department. Two professors. I went there as an assistant professor. And there was one instructor. That was it.

DeGroot: Have you maintained any contact with Howard through the years?

Blackwell: Oh yes. I guess the last time I gave a lecture there was about three years ago, but I visited many times during the years.

DeGroot: Do you see much change in the place through the years?

Blackwell: Yes, the math department now is a livelier place than it was when I was there. It's much bigger and the current chairman, Jim Donaldson, is very good and very active. There are some interesting things going on there.

DeGroot: Did you feel or find that discrimination against blacks affected your education or your career after your Ph.D.?

Blackwell: It never bothered me. I'll put it that way. It surely shaped my expectations from the very beginning. It never occurred to me to think about teaching in a major university since it wasn't in my horizon at all.

DeGroot: Even in your graduate-student days at Illinois?

Blackwell: That's right. I just assumed that I would get a job teaching in one of the black colleges. There were 105 black colleges at that time, and I wrote 105 letters of application.

DeGroot: And got 105 offers, I suppose.

Blackwell: No, I eventually got three offers, but I accepted the first one that I got. From Southern University.

DeGroot: Let's move a little further back in time. You grew up in Illinois?

Blackwell: In Centralia, Illinois. Did you ever get down to Centralia or that part of Illinois when you were in Chicago?

DeGroot: No, I didn't.

Blackwell: Well, it's a rather different part of the world from northern Illinois. It's quite southern. Centralia in fact was right on the border line of segregation. If you went south of Centralia to the southern tip of Illinois, the schools were completely segregated in those days. Centralia had one completely black school, one completely white school, and five "mixed" schools.

DeGroot: Well that sounds like the boundary all right. Which one did you go to?

Blackwell: I went to one of the mixed schools, because of the part of town I lived in. It's a small town. The population was about 12,000 then and it's still about 12,000. The high school had about 1,000 students. I had very good high school teachers in mathematics. One of my high school teachers organized a mathematics club and used to give us problems to work. Whenever we would come up with something that had the idea for a solution, he would write up the solution for us, and send it in our name to a journal called *School Science and Mathematics*. It was a great thrill to see your name in the magazine. I think my name got in there three times. And once my *solution* got printed. As I say, it was really Mr. Huck's write-up based on my idea. [Laughs].

DeGroot: Was your family encouraging about your education?

David Blackwell (lower left), 1930, probably sixth grade.

Blackwell: It was just sort of assumed that I would go to college. There was no "Now be sure to study hard" or anything like that. It was just taken for granted that I was going to go to college. They were very, very supportive.

Some Favorite Papers

DeGroot: You were quite young when you received your Ph.D. You were 21 or so?

Blackwell: 22. There wasn't any big jump. I just sort of did everything a little faster than normal.

DeGroot: And you've been doing it that way ever since. You've published about 80 papers since that time. Do you have any favorites in that list that you particularly like or that you feel were particularly important or influential?

Blackwell: Oh, I'm sure that I do, but I'd have to look at the list and think about that. May I look?

DeGroot: Sure. This is an open-book exam.

Blackwell: Good. Let's see... Well, my first statistical paper, called "On an equation of Wald" (*Ann. Math. Statist.* **17**, 84–87, 1946) grew out of that original conversation with Abe Girshick. That's a paper that I am still really very proud of. It just gives me pleasant feelings every time I think about it.

DeGroot: Remind me what the main idea was.

Blackwell: For one thing it was a proof of Wald's theorem under, I think, weaker conditions than it had been proved before, under sort of *natural* conditions. And the proof is *neat*. Let me show it to you. [Goes to blackboard.]

Suppose that $X_1, X_2, \ldots$ are i.i.d. and you have a stopping rule N, which is a random variable. You want to prove that $E(X_1 + \cdots + X_N) = E(X_1)E(N)$. Well, here's my idea. Do it over and over again. So you have stopping times $N_1, N_2, \ldots$, and you get

$$S_1 = X_1 + \cdots + X_{N_1},$$
$$S_2 = X_{N_1+1} + \cdots + X_{N_1+N_2},$$
$$\ldots$$

Consider $S_1 + \cdots + S_k = X_1 + \cdots + X_{N_1+\cdots+N_k}$. We can write this equation as

$$\frac{S_1 + \cdots + S_k}{k} = \left(\frac{X_1 + \cdots + X_{N_1+\cdots+N_k}}{N_1 + \cdots + N_k}\right)\left(\frac{N_1 + \cdots + N_k}{k}\right).$$

Now let $k \to \infty$. The first term on the right is a subsequence of the X averages. By the strong law of large numbers, this converges to $E(X_1)$. The second term on the right is the average of $N_1, \ldots, N_k$. We are assuming that they have a finite expectation, so this converges to that expectation $E(N)$. Therefore, the sequence

$$\frac{S_1 + \cdots + S_k}{k}$$

converges a.e. Then the converse of the strong law of large numbers says that the expected value of each S_i must be finite, and that

$$\frac{S_1 + \cdots + S_k}{k}$$

must converge to that expectation $E(S_1)$. Isn't that neat?

DeGroot: Beautiful, beautiful.

Blackwell: So that's the proof of Wald's equations just by invoking the strong law of large numbers and its converse. I think I like that because that was the first time that *I* decided that I could do something original. The papers based on my thesis were nice, but those were really Doob's ideas that I was just carrying out. But here I had a really original idea, so I was very pleased with that paper. Then I guess I like my paper with Ken Arrow and Abe Girshick, "Bayes and minimax solutions of sequential decision problems" (*Econometrica* **17**, 213–244, 1949).

DeGroot: That was certainly a very influential paper.

Blackwell: That was a serious paper, yes.

DeGroot: There was some controversy about that paper, wasn't there? Wald and Wolfowitz were doing similar things at more or less the same time.

Blackwell: Yes, they had priority. There was no question about that, and I think we did give inadequate acknowledgment to them in our work. So they were very much disturbed about it, especially Wolfowitz. In fact, Wolfowitz was cool to me for more than 20 years.

DeGroot: But certainly your paper was different from theirs.

Blackwell: We had things that they didn't have, there was no doubt about that. For instance, induction backward—calculation backward—that was in our paper and I don't think there is any hint of it in their work. We did go beyond what they had done. Our paper didn't seem to bother Wald too much, but Wolfowitz was annoyed.

DeGroot: Did you know Wald very well or have much contact with him?

Blackwell: Not very well. I had just three or four conversations with him.

IMPORTANT INFLUENCES

DeGroot: I gather from what you said that Girshick was a primary influence on you in the field of statistics.

Blackwell: Oh yes.

DeGroot: Were there other people that you felt had a strong influence on you? Neyman, for example?

Blackwell: Not in my statistical thinking. Girshick was certainly *the* most important influence on me. The other person who had just one influence, but it was a very big one, was Jimmie Savage.

DeGroot: What was that one influence?

Blackwell: Well, he explained to me that the Bayes approach was the right way to do statistical inference. Let me tell you how that happened. I was at Rand, and an economist came in one day to talk to me. He said that he had a problem. They were preparing a recommendation to the Air Force on how to divide their research budget over the next five years and, in particular, they had to decide what fraction of it should be devoted to long-range research and what fraction of it should be devoted to more immediate developmental research.

"Now," he said, "one of the things that this depends on is the probability of a major war in the next five years. If it's large then, of course, that would shift the emphasis toward developing what we already know how to do, and if it's small then there would be more emphasis on long-range research. I'm not going to ask you to tell me a number, but if you could give me any guide as to how I could go about finding such a number I would be grateful." Oh, I said to him, that question just doesn't make sense. Probability applies to a long sequence of repeatable events, and this is clearly a unique situation. The probability is either 0 or 1, but we won't know for five years, I pontificated. [Laughs] So the economist looked at me and nodded and said, "I was afraid you were going to say that. I have spoken to several other statisticians and they have all told me the same thing. Thank you very much." And he left.

Well, that conversation bothered me. The fellow had asked me a reasonable, serious question and I had given him a frivolous, sort of flip, answer, and I wasn't happy. A couple of weeks later Jimmie Savage came to visit Rand, and I went in and said hello to him. I happened to mention this conversation that I had had, and then he started telling me about deFinetti and personal probability. Anyway, I walked out of his office half an hour later with a completely different view on things. I now understood what was the right way to do statistical inference.

DeGroot: What year was that?

Blackwell: About 1950, maybe 1951, somewhere around there. Looking back on it, I can see that I was emotionally and intellectually prepared for Jimmie's message because I had been thinking in a Bayesian way about sequential analysis, hypothesis testing, and other statistical problems for some years.

DeGroot: What do you mean by thinking in a Bayesian way? In terms of prior distributions?

Blackwell: Yes.

DeGroot: Wald used them as a mathematical device.

Blackwell: That's right. It just turned out to be clearly a very natural way to think about problems and it was mathematically beautiful. I simply regretted that it didn't correspond with reality. [Laughs] But then what Jimmie was telling me was that the way that I had been thinking all the time was really the right way to think, and not to worry so much about empirical frequencies. Anyway, as I say, that was just one very big influence on me.

DeGroot: Would you say that your statistical work has mainly used the Bayesian approach since that time?

Blackwell: Yes. I simply have not worked on problems where that approach could not be used. For instance, all my work in dynamic programming just has that Bayes approach in it. That is *the* standard way of doing dynamic programming.

DeGroot: You wrote a beautiful book called *Basic Statistics* (New York, McGraw-Hill, 1970) that was really based on the Bayesian approach, but as I recall you never once mentioned the word "Bayes" in that book. Was that intentional?

Blackwell: No, it was not intentional.

DeGroot: Was it that the terminology was irrelevant to the concepts that you were trying to get across?

Blackwell: I doubt if the word "theorem" was ever mentioned in that book. That was not originally intended as a book, by the way. It was simply intended as a set of notes to give my students in connection with lectures in this elementary statistics course. But the students suggested that it should be published and a McGraw-Hill man said that he would be interested. It's just a set of notes. It's short; I think it's less than 150 pages.

DeGroot: It's beautiful. There are a lot of wonderful gems in those 150 pages.

Blackwell: Well, I enjoyed teaching the course.

DeGroot: Do you enjoy teaching from your own books?

Blackwell: No, not after a while. I think about five years after the book was published, I stopped using it. Just because I got bored with it. When you reach the point where *you're* not learning anything, then it's probably time to change something.

DeGroot: Are you working on other books at the present time?

Blackwell: No, except that I am *thinking about* writing a more elementary version of parts of your book on optimal statistical decisions because I have been using it in a course and the undergraduate students say that it's too hard.

DeGroot: Uh oh. I've been thinking of doing the same thing. [Laughs] Well, I am just thinking generally in terms of an introduction to Bayesian statistics for undergraduates.

Blackwell: Very good. I really hope you do it, Morrie. It's needed.

DeGroot: Well, I really hope you do it, too. It would be interesting. Are there courses that you particularly enjoy teaching?

Blackwell: I like the course in Bayesian statistics using your book. I like to teach game theory. I haven't taught it in some years, but I like to teach that course. I also like to teach, and I'm teaching right now, a course in information theory.

DeGroot: Are you using a text?

Blackwell: I'm not using any one book. Pat Billingsley's book *Ergodic Theory and Information* comes closest to what I'm doing. I like to teach measure theory. I regard measure theory as a kind of hobby, because to do probability and statistics you don't really need very much measure theory. But there are these fine, nit-picking points that most people ignore, and rightly so, but that I sort of like to worry about. [Laughs] I know that it is not important, but it is interesting to me to worry about regular conditional probabilities and such things. I think I'm one of only three people in our department who really takes measure theory seriously. Lester [Dubins] takes it fairly seriously, and so does Jim Pitman. But the rest of the people just sort of ignore it. [Laughs]

"I Would Like to see More Emphasis on Bayesian Statistics"

DeGroot: Let's talk a little bit about the current state of statistics. What areas do you think are particularly important these days? Where do you see the field going?

Blackwell: I can tell you what I'd like to see happen. First, of course, I would like to see more emphasis on Bayesian statistics. Within that area it seems to me that one promising direction which hasn't been explored at all is Bayesian experimental design. In a way, Bayesian statistics is much simpler than classical statistics in that once you're given a sample, all you have to do are calculations based on that sample. Now, of course, I say "all you have to do"—sometimes those calculations can be horrible. But if you are trying to design an experiment, that's not all you have to do. In that case, you have to look at all the different samples you might get and evaluate every one of them in order to calculate an overall risk, to decide whether the experiment is worth doing and to choose among the experiments. Except in very special situations, such as when to stop sampling, I don't think a lot of work has been done in that area.

Kenneth Arrow, David Blackwell, and M. A. Girshick,
Santa Monica, September 1948.

David Blackwell, 1984.

DeGroot: I think the reason there hasn't been very much done is because the problems are so hard. It's really hard to do explicitly the calculations that are required to find *the* optimal experiment. Do you think that perhaps the computing power that is now available would be helpful in this kind of problem?

Blackwell: That's certainly going to make a difference. Let me give you a simple example that I have never seen worked out but I am sure could be worked out. Suppose that you have two independent Bernoulli variables, say, a proportion among males and a proportion among females. They are independent, and you are interested in estimating the sum of those proportions or some linear combination of those proportions. You are going to take a sample in two stages. First of all, you can ask how large should the first sample be? And then, based on the first sample, how should you allocate proportions in the second sample?

DeGroot: Are you going to draw the first sample from the total population?

Blackwell: No. you have males and you have females, and you have a total sample effort of size N. Now you can pick some number $n \leq N$ to be your sample size. And you can allocate those n observations among males and females. Then based on how that sample comes out, you can allocate your second sample. What is the best initial allocation, and how much better is it than just doing it all in one stage? Well, I haven't done that calculation but I'm sure that it can be done. It would be an interesting kind of thing and it could be extended to more than two categories. That's an example of the sort of thing on which I would like to see a lot of work done—Bayesian experimental design.

One of the things that I worry about a little is that I don't see theoretical statisticians having as much contact with people in other areas as I would like to see. I notice here at Berkeley, for example, that the people in Operations Research seem to have much closer contact with industry than the people in our department do. I think we might find more interesting problems if we did have closer contact.

DeGroot: Do you think that the distinctions between applied and theoretical statistics are still as rigid as they were years ago or do you think that the field is blending more into a unified field of statistics in which such distinctions are not particularly meaningful? I see the emphasis on data analysis which is coming about, and the development of theory for data analysis and so on, blurring these distinctions between theoretical and applied statistics in a healthy way.

Blackwell: I guess I'm not familiar enough with data analysis and what computers have done to have any interesting comments on that. I see what some of our people and people at Stanford are doing in looking at large-dimensional data sets and rotating them so that you can see lots of three-dimensional projections and such things, but I don't know whether that suggests interesting theoretical questions or not. Maybe that's not important, whether it suggests interesting theoretical questions. Maybe the important thing is that it helps contribute to the solution of practical problems.

Infinite Games

DeGroot: What kind of things are you working on these days?

Blackwell: Right now I am working on some things in information theory, and still trying to understand some things about infinite games of perfect information.

DeGroot: What do you mean by an infinite game?

Blackwell: A game with an infinite number of moves. Here's an example. I write down a 0 or a 1, and you write down a 0 or a 1, and we keep going indefinitely. If the sequence we produce has a limiting frequency, I win. If not, you win. That's a trivial game because I can force it to have a limiting frequency just by doing the opposite of whatever you do. But that's a simple example of an infinite game.

DeGroot: Fortunately, it's one in which I'll never have to pay off to you.

Blackwell: Well, we can play it in such a way that you would have to pay off.

DeGroot: How do we do that?

Blackwell: You must specify a strategy. Let me give you an example. You know how to play chess in just one move: You prepare a complete set of instructions so that for every situation on the chess board you specify a possible response. Your one move is to prepare that complete set of instructions. If you have a complete set and I have a complete set, then we can just play the game out according to those instructions. It's just one move. So in the same way, you can specify a strategy in this infinite game. For every finite sequence that you might see up to a given time as past history, you specify your next move. So you can define this function once and for all, and I can define a function, and then we can mathematically assess those functions. I can prove that there is a specific function of mine such that no matter what function you specify, the set will have a limiting frequency.

DeGroot: So you could extract money from me in a finite amount of time. [Laughs]

Blackwell: Right. Anyway it's been proved that all such infinite games with Borel payoffs are determined, and I've been trying to understand the proof for several years now. I'm still working on it, hoping to understand it and simplify it.

DeGroot: Have you published papers on that topic?

Blackwell: Just one paper many years ago. Let me remind myself of the title [checking his files], "Infinite games and analytic sets" (*Proc. Natl. Acad. Sci. U.S.A.* **58**, 1836–1837, 1967). This is the only paper I've published on infinite games; and that's one of my papers that I like very much, by the way. It's an application of games to prove a theorem in topology. I sort of like the idea of connecting those two apparently not closely related fields.

DeGroot: Have you been involved in applied projects or applied problems through the years, at Rand or elsewhere, that you have found interesting and that have stimulated research of your own?

Blackwell: I guess so. My impression though is this: When I have looked at real problems, interesting theorems have sometimes come out of it. But never anything that was helpful to the person who had the problem. [Laughs]

DeGroot: But possibly to somebody else at another time.

Blackwell: Well, my work on comparison of experiments was stimulated by some work by Bohnenblust, Sherman, and Shapley. We were all at Rand. They called their original paper "Comparison of reconnaissances," and it was *classified* because it arose out of some question that somebody had asked them. I recognized a relation between what they were doing and sufficient statistics, and proved that they were the same in a special case. Anyway, that led to this development which I think is interesting theoretically, and to which you have contributed.

DeGroot: Well, I have certainly used your work in that area. And it has spread into diverse other areas. It is used in economics in comparing distributions of income, and I used it in some work on comparing probability forecasters.

Blackwell: And apparently people in accounting have made some use of these ideas. But anyway, as I say, nothing that I have done has ever helped the person who raised the question. But there is no doubt in my mind that you do get interesting problems by looking at the real world.

"I Don't Have Any Difficulties with Randomization"

DeGroot: One of the interesting topics that comes out of a Bayesian view of statistics is the notion of randomization and the role that it should play in statistics. Just this little example you were talking about before with two proportions made me think about that. We just assume that we are drawing the observations at random from within each subpopulation in that example, but perhaps basically because we don't have much choice. Do you have any thoughts about whether one should be drawing observations at random?

Blackwell: I don't have any difficulties with randomization. I think it's probably a good idea. The strict theoretical idealized Bayesian would of course never need to randomize. But randomization probably protects us against our own biases. There are just lots of ways in which people differ from the ideal Bayesian. I guess the ideal Bayesian, for example, could not think about a theorem as being probably true. For him, presumably, all true theorems have probability 1 and all false ones have probability 0. But you and I know that's not the way we think. I think of randomization as being a protection against your own imperfect thinking.

DeGroot: It is also to some extent a protection against others. Protection for you as a statistician in presenting your work to the scientific community, in the sense that they can have more belief in your conclusions if you use some randomization procedure rather than your own selection of a sample. So I see it as involved with the sociology of science in some way.

Blackwell: Yes, that's an important virtue of randomization. That reminds me of something else though. We tend to think of evidence as being valid only when it comes from random samples or samples selected in a probabilistically specified way. That's wrong, in my view. Most of what we have learned, we have learned just by observing what happens to come along, rather than from carefully controlled experiments. Sometimes statisticians have made a mistake in throwing away experiments because they were not properly controlled. That is not to say that randomization isn't a good idea, but it is to say that you should not reject data just because they have been obtained under uncontrolled conditions.

DeGroot: You were the Rouse Ball Lecturer at Cambridge in 1974. How did that come about and what did it involve?

Blackwell: Well, I was in England for two years, 1973–1975, as the director of the education-abroad program in Great Britain and Ireland for the University of California. I think that award was just either Peter Whittle's or David Kendall's idea of how to get me to come up to Cambridge to give a lecture. One of the things which delighted me was that it was named the Rouse Ball Lecture because it gave me an opportunity to say something at Cambridge that I liked—namely, that I had heard of Rouse Ball long before I had heard of Cambridge. [Laughs]

DeGroot: Well, tell me about Rouse Ball.

Blackwell: He wrote a book called *Mathematical Recreations and Essays*. You may have seen the book. I first came across it when I was a high school student. It was one of the few mathematics books in our library. I was fascinated by that book. I can still picture it. Rouse Ball was a 19th century mathematician, I think. [Walter William Rouse Ball, 1850–1925] Anyway, this is a lectureship that they have named after him.

DeGroot: I guess there aren't too many Bayesians on the statistics faculty here at Berkeley.

Blackwell: No. I'd say, Lester and I are the only ones in our department. Of course, over in Operations Research, Dick Barlow and Bill Jewell are certainly sympathetic to the Bayesian approach.

DeGroot: Is it a topic that gets discussed much?

Blackwell: Not really, It used to be discussed here but you very soon discover that it's sort of like religion; that it has an appeal for some people and not for other people, and you're not going to change anybody's mind by discussing it. So people just go their own ways. What has happened to Bayesian statistics surprised me. I expected it either to catch on and just sweep the field or to die. And I was rather confident that it would die. Even though to me it was the right way to think, I just didn't think that it would have a chance to survive. But I thought that if it did, then it would sweep things. Of course, neither one of those things has happened. Sort of a steady 5-10% of all the work in statistical inference is done from a Bayesian point of view. Is that what you would have expected 20 years ago?

DeGroot: No, it certainly doesn't seem as though that would be a stable equilibrium. And maybe the system is still not in equilibrium. I see the Bayesian approach growing, but it certainly is not sweeping the field by any means.

Blackwell: I'm glad to hear that you see it growing.

DeGroot: Well, there seem to be more and more meetings of the Bayesians, anyway. The actuarial group that met here at Berkeley over the last couple of days to discuss credibility theory seems to be a group that just naturally accepts the Bayesian approach in their work in the real world. So there seem to be some pockets of users out there in the world, and I think maybe that's what has kept the Bayesian approach alive.

Blackwell: There's no question in my mind that if the Bayesian approach does grow in the statistical world it will not be because of the influence of other statisticians but because of the influence of actuaries, engineers, business people, and others who actually like the Bayesian approach and use it.

DeGroot: Do you get a chance to talk much to researchers outside of statistics on campus, researchers in substantive areas?

Blackwell: No, I talk mainly to people in Operations Research and Mathematics, and occasionally Electrical Engineering. But the things in Electrical Engineering are theoretical and abstract.

"THE WORD 'SCIENCE' IN THE TITLE BOTHERS ME A LITTLE"

DeGroot: What do you think about the idea of this new journal, *Statistical Science*, in which this conversation will appear? I have the impression that you think the IMS is a good organization doing useful things, and there is really no need to mess with it.

Blackwell: That is the way I feel. On the other hand, I must say that I felt exactly the same way about splitting the *Annals of Mathematical Statistics* into two journals, and that split seems to be working. So I'm hoping that the new journal will add something. I guess the word "science" in the title bothers me a little. It's not clear what the word is intended to convey there, and you sort of have the feeling that it's there more to contribute a tone than anything else.

DeGroot: My impression is that it *is* intended to contribute a tone. To give a flavor of something broader than just what we would think of as theoretical statistics. That is, to reach out and talk about the impact of statistics on the sciences and the interrelationship of statistics with the sciences, all kinds of sciences.

Blackwell: Now I'm all in favor of that. For example, the relation of statistics to the law is to me a quite appropriate topic for articles in this journal. But somehow calling it "science" doesn't emphasize that direction. In fact, it rather suggests that that's *not* the direction. It sounds as though it's tied in with things that are supported by the National Science Foundation and to me that restricts it.

DeGroot: The intention of that title was to convey a broad impression rather than a restricted one. To give a broader impression than just statistics and probability, to convey an applied flavor and to suggest links to all areas.

Blackwell: Yes. It's analogous to computer science, I guess. I think *that* term was rather deliberately chosen. My feeling is that the IMS is just a beautiful organization. It's about the right size. It's been successful for a good many years. I don't like to see us become ambitious. I like the idea of just sort of staying the way we are, an organization run essentially by amateurs.

DeGroot: Do you have the feeling that the field of statistics is moving away from the IMS in any way? That was one of the motivations for starting this journal.

Blackwell: Well, of course, statistics has always been substantially bigger than the IMS. But you're suggesting that the IMS represents a smaller and smaller fraction of statistical activity.

DeGroot: Yes, I think that might be right.

Blackwell: You know, Morrie, I see what you're talking about happening in mathematics. It's less and less true that all mathematics is done in mathematics departments. On the Berkeley campus, I see lots of interesting mathematics being done in our department, in Operations Research, in Electrical Engineering, in Mechanical Engineering, some in Business Administration, a lot in the Economics Department by Gerard Debreu and his colleagues; a lot of really interesting, high class mathematics is being done outside mathematics departments. What you're suggesting is that statistics departments and the journals in which they publish are not necessarily the centers of statistics the way they used to be, that a lot of work is being done outside. I'm sure that's right.

DeGroot: And perhaps *should* be done outside statistics departments. That used to be an unhealthy sign in the field, and we worked hard in statistics departments to collect up the statistics that was being done around the campus. But I think now that the field has grown and matured, that it is probably a healthy thing to have some interesting statistics being done outside.

Blackwell: Yes. Consider the old problem of pattern recognition. That's a statistical problem. But to the extent that it gets solved, it's not going to be solved by people in statistics departments. It's going to be solved by people working for banks and people working for other organizations who really need to have a device that can look at a person and recognize him in lots of different configurations. That's just one example of the cases where we're somehow too narrow to work on a lot of serious statistical problems.

DeGroot: I think that's right, and yet we have something important to contribute to those problems.

Blackwell: I would say that we *are* contributing, but indirectly. That is, people who are working on the problems have studied statistics. It seems to me that a lot of the engineers I talk to are very familiar with the basic concepts of decision theory. They know about loss functions and minimizing expected risks and such things. So, we have contributed, but just indirectly.

DeGroot: You are in the National Academy of Sciences...

Blackwell: Yes, but I'm very inactive.

DeGroot: You haven't been involved in any of their committees or panels?

Blackwell: No, and I'm not sure that I would want to be. I guess I don't like the idea of an official committee making scientific pronouncements. I like people to form opinions about scientific matters just on the basis of listening to individual scientists. To have one group with such overwhelming prestige bothers me a little.

DeGroot: And it is precisely the prestige of the Academy that they rely on when reports get issued by these committees.

Blackwell: Yes. So I think it's just great as a purely honorific organization, so to speak. To meet just once a year, and elect people more or less at random. I think everybody that's in it has done something reasonable and even pretty good, in fact. But on the other hand, there are at least as many people *not* in it who have done good things as there are in it. It's kind of a random selection process.

DeGroot: So you think it's a good organization as long as it doesn't do anything.

Blackwell: Right. I'm proud to be in it, but I haven't been active. It's sort of like getting elected to Phi Beta Kappa—it's nice if it happens to you...

"I Play with this Computer"

DeGroot: Do you feel any relationship between your professional work and the rest of your life, your interests outside of statistics? Is there any influence of the outside on what you do professionally, or are they just sort of separate parts of your life?

Blackwell: Separate, except my friends are also my colleagues. It's only through the people with whom I associate outside that there's any connection. It's hard to think of any other real connection.

DeGroot: It's not obvious what these connections might be for anyone. One's political views or social views seem to be pretty much independent of the technical problems we work on.

Blackwell: Yes. Although it's hard to see how it could *not* have an influence, isn't it? I guess my life seems all of a piece to me but yet it's hard to see where the connections are. [Laughs]

DeGroot: What do you see for your future?

Blackwell: Well, just gradually to wind down, gracefully I hope. I expect to get more interested in computing. I have a little computer at home, and it's a lot of fun just to play with it. In fact, I'd say that I play with this computer here in my office at least as much as I do serious work with it.

DeGroot: What do you mean by play?

Blackwell: Let me give you an example. You know the algorithm for calculating square roots. You start with a guess and then you divide the number by your guess and take the average of the two. That's your next guess. That's actually Newton's method for finding square roots, and it works very well. Sometimes doing statistical work, you want to take the square root of a positive definite matrix. It occurred to me to ask whether that algorithm works for finding the square root of a positive definite matrix. Before I got interested in computing, I would have tried to solve it theoretically. But what did I do? I just wrote up a program and put it on the computer to see if it worked. [Goes to blackboard]

Suppose that you are given the matrix M and want to find $M^{1/2}$. Let G be your guess of $M^{1/2}$. Then you new guess is $1/2(G+MG^{-1})$. You just iterate this and see if it converges to $M^{1/2}$. Now, Morrie, I want to show you what happens. [Goes to terminal]

Let's do it for a 3×3 matrix. We're going to find the square root of a positive definite 3×3 matrix. Now, if you happen to have in mind a particular 3×3 positive definite matrix whose square root you want, you could enter it directly. I don't happen to have one in mind, but I do know a theorem: If you take any nonsingular 3×3 matrix A, then AA' is going to be positive definite. So I'm just going to enter any 3×3 nonsingular matrix [putting some numbers into the terminal] and let $M = AA'$. Now, to see how far off your guess G is at any stage, you calculate the Euclidean norm of the 3×3 matrix $M - G^2$. That's what I call the error. Let's start out with the identity matrix I as our initial guess. We get a big error, 29 million. Now let's iterate. Now the error has dropped down to 7 million. It's going to keep being divided by 4 for a long time. [Continuing the iterations for a while] Now notice, we're not bad. There's our guess, there's its square, there's what we're trying to get. It's pretty close. In fact the error is less than one. [Continuing] Now the error is really small. Look at that, isn't that beautiful? So there's just no question about it. If you enter a matrix at random and it works, then that sort of settles it.

But now wait a minute, the story isn't quite finished yet. Let me just continue these iterations... Look at that! The error got bigger, and it keeps getting bigger. [Continuing] Isn't that lovely stuff?

DeGroot: What happened?

Blackwell: Isn't that an interesting question, what happened? Well, let me tell you what happened. Now you can study it theoretically and ask, should it converge? And it turns out that it will converge if, and essentially only if, your first guess commutes with the matrix M. That's what the theory gives you. Well, my first guess was I. It commutes with everything. So the procedure theoretically converges. However, when you calculate, you get round-off errors. By the way, if your first guess commutes, then all subsequent guesses will commute. However, because of round-off errors, the matrices that you actually get don't quite commute. There are two ways to do this. We could take MG^{-1} or we could have taken $G^{-1}M$. Of course, if M commutes with G, then it commutes with G^{-1} and it doesn't matter which way you do it. But if you don't calculate G exactly at some stage, then it will not quite commute. And in fact, what I have here on the computer is a calculation at each stage of the noncommutativity norm. That shows you how different MG^{-1} is from $G^{-1}M$. I didn't point those values out to you, but they started out as essentially 0, and then there was a 1 in the 15th place, and then a 1 in the 14th place, and so on. By this stage, the noncommutativity norm has built up to the point where it's having a sizable influence on the thing.

DeGroot: Is it going to diverge or will it come back down after some time?

Blackwell: It won't come back down. It will reach a certain size, and sometimes it will stay there and sometimes it will oscillate. That is, one G will go into a quite different G, but then that G will come back to the first one. You get periods, neither one of them near the truth. So that's what I mean by just playing, instead of sitting down like a serious mathematician and trying to prove a theorem. Just try it out on the computer and see if it works. [Laughs]

DeGroot: You can save a lot of time and trouble that way.

Blackwell: Yes. I expect to do more and more of that kind of playing. Maybe I get lazier as I get older. It's fun, and it's an interesting toy.

DeGroot: Do you find yourself growing less rigorous in your mathematical work?

Blackwell: Oh yes. I'm much more interested in the ideas, and in truth under not-completely-specified hypotheses. I think that has happened to me over the last 20 years. I can certainly notice it now. Jim MacQueen was telling me about something that he had discovered. If you take a vector and calculate the squared correlation between that vector and some permutation of itself, then the average of that squared correlation over all possible permutations is some simple number. Also, there was some extension of this result to k vectors. He has an interesting algebraic identity. He told me about it, but instead of my trying to prove it, I just selected some numbers at random and checked it on the computer. Also, I had a conjecture that some stronger result was true. I checked it for some numbers selected at random and it turned out to be true for him and *not* true for what I had said. Well, that just settles it. Because suppose you have an algebraic function $f(x_1, \ldots, x_n)$ and you want to find out if it is identically 0. Well, I think it's true that any algebraic function of n variables is either identically 0 or the set of x's for which it is 0 is a set that has measure 0. So you can just select x's at random and evaluate f. If you get 0, it's identically 0. [Laughs]

DeGroot: You wouldn't try even a second set of x's?

Blackwell: I did. [Laughs]

DeGroot: Getting more conservative in your old age.

Blackwell: Yes. [Laughs] I've been wondering whether in teaching statistics the typical set-up will be a lot of terminals connected to be a big central computer or a lot of small personal computers. Let me turn the interview around. Do you have any thoughts about which way that is going or which way it ought to go?

DeGroot: No, I don't know. At Carnegie-Mellon we are trying to have both worlds by having personal computers but having them networked with each other. There's a plan at Carnegie-Mellon that each student will have to have a personal computer.

Blackwell: Now when you say each student will have to have a personal computer, where will it be physically located?

DeGroot: Wherever he lives.

Blackwell: So that they would not actually use computers in class on the campus?

DeGroot: Well, this will certainly lessen the burden on the computers that are on campus, but in a class you would have to have either terminals or personal computers for them.

Blackwell: Yes. I'm pretty sure that in our department in five years we'll have several classrooms in which each seat will be a work station for a student, and in front of him will be either a personal computer or a terminal. I'm not sure which, but that's the way we're going to be in five years.

"I Wouldn't Dream of Talking about a Theorem Like That Now"

DeGroot: A lot of people have seen you lecture on film. I know of at least one film you made for the American Mathematical Society that I've seen a few times. That's a beautiful film, "Guessing at Random."

Blackwell: Yes. I now, of course, don't think much of those ideas. [Laughs]

DeGroot: There were some *minimax* ideas in there...

Blackwell: Yes, that's right. That was some work that I did before I became such a committed Bayesian. I wouldn't dream of talking about a theorem like that now. But it's a nice result...

DeGroot: It's a nice result and it's a beautiful film. Delivered so well.

Blackwell: Let's see... How does it go? If I were doing it now I would do a weaker and easier Bayesian form of the theorem. You were given an arbitrary sequence of 0's and 1's, and you were going to observe successive values and you had to predict the next one. I proved certain theorems about how well you could do against every possible sequence. Well, *now* I would say that you have a probability distribution on the set of all sequences. It's a general fact that if you're a Bayesian, you don't have to be clever. You just *calculate.* Suppose that somebody generates an arbitrary sequence of 0's and 1's and it's your job after seeing each finite segment to predict the next coordinate, 0 or 1, and we keep track of how well you do. Then I have to be clever and invoke the minimax theorem to devise a procedure that asymptotically does very well in a certain sense. But now if you just put a prior distribution on the set of sequences, any Bayesian knows what to do. You just calculate the probability of the next term being a 1 given the past history. If it's more than 1/2 you predict a 1, if it's less than 1/2 you predict a 0. And that simple procedure has the corresponding Bayesian version of all the things that I talked about in that film. You just know what is the right thing to do.

DeGroot: But how do you know that you'll be doing well in relation to the reality of the sequence?

Blackwell: Well, the theorem of course says that you'll do well for all sequences except a set of measure zero according to your own prior distribution, and that's all a Bayesian can hope for. That is, you have to give up something, but it just makes life so much *neater*. You just know that this is the right thing to do.

I encountered the same phenomenon in information theory. There is a very good theory about how to transmit over a channel, or how to transmit over a sequence of channels. The channel may change from day to day, but if you know what it is every day, then you can transmit over it. Now suppose that the channel varies in an arbitrary way. That is, you have one of a finite set of channels, and every day you're going to be faced with one of these channels. You have to put in the input and a guy at the other end gets an output. The question is, how well can you do against all possible channel sequences?

You don't really know what the weather is out there, so you don't know what the interference is going to be. But you want to have a code that transmits well for all possible weather sequences. If you just analyze the problem crudely, it turns out that you can't do *anything* against all possible sequences. However, if you select the code in a certain random way, your overall error probability will be small for each weather sequence. So you see, it's a nice theoretical result but it's unappealing. However, you can get exactly the same result if you just put a probability distribution on the sequences. Well, the weather could be any sequence, but you expect it to be sort of this way or that. Once you put a probability distribution on the set of sequences, you no longer need random codes. And there is a deterministic code that gives you that same result that you got before. So either you must behave in a random way, or you must put a probability distribution on nature.

[Looking over a copy of his paper, BLACKWELL, D., BREIMAN, L. and THOMASIAN, A. J., "The capacities of certain channel classes under random coding," *Ann. Math. Statist.* **31**, 558–567, 1960] I don't think we did the nice easy part. We behaved the way Wald behaved. You see, the minimax theorem says that if for every prior distribution you can achieve a certain gain, then there is a random way of behaving that achieves that gain for every parameter value. You don't need the prior distribution; you can throw it away. Well, I'm afraid that in this paper, we invoked the minimax theorem. We said, take any prior distribution on the set of channel sequences. Then you can achieve a certain rate of transmission for that prior distribution. Now you invoke the minimax theorem and say, therefore, there is a randomized way of behaving which enables you to achieve that rate against every possible sequence. I now wish that we had *stopped* at the earlier point. [Laughs] For us, the Bayesian analysis was just a preliminary which, with the aid of the minimax theorem, enabled us to reach the conclusions we were seeking. That was Wald's view and that's the view that we took in that paper. I'm sure I was already convinced that the Bayes approach was the right approach, but perhaps I deferred to my colleagues.

DeGroot: That's a very mild compromise. Going *beyond* what was necessary for a Bayesian resolution of the problem.

Blackwell: That's right. Also, I suspect that I had Wolfowitz in mind. He was a real expert in information theory, but he wouldn't have been interested in anything Bayesian.

DeGroot: What about the problem of putting prior distributions on spaces of infinite sequences or function spaces? Is that a practical problem and is there a practical solution to the problem?

Blackwell: I wouldn't say for infinite sequences, but I think it's a very important practical problem for large finite sequences and I have no idea how to solve it. For example, you could think that the pattern recognition problem that I was talking about before is like that. You see an image on a TV screen. That's just a long finite sequence of 0's and 1's. And now you can ask how likely it is that that sequence of 0's and 1's is intended to be the figure 7, say. Well, with some you're certain that it is and some you're certain that it isn't, and with others there's a certain probability that it is and a probability that it isn't. The problem of describing that probability distribution is a very important problem. And we're just not close to knowing how to describe probability distributions over long finite sequences that correspond to our opinions.

DeGroot: Is there hope for getting such descriptions?

Blackwell: I don't know. But again it's a statistical problem that is not going to be solved by professors of statistics in universities. It might be solved by people in artificial intelligence, or by researchers outside universities.

"Just Tell Me One or Two Interesting Things"

DeGroot: There's an argument that says that under the Bayesian approach, you have to seek the optimal decision and that's often just too hard to find. Why not settle for some other approach that requires much less structure and get a reasonably good answer out of it, rather than an optimal answer?

Especially in these kinds of problems where we don't know how to find the optimal answer.

Blackwell: Oh, I think everybody would be satisfied with a reasonable answer. I don't see that there's more of an emphasis in the Bayesian approach on optimal decisions than in other approaches. I separate Bayesian inference from Bayesian decision. the inference problem is just calculating a posterior distribution, and that has nothing to do with the particular decision that you're going to make. The same posterior distribution could be used by many different people making different decisions. Even in calculating the posterior distribution, there is a lot of approximation. It just can't be done precisely in interesting and important cases. And I don't think anybody who is interested in applying Bayes method would insist on something that's precise to the fifth decimal place. That's just the conceptual framework in which you want to work, and which you want to approximate.

DeGroot: That same spirit can be carried over into the decision problem, too. If you can't find the optimum decision, you settle for an approximation to it.

Blackwell: Right.

DeGroot: In your opinion, what have been the major breakthroughs in the field of statistics or probability through the years?

Blackwell: It's hard to say... I think that theoretical statistical thinking was just completely dominated by Wald's ideas for a long time. Charles Stein's discovery that $\overline{X}$ is inadmissible was certainly important. Herb Robbin's work on empirical Bayes was also a big step, but possibly in the wrong direction.

You know, I don't view myself as a statesman or a guy with a broad view of the field or anything like that. I just picked directions that interested me and worked in them. And I have had fun.

DeGroot: Well, despite the fact that you didn't choose the problems for their impact or because of their importance, a lot of people have gained a lot from your work.

Blackwell: I guess that's the way scholars *should* work. Don't worry about the overall importance of the problem; work on it if it looks interesting. I think there's probably a sufficient correlation between interest and importance.

DeGroot: One component of the interest is probably that others are interested in it, anyway.

Blackwell: That's a big component. You want to tell somebody about it after you've done it.

DeGroot: It has not always been clear that the published papers in our more abstract journals did succeed in telling anybody about it.

Blackwell: That's true. But if you get the fellow to give a lecture on it, he'll probably be able to tell you something about it. Especially if you try to restrict him: Look, don't tell me everything. Just tell me *one or two* interesting things.

DeGroot: You have a reputation as one of the finest lecturers in the field. Is that your style of lecturing?

Blackwell: I guess it is. I try to emphasize that with students. I notice that when students are talking about their theses or about their work, they want to tell you everything they know. So I say to them: You know much more about this topic than anybody else. We'll never understand it if you tell it all to us. Pick just one interesting thing. Maybe two.

DeGroot: Thank you, David.

A native of Canada, Cecil J. Nesbitt did his undergraduate and graduate work at the University of Toronto, where he received his Ph.D. in 1937 as a student of Richard Brauer. After a postdoctoral year at the Institute for Advanced Study, he took a position at the University of Michigan and remained there until his retirement in 1980. His early research was in algebra, but at both Toronto and Michigan his primary bent was to actuarial mathematics. With Carl H. Fischer he led a flourishing actuarial program in the Mathematics Department at Michigan, while publishing actively and serving the Society of Actuaries in various capacities. His main work has been in the areas of pension funding and social insurance.

Personal Reflections on Actuarial Science in North America from 1900

CECIL J. NESBITT

1. INTRODUCTION

At the outset, it should be made clear that this article does not pretend to be a definitive history of actuarial science developments in North America since the beginning of the century. Deadlines, and my own available time and energy, do not permit such an undertaking, worthy as it may be. Instead I shall draw on memories of almost 60 years as an actuarial student, teacher, practitioner, and researcher, to indicate actuarial highlights of that period, and also sources for further review if readers become so inclined. Such readers should turn first to *Actuarial Mathematics* (Proc. Symp. Appl. Math. 35, 1986) and peruse it alongside this article to gain detailed, introductory overviews of the diverse actuarial models that will be mentioned here. The non-exhaustive list of references at the end of the article is selected to aid such review by pointing the way to other more complete lists in regard to various topics mentioned herein. The body of ideas, known and unknown, is infinite, and even in one special area, such as the intersection of actuarial science and mathematics, can be covered only by broad strokes.

Actuarial science has a major role in the guidance of financial security systems, developed to protect individuals and groups against a multiplicity of risks such as impairment of health, premature death, destruction of property, and extended old age. The systems may range from self-insured groups to national programs of social security. Some of these systems operate on an international basis, and more such development may lie in the future. These systems during my life have made much progress despite economic, financial, and political disturbances and disasters. The systems have been facing fast-growing environmental hazards, and military potentialities of incredible magnitude. Actuarial science has a role to play, as do all fields, in finding viable equilibria in a fast-changing world.

In the following section, there will be brief discussion of the main fields of knowledge on which actuarial science draws. Those to be mentioned are mathematics, statistics, probability, accounting, computer science, demography, economics, finance theory, law and medical science. Some of these fields were relatively undeveloped at the beginning of this century. The section on sources will be followed by one on distinctively (although not exclusively) actuarial theories. These are: estimation of mortality and other rates, life tables (now broadened to survival models), graduation theory, risk theory, credibility theory, actuarial finance theory, life insurance mathematics including growth models and stochastic models of life contingencies. The application of these theories to various fields of practice will come next, with a final summary overview.

2. Sources of Actuarial Science

In a broad sense, all portions of actuarial science relate to some form of mathematical theory or application. The theory may be relatively elementary, but the application may be extremely detailed and numerical. Of prime importance are the actuarial assumptions from which the mathematical model is developed. For short-term insurances, there may be a large volume of current data for statistical and probability analysis. Such current data may also be available for long-term insurances, as for example, whole life insurance, pension systems, and social security, but must be extended by projection factors to guide the future growth of the financial security system.

From the data analysis, one may estimate probabilities needed for the model of the system. The mathematical model may draw heavily on probability theory for its structure, or for the longer term it may be deterministic in character, following out the consequences of assumed rates of growth and eligibility for benefits. Statistical and probability theories, which have grown rapidly in this century, are playing an expanding role in actuarial science.

Another main source of actuarial science is the mathematics of finance. Until recent years, this has been an elementary theory, defining various discrete and continuous rates of interest, and utilizing a constant rate compound interest model to calculate present and accumulated values of series of payments. Still more recently, the turbulence of financial markets, has led to simulation studies being conducted by committees of the Society of Actuaries under the leadership of C. L. Trowbridge and D. D. Cody (Cody, 1987). About the same time there appeared Phelim Boyle's "Immunization under stochastic models of the term structure" (Boyle, 1978). Also, finance theory, with application to the pricing of options, has been advancing strongly. (For a comprehensive view of this last work, see Pedersen, Shiu and Thorlacius, forthcoming, and D'Arcy and Doherty, 1988.)

Computer science has greatly empowered actuaries in regard to: estimation of rates or probabilities, the calculation of premiums or contributions, the projection of future benefit outgo and of premium or contribution income, and the corresponding accumulation of reserves. A notable example is provided by the annual actuarial projections for old-age, survivors and disability insurance in the United States (see Andrews and Beekman, 1987).

Other bodies of knowledge or practice which impinge on actuarial practice are indicated by the examples below:

Accounting. To get a feel for some of the discussion preceding (Financial Accounting Standards No. 87, 1985), see E. L. Hicks and C. L. Trowbridge, *Employer accounting for pensions* (Hicks and Trowbridge, 1985).

Demography. A major actuarial concern here is in regard to the development of national life tables. An early reference was H. H. Wolfenden's *Population statistics and their compilation* (Wolfenden, 1925). This was followed by M. Spiegelman's *Introduction to demography* (Spiegelman, 1955). One of the current references in the actuarial education syllabus is *Demography through problems* (Keyfitz and Beekman, 1984). See also A. Wade's *Social security area population projections* (Wade, 1988) and J. Wilkins' *OASDI long-range beneficiary projection, 1987* (Wilkins, 1988).

Economics. Recently, the Office of the Actuary, Social Security Administration, has published Actuarial Study No. 101, *Economics projections for OASDHI cost and income estimates*: 1987 (Goss, 1988).

Law and Regulation. Life insurance companies are supervised by the State Insurance Departments in the United States, and by federal and provincial departments in Canada. In the United States, the Employee Retirement Income Security Act of 1974 and subsequent legislation have been a major influence on pension funds.

Medicine. This last comes to the fore in the underwriting of life insurance, in health insurance, in the projection of future mortality improvements, and in regard to current epidemics such as AIDS.

3. Some Actuarial Theories

This section overviews some of the theories used by actuaries in their professional practices.

3.1. Estimation of Mortality and Other Rates

From chapter III of J. S. Elston's *Sources and characteristics of the principal mortality tables* (Elston, 1932), we have the quotation:

"United States Life Tables, 1910

United States Life Tables 1890, 1901, 1910 and 1901–1910

These tables are the first of any scientific value prepared by the U. S. Government from census returns. When the census of 1910 was taken, the Bureau of the Census called into consultation a committee of The Actuarial Society of America, and this committee gave general advice with reference to the taking of the census, the tabulation of the data, and the preparation of life tables. Although not all the committee's recommendations were followed, these tables, which were prepared under the supervision of Professor James W. Glover, mark a notable epoch in the history of mortality."

In all, 69 life tables were prepared from the censuses, and the death statistics of the ten original registration states and the District of Columbia (Glover, 1916 and 1921). In view of the status of computing facilities in the 1910–1920 decade, the preparation and publication of these tables was a monumental task.

Another mathematician who has been a principal and innovative consultant for U. S. Life Tables for 1939–1941, and for subsequent intervals around the decennial censuses, is T. N. E. Greville (Greville, 1946). The problems he met and solved for the 1939–1941 Tables led him to many later developments in theories of interpolation, graduation, splines, generalized inverses of matrices, and life tables, as indicated in bibliographies of (Bowers et al, 1986), (London, 1985), and (Shiu, 1984 and 1986).

Turning now to insured lives mortality, we have as an early reference, *Construction of mortality tables from the records of insured lives* (Murphy and Papps, 1922). This was followed by a number of papers in the actuarial literature which plateaued in the texts (Gershenson, 1961) and (Batten, 1978). Concurrently, beginning in 1934, committees of the Society of Actuaries, or its predecessors, have published a series of annual *Reports of mortality, morbidity and other experience* based on data mainly contributed by a number of life insurance companies (Society of Actuaries, 1984).

In regard to annuitant mortality, a landmark paper by W. A. Jenkins and E. A. Lew developed the idea of scales of projection factors to allow for future mortality improvements (Jenkins and Lew, 1948).

Much statistical work has gone into the estimation of mortality rates from clinical data, and the subject has broadened to that of survival models. Simultaneously, the computer evolution has greatly facilitated the calculation of exposed to risk from the records of the individuals observed in the estimation process. These new approaches, are presented in Dick London's text, *Survival models and their estimation* (London, 1988). See also J. D. Broffitt's paper "Maximum likelihood alternatives to actuarial estimators of mortality rates" (Broffitt, 1984). It should be added that census methods used in the estimation of population mortality also have application to estimation of mortality of insured lives or of pension fund participants. There are, of course, distinctive differences in the data for the various studies.

3.2. Graduation Theory

This topic is concerned with the systematic revision of estimates of series of rates, in particular, those to be used as bases for survival models. A fine survey is given by E. S. Shiu in the 1986 *Proceedings*, volume 35. His abstract is as follows: "Graduation is the process of obtaining from an irregular set of observed values, a corresponding smooth set of values consistent in a general way with the observed values. This is a survey of various methods of graduation used by actuaries."

Some early work goes back to E. L. DeForest in the 1870s which was later brought to life by H. H. Wolfenden in (Wolfenden, 1925). R. Henderson, who was prominent in the early history of both the actuarial and the mathematical professions, prepared the monograph *Mathematical theory of graduation* (Henderson, 1938). Let us pause for a moment to pay tribute to this distinguished man.

His life spanned from 1871 to 1942. He graduated from the honors mathematics program of the University of Toronto in 1891. He became a Fellow of the Actuarial Society of America and of the Casualty Actuarial Society, and was elected president of the former organization. Robert Henderson rose to become actuary of the Equitable Life Assurance Society. He served as trustee of the Mathematical Association of America and of the American Statistical Association. In addition, "For a number of years, Mr. Henderson served as a member of the Board of Trustees of the American Mathematical Society. He felt the keenest interest in the place which the Society was taking in scientific progress and lent earnest assistance to the raising of funds in order that its work might continue unimpaired despite the economic difficulties of recent years. He also served from 1935 until shortly before his death as a Director of the Teachers Insurance and Annuity Association." (Quotation from the obituary for Robert Henderson in *Transactions of the Actuarial Society of America*, **43** (1942)).

We should also note that Robert Henderson delivered the second Gibbs Lecture on "Life insurance as a social science and as a mathematical problem" (Henderson, 1925). The entire principal of his estate was received by the American Mathematical Society in 1961 for its Endowment Fund.

Another notable author was C. A. Spoerl, a summa cum laude graduate of Harvard University. See, for instance, his paper "Whittaker-Henderson graduation formula A" (Spoerl, 1937).

For a number of years following 1950, a new monograph, *Elements of graduation*, by M. D. Miller served as education reference (Miller, 1946). Meanwhile, a succession of papers were coming from the pen of T. N. E. Greville, which may be well seen in the book, *Selected papers of T. N. E. Greville, 1984*. These have influenced the work of G. S. Kimeldorf and D. A. Jones in "Bayesian graduation" (Kimeldorf and Jones, 1967), and E. S. Shiu in "Minimum-R_z moving-weighted-average formulas" (Shiu, 1984). Another approach is exemplified by D. R. Schuette's "A linear programming approach to graduation", (Schuette, 1978).

3.3. Risk Theory

We consider first the simpler case of short-term insurances. Here one may be concerned with the distribution of total claims in a given period for a given portfolio of insurance policies. The approach in individual risk theory is to set up a random variable

$$Y_j = I_j B_j \quad j = 1, 2, \ldots, n \tag{3.3.1}$$

for each of the n insurance policies in the portfolio under consideration. Here I_j is 1 if policy j leads to a claim and is 0 otherwise; B_j is the amount of such a claim, given that it occurs. On the assumption that I_j, B_j, $j = 1, 2, \ldots, n$ are mutually independent, one proceeds to approximate the distribution of aggregate claims for the period, that is, the distribution of

$$S_{IR} = \sum_{j=1}^{n} Y_j. \tag{3.3.2}$$

For this we need knowledge of the probability that $I_j = 1$ and of the distribution of B_j, for each j.

In the collective risk model, the basic concept is that of a random process that generates claims for a portfolio of policies. This process is in terms of a portfolio as the whole rather than in terms of the individual policies. Let N be the random number of claims for a portfolio of policies in the given period. If X_1 is the random amount of the first claim, X_2, the random amount of the second claim, and so on, then

$$S_{CR} = X_1 + X_2 + \cdots + X_N \tag{3.3.3}$$

is the random amount of aggregate claims. The random variable N is referred to as frequency of claims and the random variables X_j measures the size of claims. In order to proceed, one makes the assumptions that:

1. $X_1, X_2 \ldots$ are identically distributed.
2. The random variables $N, X_1, X_2, \ldots$ are mutually independent.

An overview of risk theory, with emphasis on the collective theory, is given by H. Panjer's, "Models in risk theory" (Panjer, 1986). See also H. Gerber's *An introduction to mathematical risk theory* (Gerber, 1979). Both of these references provide bibliographies which indicate the historical development of risk theory. A major figure is H. L. Seal, as the bibliographies attest (see Seal, 1969). The reader interested in connecting the two approaches to risk theory for short-term insurances is referred to Section 13.5 of (Bowers et al, 1986). For information about estimating the probability distribution of the X_j's one can refer to S. A. Klugman's "Loss Distributions" (Klugman, 1986), or to the book by R. V. Hogg and S. A. Klugman with the same title (Hogg and Klugman, 1984).

An early discussion of risk theory for individual insureds under long-term life insurance and annuity contracts was given by W. O. Menge, a later-year colleague of J. W. Glover, in the paper "A statistical treatment of actuarial functions" (Menge, 1937). An extensive development of individual risk theory for such contracts is a major theme of (Bowers et al, 1986).

3.4. Credibility Theory

Since the early papers of F. A. Perryman (Perryman, 1937) and A. L. Bailey (Bailey, 1950), an extensive literature has grown up. Successive overviews of this literature have been presented by P. M. Kahn in 1967, 1968, 1975 and 1986. The reader is referred to this last paper, and its bibliography (Kahn, 1986).

In brief, credibility theory applies mainly to short-term insurances such as group life insurance, or those in various casualty lines, or the year-to-year risks under individual life insurances. The theory studies the revision of premium rates in the light of current claim experience. To quote from (Kahn, 1986):

"In the classical approach the actuary must first determine the size of the experience which warrants full credibility, i.e. a credibility factor $Z(t)$ of 1, where t measures the size of the exposure or insurance experience which generated the level of chaims x. The next step is to determine partial weights, or partial credibility factors for some smaller groups. Then the adjusted estimate of claims may be expressed as

$$Z(t)x + [1 - Z(t)]m(t) \tag{3.4.1}$$

where $m(t)$ is the prior estimate of expected claims."

A. L. Mayerson's paper "A Bayesian view of credibility" (Mayerson, 1964) was a stimulus for much further research. In 1975, J. C. Hickman drew a distinction between classical theory where the parameters of the claims process are considered as fixed constants, and the newer theories where the parameters are themselves random variables (Hickman, 1975). As with much actuarial theory, the newer concepts of credibility must undergo validation and refinement in actual insurance experience.

3.5. Mathematics of Compound Interest

In Section 2, Sources of Actuarial Science, reference has been made already to the mathematics of finance and the direction in which it is headed. Here, and in the next section, we refer to some of the classical actuarial mathematics texts. For further information on these texts, and how they became incorporated into the education and examination processes of the profession in North America, the reader is referred to the chapter on actuarial education in E. J. Moorhead's forthcoming 1809–1979 history of the actuarial profession, entitled *Our yesterdays* (Moorhead, forthcoming). This chapter, from a different viewpoint, gives insight about the professors and universities that have contributed to actuaral education and science.

From the University of Toronto, we have had M. A. Mackenzie's *Interest and bond values* (Mackenzie, 1917), and N. E. Sheppard and D. C. Baillie's *Compound interest* (Sheppard and Baillie, 1960). From the University of Michigan, there has appeared M. V. Butcher and C. J. Nesbitt's *Mathematics of compound interest* and, as one of the leading more elementary texts, P. R. Rider and C. H. Fischer's *Mathematics of investment* (Rider and Fischer, 1951). Since 1970, the Society of Actuaries has benefitted from S. G. Kellison's *The Theory of Interest*. The newest text in the English language is J. J. McCutcheon and W. F. Scott's *An introduction to the mathematics of finance* (McCutcheon and Scott, 1986). These texts treat basic finance concepts which go far back into the mists of history of civilization.

3.6. Mathematics of Life Contingencies

As a major part of the core of actuarial mathematics is the subject of this subsection, it is in order to discuss the principal textbooks that have appeared from time to time in the English language. If one refers to (Moorhead, forthcoming), one reads about such early works as R. Price's *Observations on reversionary payments* (1771), William Morgan's *The doctrine of annuities and assurances on lives and survivorships* (1779), Francis Bailey's *Doctrine* (1812–1813), Joshua Milne's *Treatise* (1815), and David Jones' *Value of annuity and reversionary payments* (1843). My own acquaintance goes back to G. King's *Institute of actuaries textbook, Part II* (King, 1902), and I endured through examinations on E. F. Spurgeon's *Life contingencies* (Spurgeon, 1922).

It is probably little known by now that C. H. Fischer and myself were invited in the late 1940s to undertake for the Society of Actuaries a new textbook. At that time concepts about the probability distributions of random variables were not well organized, at least in my mind, but nevertheless, it seemed to me then to be the way to proceed. The Society was not ready for what appeared to be a novel approach, and turned the project over to C. W. Jordan who by 1952 produced a book which served the profession well for over thirty years (Jordan, 1952). His book began with the notion of survival function but soon settled down to deterministic formulas. This was followed by P. F. Hooker and L. H. Longley-Cook's two-volume text, *Life and other contingencies* (Hooker and Longley-Cook, 1953, 1957). This text had brief discussion of variance around the expected values, as did also the successor book, A. Neill's *Life contingencies* (Neill, 1977). In 1978, the author team of N. L. Bowers, H. U. Gerber, J. C. Hickman, D. A. Jones and C. J. Nesbitt began work on a new textbook, entitled *Actuarial mathematics*, which emerged in final form by 1986 (Bowers et al, 1986). An enlightening overview is given by J. C. Hickman's paper "Updating life contingencies" (Hickman, 1988). This textbook goes way beyond what Fischer and I attempted forty years earlier. It intertwines individual risk theory and individual life insurance mathematics, and introduces collective risk theory, with various practical applications in group insurance and reinsurance. It ends with a chapter on "Theory of pension funding," using a mathematical deterministic model, generalizing the work of C. L. Trowbridge in "Fundamentals of pension funding" (Trowbridge, 1952). The extensive bibliography lists the many authors whose works have helped to shape the text.

For some time, a new direction in actuarial mathematics has been appearing in Europe. This is exemplified by J. Hoem's "The versatility of the Markov chain as a tool in the mathematics of life insurance" (Hoem, 1988), and by H. Wolthuis' doctoral thesis, *Savings and risk processes in life contingencies* (Wolthuis, 1988). To a considerable extent, this direction runs counter to American practice which models separately each state that an insured may enter, for example, the state of disability, rather than use an integrated model covering all states, and transfers among them. It remains for the future to determine the usefulness of the integrated models. One indication is that the work of M. J. Cowell and W. H. Hoskins (Cowell and Hoskins, 1987), and of H. J. Panjer (Panjer, 1988) on projections regarding the AIDS epidemic, and recent work of J. Beekman in modeling decline of activity of the aged, are related thereto.

Meanwhile, actuaries like myself who are interested in the long-term guidance of pension funds and social security, are prone to use what I term mathematical deterministic (or growth) models, and to utilize a range of actuarial assumptions which are monitored regularly. This viewpoint is reflected in B. N. Berin's *The Fundamentals of pension mathematics* (Berin, 1978), and in the long-range projections for U.S. Social Security (Andrews and Beekman, 1987). This approach is also exploited in A. W. Anderson's *Pension mathematics for actuaries* (Anderson, 1985).

Another example of theory developments which have not gained much usage yet in practice is given by W. S. Bicknell's thesis "Premiums and reserves in multiple decrement theory (Bicknell and Nesbitt, 1956). This discusses three systems for premiums and reserves for the case of multiple forms of termination and benefits related thereto, as in pension plans. The second and third systems involved somewhat complex composition of the actuarial bases for the several benefits. This, we have noted, is not the American way in practice. The third system, which goes back to Alfred Loewy, has considerable possibilities, but has practical and throretical subtleties which have been explored by (Schuette and Nesbitt, forth coming in *ARCH*).

It seems fitting to end this subsection with a tribute to H. L. Rietz who was from 1918 to 1962 influential in the development of mathematical statistics and actuarial science at the University of Iowa. He served as vice president of the American Institute of Actuaries, 1919–1920, as president of the Mathematical Association of America in 1924, as vice president of the American Statistical Association in 1925, and as vice president of the American Association for the Advancement of Science in 1929. He was the first president of the Institute of Mathematical Statistics, organized in 1935, and the 1943 volume of the *Annals of Mathematical Statistics* was dedicated to him on the occasion of his retirement. Among his doctoral students was C. H. Fischer, my long-time colleague at the University of Michigan.

4. Applications

From 1900 through 1987, life insurance in force in the United States has grown from a little over \$7.5 billion to almost \$7.5 trillion. Some \$3 trillion of this latter amount is classified as group insurance, a form which did not exist in 1900. This period saw the development of retirement income policies, variable life insurance and several forms of flexible life insurance. Discussions of these may be found in chapter 16 of (Bowers et al, 1986). Some initial papers were authored by E. G. Fassel (1930), J. C. Fraser, W. N. Miller and C. M. Sternhell (1969), W. L. Chapin (1976), and S. A. Chalke and M. F. Davlin (1983). (See bibliography in Bowers et al, 1986 for references.)

During the same period, the growth of pension funds is indicated by the increase from about $20 billion of assets in 1900 to about $2 trillion in 1986. A notable development during this period was the concept of variable annuities and the formation in 1952 of the College Retirement Equities Fund (CREF). The actuarial basis for that fund was pioneered by R. M. Duncan's "A retirement system granting unit annuities and investing in equities" (Duncan, 1952). I recall one lunchtime where Carl Fischer and I pressed Robert Duncan on the theory of dollar averaging for accumulating purchases of units by a series of regular contributions to CREF. When asked what would happen if the stock market collapsed, he thought for a moment and then with a smile said "They might not be worth very much, but you would have a lot of accumulation units."

For further information about the Teachers Insurance and Annuity Association (TIAA) and CREF, see my paper, "On the performance of pension plans" (Nesbitt, 1986). In particular, note the graded benefit annuity option which has in recent years become available from TIAA.

This section concludes with a few comments on the Old Age, Survivors and Disability Insurance (OASDI) system, popularly called Social Security but this latter also embraces the insurances under Medicare. OASDI is an extremely large system with annual benefit outgo now at the level of $235 billion, and projected level of $8 trillion by year 2045 under moderate growth assumptions (Annual Report, 1988). With good reason, the actuaries of the System prefer to project benefit outgo as a percent of projected taxable payroll for the System. On this basis, projected OASDI outgo in 2045 is 16.25 percent of taxable payroll. The actuarial guidance of this huge system is a major challenge for the actuarial profession.

An ackowledged leader in such guidance has been R. J. Myers who has written very extensively on Social Security (see, for instance, Myers, 1985). He set the pattern for the short-range and long-range projections, the processes for which are continuously evolving. An overview of these processes is given in (Andrews and Beekman, 1987) and (Annual Report, 1988). A recently formed National Academy of Social Insurance, with Alicia Munnell of the Federal Reserve Bank of Boston as president, will form a common ground for persons from different fields who are interested in Social Security. OASDI developments over the past fifty years have been of major importance as a foundation of benefits to be supplemented by nonfederal life insurance and pension-funding, and should remain so.

5. Concluding Comments

This paper has been written mainly as personal reactions to actuarial developments of the last fifty years. As I got further into it, I had more and more occasion to refer to (Bowers et al, 1986) and (*Proceedings*, 1986). For the reader interested in going beyond this paper, I recommend a perusal of the latter reference. I draw special attention to J. C. Hickman's introduction, and to his paper "Updating life contingencies" which enlarges on the concepts underlying the textbook (Bowers et al, 1986).

I have tried to make at least one reference to many, but by no means all, contributors to actuarial science in North America. Most of the omitted names may be found in the bibliographies of (Bowers et al, 1986) and (*Proc. Symp. Appl. Math.*, 1986). Some omissions relate to young men whose work is in process of recognition, and some to special areas of expertise. I have depended on the useful bibliographies in the various references cited to provide a more complete picture of developments, including much work not cited here.

Two references I wish to add here are to W. O. Menge and J. W. Glover's *An introduction to the mathematics of life insurance* (Menge and Glover, 1935), and its later revision, *Mathematics of life insurance* (Menge and Fischer, 1965). In a very real sense, these helped to clarify life insurance actuarial practice.

While risk theory and credibility theory are major elements in the actuarial mathematics of nonlife insurance, beyond these two theories and reference to (D'Arcy and Doherty, 1988) no attempt has been made to cover that field further, as it has not been part of my experience. A similar remark applies for the large field of health insurance on an individual, group or national basis.

It should be recorded that there are some actuaries who have realized that we have undergone in the last forty years the risk of incredible destruction by nuclear war. Among these was Edmund C. Berkeley who included this topic in his address "Society, computers, thinking and actuaries" to the 16th Annual Actuarial Research Conference, University of Manitoba (see Berkeley, 1982). In papers presented to the 22nd and 23rd annual actuarial research conferences, I have indicated a simple model for recognising nuclear holocaust hazard, and its pervasive effect on all longterm actuarial calculations such as those regarding average length of life, or for mortgage amortization over a term of years (Nesbitt, 1987 and 1988). This is the actuarial science that must be communicated to protect life, and to counter the weight of science that could destroy life. These tasks really fall upon teachers in all fields, but actuarial science should do its part.

Finally, this review, which was mainly retrospective but had some updating and prospective aspects, has encouraged me about this century's progress in actuarial science, and has increased my awareness of developments to come.

ADDENDUM

Here I present some information and reflections on the education and examination of actuaries. This could be a very large assignment, and I shall resolve it mainly by pointing to sources of information. An immediate problem is that there is not just one, but four, actuarial organizations directly involved in actuarial education and examination, and three others that cosponsor some or all of the examinations. The two oldest organizations, the Society of Actuaries and the Casualty Actuarial Society, have leading roles but have been supplemented by the Joint Board for the Enrollment of Actuaries (a unit of the U.S. Department of the Treasury), and the American Society of Pension Actuaries. Cosponsoring organizations are the American Institute of Actuaries, the Canadian Institute of Actuaries (CIA has a different connotation in Canada than in the U.S.), and the Conference of Actuaries in Public Practice. This may seem confusing but there is considerable coordination among the seven bodies through the Council of Presidents, and also through overlapping memberships. My own experience has been mainly with the Society of Actuaries, and I shall use the Society as my information source. The catalogs of the Society list the addresses of all seven organizations.

A second problem is that culminating in the years since 1985, there has been a restructuring of the Society's education process into a Flexible Education System (FES), and a follow-up by proposed Future Education Methods (FEM). The multi-membered Education and Examination Committee distributed two white papers, on FES in 1986, and on FEM in 1987, setting forth the proposed changes and their rationale. As Vice President for Research and Studies in 1985–1987, I witnessed the presentation of these documents and both the general support and the counter-reactions that they gathered. The new emphasis is on education that can adapt itself to our fast-changing world and that achieves a better balance with the discipline of the actuarial examinations.

FES is now in place and some steps have been taken in regard to FEM. These are reflected in the booklets, 1989 Associateship Catalog, and Spring 1989 Fellowship Catalog, where associateship is the first level and fellowship is the second level of qualification for membership in the Society. Both booklets state the following:

Principles Underlying the Education and Examination System

The Society of Actuaries administers a series of self-study courses and examinations leading to Associateship and Fellowship. The principles underlying the Society's education and examination system are the following:

(1) To provide the actuary with an understanding of fundamental mathematical concepts and how they are applied, with recognition of the dynamic nature of these fundamental concepts in that they must remain up-to-date with developments in mathematics and statistics;

(2) To provide the actuary with an accurate picture of the socio-demographic, political, legal and economic environments within which financial arrangements operate, along with an understanding of the changing nature and potential future directions of these environments;

(3) To expose a broad range of techniques that the actuary can recognize and identify as to their application and as to their inherent limitations, with appropriate new techniques introduced into this range as they are developed;

(4) To expose a broad range of relevant actuarial practice, including current and potential application of mathematical concepts and techniques to the various and specialized areas of actuarial practice; and

(5) To develop the actuary's sense of inquisitiveness so as to encourage exploration into areas where traditional methods and practice do not appear to work effectively."

Under FES, a number of self-study courses are available, each providing a certain number of credits. Completion of the Series 100 requirements now satisfies the education requirements for the Associate of the Society of Actuaries (ASA) designation. A candidate must obtain 200 units of credit prior to 1995 for courses listed in Table A to satisfy the Series 100 requirements.

Table A. Course Description

Course	Description	Credits	Type
100	Calculus and Linear Algebra	30	Required
110	Probability and Statistics	30	Required
120	Applied Statistical Methods	15	Required
130	Operations Research	15	Elective
135	Numerical Methods	10	Elective
140	Mathematics of Compound Interest	10	Required*
141	EA-1, Segment A	10	Required*
150	Actuarial Mathematics	40	Required
151	Risk Theory	15	Required
160	Survival Models	15	Required
161	Mathematics of Demography	10	Elective
162	Construction of Actuarial Tables	10	Elective
165	Mathematics of Graduation	10	Elective

Each course is designated as required or elective. A candidate must obtain 155 credits in "required" courses and 45 credits in "elective" courses to satisfy the Series 100 requirements for this catalog.

Credit for courses 140, 150, and 151 must be obtained by examinations offered by the Society of Actuaries. Credit for course 141 must be obtained by passing EA-1, Segment A of the Enrolled Actuary (EA) Examinations. Credit for all other courses must be obtained by examinations offered by the Society of Actuaries or by an alternative method which has been approved by the Board of Governors. For fall 1988 and spring 1989, credit for course 100 may be obtained by an alternative method (an appropriate score on the Graduate Record Examination Mathematics Test)."

While each 10 credits usually implies one hour of multiple-choice examination, there are exceptions. Course 140 has a one-and-one-half-hour examination, and course 150 has a four-and-one-half-hour examination split into two sessions, and including some written-answer questions.

The written-answer examinations I took years ago had algebra based on the classical Hall and Knight textbook, had analytic geometry and calculus together, and scarcely touched linear algebra. Probability was mainly combinatorics based on Whitworth's Choice and Chance, and statistics was at a precalculus descriptive level. Now course 110 includes topics among those proposed for a one-year college course in probability and statistics by the Committee on the Undergraduate Program in Mathematics. Course 120 covers analysis of variance, regression analysis and time-series analysis which were largely omitted from the syllabus of my examination-writing years. Course 130 on linear programming, project scheduling, dynamic programming, relates to topics that came to the fore during World War II.

Course 135, Numerical Methods, replaces the former examination on finite differences. The finite (as opposed to the infinitesimal) calculus was one of my teaching joys. It was always a pleasure to define divided differences, proceed to the Lagrange interpolation formula with remainder, relate divided differences under prescribed conditions to derivatives at intermediate points, and pull out Newton's divided difference interpolation formula with remainder, and as special cases obtain Taylor's series and the various classical polynomial interpolation formulas, all with remainders. One then was set to make applications to summation, approximate integration, and difference equations. Now, the impact of computers has greatly expanded the subject to modern numerical analysis with its algorithmic approach. This is reflected in course 135 which covers iteration, interpolation, numerical integration and linear systems.

---

*Candidates must receive credit for either course 140 or 141 but will not receive credit for both.

Compound interest theory, the subject of course 140, while benefitting from some refinement of basic concepts, and from the enormous improvement in computing facilities, has been well established for many decades. Course 141, administered by the Joint Board for the Enrollment of Actuaries, has a two-and-one-half-hour examination covering the mathematics of compound interest and of life contingencies.

Course 150 is an extensive coverage of the mathematics of life contingencies. It is based on the new textbook (Bowers et al, 1986) which employs future lifetime as the underlying random variable. For the development of this central subject of life actuarial science over the past two centuries, and its updated setting in *Actuarial mathematics*, see Section 3.6 of my foregoing Reflections and also (Hickman, 1986).

While "Economics of insurance", and "Individual risk models for a short term," (chapters 1 and 2 of *Actuarial mathematics*) are recommended as background readings for course 150, these topics together with "Collective risk models for an extended period", and "Applications of risk theory," comprise the examination subjects for course 151. Random variables were only vaguely elaborated in the syllabus when I was a student, and much of this theory has developed since.

The subjects of courses 160, 161, 162 and 165 have been touched upon in Sections 3.1 and 3.2 of my foregoing paper. Only course 160 is required, the others are elective, but 45 credits, as of now, must be chosen from 55 available. In many cases, actuaries have very large amounts of data available (relative to insureds and deaths) in the form of policies of insurance, amounts of insurance, annuity or pension incomes, census counts, and vital statistics of births and deaths. Methods for analyzing such data may then differ in some degree from those for smaller, more detailed studies of clinical data, or of impaired lives. In any case, the actuarial profession seeks reasonable understanding of the various estimation procedures that are feasible and available.

After attaining ASA designation, many actuarial students aspire to complete the education requirements for the Fellow of the Society of Actuaries (FSA) designation. To do so, they must undertake the Series 200–500 courses. In the Fellowship Catalog, we read these Series "are divided into four groups; the common Core and three specialty tracks: the group Benefits (GB) Track, the Individual Life and Annuity (ILA) Track and the Pension (P) Track. All candidates must earn 100 credits from the core courses, 90 credits from the required courses in one of the tracks with a single national emphasis, and 60 credits from other Fellowship courses. Within a track, some courses are designed to be national in emphasis (either Canada or U.S.)".

To give a little more insight to the nature of these requirements, I quote from the Fellowship Catalog.

Course 200.
Introduction to Financial Security Programs

(40 Credits) Required

The examination for this course is a four-hour multiple-choice and written-answer examination. The course covers: design, regulation and taxation of the major voluntary financial security programs involving life insurance, health insurance, property and casualty insurance, and employee benefit and pension programs; characteristics of the major social insurance programs in Canada and the U.S.; description of the providers of financial security programs; and an introduction to taxation of insurance companies in both Canada and the U.S."

Some hardy souls, after attaining the FSA designation proceed to the ACAS and FCAS designations of the Casualty Actuarial Society. Canadian FSAs take whatever additional steps may be needed for the Fellow of the Canadian Institute of Acturies (FCIA) designation.

Future Education Methods (FEM) are in progress. In October 1987, the Board of Governors of the Society approved implementation of five programs, namely:

(i) a Fellowship Admissions Course, a two-and-one-half-day course focusing on professional ethics and integration of syllabus material.
(ii) a research paper option for 30 elective Fellowship credits (details in the Fellowship Catalogs).
(iii) credit for examinations of other actuarial organizations and complete designations of non-actuarial organizations.
(iv) elective credit for an Intensive Seminar at the Associate level.
(v) an experiment in allowing credit for college courses, approved by the Society of Actuaries Education and Examination Committee, covering the topics of applied statistics, operations research and numerical methods.

These programs are at various stages of implementation.

In my fellowship student days, I was required to pass three six-hour examinations, each of which had a number of subjects. Fortunately, one examination was in my special fields of interest of pensions and of social insurance. The current requirement of 250 credits may require more examination hours, although FEM programs may effect such increase.

Of 112 members of the Society of Actuaries who hold appointments in U.S. and Canadian colleges and universities, 40 are in departments of mathematics or mathematical sciences; 27, in departments of statistics and actuarial science; 22, in schools of business administration; and 10, in actuarial science and insurance programs. The remaining 13 are in miscellaneous or unstated units. An additional 8 members of the Society are in foreign colleges and universities. As of November 1, 1988, total membership of the Society was 11,157, consisting of 6,039 Fellows and 5,118 Associates. Some 721 members reside outside Canada and the U.S.A. Additional membership statistics can be found in the 1989 Yearbook of the Society.

Much detailed information about the Society of Actuaries courses and examinations can be obtained by writing to:

Society of Actuaries
475 N. Martingale Road
Schaumburg, IL 60173
(312) 706-3500

and requesting a copy of the 1989 Associateship Catalog and of the spring 1989 Fellowship Catalog.

Education in the topics of courses 100–135 can be acquired at many colleges and universities in the United States and Canada. A list of schools which offer degree programs covering much of courses 150–165 can be obtained from the Society of Actuaries. In addition, the catalogs give information about study manuals and study groups that a student may wish to utilize.

Throughout my teaching career, and in following years, there has been a strong demand for actuarial students. These may find employment in insurance companies, consulting firms, state and federal government agencies (including insurance departments, the Internal Revenue Service and the Social Security Administration). Such organizations cover the tremendous range of financial security systems such as property-liability insurance, health insurance, life insurance, annuities and social insurance.

Actuarial education and examinations provide a rigorous but equitable process, manned by many dedicated volunteers who are complemented by an able, growing staff. The process provides a pathway to a challenging life devoted to making our financial security systems truly effective. My part therein has been a major satisfaction of my life.

References

[1] *Actuarial Research Clearing House (ARCH)*, C. S. Fuhrer and A. S. Shapiro, coeditors, distributed by the Society of Actuaries in 2 or 3 issues per year.

[2] A. W. Anderson, *Pension mathematics for actuaries*, Windsor Press, P. O. Box 87, Wellesley Hills, Massachusetts 1985.

[3] G. H. Andrews and J. A. Beekman, *Actuarial projections for the Old Age, Survivors, and Disability Insurance Program of Social Security in the United States of America*, Actuarial Education and Research Fund, 475 N. Martingale Rd., Schaumburg, IL 60173, 1987.

[4] Annual Report, Board of Trustees, Federal Old-Age, Survivors and Disability Insurance Trust Funds, 1988.

[5] A. L. Bailey, "Credibility procedures", *Proc. Casualty Actuar. Soc.* **37** (1950), 7–28.

[6] R. W. Batten, *Mortality table construction*, Prentice-Hall, Inc., Englewood Cliffs, New Jersey, 1978.

[7] J. A. Beekman, "Actuarial assumptions and models for social security projections", *Proc. Symp. Appl. Math.* **35**, 1986, 85–104.

[8] B. N. Berin, *The fundamentals of pension mathematics*, William M. Mercer, Inc. 1978.

[9] B. C. Berkeley, "Society, computers, thinking and actuaries", *ARCH*, New York, 1982. 1, 11–20.

[10] W. S. Bicknell and C. J. Nesbitt, "Premiums and reserves in multiple decrement theory", *Trans. Soc. Actuar.* **8** (1956), 344–377.

[11] P. P. Boyle, "Immunization under stochastic models of the term structure", *Jour. Inst. Actuar.* **105** (1978), 177–187. Also, *ARCH*, 1980.1, 19–29.

[12] N. L. Bowers, H. U. Gerber, J. C. Hickman, D. A. Jones, C. J. Nesbitt, *Actuarial Mathematics*, Society of Actuaries, 1986.

[13] J. D. Broffitt, "Maximum likelihood alternatives to actuarial estimators of mortality rates", *Trans. Soc. Actuar.* **36** (1984), 77–142.

[14] M. V. Butcher and C. J. Nesbitt, *Mathematics of compound interest*, Ulrich's Books, Inc. Ann Arbor, Michigan, 1971.

[15] D. D. Cody, "Discoveries to-date on risk, valuation and surplus", *Actuarial Research Clearing House*, 1987.2, 35–44.

[16] M. J. Cowell and W. H. Hoskins, *AIDS, HIV mortality and life insurance, parts* 1 *and* 2, Special Report, Society of Actuaries, 1987.

[17] S. P. D'Arcy and N. A. Doherty, *The financial theory of pricing property-liability insurance contracts*, Richard D. Irwin, Inc., Homewood, Illinois, 1988.

[18] R. M. Duncan, "A retirement system granting unit annuities and investing in equities", *Trans. Soc. Actuar.* **4** (1952), 317–344.

[19] J. S. Elston, *Sources and characteristics of the principal mortality tables*, The Actuarial Society of America, 1932.

[20.] Financial Accounting Standards Board, *Statement of financial accountig standards No. 87*, P. O. Box 3821, Stamford, CN 06905-0821, 1985.

[21] H. Gerber, *An introduction to mathematical risk theory*, Richard D. Irwin, Inc., Homewood, Illinois, 1979.

[22] H. Gershenson, *Measurement of mortality*, Society of Actuaries, 1961.

[23] J. W. Glover, *United States life tables*: 1910, Bureau of the Census, 1916.

[24] ——, *United States life tables*: 1890, 1901, 1910, and 1901–1910, Bureau of the Census, 1921.

[25] S. C. Goss, M. P. Glanz, E. Lopez, *Economic projections for OASDHI cost and income estimates*: 1987, Actuarial Study No. 101, Social Security Administration, Baltimore, MD, 1988.

[26] T. N. E. Greville, *United States life tables and actuarial tables*, 1939–1941, Bureau of the Census, 1946.

[27.] ——, *Selected papers of T. N. E. Greville*, edited by D. S. Meek, R. G. Stanton, Winnipeg, Manitoba, 1984.

[28] R. Henderson, "Life insurance as a social science and as a mathematical problem", *Bull. Amer. Math. Soc.* **31** (1925), 227–252.

[29] ——, *Mathematical theory of graduation*, The Actuarial Society of America, 1938.

[30] E. L. Hicks and C. L. Trowbridge, *Employer accounting for pensions*, Pension Research Council, Wharton School, University of Pennsylvania, 1985.

[31] J. C. Hickman, "Introduction and historical overview of credibility", in *Credibility: theory and applications*, edited by P. M. Kahn, Academic Press, New York, 1975.

[32] ——, "Updating life contingencies", *Proc. Symp. Appl. Math.* **35**, 1986, 5–15.

[33] Jan M. Hoem, *The versatility of the Markov chain as a tool in the mathematics of life insurance transactions*, 23rd. International Congress of Actuaries, Helsinki, 1988.

[34] R. V. Hogg and S. A. Klugman, *Loss distributions*, Wiley, New York, 1984.

[35] D. F. Hooker and L. H. Longley-Cook, *Life and other contingencies*, Vol. I, Cambridge University Press, 1953, and *Life and other contingencies*, Vol. II, Cambridge University Press, 1957.

[36] W. A. Jenkins and E. A. Lew, "A new mortality basis for annuities", *Trans. Soc. Actuar.* **1** (1949), 369–466.

[37] C. W. Jordon, *Life contingencies*, Society of Actuaries, 1952 lst ed., 1967 2nd ed.

[38] P. M. Kahn, "Overview of credibility theory", *Proc. Symp. Appl. Math.* **35**, 1986, 57–66.

[39] S. G. Kellison, *The theory of interest*, Richard D. Irwin, Inc., Homewood, Illinois, 1970.

[40] N. Keyfitz and J. A. Beekman, *Demography through problems*, Springer-Verlag, New York, 1984.

[41] G. S. Kimeldorf and D. A. Jones, "Bayesian graduation", *Trans. Soc. Actuar.* **19** (1967) 66–112.

[42] G. King, *Institute of actuaries textbook, part II*, Charles and Edwin Layton, London, 1887 1st ed., 1902 2nd ed.

[43] S. A. Klugman, "Loss distributions", *Proc. Symp. Appl. Math.* **35**, 1986, 31–55.

[44] R. London, *Survival models and their estimation*, ACTEX Publications, Winsted, Connecticut, 1988.

[45] M. A. Mackenzie, *Interest and bond values*, University of Toronto Press, Toronto, 1917.

[46] A. L. Mayerson, "A Bayesian view of credibility", *Proc. Casualty Actuar. Soc.* **51** (1964), 85–104.

[47] J. J. McCutcheon and W. F. Scott, *An introduction to the mathematics of finance*, Heinemann, London, 1986.

[48] W. O. Menge, "A statistical treatment of actuarial functions", *Record Amer. Inst. Actuar.* **26** (1937), 65–88.

[49] W. O. Menge and J. W. Glover, *An introduction to the mathematics of life insurance*, MacMillan, 1935.

[50] W. O. Menge and C. H. Fischer, *The mathematics of life insurance*, MacMillan, New York, 1965.

[51] M. D. Miller, *Elements of graduation*, The Actuarial Society of America and American Institute of Actuaries, 1946.

[52] E. J. Moorhead, *Our yesterdays*, a history of the actuarial profession 1809–1979 in North America, prepared for the Centennial Celebration of the actuarial profession, forthcoming. See also Actuarial History 1889–1989 in 1989 Year book, Society of Actuaries.

[53] R. D. Murphy and P. C. H. Papps, *Construction of mortality tables from the records of insured lives*, The Actuarial Society of America, 1922.

[54] Robert J. Myers, *Social Security*, Richard D. Irwin, Inc., Homewood, Illinois, 1985.

[55] A. Neill, *Life contingencies*, Heinemann, London, 1977.

[56] C. J. Nesbitt, "On the performance of pension plans", *Proc. Symp. Appl. Math.* **35**, 1986, 113–129.

[57] ——, "Exploration of actuarial mathematics with recognition of nuclear holocaust hazard", 22nd Actuarial Research Conference, University of Toronto, 1987. Forthcoming in ARCH.

[58] ——, "Further reflections on actuarial recognition of nuclear holocaust hazard", 23rd Actuarial Research Conference, University of Connecticut, 1988. To appear in *Insurance, Mathematics and Economics* and in ARCH.

[59] H. Panjer, "Models in risk theory", *Proc. Symp. Appl. Math.* **35**, 1986, 17–30.

[60] H. J. Panjer, *AIDS: survival analysis of persons testing HIV+, Trans. Soc. Actuar.* **40** (1988), forthcoming.

[61] H. W. Pedersen, E. S. Shiu and A. E. Thorlacius, "Arbitrage-free pricing of interest-rate contingent chains", *Trans. Soc. Actuar.*, forthcoming.

[62] F. W. Perryman, "Experience rating plan credibilities", *Proc. Casualty Actuar. Soc.* **24** (1937), 60–125.

[63] Proceedings of Symposia in Applied Mathematics, Volume 35, *Actuarial Mathematics*, American Mathematical Society, 1986.

[64] P. R. Rider and C. H. Fischer, *Mathematics of investment*, Rinehart, New York, 1951.

[65] D. R. Schuette, "A linear programming approach to graduation", *Trans. Soc. Actuar.* **30** (1978), 407–431.

[66] D. R. Schuette and C. J. Nesbitt, "Withdrawal benefit equal to reserve: non-neutrality in the discrete case", 22nd Actuarial Research conference, University of Toronto, 1987. Forthcoming in *ARCH*.

[67] H. L. Seal, *Stochastic theory of a risk business*, Wiley and Sons, New York, 1969.

[68] N. E. Sheppard and D. C. Baillie, *Compound interest*, University of Toronto Press, Toronto, 1960.

[69] E. S. Shiu, "Minimum-R_z moving-weighted-average formulas", *Trans. Soc. Actuar.* **36** (1984), p. 489.

[70] ——, "A survey of graduation theory", *Proc. Symp. Appl. Math.* **35** (1986), 67–84.

[71] Society of Actuaries, 1984 Reports of Mortality, Morbidity and Other Experience, 1988.

[72] M. Spiegelman, *Introduction to demography*, Harvard University Press, Cambridge, MA., 1955 and 1968.

[73] C. A. Spoerl, "Whittaker-Henderson graduation formula A", *Trans. Actuar. Soc. Amer.* **38** (1937), 403–462.

[74] E. F. Spurgeon, *Life contingencies*, Cambridge University Press, 1922 1st ed., 1929 2nd ed., 1932 3rd ed.

[75] C. L. Trowbridge, "Fundamentals of pension funding", *Trans. Soc. Actuar.* **4** (1952), 17–43.

[76] A. Wade, *Social Security area population projections*: 1988, Actuarial Study, No. 102, Social Security Administration, Baltimore, MD, 1988.

[77] J. C. Wilkin, OASDI, *long-range beneficiary projection*: 1987, Actuarial Study No. 100, Social Security Administration, Baltimore, MD, 1988.

[78] H. H. Wolfenden, *Population statistics and their compilation*, The Actuarial Society of America, New York, 1925.

[79] ——, "On the development of formulae for graduation by linear compounding, with special reference to the work of Erastus L. DeForest", *Trans. Actuar. Soc. Amer.* **26** (1925), 81–121.

[80] H. Wolthuis, *Savings and risk processes in life contingencies*, Universiteit van Amsterdam, Institut voor Actuariaat en Econometrie, 1988.

Uta C. Merzbach studied mathematics at the University of Texas at Austin and at Harvard University before becoming involved in research projects that introduced her to the history of science. She received her Ph.D. in mathematics and the history of science at Harvard (Radcliffe College) in 1965. Garrett Birkhoff and I. Bernard Cohen supervised the writing of her dissertation. She has spent much of her professional career as a teacher and curator of mathematics. She is Curator Emeritus of Mathematics at the Smithsonian Institution and Director of the LHM Institute, a research organization providing support services in the history of mathematics.

The Study of the History of Mathematics in America: A Centennial Sketch

UTA C. MERZBACH

0. Introduction

Anniversaries provoke reflections on the past. The centennial of the American Mathematical Society (AMS) provided an appropriate opportunity to review past contributions made in the United States to the study of the history of mathematics and to reflect on the changing status of the subject over the years. History of mathematics has been especially closely tied to mathematics, to mathematics education, and to history of science. But the alliances have been uneasy and have shifted over the years. Early in this century, history of mathematics in this country derived its strength from the singular energy of a few men, their close affiliation with the mathematical community here and abroad, and a tradition of history as a literary form. Its chief supporters tended to share a belief in progress and rationalism. In its current renewal, there are many workers in the field, a multiplicity of methodologies, and numerous motivations for promoting the subject. It is too soon to tell whether contemporary collective efforts will result in the desirable balance of sound exposition and rigorous research for which our predecessors prepared us.

Florian Cajori

D. E. Smith

R. C. Archibald

D. J. Struik

(Photograph of F. Cajori courtesy of Special Collections, The Colorado College Library; photograph of D. E. Smith courtesy of Special Collections, Teachers College, Columbia University; and photograph of R. C. Archibald courtesy of Boston University, Archives.)

This cursory sketch is not an attempt to present a history of history of mathematics in America. Rather, it is intended to call attention to a few of the currents that affected the subject over the past century, to note something of the relationship of past American historians of mathematics to the AMS and to the community of scholars at large, and to recall a few of the major contributions to the subject made in this country. The discussion focuses on the early part of this century, when America's leading historians of mathematics were active in the mathematical community. The post-1945 period is characterized only briefly, and very little is said about the contemporary scene. Although no specific references are made to publications of the last twenty years, it should be noted that these include some of the most interesting contributions to history of modern mathematics that have been made in the United States.

1. The Pre-World War I Period

In 1890 the United States Bureau of Education issued a monograph entitled *The Teaching and History of Mathematics in the United States.* The author was Florian Cajori (1859–1930), a native of Switzerland, trained in the United States, who had recently assumed a professorship for physics in Colorado. The work not only was the first comprehensive history of mathematics in the United States, but the first major work dealing with history of mathematics to be published in this country. Its publication marks the beginning in America of the organized study of the history of mathematics, and the life of its author spans the formative years of history of mathematics in the United States.

Cajori's pioneering study was published in a period during which there were widespread attempts to expand the intellectual life of the country. These took many forms; in most areas they were accompanied by efforts at control on the part of increasing groups of "professionals." It is symptomatic of the period that between 1870 and 1890 more than 200 "learned societies" were founded; these included the New York Mathematical Society (1888), the American Historical Association (1884), and the National Education Association (1870). Other factors, of special relevance to history of mathematics, include the rise of graduate education, the establishment of professional schools in engineering and business, and the conversion of the nineteenth century teachers' training institutes and normal schools to graduate "schools of education." Educated Americans banded together in professional groups not only to exercise control over the future development of their subject, but to communicate with colleagues in their fields of research, and to enhance the resources available for study and research. In conjunction with these aims, many strove to enlarge awareness of their fields of study among laymen; a historical approach was generally regarded as a useful means of achieving these ends.

Just as the celebration of the nation's centennial in 1876 had fanned historical sparks in the United States, so the closing of the century and the self-consciousness of the "new" professionals in many academic disciplines in the 1890s led to orations and papers on "Review of Progress in Subject x" or "History and Future Outlook of Topic y." In addition, national magazines such as *The Nation* or *Century* not only guided their readers in matters pertaining to literature, politics, and the arts, but also sprinkled their widely read issues with occasional essays or biographies pertaining to science or mathematics. Since it was still fashionable at the time to discuss most topics against a historical framework, it is not surprising that such articles, too, frequently added to the literature of mathematical history.

In examining American contributions to history of mathematics prior to World War I, one observes that, with one notable exception, they came from those trained in mathematics and allied fields rather than from historians. American historians in the 1890s tended to be preoccupied with the shift from episodic narrative history to a more unified approach to political and, occasionally, military history. While anxious to alter the pattern of amateur elder statesmen being the chief writers of history, many of the new professional historians, in fact, attempted to invert this pattern and were busy convincing the country that their insight was needed to lead the nation into the twentieth century [Higham 1965]. Despite this divergence of priorities, work done by historians in developing generally available research materials helped the efforts of those working in the history of mathematics. This is especially true of the part played by historians in rescuing from destruction archival records and other forms of primary research materials, in encouraging the proliferation of libraries, and in expanding the means of publication for various disciplines. For example, the fact that Cajori's 1890 work was published by the Bureau of Education serves as a reminder that American reform educators for half a century had published reports and essays describing the history leading to whatever state of their subject they were discussing; it must also be noted, however, that at this time the connection between the Bureau and the historical community was being strengthened through the work of the historian Herbert Baxter Adams of the Johns Hopkins University, who prepared a series of monographs for the Bureau.

History of mathematics turned up in a variety of forms and places. Most American contributions to the subject prior to World War I were expository in nature, rather than devoted to conveying new research results. Depending largely on the publication outlet, the expositions could be research-oriented or general. Most expository history designed for the mathematician was found in the *Bulletin of the American Mathematical Society*; this was consonant with the *Bulletin's* purpose, iterated on its masthead, to provide "a historical and critical review of mathematical science." Less technical articles appeared in the *American Mathematical Monthly*, although, in time, some of these became more research-oriented. Research results in history occasionally appeared in the *Bulletin*. Most of America's leading historians of mathematics also had papers in *Bibliotheca Mathematica*, an international journal for the history of mathematics founded and edited by the Swedish mathematician Gustav Eneström, who maintained high standards of scholarship; it was published by Teubner in Germany.

Maxime Bôcher, in his AMS presidential address of 1911 dealing with the early history of Sturm-Liouville theory, called attention to an interesting example of the occasional integration of historical study with mathematical research [Bôcher 1911]. Years earlier Bôcher had noticed that there was a lacuna among Sturm's extant papers. Bôcher's student, M. B. Porter, set about to reconstruct the missing paper; his partial reconstruction appeared in the *Annals of Mathematics* [Porter 1902]. Porter, in turn, interested Helen A. Merrill in the subject, resulting in her paper "On Solutions of Differential Equations which Possess an Oscillation Theorem," published in the *Transactions of the AMS* [Merrill 1903].

Another variety of historical publications was designed to facilitate research. This included bibliographic work, translations, and some book reviews. Bibliographies and translations often appeared as separate monographs; however, the *Bulletin*, especially in its early years, frequently carried translations of historically slanted articles by contemporary European mathematicians; for the most part, the authors were men like Felix Klein or Emile Picard who had connections with American mathematicians. The *Bulletin* consistently presented book reviews dealing with historical topics.

The most pervasive and traditional historical papers dealt with the life or work of an individual. Earlier in the nineteenth century, obituaries of mathematicians had appeared in a few serial publications such as the *American Journal of Science* or in privately printed memorial volumes such as that issued by the Bowditches upon the death of Nathaniel in 1839. By the turn of the century, biographic sketches might appear in one of this country's mathematical journals such as the *American Mathematical Monthly*, a general scientific journal like *Science*, or in a nontechnical magazine like *Century*. In general, accounts of individual mathematicians ranged from simple declarations that the person had been a scholar and a gentleman to incisive mathematical evaluations. The *Bulletin* usually carried articles stressing the mathematical contributions of the individual more than the details of his life history. Occasionally, there was an article such as Wilczynski's paper on Lazarus Fuchs [Wilczynski 1902] that managed to convey a great deal about the individual in a few lines, while presenting a clear mathematical exposition with historical perspective of the person's contribution to a research area.

Nontechnical expository articles dealing with the role of mathematics in civilization, or with special mathematical topics, usually were intended to provide motivation for the lay person and were frequently education-oriented. They might appear in *Science*, *Popular Science Monthly*, or one of the general magazines. There were also expository articles dealing with mathematical history geared to the mathematical or larger scientific community; presidential addresses constituted a special sub-genre of this type of history.

On the border line between general exposition and original research were historical summaries of specific subjects. These could take the form of brief historical references serving to introduce a subject or to support a form of mathematical argument, or they might be detailed subject reviews that traced the roots of then current research. The American Association for the Advancement of Science (AAAS), following the pattern established by the British Association, produced periodic historical reviews. Thus we find E. W. Brown giving a "Report on the Recent Progress of Solids and Fluids" in the *Proceedings of the AAAS for 1897*, which essentially outlines work in hydrodynamics during the preceding fifteen-year span. The following year, A. G. Webster provided a similar report "On the Mathematical Theory of Electricity and Magnetism" and, in 1899, G. B. Halsted reported on "Progress in NonEuclidean Geometry."

Histories of specific mathematical topics were provided by a number of mathematicians. Examples are in the work found of Halsted on non-Euclidean geometry, Miller and his student Josephine Burns on group theory, and Emch on geometry. Occasionally one might also find an essay dealing with aspects of institutional history.

Narrative historical articles or large-scale histories of mathematics based on existing histories began to appear before the turn of the century as well. These tended to be carefully crafted and provided a valuable resource for students, mathematicians, and lay people. Their flaws were those of their nineteenth century predecessors, on whom they improved in many instances. For many years, the chief American authors of such general historical surveys were Florian Cajori and David Eugene Smith.

Cajori had come to this country when he was sixteen. He attended Whitewater Normal School in Wisconsin and taught school before matriculating at the University of Wisconsin, where he obtained a B.S. degree in 1883. After a year's graduate study at the Johns Hopkins University from 1884 to 1885, he spent three years as an assistant professor of mathematics and professor of applied mathematics at Tulane. During this period, he contributed several papers in the history of mathematics to the *Journal of Education*, published in New Orleans, and to the New Orleans Academy of Science; in 1886, he was granted an M. S. degree from Wisconsin. A year's stay in Washington, during which he was a researcher at the United States Bureau of Education, resulted not only in [Cajori 1890] but in several smaller articles. He spent the next three decades at Colorado College, serving as professor of physics from 1889 to 1898, and as professor of mathematics from 1898 to 1918. In addition, he was dean of the department of engineering at the college from 1903 to 1918. Despite a heavy teaching and administrative load and lack of major regional research facilities, while at Colorado College Cajori produced nearly 100 research and expository papers, two dozen book reviews, and several books and monographs, not counting reprints and translations into foreign languages. His nonhistorical work included elementary textbooks of mathematics and some research on semiconvergent series. In [Cajori 1890], which broke new ground in dealing with American mathematics, he relied heavily on the use of questionnaires and letters to gather his data. The fact that he was asked to contribute [Cajori 1908] to the volume of Cantor's *Geschichte* dealing with the late eighteenth century attests to his international reputation early in the century.

A Ph.D. degree awarded to Cajori by Tulane University in the 1890s was apparently honorary. Evidence of the regard that American mathematicians had for him at this time, when he worked in relative isolation, is provided by the fact that in 1903 he ranked 36th in Cattell's survey of mathematicians and was elected to the council of the AMS, serving from 1904 to 1906.

Cajori's contemporary, David Eugene Smith (1860–1943), was the most prolific historian of mathematics America has produced. Because of his long-lasting influence on history of mathematics and mathematics education, it is frequently overlooked that he was only a year younger than Cajori. In many ways, he had more kinship with the nineteenth century historical scholarship that was rooted in literature and philology than he did with higher mathematics or modern historical research techniques. Without specialized formal training in either history or mathematics, his strength lay in the classical training he had received as a child. He studied art and classical languages at Syracuse University, where he graduated in 1881. Although admitted to the bar in New York state three years later, he preferred to teach mathematics at the State Normal School in Cortland, while pursuing further graduate work at Syracuse. He received a Ph.D. degree from Syracuse with a thesis on classical art. In 1891 he became professor of mathematics in the State Normal College at Ypsilanti, Michigan. He collected a degree in pedagogy there before becoming principal of the State Normal School, which, subsequently, led to his assuming the professorship of mathematics at Teachers College of Columbia University.

Among his major works of the pre-World War I period one must single out *Rara Arithmetica, a Catalogue of the Arithmetics Written before the Year MDCI with a Description of Those in the Library of George Arthur Plimpton of New York.* (1908). Not only has this work remained a standard reference among bibliographers, book collectors, and historians of early modern mathematics, but it served to cement his friendship with Plimpton, who was the chairman of Ginn and Company from 1914 to 1931. Both men were collectors; for years, Smith assisted Plimpton in developing the mathematical parts of his library. Not surprisingly, Ginn published many of Smith's books.

Another publication of this period, [Smith and Mikami 1914] is representative of the great service Smith performed in calling attention to the history of mathematics in the Far East. He was instrumental in promoting research in the history of mathematics in China, India and Japan, and encouraged many of the contributions on the subject that appeared in American journals prior to World War II.

In the AMS, which he had joined in 1893, Smith served on the Committee on Publication from 1903 to 1909; he was an editor of the *Bulletin* from 1910 to 1920, having assisted briefly in 1902. Earlier, in 1896–1897, he had been part of the group instrumental in forming the Chicago Section of the Society. From 1902 to 1920 he was librarian of the AMS. His contributions in that capacity have been noted by Archibald [1938:90–92]. It seems appropriate that his first major acquisitions for the Society should have come from the library of G. W. Hill. For it was Hill who, in his presidential address before the AMS in 1895, had commented on the difficulties American mathematicians faced in trying to do historical research without proper library resources [Hill 1896].

The offices Smith held in national and international organizations that he used to promote the history of mathematics are too numerous to recount here. Suffice it to note that, as member of the International Commission on the Teaching of Mathematics, Smith collaborated with the other two American commissioners, William Fogg Osgood and J. W. A. Young, in writing and editing numerous reports of the Commission. His influence is seen in journals such as *School Science and Mathematics*, which he served as associate editor; it published historical articles such as [Benedict 1909].

Aside from his publications and his organizational activities, Smith exerted strong influence on mathematics education and history of mathematics through his teaching. His courses were extremely popular. A survey of 113 schools, published in 1915, indicated that Smith's course had the largest enrollment among 47 courses in the history of mathematics; in enrollment among over 175 courses in the history of science, mathematics, and psychology, it ran second only to the course in the history of chemistry taught by Theodore W. Richards of Harvard. [Brasch 1915]. Usually, the history of mathematics was introduced into the curriculum either as part of a course in the history of science or as a course by itself. A separate course had been taught at Yale, where James Pierpont was interested in history, since 1892. A history of science course was introduced by Tyler and Sedgwick at MIT in 1905. [Tyler 1910–1911]. Smith's course outlasted these and most others like them. Like other such courses, his had a reputation for being easy;

but, unlike many others, it conveyed to the students knowledge and appreciation for the subject. Although his strongest influence was exerted on the students enrolled in Teachers College, it was not limited to these. Numerous students from the mathematics department in Columbia College attended Smith's courses; for example, E. T. Bell [1945] described his experience when sent there by Cassius Jackson Keyser (1862–1947), long-time member of the Columbia mathematics faculty, who taught history of mathematics himself at times and steered students to Smith.

For twenty years, Smith's graduate students at Teachers College produced respectable theses devoted to the history of mathematics education. Typical of these are two of the earliest, [Jackson 1906] and [Stamper 1906]. Smith's influence and collaborations extended beyond his regular graduate students, however.

In the academic year 1909–1910, an instructor from the University of Michigan spent a year's leave of absence at Teachers College. The stay resulted in a joint publication by Smith and Louis Charles Karpinski (1878–1956) on *The Hindu-Arabic Numerals*, which was widely hailed as the best exposition on this frequently treated topic. Karpinski was a graduate of Cornell University who had presented a dissertation on distributions of quadratic residues to obtain his Ph.D. degree from the University of Strassburg in 1903. He, too, had gained his first teaching experience as a young man, when he had taught mathematics at Berea College in Kentucky. After his return from Strassburg, he spent a year as instructor at the New York State Normal School in Oswego, after which he joined the faculty of the University of Michigan, where he remained the rest of his life. His interest and competence in the medieval period was demonstrated further in 1912 when his paper on "The 'Algebra' of Abu Kamil Shoja' ben Aslam" appeared in Eneström's *Bibliotheca Mathematica.*

Soon after the turn of the century another member of the Columbia University faculty was placed in a position to provide substantial support to the history of mathematics. This was Robert Simpson Woodward (1849–1924), who became president of the Carnegie Institution in 1904, two years after Andrew Carnegie had provided the funding for "an Institution to promote study and research." Woodward, an applied mathematician with special research interests in geophysics, geodesy and astronomy, had served with the U.S. Geological Service and the U.S. Coast and Geodetic Survey before joining the Columbia faculty in 1893; in 1895 he had become dean of the school of pure science at Columbia. His expository and historical skills and interests had been apparent for some time; his vice-presidential address on "The Mathematical Theories of the Earth" presented to the American Association for the Advancement of Science in 1889, met with sufficient interest to be printed in several of the leading scientific publications of the day. He produced a

"Historical Survey of the Science of Mechanics," which was published in *Science* in 1895. As co-editor, with Mansfield Merriman, of a volume on *Higher Mathematics*, it was he who had asked David Eugene Smith to contribute an article on the history of modern mathematics; this project fixed Smith's historical direction and renewed his contact with Felix Klein, whom he asked for assistance. In the meantime, Woodward himself was preparing the article on mathematics for the *History of the Smithsonian Institution, 1846–1896*, which appeared in 1897. His awareness of history may have received some reinforcement through the fact that he had been elected president of the AMS for 1899–1900 and president of the AAAS for 1901. At any rate, he did not miss the opportunity of welcoming the new century in the properly progressive spirit when he chose the topics "The Century's Progress in Applied Mathematics" for his presidential AMS address and "The Progress of Science" for his presidential AAAS address. Both addresses were widely read; the first was published in the *Bulletin*, the second in *Science*, as one would expect; but, in addition, there was a reprint in *The Scientific Monthly*, and there were more reprints and translations in England, Germany, and Poland. During his first ten years as president of the Carnegie Institution, that body supported the publication (1907–1909) of the *Collected Mathematical Works* of George William Hill, and of Derrick Lehmer's factor tables and tables of primes.

There was another man in the pre-World War I period who, aided by substantial financial support, was to further publications in this country in the history of mathematics. This was Paul Carus (1852–1919). He came to the United States from Germany, where he had obtained a Ph.D. in Tübingen in 1876. He was a believer in the monism of mind and matter, convinced that philosophy could be put on a scientific basis. He soon became the editor of the *Open Court*, a journal founded in 1887 by the zinc mogul A. Hegeler of Chicago, who was equally convinced that religion could be put on a scientific basis. In 1888, Carus had married Hegeler's daughter, Mary, subsequently a major benefactor to mathematical expository writing. The following year Hegeler founded a new, more technically oriented journal, the *Monist*, and Carus became its editor. Carus expanded the orientation of the journal. It was a general journal of philosophy, and one of the directions of special interest to the editor was the philosophy of mathematics along with related history.

As a result, the *Monist* featured articles such as G. B. Halsted's [1902/1903] translation of an extract from Gino Loria on the history of geometry prior to 1850, and L. Robinson's translation [Robinson 1909] with a commentary by D. E. Smith of Heiberg's account of the palimpsest on Archimedes' *Method*, discovered in Constantinople in 1906. Before the turn of the century, Hegeler and Carus expanded their activity even further, establishing the Open Court Publishing Company.

Under its imprint appeared numerous monographs in the history of mathematics. Many were translations into English of classics in modern mathematics. Among the first such Open Court monographs were [McCormack 1898], a translation of Lagrange's elementary lectures at the *Ecole Normale*, and [Beman 1901], a translation of Dedekind's "*Was sind und was sollen die Zahlen*" and "*Stetigkeit und Irrationalzahlen.*"

The year that Karpinski spent at Teachers College with Smith, another Ph.D. from Strassburg, Raymond Clare Archibald (1875–1955), was doing postgraduate work at the Sorbonne, where he came under the influence of Jules Tannery, but also heard Borel, Darboux, Goursat, Picard, and Poincaré [Sarton 1956]. A native of Nova Scotia, Archibald had returned to Canada after obtaining his doctorate in 1900 and spent seven years at Mount Allison Ladies' College at Sackville. There he furthered the three areas that most interested him for the rest of his life: mathematics, music, and the study of books. In 1908, he received an appointment at Brown University, where he remained more than three decades. During the pre-World War I period he came to attention with a biobibliography of Simon Newcomb, published in the *Transactions of the Royal Society of Canada*, and with numerous minor historical notes and reviews that appeared in the *Bulletin of the AMS*, the *Proceedings of the Edinburgh Mathematical Society*, and the *Mathematical Gazette*.

By 1914, these individuals and others formed an active group, promoting the history of mathematics as an independent research field, as a motivating subject for teachers of mathematics, as a stimulus for mathematical research, and as a source of general edification and pleasure. Conscious of the limited availability of reference materials and libraries, they collaborated in making requisite primary and secondary source materials more easily available, be it through book purchases, through translations, through bibliographies, through text editions and analyses, or simply through reviews.

2. World War I to 1930

During the post-World War I period, history of mathematics grew steadily in America and flourished within the mathematical community. It is true that World War I markedly affected research and international collaboration in history of mathematics as it did in other fields. For example, *Bibliotheca Mathematica*, whose rigorous editor had featured research contributions by Cajori, Karpinski, Miller, and Smith, ceased publication after 1914. The International Commission on the Teaching of Mathematics, in which David Eugene Smith had become of considerable influence, suspended its operations as well. Yet, during a period of institutional growth in this country, American historians of mathematics reached a peak of professional involvement and sharpened their research.

The pioneer historians of mathematics continued to be involved in the activities of the American Mathematical Society. Both Cajori (1919) and D. E. Smith (1922) served as vice-presidents of the Society. Archibald had a term on the council (1918–1920) and in 1921 succeeded Smith as librarian, a position he retained for twenty years; in 1925 he edited an expanded catalogue of the Society's library. In 1928 the Society appointed its Committee on the Semicentennial Celebration. Archibald ended up as its vice-chairman and as chairman of the program subcommittee. D. E. Smith chaired the subcommittee on exhibits, and Archibald, together with T. S. Fiske, was put in charge of the history of the Society. The result was [Archibald 1938]. Cajori was an active member of the California section of the AMS, serving twice as its chairman (1918–1919 and 1922–1923); he was an invited AMS-MAA speaker in 1922.

Upon the founding of the Mathematical Association of America (MAA) in 1915, Cajori, Smith, and Archibald became even more involved in organizational activities. All three, as well as Karpinski, Miller, and others sympathetic to mathematical history, were charter members of the Association. Cajori served as president (1917–1918) and was a member of several committees which prepared lists of suggested mathematical books for college and junior college libraries. Archibald, a member of the International Mathematical Union, served on its International Commission of Mathematical Bibliography. Archibald, Cajori, and Karpinski were elected vice-presidents of the American Association for the Advancement of Science (AAAS); Archibald served as chairman of section A (mathematics), Cajori and Karpinski, of the recently (1921) established section L (history).

The pioneers continued to be productive in research and publications as well. Cajori, who left Colorado Springs for Berkeley in 1918 to assume a newly created professorship for the history of mathematics at the University of California, maintained a steady output of research articles leading to new textbooks and monographs. These included his study of the seventeenth century English mathematician William Oughtred (1916), his *History of Elementary Mathematics, with Hints on Methods of Teaching* (1917), and his *History of the Conceptions of Limits and Fluxions in Great Britain* (1919). In 1919 he also produced a second edition of his *History of Mathematics*, which for three decades was the only English-language history that provided at least a cursory treatment of nineteenth-century mathematics. His major achievement was the publication in 1928 of the two-volume *History of Mathematical Notations*, which remains the standard reference work on the subject.

David Eugene Smith continued his bibliographic contributions with a "Union List of Mathematical Periodicals," which he produced in collaboration with Caroline E. Seely, a mathematician who served as secretary of the AMS for many years. The work was published as a *Bulletin* of the U.S. Bureau of Education in 1918. It was followed in 1923 with a "Bibliography of the Teaching of Mathematics, 1911–1921," compiled in collaboration with J. A. Foberg. A small monograph on computing jetons (counters), published by the American Numismatic Society in 1921, became popular with collectors and students of the subject. A *History of Mathematics*, published by Ginn, appeared in two volumes between 1923 and 1925; re-issued as a Dover paperback, this work dealing with elementary mathematics has remained a favorite with many teachers of mathematics. In addition to his books and monographs, which included successful textbooks in elementary mathematics beside his historical works, Smith published over a hundred journal articles and another hundred book reviews during the period 1914–1930.

Among the younger men, Archibald gained international attention with his English edition of Euclid's *Division of Figures* [Archibald 1915] and continued to contribute articles to the *Monthly*. In 1925, he edited a volume on Benjamin Peirce, which demonstrates his biobibliographic skills. The previous year, his memoir on Simon Newcomb for the National Academy had involved similar skills; his subject, in this case, was a man he had known in his youth, who had been the subject of his first historical publication. His major research achievement of the 1920s was his contribution to the edition of the Rhind Mathematical Papyrus produced by A. B. Chace, the Chancellor of Brown University; Archibald's 102-page bibliography on Egyptian mathematics appended to this work remains a reference source in this area of study. Minor research products by Archibald in the 1920s appeared in foreign journals like the *Mathematical Gazette* and *Nature*; after 1928, he also contributed to the *Dictionary of American Biography*, to the *Encyclopaedia Britannica* and to Smith's *Sourcebook of Mathematics*, to name but a few of his varied projects.

Karpinski made a notable contribution to the history of medieval mathematics with his translation and edition of Robert of Chester's algebra of al-Khwarizmi. [Karpinski 1915]. In the 1920s he continued to call attention to medieval sources available for study in this country. Halsted added to his editions of geometric classics with a translation of Saccheri's *Euclides Vindicatus*, published by Open Court in 1920. George Abram Miller in 1916 had published a widely noted *Historical Introduction to Mathematical Literature*. He continued to contribute historical articles to a range of journals. They tended to fall into three groups. There were technical contributions to the history of algebra; these appeared in the *Monthly*. He produced lists of errors in the literature, which he published in several journals. Finally, there was a variety of expository articles, often directed to teachers of mathematics.

Among other mathematicians making regular historical contributions was R. B. McClenon of Grinnell College, who had succeeded Karpinski as librarian of the MAA. His articles in the *Monthly* included a study on Leibniz and complex numbers, and a discussion of Leonard of Pisa and his *Liber quadratorum*. Lao Simons, who had studied mathematics at Vassar in the

1890s, been a member of the faculty at Hunter College since 1895, and extended her training in pedagogy and the history of mathematics at Teachers College, in the 1920s systematically produced research reports on the history of American textbooks. In the process, she obtained a Ph.D. degree from Teachers College. Vera Sanford, of the State Teachers College at Oneonta, N.Y., produced brief articles of high quality in the *Mathematics Teacher* in the twenties, and wrote a *Short History of Mathematics*, published in 1930. This became a successful textbook in courses designed for teachers of mathematics in the 1930s.

The *Mathematics Teacher*, which had been the journal of the Association of Teachers of Mathematics in the Middle States and Maryland, became the official journal of the National Council of Teachers of Mathematics (NCTM) in 1921. From that year on, brief expository contributions on historical topics appeared in that journal with some regularity, reaching a peak in quantity and quality in the late twenties.

Under the leadership of R. S. Woodward, the Carnegie Institution continued to support research in the history of mathematics. Two Carnegie projects should be cited in any review of history of mathematics in the 1920s: One was Leonard Eugene Dickson's three-volume publication entitled *A History of Number Theory*, which appeared between 1919 and 1923. It has remained a standard reference work, largely because it is not the narrative history the title might suggest, but instead a reasonably reliable bibliographic guide through the history of number theory.

The other Woodward-Carnegie project relates to the coming to this country of the Belgian historian of science George Sarton (1884–1956). Sarton, an admirer of Poincaré, had turned from philosophy to study chemistry and mathematics. After obtaining his doctorate with a thesis on the mechanics of Newton in 1911, he founded the journal *Isis*, the first volume of which appeared in 1912. It was conceived as an international journal for the history of science; Sarton edited it from his home outside Ghent. Upon the invasion of Belgium at the beginning of World War I, Sarton buried many of his research notes in his garden and fled to England. [Sarton 1927:45]. In 1915, Sarton came to the United States, assisted by Smith. The following year he held an appointment at the philosophy department at Harvard. Woodward created for him a position as associate in the history of science at the Carnegie, which became effective in 1918. It was this appointment that fed Sarton for many years, even after Harvard offered him research space and library facilities. In addition, the Carnegie Institution sponsored the publication of Sarton's monumental *Introduction to the History of Science*, a major bibliographic work; the scope of the project for exceeded available resources, however, and only three volumes could be completed.

Sarton is justly credited with establishing the history of science as an academic discipline in the United States and with shaping the basic research tools needed by workers in the field. Because his journal *Isis* has been the official journal of the History of Science Society since the founding of that society in 1924, it is often assumed that the Society was his sole creation as well. In fact, however, the establishment of the Society involved several American historians of mathematics, notably the indefatigable David Eugene Smith. In 1915, Smith had called attention to *Isis* through a note published in *Science*. In December 1923, Smith sent a letter to 45 individuals, suggesting a meeting in Boston. As a result, the next month, 37 individuals met and founded the History of Science Society: besides Sarton, the organizing committee included Archibald, E. W. Brown, Karpinski, Smith, and H. W. Tyler, among others [Isis 6:6–7]. Cajori served as one of the History of Science Society's two vice-presidents for the first two years. Smith was the first secretary of the Society, to be succeeded by L. Leland Locke, a mathematics teacher from Brooklyn with a special interest in the history of calculating machines. Archibald served as one of the associate editors of *Isis*, a position he retained for the rest of his life.

Florian Cajori died in 1930. His death marked the end of an era during which Americans had developed impressive bibliographic skills and resources in the history of mathematics, had produced notable translations and editions of mathematical works, had demonstrated critical judgment in the analysis of ancient and medieval mathematical texts, and had authored exemplary textbooks in the history of elementary mathematics, along with other useful materials for teachers of mathematics. It would appear that they had paved the road for the next generation of American scholars to make its mark by examining conceptual developments in modern mathematics, and by subjecting to deeper analysis social, economic, and cultural issues affecting the subject. However, this road was soon covered with obstacles.

3. The 1930s and World War II

Through hindsight it is possible to detect the beginning of a decline in mathematical history in the late twenties. By the end of World War II, American research results in the history of mathematics were becoming scarce, and the most widely read expository presentations sacrificed accuracy for literary bon mots or philosophic preconceptions. The best research work being done was no longer published in the mathematical journals; and the occasional expository article dealing with history tended to be chatty. The quality of courses in the history of mathematics, never very demanding, sank further. In the minds of most mathematicians, history of mathematics had lost any claim to status as a legitimate field of mathematical specialization.

In the early thirties, the change was not obvious. To be sure, in 1931 the *Bulletin* dropped from its masthead the reference to being "a historical and critical review of mathematical science." It still carried an occasional historical book review; but neither in the *Bulletin* nor in other American research journals could one find the historical framework that had once surrounded many research articles. Yet there appeared to be other outlets for historical articles.

The *Monthly* continued to publish a variety of readable articles, covering a wide range of mathematical history, which occasionally included original research.

In 1932, a new journal was founded, entitled *Scripta Mathematica.* Its masthead proclaimed that it was "A Quarterly Journal Devoted to the Philosophy, History, and Expository Treatment of Mathematics." The editor-in-chief of the journal was Jekuthiel Ginsburg of Yeshiva University; the listing of the editorial board read like a Who's Who in the History and Philosophy of Mathematics: it consisted of Archibald, Karpinski, Keyser, Loria, Simons, and Smith. Although this board was expanded several times, eventually the burden of editorship rested almost exclusively on the hard-working Ginsburg. There was a pleasant mixture of original research and exposition in the historical articles; most active American historians of mathematics contributed to *Scripta* at some time during the thirties and forties.

The *Mathematics Teacher* brought out numerous solid articles on history of mathematics during the early thirties, thanks largely to the efforts of Vera Sanford; however, these gradually declined in quantity and quality. Yet, another new magazine oriented to mathematics teachers featured interesting research notes in the history of mathematics. This was *The National Mathematics Magazine*, published by Louisiana State University. It had been established in the late twenties as *Mathematics News Letter*, and had contained an occasional note on history. After its reorganization, contributions on history of mathematics increased substantially and from the mid-thirties to the mid-forties it featured historical notes and articles by most contributors to the field. Much of the effort was due to G. W. Dunnington, the Gauss biographer with close ties to G. A. Miller.

Among publications designed for a wider, nonmathematical, readership, *Science and Society*, a Marxian journal founded in the 1930s, contained occasional stimulating reflections or reviews on the history of mathematics. As an example of a specialized journal, which, thanks to Alonzo Church, began with a major contribution to historical bibliography, one must note the *Journal of Symbolic Logic.*

Periodically, research results in the history of mathematics appeared in *Isis.* Mathematical items also could be found in the series of companion volumes entitled *Osiris* that George Sarton began to issue in the 1930s. It, too, was published in Belgium. Each volume was dedicated to a leading figure in the history of science, and contributions to that volume usually, though not always, related to that individual's interests. The first volume in the series was dedicated to David Eugene Smith. It, and several subsequent volumes, contained substantial articles in the history of mathematics; American authors were well represented.

Another new publication, sponsored by Julius Springer in Germany and founded in 1929 by Otto Neugebauer, J. Stenzel, and Otto Toeplitz, provided an international outlet for solid research contributions and promised to become the scholarly vehicle that the history of mathematics needed at this stage. This was *Quellen und Studien zur Geschichte der Mathematik.* Divided into two parts, one for "Quellen" the other for "Studien," it began propitiously, with articles of high quality, including some by American authors. Its life was cut short by the advent of the Third Reich, however.

The aging pioneers continued their work. Smith collaborated with Ginsburg on a *A History of Mathematics in America before 1900*, which appeared as a Carus Mathematical Monograph in 1934. In addition, he provided dozens of book reviews, encyclopedia articles, and occasional pieces, served on committees, and organized his collections. In 1931, he donated his library to Columbia University; this gift, along with the University's previous holdings, the Plimpton collection, Smith's collection of mathematical objects, and the library of the AMS, all housed on the Columbia campus, provided a substantial resource for mathematicians and historians of mathematics.

Karpinski continued his bibliographic work, culminating in his *Bibliography of Works Published in America Prior to 1850*, which appeared in 1940. Before his retirement in 1948, he had built up the mathematical rare book and manuscript collection at the University of Michigan, produced bibliographies on cartography, sold his map collection to Yale University, served a term as chairman of Section L of the AAAS and, in 1941, been elected president of the History of Science Society.

Archibald remained active throughout the 1930s and World War II. He made substantial contributions to each volume of *Scripta.* He issued another edition of the Catalogue of the library of the AMS in 1932; he brought to notice, through knowledgeable reviews, the work of Neugebauer; he contributed to the *Dictionary of American Biography*; and he produced successive editions of his *Outline of the History of Mathematics.* In 1938 his *Semicentennial History of the AMS* was published. During World War II he served as chairman of the National Research Council's Committee on Mathematical Tables and Aids to Computation, and founded the journal by the same name. Meanwhile he had watched the passing of three generations of historians of mathematics; he eulogized Cajori and Chace, Brown and Heath, and finally H. W. Tyler.

Among other mathematicians who had made earlier contributions to the history of mathematics, Julian Lowell Coolidge produced a number of works on algebraic geometry with historical overtones; his *History of Geometric*

Methods was published by Oxford in 1940. Emch contributed another research piece on Saccheri in *Scripta.* Miller still wrote occasional articles in *National Mathematics Magazine*, the *Monthly*, *School Science and Mathematics*, and *Science.* Sanford contributed to the *Mathematics Teacher.*

Although it is striking that most of those who produced more than an occasional piece in the 1930s and 1940s were near or past retirement, there was a scattering of younger mathematicians and historians who displayed interest in history of mathematics. For example, among mathematicians who played a significant part in the resurgence of history of mathematics after the 1950s, Garrett Birkhoff made a contribution to the third volume of *Osiris* [Birkhoff 1935], and Dirk Struik wrote thoughtful historical articles and reviews; after the establishment of *Mathematical Reviews*, Struik handled a large portion of reviews on modern topics for the section on history.

The individual who in the latter part of this period was most successful in bringing aspects of the history of mathematics to the attention of mathematicians as well as the general public was Eric Temple Bell (1883–1960). Born in Scotland, he came to the United States at age nineteen. During a decade spent in the western United States, he studied at Stanford and the University of Washington, taught school, and, according to his own account, was a mule skinner, surveyor, lumberjack, and minor entrepreneur. He spent a brief period in New York, where he obtained a Ph.D. at Columbia in 1912. A specialist in number theory, he returned to the University of Washington, serving on its faculty until his appointment to the California Institute of Technology in 1926. In the meantime, he had won the Bôcher prize in 1921, and was invited to give the Colloquium Lecture for 1927. He served as president of the MAA from 1931–1933. A prolific author, he devoted much of his time after 1930 to popularization and to history, besides writing science fiction novels under the pseudonym John Taine. His two works most widely read by students of the history of mathematics have been *Men of Mathematics*, which first appeared in 1937, and *Development of Modern Mathematics*, the first edition of which was published in 1940. The fluidity of Bell's prose often obscures the lack of evidence for his assertions. Struik characterized Bell's style in gentlemanly fashion by noting "the experience of the author as a creative mathematician, a teacher and interested colleague has made it possible to place lively comment, pithy summaries and challenging outlooks between an otherwise factual survey of achievements." [Struik 1940].

The interest in popularization and biography during the late thirties and forties was not peculiar to the history of mathematics. Among American historians this was a time of major controversy concerning these two issues [Higham 1965]. American historians had come a long way since the turn of the century in developing research strength and generational continuity and perhaps benefitted from vigorous debates. American historians of mathematics, however, had just begun to show their research potential. Their limited publication outlets for serious research contributions were being shut down by the spread of National Socialism on the European continent and the competition for increasingly limited resources at home. The disdain for historians expressed by men like Bell was hardly designed to encourage young people with an interest in mathematics to take up the study of history. All of this exacerbated the major problem, which was that there was no new generation of historians of mathematics to take the place of those who were retiring in the 1930s. Smith and Karpinski had had substantial numbers of graduate students; but even those among them who made sound contributions to the history of mathematics had been prepared to become mathematics "educators," not historians of mathematics. Their careers were in teaching and administration; the time people like Sanford and Simons found to edit and produce historical articles is a testament to their devotion to the subject. Neither economic conditions nor the academic climate, said to have produced a "schism in scholarship" [Higham 1970], could be expected to encourage many to take up a research career in a fading field.

4. The Post-World War II Period: 1945–1968

In the post-war period, the history of science emerged as an independent discipline. As courses in the subject multiplied, and graduate programs and departments grew, certain areas within the history of mathematics also assumed new strength; this was especially true of some aspects of the history of ancient and medieval mathematics. Yet, for the most part, history of modern mathematics existed on the fringes of history of science and became practically defunct within mathematics, where it was promoted, albeit at a rather light level, only by those concerned with mathematics education. There were spin-offs from activities in history and history of science that were to prove beneficial to the history of mathematics. But these could be utilized fully only after renewed interest among mathematicians in history and a resurgence of research activity in the history of modern mathematics became apparent in the late sixties.

The post-war period started with promise. In 1948, Struik's *Concise History of Mathematics* made its appearance. For the first time, an American had produced a general history that introduced intellectual, social, and economic factors while remaining mathematically sound and historically perceptive. It was followed the next year by J. L. Coolidge's charming *The Mathematics of Great Amateurs*, and by Boyer's *Concepts of the Calculus*, about which more will be said below.

An observer of the scene in the early 1950s was still receiving mixed messages: The AMS sold its library to the highest bidder, the University of Georgia; a note in *Isis* called attention to a seminar on the history of mathematics conducted by Otto Neugebauer at Brown; at the International Congress

Otto Neugebauer

of Mathematicians in 1950, George Sarton expressed grave concern at the lumping of history of mathematics with mathematics education; carefully executed editions and analyses of medieval mechanics began to appear from Marshall Clagett and his students.

It soon became clear, however, that a revival was overdue. Smith had died in 1946, Simons in 1949. Solomon Gandz followed in 1954, Archibald in 1955, Karpinski and Sarton in 1956, and Ginsburg in 1957. Among younger mathematicians, Carolyn Eisele began her championship of the mathematics of C. S. Peirce. Phillip S. Jones promoted history among mathematics teachers. Morris Kline called attention to the relationship of mathematics to other disciplines in the history of Western culture and encouraged mathematicians with historical interests. Still, there was little fresh research in the history of modern mathematics. Popular accounts continued and were well received. *The World of Mathematics*, a successful four-volume anthology, found favor with many lay people. Books such as [Bochner 1966] interested the scientifically oriented. [Kline 1953] sold well. [Eves 1964] became an unusually successful textbook. [Aaboe 1964] was a rare example of a technical, intelligible, mathematically and historically sound introduction to topics in Greek mathematics. But research outlets for history of mathematics continued to shrink. Upon the death of Ginsburg, *Scripta Mathematica* in 1957 dropped the reference to philosophy and history from its masthead, replacing it with the statement that it was "devoted to the Expository and Research Aspects of Mathematics." *Osiris* ceased. *Isis* carried a few articles pertaining to mathematics, and began to discourage any that were technically oriented.

There was increased interest in mathematical classics as collectibles; perhaps this would have pleased Karpinski, who had retired to Florida to spend his remaining years as a book dealer. What might have surprised him and his contemporaries was the flood of reprints that appeared on the book market in the 1960s; perhaps even more surprising might have been the fact that not only was no editing or updating done on these publications, but that they occasionally appeared without the scholarly apparatus that had made the original edition so valuable to the student ([Gauss 1902] vs. [Gauss 1965], for example). Nevertheless, the availability of so much historical material in libraries and bookstores compensated the interested student for the scarcity of courses, seminars, and other organized activity in history of mathematics. Occasionally, the lucky browser encountering a genetic introduction to the calculus [Toeplitz 1963] next to a history of its concepts [Boyer 1939], both stacked on top of a good traditional calculus textbook, might gain an appreciation for an aspect of mathematics and its history not necessarily attainable through the single-minded approach of the lecture room.

Members of the History of Science Society from time to time conducted surveys dealing with their growing field. The results of one such survey [Price 1967], although not altogether reliable in its details, reflected the scarcity of research opportunities for graduate students in the history of mathematics. There was one department for the history of mathematics; it had an exceptionally strong base in the ancient, medieval, and Renaissance periods. In other departments, the programs that came closest to providing guidance for work in the history of mathematics were those listing the history of the exact sciences as fields of specialty. Almost all of these had strength in the ancient and medieval areas, a testimony to the influence of Neugebauer and Clagett. The only listings for study in the history of the exact sciences in the seventeenth and eighteenth centuries were attributable to I. Bernard Cohen, whose influence in encouraging mathematicians and historians interested in the modern period became obvious only recently.

The one department in the history of mathematics that existed in the 1960s was that of Otto Neugebauer at Brown. Neugebauer's contributions to the history of mathematics go back to the 1920s. As a graduate student in Göttingen, he assisted Richard Courant in editing the posthumous second volume of Felix Klein's *Vorlesungen über die Geschichte der Mathematik im 19. Jahrhundert* and "was introduced [by Courant] to modern mathematics and physics as a part of intellectual endeavor, never isolated from each other nor from any other field of our civilization." [Neugebauer 1957:vii]. But beyond this, he has credited Courant with encouraging him in his study of the mathematical sciences of the Near East. Beginning with the publication by Springer in 1926 of his *Grundlagen der ägyptischen Bruchrechnung*, Neugebauer has produced a steady stream of scholarly and ground-breaking contributions to the history of ancient Egyptian and Babylonian mathematics and astronomy. As noted above, in the United States, Archibald had called attention to the significance of this work since the early thirties. Neugebauer came to this country in 1939, to serve on the faculty of Brown University and to edit the newly established *Mathematical Reviews* [Pitcher 1988:69–85]. His research production did not flag. *Mathematical Cuneiform Texts* in 1945 was his first major work published in this country, in collaboration with A. Sachs. It brought to light and interpreted a substantial collection of mathematical tablets in the United States that had not been previously analyzed. Incidentally, this publication, dedicated to Archibald, was supported in part by the MAA's Chace Fund. Neugebauer and his school continued their systematic research activities throughout the period under discussion. Their publications appeared in mathematical, historical, and philological journals. As time passed, *Centaurus* rather than *Isis* became the transmitting journal for many research results of the Neugebauer school. Neugebauer himself placed more distance between himself and historians of science. Typically, in a review written in 1955, he expressed his opposition to the dominant historical direction by commenting that "the trend toward 'synthesis' in historical studies at the expense of factual, detailed analysis shows its detrimental effects." [Neugebauer 1955].

In view of the important role played by historians of mathematics in the early days of the History of Science Society, the extent to which mathematics receded from the center of action in the history of science may seem surprising. It is explained, in part, by the growing proximity of history of science to the general field of history. In the 1920s, when the History of Science Society was founded, its connections to the scientific community were stronger than its ties to American historians. Intellectual history in America was in its youth; just twenty years had passed since James Harvey Robinson at Columbia had introduced a course in intellectual history. It was only in 1924 that Arthur Schlesinger introduced his course on social and intellectual history. By contrast, in the 1960s, intellectual history had peaked, and socioeconomic history was in the ascendant. Freshly trained historians of science found most of their jobs in history departments, although the jobs had been frequently created at the urging of the science sector, and most of the students were science majors [Kuhn 1971]. If the scenario described by Kuhn in 1971 is accurate, it is not surprising that economic and intellectual pressures moved historians of science more closely into the history camp. For the history of mathematics, there were special problems. The work done by most of the earlier pioneers in the field was pronounced methodologically worthless by various historians' spokesmen. In addition, there seemed even less reason for either the intellectual or the socioeconomic historian to be concerned with the history of mathematics than with the history of the experimental sciences. For Kuhn [1971] hypothesized that "after a science has become thoroughly technical, particularly mathematically technical, its role as a force in intellectual history becomes relatively insignificant." This notion, coupled with the statement that "Science, when it affects socioeconomic development at all, does so through technology," left little room for mathematics in two areas of history especially relevant to the history of science. Whatever the reasons, it is true that by the end of the sixties the new generation of historians of science excelled in areas where training in history was more important than training in mathematics.

It was a man trained as an intellectual historian, Carl Boyer (1906–1976), of the mathematics department of Brooklyn College, who bridged the gap, and was the primary representative in this period of the history of mathematics among historians of science. Trained in mathematics at Brooklyn College and Columbia University, he had received a Ph.D. degree in intellectual history in 1939 with a dissertation on "The Concepts of the Calculus." Originally published as a hardback, this work became better known in its Dover edition, especially after the title was changed to read *The History of the Calculus and its Conceptual Development*. Prior to the end of World War II, Boyer had contributed about a dozen notes and articles ranging over topics as diverse as "A Vestige of Babylonian Influence in Thermometry" to "Fundamental Steps in the Development of Numeration." These appeared in the journals mentioned in the previous section—*Scripta Mathematica*, *Isis*, the *American Mathematical Monthly*, the *Mathematics Teacher*, and the *National Mathematics Magazine*—as well as more broadly based publications such as *Science* and the *Scientific Monthly*. He continued to contribute to these and other journals and produced three more books: *A History of Analytic Geometry*, published in the *Scripta Mathematica Studies* series in 1956, followed three years later by *The Rainbow: from Myth to Mathematics* (Yoseloff), and, in 1968, a *History of Mathematics* (Wiley). It was Boyer who, as book review editor of *Scripta Mathematica* from 1947 to 1970, sustained its historic component; as member of the Editorial Committee of *Isis* from 1954 to 1970 he represented mathematical interests in that group. In 1960, when the *Archive for History of Exact Sciences* was established, Boyer was appointed to its editorial board; he assumed editorial responsibilities for mathematics on the editorial board of the *Dictionary of Scientific Biography* in 1960 as well. He also served his professional organizations in other capacities: He was elected to the council of the History of Science Society (1943–1945 and 1950–1953), served as its vice president (1957–1958), and as vice-president of the American Association for the Advancement of Science (1958–1959); he contributed to the MAA and the NCTM, serving as secretary of the Metropolitan New York Section of the MAA from 1945 to 1947, and supporting publications in history in the NCTM yearbooks.

In the 1960s, two mathematicians, concerned about the decline of the level of research in the history of mathematics and the lack of appropriate journals for the field, set about to remedy the situation. The first was Clifford Truesdell, who established the *Archive for History of Exact Sciences*, published by Springer. It quickly gained a reputation for solid scholarship. In addition, it provided opportunity for publication of research papers exceeding the length allotted by most journals. The second man was Kenneth O. May. His goal was broader than the establishment of a research journal alone; he wished also to establish better communications among historians of mathematics about other matters of common interest. In 1971, the first volume of *Historia Mathematica*, now sponsored by the International Commission on the History of Mathematics, appeared. Bringing together among its editors and contributors individuals with primary strengths in mathematics, history, philosophy, and education, it set a new direction.

5. The Contemporary Scene: 1969–1989

During the last twenty years there has been a resurgence in the history of mathematics. This has involved a large number of individuals. What is more interesting than the number of efforts is the fact that while few of the contributors consider themselves historians of mathematics exclusively, they include mathermaticians as well as historians, historians of science, mathematics educators and philosophers. In contrast to the early decades of this century, one can no longer single out two or three "leaders" of the field. Instead, one finds numerous men and women pursuing research, teaching, and expository writing in many areas of the history of mathematics.

The resurgence has both positive and negative aspects. On the positive side, history is again finding a niche in the activities of the AMS and of other mathematical organizations. Since the mid-seventies, the annual meetings of the Society have featured historical topics with some regularity, whether in special sessions, in sessions of contributed papers, or as the choice of invited one-hour speakers. These contributions to the annual programs have drawn large audiences. Both the Society and the Association have re-established committees on history. The Society, the Association, and the Association for Women in Mathematics (AWM) have sponsored special programs and book-length publications in history. The return of the *Bulletin* to expository surveys, recent *Mathematics Magazine* policy statements stressing the desirability of historical articles, the popularity of historical notes and articles in the *College Mathematics Journal* and the *Mathematical Intelligencer* all suggest a renewed acceptance of mathematical history. *Historia Mathematica*, the *Archive for History of Exact Sciences*, and *Centaurus* provide the outlets for research articles that were in short supply in preceding decades. In recent years, *Isis* and other journals of the history of science community have had an increased number of articles dealing with post–1750 history of mathematics, largely with emphasis on surrounding social and cultural factors. Aside from numerous, sound biographic studies and historical analyses of mathematical topics, there has been a spate of collected papers, autobiographies, source books, and other book-length publications. The bibliographic abstracts in *Historia* and the *Mathematical Reviews* have greatly facilitated following the growing literature. Besides the Archives of American Mathematics at Texas described elsewhere in this volume, there are accessible collections of mathematical manuscripts and rare books at centers such as the Library of Congress, Brown, Columbia, Chicago, California, Cal Tech, Harvard, MIT, Michigan, and Wisconsin, to name only some of the major repositories. All of this, and the return of history of mathematics courses to the curriculum, especially in mathematics education, suggests a new growth phase in history of mathematics.

There are some other factors to be considered, however. Expository history has been encouraged in large part as a "royal road" to mathematics. At the same time, there is still a lack of institutional training grounds for research in history of mathematics. Students who have located a mathematics or a history of science department, or an individual mentor, willing to support such research too often are hemmed in by methodological dogma. They might benefit from the wisdom of the French historian Marc Bloch (shot in 1944) who observed that "history seeks for causal wave-trains and is not afraid, since life shows them to be so, to find them multiple." [Bloch 1964]. To avoid a repetition of the decline that history of mathematics faced fifty years ago, those working in the field may need to beware of both its popularity and its methodological champions. Perhaps it is not necessary to exclude historical references from the mathematics classroom in order to help eradicate the perpetuation of myths—a suggestion attributed to Gustav Eneström. But it may still serve us well be be guided by the spirit of that dedicated Swedish historian of mathematics, of whom Sarton said that "the very presence of Eneström obliged every scholar devoting himself to the history of mathematics to increase his circumspection and improve his work." [Sarton 1923].

References

Aaboe, Asger. 1964. *Episodes from the early history of mathematics.* New York: Random House.

Archibald, Raymond Clare. 1915. *Euclid's book on* Divisions of figures *with a restauration based on Woepcke's text and on the* Practica Geometriae *of Leonardo Pisano.* Cambridge: University Press.

——, 1932. Florian Cajori. *Isis* **17**:384–407.

——, 1938. *A semicentennial history of the American Mathematical Society 1888–1938.* New York: American Mathematical Society.

Bell, Eric Temple. 1945. Possible projects in the history of mathematics. *Scripta Mathematica* **11**:308–316.

——, 1948. Cassius Jackson Keyser. *Scripta Mathematica* **14**:27–33.

Beman, W. W., trans. 1901. *Essays on the theory of numbers.* By Richard Dedekind. Chicago: Open Court.

Benedict, Suzan R. 1909. The development of algebraic symbolism from Paciuolo to Newton. *School Science and Mathematics* **9**:375–384.

Birkhoff, Garrett. 1935. Galois and group theory. *Osiris* **3**:260–268.

Bloch, Marc. 1964. *The historian's craft.* New York: Vintage Books. p. 194.

Bôcher, Maxime. 1911. The published and unpublished work of Charles Sturm on algebraic and differential equations. *Bull. Amer. Math. Soc.* (2) **18**:1–18.

Bochner, Salomon. 1966. *The role of mathematics in the rise of science.* Princeton: Princeton University Press.

Boyer, Carl B. 1939. *The concepts of the calculus.* New York: Columbia University Press.

Brasch, Frederick E. 1915. The teaching of the history of science. Its present status in our universities, colleges and technical schools. *Science* (ns) **42**:746–760.

Cajori, Florian, 1890. *The teaching and history of mathematics in the United States.* Washington, D.C.: Government Printing Office.

——, 1908. Arithmetik, Algebra, Zahlentheorie. In: *Vorlesungen über Geschichte der Mathematik.* Ed. by Moritz Cantor. **4**:37–198.

Eisele, Carolyn, 1950. Lao Genevra Simons. *Scripta Mathematica.* **16**:22–30.

Eves, Howard, 1964. *An introduction to the history of mathematics.* Rev. ed. New York: Holt, Rinehart and Winston.

Gauss, Carl Friedrich. 1902. *General investigations of curved surfaces.* Trans. by A. Hiltebeitel and J. Morehead. Princeton: University Library.

——, 1965. *General investigations of curved surfaces.* Trans. by A. Hiltebeitel and J. Morehead. New York: Raven Press.

Halsted, G. B., trans. 1902/3. Extract of 'Sketch of the origin and development of geometry prior to 1850' by G. Loria. *The Monist* **13**:80–120 and 218–234.

Higham, John with Leonard Krieger and Felix Gilbert. 1965. *History: The development of historical studies in the United States.* Englewood Cliffs, N.J.: Prentice-Hall.

Higham, John. 1970. The schism in American scholarship. Pp. 1f. in: *Writing American history. Essays on modern scholarship.* Bloomington, Ind.: Indiana University Press.

Hill, G. W., 1896. Remarks on the progress of celestial mechanics since the middle of the century. *Bull. Amer. Math. Soc.* (2) **2**:125–136.

Jackson, L. L. 1906. *The Educational significance of sixteenth century arithmetic from the point of view of the present time.* New York: Columbia University.

Jones, Phillip S. 1976. Louis Charles Karpinski, historian of mathematics and cartography. *Historia Mathematica* **3**:185–202.

Kline, Morris. 1953. *Mathematics in Western culture.* New York: Oxford University Press.

Kline, Morris. 1976. Carl B. Boyer. *Historia Mathematica* **3**:387–394.

Kuhn, Thomas S. 1971. The relations between history and the history of science. *Daedalus* **100** (Spring): 271–304.

Levey, Martin. 1955. Solomon Gandz. *Isis* **46**:107–110.

May, Kenneth O., ed. 1972. *The Mathematical Association of America. Its first fifty years.* np.: The Mathematical Association of America.

McCormack, Th. J., trans. 1898. *Lagrange's Lectures on elementary mathematics.* Chicago: Open Court.

Merrill, Helen A. 1903. "On solutions of differential equations which possess an oscillation theorem." *Trans. Amer. Math. Soc.* **4**:423–433.

Neugebauer, Otto. 1957. *The exact sciences in antiquity.* Second rev. ed. Providence, R.I.: Brown.

——, 1955. [Review]. *Isis* **46**:71.

Pitcher, Everett. 1988. *A history of the second fifty years. American Mathematical Society 1939–1988.* Providence: American Mathematical Society.

Porter, M. B. 1902. On the roots of functions connected by a linear recurrent relation of the second order. *Annals of Mathematics* (2) **3**:55–70.

Price, Derek J. de Solla. 1967. A guide to graduate study and research in the history of science and medicine. *Isis* **58**:385–395.

Robinson, Lydia G., trans. 1909. *Archimedes. Geometrical solutions derived from mechanics.* From the German translation by J. L. Heiberg. Chicago: Open Court.

Sarton, George. 1923. For Gustav Eneström's seventy-first anniversary. *Isis* **5**:421.

——, 1927. *Introduction to the history of science. Volume I. From Homer to Omar Khayyam.* (Carnegie Institution of Washington. Publication no. 376). Baltimore: Williams & Wilkins.

——, 1956. Raymond Clare Archibald. *Osiris* **12**:5–31.

Smith, David Eugene and Y. Mikami. 1941. *A history of Japanese mathematics.* Chicago: Open Court.

Stamper, Alva Walker. 1906. *A history of the teaching of elementary geometry.* New York: Columbia University.

Struik, D. J. 1940. *Math. Reviews* **2** no. 4: 113.

Toeplitz, Otto. 1963. *The calculus. A genetic approach.* Chicago: University of Chicago Press.

Tyler, H. W. 1910–1911. On the course in the history of mathematics in the Massachusetts Institute of Technology. *Bibliotheca Mathematica* (3) **10-11**:48–52.

Wilczynski, E. J. 1902. Lazarus Fuchs. *Bull. Amer. Math. Soc.* (2) **9**:46–49.

Frederic F. Burchsted received a B.A. from the University of Chicago in 1971 *and a Ph.D. in zoology from the University of Wisconsin in* 1980, *with concentration in the history of science. After serving for several years as a librarian at the University of Texas, including three years in the Physics-Mathematics-Astronomy Library, in* 1986 *he received a M.L.I.S. degree and was appointed Archivist of the Archives of American Mathematics.*

Sources for the History of Mathematics in the Archives of American Mathematics

FREDERIC F. BURCHSTED

The *Archives of American Mathematics* (*University Archives, University of Texas at Austin*) is dedicated to the preservation of the American mathematical heritage for the use of mathematicians and historians of mathematics. The AAM serves as a national repository for papers of mathematicians and records of mathematical organizations for which preservation at the home institution is not available. The preservation of sets of papers, including correspondence, teaching materials, records of professional society affiliation, as well as manuscripts, is of the first importance in preserving the full record of American mathematics in its intellectual, institutional, and social contexts. The AAM is interested in hearing of collections needing preservation. Requests for information on any aspect of historical documentation in mathematics are welcome.

The AAM was initiated in 1975 with the preservation of the papers of H. S. Vandiver and Robert Lee Moore, the University of Texas number theorist and point-set topologist, at the University of Texas at Austin Humanities Research Center. The Vandiver and Moore papers were conceived as the starting point of an archival collection documenting the history of American mathematics. Papers of several prominent mathematicians were rapidly added, and in 1978 the Mathematical Association of America named the AAM as the official repository of its archival records. In 1984, custody of the AAM was transferred to the University of Texas at Austin University Archives where it now resides. Accounts of the establishment of the AAM have been published by Albert C. Lewis, the first curator (Lewis, 1978a, 1978b).

Left to right: Wilfrid Wilson, J. W. Alexander, W. L. Ayres, G. T. Whyburn, R. L. Wilder, P. M. Swingle, C. N. Reynolds, W. W. Flexner, R. L. Moore, T. C. Benton, K. Menger, S. Lefschetz. This picture was taken at the mathematical meetings in Cleveland, Ohio, December 1930.

(Photograph courtesy of the University of Texas at Austin, Archives.)

The bulk of the collection dates from after approximately 1920. Sources for earlier work include some notebooks of R. L. Moore (including diaries) from his graduate student years (1903–1905) at the University of Chicago, together with correspondence (1898–1920) with George Bruce Halsted and, in lesser amounts, with D. R. Curtiss, L. E. Dickson, E. H. Moore, E. B. Van Vleck, and others. The R. L. Moore papers include a collection of publications and clippings on G. B. Halsted. H. J. Ettlinger pursued his graduate studies at Harvard University (1910–1913), and the AAM has his student notebooks (7 in.) which record courses of G. D. Birkhoff, M. Bôcher, C. L. Bouton, W. E. Byerly, J. L. Coolidge, H. N. Davis, D. Jackson, W. F. Osgood, and B. O. Peirce. In the University of Texas University Archives, although not in the AAM, are Harvard University graduate student notebooks (1895–1898) of H. Y. Benedict, a University of Texas professor of applied mathematics and astronomy, on courses of M. Bôcher, W. E. Byerly, A. Hall, W. F. Osgood, and B. O. Peirce.

The AAM has particular strengths in several mathematical fields as follows:

Analysts represented include H. J. Ettlinger (1889–1986) and his students William T. Reid (1907–1977) and William M. Whyburn (1901–1972). The Ettlinger papers (1909–1979; 3 ft.) include teaching notes for University of Texas courses, and notes and technical reports concerning his aerodynamics work with the University of Texas Defense Research Laboratory. Also included are mimeographed lecture notes and handwritten notes on some lectures given in 1925 at the Massachusetts Institute of Technology on the operational calculus of Oliver Heaviside, a subject later used by Ettlinger in his own work. The papers (1926–1977; 28 ft.) of William T. Reid include some of Reid's notes made during his studies with Ettlinger and substantial quantities of notes and drafts related to his research on differential equations, calculus of variations, and optimal control. Also included are notes of his analysis seminars, some taught with G. M. Ewing, E. D. Hellinger, and W. T. Scott, at Northwestern University, University of Iowa, and University of Oklahoma. Reid's World War II work on ballistics and aerial photogrammetry is represented. The William M. Whyburn papers (1923–1970; 10 ft.) are dominated by his administrative work at Texas Technological College and the University of North Carolina, but also include some material on his differential equations research and his consulting work with the Air Force, Navy, and Oak Ridge National Laboratory. The papers (1951–1980; 6 ft.) of Charles B. Morrey, Jr. (1907–1984) contain records of his research in the calculus of variations and of his textbooks, written with M. H. Protter. The papers (1936–1986; 16 ft.) of W. F. Eberlein (1917–1986) include manuscripts and notes for papers on ergodic theory, mean value theorems, numerical integration, and functional and harmonic analysis, together with teaching notes. Papers (1915–1949; 5 in.) of Ernst D. Hellinger (1883–1950) are included in the William T. Reid papers and the Max Dehn papers. These are mainly research and lecture notes, including work on the theory of spectra, on Hermitian operators in Hilbert space, and on Stieltjes continued fractions. The Otton Martin Nikodým papers (see below: **Mathematical physics**) contain material on measure theory and operators in Hilbert space. In a small collection (10 in.) of Abraham Robinson's (1918–1974) manuscripts are some items on nonstandard analysis. A film of the Conference on Orthogonal Expansions and their Continuous Analogues (Southern Illinois University, Edwardsville, 1968) is held. A small collection (1950–1967; 1 in.) of the papers of Louis L. Silverman (1884–1967), whose interest centered on divergent series, includes biographical clippings and three lectures in Hebrew.

Mathematical physics is represented by the papers of the relativity theorists George Yuri Rainich (1886–1968) and Alfred Schild (1921–1977). Rainich's papers (1941–1967, bulk: 1960's; 5 ft.) are largely notes for relativity seminars at the Universities of Michigan and Notre Dame, together with work for a projected book on relativity. Schild's papers (1944–1977; 21 ft.) contain notes and drafts for lectures and publications on relativity and gravitation, in particular, algebraically special solutions, quantization, conformal techniques, Fokker action principles, and string models of particles. There are also records of the preparation of his book *Tensor Calculus* (1949),written with J. L. Synge. The papers (1925–1974; 12 ft.) of Otton Martin Nikodým (1887–1974) contain notes and drafts for his *The Mathematical Apparatus for Quantum-theories* (vol. I, 1968), which utilizes a theory of operators in Hilbert space based on abstract Boolean lattices. The nearly finished manuscript for the unpublished second volume is included. The W. F. Eberlein papers (see above: **Analysts**) contain notes and drafts for works on models of space-time, relativity, and quantum mechanics, particularly internal symmetry and spinor analysis.

Number theory is represented by the correspondence files (1942–1988; 4 ft.) of Emil Grosswald (1912–) which include letters from P. Bateman, H. Rademacher, and C. L. Siegel, among many others. Records of Grosswald's editing of Hans Rademacher's *Collected Papers* and *Topics in Analytic Number Theory* are also included. The H. S. Vandiver (1882–1973) papers (1900–1965; 17 ft.) contain over 2300 letters written between 1910 and 1965, together with notes and drafts for publications, including his unfinished book on the history of Fermat's Last Theorem. A small collection (1904–1925; 3 in.) of Albert E. Cooper's (1893–1960) papers documents his involvement with the preparation of L. E. Dickson's *History of the Theory of Numbers.* Francis L. Miksa (see below: **Amateur mathematician**) also worked in number theory.

Rational mechanics is represented by papers (1939–1984; 18 ft.) of C. Truesdell (1919–). Truesdell's papers consist largely of manuscripts of his books and articles, texts of lectures, notes on his courses, notes taken by

Truesdell during his own education, and biographical material and reminiscences. There is an annotated item list prepared by Professor Truesdell.

Topology is a particular strength of the AAM. The papers (1898–1974; 32 ft.) of Robert Lee Moore (1882–1974) contain notes and drafts for his publications, over four feet of correspondence, and notes on his teaching. The AAM also holds the papers of R. H. Bing (1914–1986), R. G. Lubben (1898–1980), and Raymond Louis Wilder (1896–1982)—all students of R. L. Moore. The Bing papers (1948–1986; 4 ft.) consist largely of material from after his return to the University of Texas in 1973, and include manuscript material for his *The Geometric Topology of 3-Manifolds* (1983), records of the organization of several topology conferences and institutes, and records of his work in preserving the memory and papers of R. L. Moore. The R. G. Lubben papers (1922–1974; 6 ft.) include mathematical notes and manuscripts, together with his teaching notes. The R. L. Wilder papers (1916–1982; 19 ft.) stress his work on the foundations and history of mathematics, but there is a substantial quantity of correspondence, manuscripts, and research notes on topological subjects. A small quantity of papers (1948–1979; 20 in.) of the algebraic topologist Norman Earl Steenrod (1910–1971) includes a draft of his *Foundations of Algebraic Topology* (1952), written with S. Eilenberg, records of the preparation of *Reviews of Papers in Algebraic and Differential Topology, Topological Groups, and Homological Algebra*, and texts of several lectures. There are small collections of items documenting the careers of Clark M. Cleveland (1892–1969, papers: 1927–1930; 1 in.), another R. L. Moore student, and Albert W. Tucker (1905– , papers: 1946–1979; 5 in.).

Mathematical logic and **foundations of mathematics** are represented by papers of Jean van Heijenoort (1912–1986) and Raymond Louis Wilder (1896–1982). The Jean van Heijenoort papers (1946–1983; 16 ft.) include his published and unpublished writings, his notes and unfinished manuscripts, and his correspondence files, which include letters from A. Church, B. Dreben, K. Gödel, S. C. Kleene, R. Martin, C. D. Parsons, and W. V. Quine, among others. The Wilder papers (1916–1982; 19 ft.) contain material on his publications and courses on the foundations of mathematics and on his later work on the application of anthropological theory to the history of mathematics. Otton Martin Nikodým's interest in logic is reflected in his papers. The Abraham Robinson papers include manuscripts on mathematical logic and model theory.

The Robert E. Greenwood papers contain a collection of printed matter on the **history of computing** and **numerical analysis**, including technical reports by H. H. Goldstine, J. von Neumann, and others.

It is well-known that many **immigrant mathematicians** have made important contributions to American mathematics. The AAM holds papers (1899–1954; 15 in.) of Max Dehn (1878–1952), which contain notes, drafts of publications and lectures, and correspondence on geometry, topology, group theory, and the history of mathematics. Dehn came to the United States in 1940. Material from Dehn's European and American years is included, with the bulk dating from 1921–1952. Otton Martin Nikodým (see above: **Mathematical physics**) came to the United States in 1948. Although little pre-1948 material is included, Nikodým maintained his ties with his European colleagues, and his correspondents include J. L. Destouches, M. Fréchet, H. Hasse, G. Ludwig, C. Y. Pauc, W. Sierpiński, and A. Tònolo. Emil Grosswald, Ernst D. Hellinger, George Yuri Rainich, Abraham Robinson, and Alfred Schild have already been mentioned. A small collection (1915–1917, 1930; 1 in.) of Eduard Helly (1884–1943) papers includes a manuscript of a published paper and some postcards dating from his World War I imprisonment in Siberia.

The papers (1937–1975; 40 ft.) of Francis L. Miksa (1901–1975), an **amateur mathematician** of Aurora, Illinois, comprise records of his work in problem-solving, magic squares, dyad squares, Pythagorean triangles, Stirling numbers, and several number theory topics. His magic squares work led to a group theory method for systematically constructing complete sets of magic squares without duplication. His correspondence documents his interaction with problem-solvers, other amateurs, and professional mathematicians including Leo Moser, with whom Miksa collaborated on several papers, and Robert E. Greenwood, who has donated his Miksa letters.

The institutional and social contexts of mathematics are important subjects for historical research which rely heavily on archival sources. The context of mathematical research is documented particularly in correspondence files, records of professional society affiliation, and departmental records.

Professional Societies. The AAM is the archival repository for the records of the Mathematical Association of America (MAA). The MAA Headquarters records (1916–1967; 7 ft.) include correspondence of, and biographical information on, past secretaries (1916–1976; bulk: 1916–1967), and records (correspondence, programs, photographs, newspaper clippings) concerning MAA conferences and annual meetings (1916–1960; bulk: 1930–1960). Records of the MAA History of American Mathematics in World War II Committee (1980–1981; 5 in.), and the William Lowell Putnam Mathematical Competition (1938–1980; 20 in.) are also held. The AAM holds records of the following MAA presidents: Raymond Louis Wilder (1965–1966), Henry L. Alder (1977–1978), and Dorothy L. Bernstein (1979–1980). There are also records of Emil Grosswald's work (1965–1977; 5 in.) with the Board of Governors and the Ford Award committee, and of Phillip S. Jones' work (1955–1967; 5 in.) with the Committee on Instructional Films and the Committee on Educational Media. Several of the sets of personal papers also contain files on MAA committee work.

Records of the American Mathematical Society include papers of the following presidents: R. L. Moore (1937–1939), R. L. Wilder (1955–1956), and Saunders Mac Lane (1973–1974).

Departmental records. Records of mathematics departments are included in several of the collections of personal papers, but the mathematics departments of the Universities of Michigan, 1928–1982 (R. L. Wilder papers), North Carolina, 1949–1971 (William M. Whyburn papers), and Texas, 1920–1986 (R. H. Bing, H. J. Ettlinger, R. G. Lubben, and R. L. Moore papers), are especially well-represented. H. J. Ettlinger and Albert C. Lewis's list of doctoral descendents of University of Texas at Austin mathematics faculty is available. The AAM has a copy of *The Princeton Mathematics Community in the* 1930*s: An Oral-History Project* (administered by Charles C. Gillespie, edited by Frederik Nebeker, 1985) which comprises transcripts of interviews with forty-three mathematicians. The C. Truesdell papers contain his reminiscences of the mathematics departments at Brown University (1942), and Princeton University (1944).

Mathematics education. Records of the MAA reflect the Association's interest in mathematics education. The AAM holds records (1958–1977; 130 ft.) of the School Mathematics Study Group (SMSG, the "New Math" movement of the 1960s) comprising the files of its director, Edward G. Begle, together with a collection of its textbooks. The records (1957–1976; 5 ft.) of the New Mathematical Library, edited by Anneli Lax, before it was taken over by the MAA are held. This was conceived as a monograph series to accompany the SMSG curriculum.

A small collection documents the founding of the Duke Mathematical Journal (1927–1934; 1 in.).

The AAM houses a considerable quantity of published and unpublished biographical sketches and portraits which are being indexed.

Fuller descriptions of most of the collections mentioned above are printed in Burchsted (1987). Complete inventories are available for several of the collections. New acquisitions will be announced in *Historia Mathematica, Focus, History of Science in America News and Views*, and the *History of Science Society Newsletter.* Inquiries are welcome and may be addressed to the archivist:

Frederic F. Burchsted, Archivist.

ARCHIVES OF AMERICAN MATHEMATICS

UNIVERSITY ARCHIVES, SRH 2.109

UNIVERSITY OF TEXAS

AUSTIN, TX 78713

(512) 471-3051

References

Burchsted, F. F., 1987. "Archives of American Mathematics," *Historia Mathematica* **14**, 366–374.

Lewis, A. C., 1978a. "Establishment of the Archives of American Mathematics," *Historia Mathematica* **5**, 340–341.

——, 1978b. "An Archive for American Mathematics," *The Mathematical Intelligencer* **1**, 175–176.

Acknowledgments

Scott, Charlotte Angas. “Edwards’ Differential Calculus.” *Bulletin of the New York Math. Soc.* v. 1 (1892) pp. 217–223.

Kenschaft, Patricia Clark. “Charlotte Angas Scott (1858–1931).” *Women in Mathematics* (Greenwood Press, 1987) pp. 193–203.

Langer, Rudolph E. and Ingraham, Mark H. “Edward Burr Van Vleck, 1863–1943.” *Biographical Memoirs* (National Academy of Sciences) v. 30 (1957) pp. 399–409.

Wilder, R. L. “The Mathematical Work of R. L. Moore: Its Background, Nature and Influence.” *Arch. Hist. Exact Sci.* v. 26 (1982), pp. 73–97.

Grinstein, Louise S. and Campbell, Paul J. “Anna Johnson Pell Wheeler (1883–1966).” *Women in Mathematics* (Greenwood Press, 1987) pp. 241–246.

Reid, Constance with Robinson, Raphael M. “Julia Bowman Robinson (1919–1985).” *Women in Mathematics* (Greenwood Press, 1987) pp. 182–189.

Stigler, Stephen M. “Mathematical Statistics in the Early States.” *Annals of Statistics* v. 6 (1978) pp. 239–265.

Doob, J. L. “William Feller and Twentieth Century Probability.” *Proceedings of the Sixth Berkeley Symposium on Mathematical Statistics and Probability* (University of California Press, Berkeley and Los Angeles, 1972) v. II pp. xv–xx.

Craig, Cecil C. “Early Days in Statistics at Michigan.” *Statistical Science.* v. 1 (1986) pp. 292–293.

DeGroot, Morris H. “A Conversation with David Blackwell.” *Statistical Science.* v. 1 (1986) pp. 40–53.

NON CIRCULATING

NATIONAL UNIVERSITY
LIBRARY SAN DIEGO